MW01518644

INSULATORS FOR ICING AND POLLUTED ENVIRONMENTS

BOOKS IN THE IEEE PRESS SERIES ON POWER ENGINEERING

Principles of Electric Machines with Power Electronic Applications, Second Edition
M. E. El-Hawary

Pulse Width Modulation for Power Converters: Principles and Practice
D. Grahame Holmes and Thomas Lipo

Analysis of Electric Machinery and Drive Systems, Second Edition
Paul C. Krause, Oleg Wasynczuk, and Scott D. Sudhoff

Risk Assessment for Power Systems: Models, Methods, and Applications
Wenyuan Li

*Optimization Principles: Practical Applications to the Operations of Markets of the
Electric Power Industry*
Narayan S. Rau

Electric Economics: Regulation and Deregulation
Geoffrey Rothwell and Tomas Gomez

Electric Power Systems: Analysis and Control
Fabio Saccomanno

*Electrical Insulation for Rotating Machines: Design, Evaluation, Aging, Testing,
and Repair*
Greg Stone, Edward A. Boulter, Ian Culbert, and Hussein Dhirani

Signal Processing of Power Quality Disturbances
Math H. J. Bollen and Irene Y. H. Gu

Instantaneous Power Theory and Applications to Power Conditioning
Hirofumi Akagi, Edson H. Watanabe, and Mauricio Aredes

Maintaining Mission Critical Systems in a 24/7 Environment
Peter M. Curtis

Elements of Tidal-Electric Engineering
Robert H. Clark

Handbook of Large Turbo-Generator Operation and Maintenance, Second Edition
Geoff Klempner and Isidor Kerszenbaum

Introduction to Electrical Power Systems
Mohamed E. El-Hawary

Modeling and Control of Fuel Cells: Distributed Generation Applications
M. Hashem Nehrir and Caisheng Wang

Power Distribution System Reliability: Practical Methods and Applications
Ali A. Chowdhury and Don O. Koval

Insulators for Icing and Polluted Environments
Masoud Farzaneh and William A. Chisholm

*FACTS Controllers: Theory, Modeling, and Applications for Electronic Transmission
Systems*
Kalyan K. Sen and Mey Ling Sen

INSULATORS FOR ICING AND POLLUTED ENVIRONMENTS

MASOUD FARZANEH
Université du Québec à Chicoutimi, Chicoutimi, Québec, Canada

WILLIAM A. CHISHOLM
Kinectrics, Toronto, Ontario, Canada

IEEE PRESS

A JOHN WILEY & SONS, INC., PUBLICATION

Library of Congress Cataloging-in-Publication Data is available.

ISBN 978-0-470-28234-2

Printed in the United States of America

10 9 8 7 6 5 4 3 2 1

CONTENTS

PREFACE   **xxi**

ACKNOWLEDGMENTS   **xxv**

1. INTRODUCTION   **1**

    1.1.   Scope and Objectives / 1

            1.1.1.   Problem Areas / 2

            1.1.2.   Problem Characteristics / 4

            1.1.3.   Intended Audience / 4

    1.2.   Power System Reliability / 6

            1.2.1.   Measures of Power System Reliability / 6

            1.2.2.   Achieving Reliability with Redundant Components / 10

            1.2.3.   Achieving Reliability with Maintenance / 10

            1.2.4.   Cost of Sustained Outages / 11

            1.2.5.   Cost of Momentary Outages / 11

            1.2.6.   Who Is Responsible for Reliable Electrical Systems? / 13

            1.2.7.   Regulation of Power System Reliability / 16

    1.3.   The Insulation Coordination Process: What Is Involved? / 16

    1.4.   Organization of the Book / 17

    1.5.   Précis / 20

          References / 20

2. INSULATORS FOR ELECTRIC POWER SYSTEMS   **23**

2.1. Terminology for Insulators / 23

    2.1.1. Electrical Flashover / 24

    2.1.2. Mechanical Support / 24

    2.1.3. Insulator Dimensions / 24

    2.1.4. Interpretation of Terminology for Winter Conditions / 29

2.2. Classification of Insulators / 30

    2.2.1. Classification by Ceramic or Polymeric Material / 31

    2.2.2. Classification by Station or Line Application / 32

    2.2.3. Classification by Nature of Mechanical Load / 34

2.3. Insulator Construction / 35

    2.3.1. Ceramic Materials / 35

    2.3.2. Polymeric Materials / 44

    2.3.3. End Fittings / 46

    2.3.4. Other Materials in Series or Parallel with Insulators / 47

2.4. Electrical Stresses on Insulators / 48

    2.4.1   Power Frequency Electrical Stresses / 49

    2.4.2. Impulse Electrical Stresses / 50

    2.4.3. Major Electrical Factors in Freezing Conditions / 51

2.5. Environmental Stresses on Insulators / 52

    2.5.1. Major Environmental Factors in Temperate Conditions / 52

    2.5.2. Major Environmental Factors in Freezing Conditions / 53

2.6. Mechanical Stresses / 54

    2.6.1. Important Factors in Temperate Conditions / 54

    2.6.2. Important Factors in Freezing Conditions / 54

2.7. Précis / 55

    References / 56

3. ENVIRONMENTAL EXPOSURE OF INSULATORS   **59**

3.1. Pollution: What It Is / 59

3.2. Pollution Deposits on Power System Insulators / 62

    3.2.1. Typical Sources / 62

    3.2.2. Deposit Processes / 63

    3.2.3. Monitoring Methods for Site Pollution Severity / 63

3.2.4. Short-Term Changes in Pollution Levels / 65

3.2.5. Cleaning Processes and Rates / 69

3.2.6. Long-Term Changes in Pollution Levels / 70

3.2.7. Other Factors in Pollution Problems / 71

3.3. Nonsoluble Electrically Inert Deposits / 72

3.3.1. Sources and Nature of Nonsoluble Deposits / 72

3.3.2. Direct Measurement Method for NSDD / 72

3.3.3. Indirect Measurement Methods for NSDD / 73

3.3.4. Role of NSDD in Insulator Surface Resistance / 75

3.3.5. Case Studies: NSDD Measurements / 75

3.4. Soluble Electrically Conductive Pollution / 78

3.4.1. Electrical Utility Sources / 78

3.4.2. Other Fixed Sources / 83

3.4.3. Conductance of Electrolytes / 91

3.5. Effects of Temperature on Electrical Conductivity / 94

3.5.1. Equivalent Conductance of Ions / 94

3.5.2. Effect of Temperature on Liquid Water Conductivity / 95

3.5.3. Effect of Temperature on Ice Conductivity / 97

3.6. Conversion to Equivalent Salt Deposit Density / 100

3.6.1. Insulator Case Study: Mexico / 101

3.6.2. Insulator Case Study: Algeria / 103

3.6.3. Insulator Case Study: Japan / 103

3.6.4. Surface Resistance of Insulator / 103

3.6.5. Insulator Leakage Current: Case Studies / 106

3.6.6. Estimating ESDD from Environmental Measures for Corrosion / 106

3.6.7. Statistical Distribution of the Conductivity of Natural Precipitation / 110

3.6.8. Mobile Sources / 112

3.7. Self-Wetting of Contaminated Surfaces / 122

3.8. Surface Wetting by Fog Accretion / 124

3.8.1. Fog Measurement Methods / 124

3.8.2. Typical Observations of Fog Parameters / 125

3.8.3. Fog Climatology / 127

3.8.4. Fog Deposition on Insulators / 128

3.8.5. Heat Balance Between Fog Accretion and Evaporation / 130

3.8.6. Critical Wetting Conditions in Fog / 131

3.9.   Surface Wetting by Natural Precipitation / 132

    3.9.1.   Measurement Methods and Units / 133

    3.9.2.   Droplet Size and Precipitation Conductivity / 136

    3.9.3.   Effects of Washing on Surface Conductivity / 137

    3.9.4.   Rain Climatology / 137

3.10.   Surface Wetting by Artificial Precipitation / 139

    3.10.1.   Tower Paint / 139

    3.10.2.   Bird Streamers / 139

    3.10.3.   Dam Spray / 139

    3.10.4.   Irrigation with Recycled Water / 141

    3.10.5.   Cooling Pond Overspray: Freshwater Makeup / 142

    3.10.6.   Cooling Tower Drift Effluent / 144

    3.10.7.   Cooling Water Overspray: Brackish or Saltwater Makeup / 145

    3.10.8.   Manurigation / 145

3.11.   Précis / 147

References / 148

4. INSULATOR ELECTRICAL PERFORMANCE IN POLLUTION CONDITIONS   **155**

4.1.   Terminology for Electrical Performance in Pollution Conditions / 155

    4.1.1.   Terms Related to Pollution and Its Characterization / 155

    4.1.2.   Terms Related to In-Service Environment / 157

    4.1.3.   High-Voltage Measurement Terminology / 157

4.2.   Air Gap Breakdown / 159

    4.2.1.   Air Breakdown in Uniform Field / 159

    4.2.2.   Air Breakdown in Nonuniform Field / 161

    4.2.3.   Breakdown of Clean and Dry Insulators / 162

    4.2.4.   Breakdown of Clean and Wet Insulators / 164

4.3.   Breakdown of Polluted Insulators / 165

    4.3.1.   Breakdown Process on a Contaminated Hydrophilic Surface / 165

    4.3.2.   Breakdown Process on a Contaminated Hydrophobic Surface / 167

    4.3.3.   Complications in the Process for Real Insulators / 168

4.4.   Outdoor Exposure Test Methods / 169

    4.4.1.   Field Observations of Leakage Current Activity / 169

4.4.2. Field Observations of Flashover Performance / 170

4.4.3. Field Observations of Other Variables / 171

4.4.4. Observations at Croydon, United Kingdom, 1934–1936 / 171

4.4.5. Observations at Croydon, United Kingdom, 1942–1958 / 172

4.4.6. Observations at Brighton, United Kingdom / 175

4.4.7. Observations at Martigues, France / 176

4.4.8. Observations at Enel Sites in Italy / 176

4.4.9. Observations at Noto, Akita, and Takeyama, Japan / 176

4.5. Indoor Test Methods for Pollution Flashovers / 178

4.5.1. Comparison of Natural and Artificial Pollution Tests / 180

4.5.2. Power Supply Characteristics for Pollution Tests / 181

4.5.3. Electrical Clearances in Test Chamber / 184

4.6. Salt-Fog Test / 185

4.6.1. Description of Salt-Fog Test Method / 185

4.6.2. Validation of Salt-Fog Test Method / 185

4.6.3. Quick Flashover Voltage Technique / 187

4.7. Clean-Fog Test Method / 187

4.7.1. Precontamination Process for Ceramic Insulators / 188

4.7.2. Precontamination Process for Nonceramic Insulators / 190

4.7.3. Artificial Wetting Processes / 192

4.7.4. Validation of Clean-Fog Test Method / 193

4.7.5. Rapid Flashover Voltage Technique / 195

4.8. Other Test Procedures / 196

4.8.1. Naturally Polluted Insulators / 197

4.8.2. Liquid Pollution Method / 197

4.8.3. Dust Cycle Method / 198

4.8.4. Dry Salt Layer Method / 198

4.8.5. Cold-Fog Test Method / 199

4.8.6. Tests of Polymeric Insulator Material Endurance / 200

4.8.7. Summary of Pollution Test Methods / 201

4.9. Salt-Fog Test Results / 203

4.9.1. ac Salt-Fog Test Results / 203

4.9.2. dc Salt-Fog Test Results / 204

4.10. Clean-Fog Test Results / 205

   4.10.1. ac Clean-Fog Tests / 205

   4.10.2. dc Clean-Fog Tests / 207

   4.10.3. Impulse Voltage Clean-Fog Tests / 209

4.11. Effects of Insulator Parameters / 211

   4.11.1. Leakage Distance and Profile / 211

   4.11.2. Effect of Small Diameter : Monofilaments and ADSS / 212

   4.11.3. Influence of Average Insulator Diameter / 214

   4.11.4. Influence of Insulator Form Factor / 219

   4.11.5. Influence of Surface Material / 221

4.12. Effects of Nonsoluble Deposit Density / 223

4.13. Pressure Effects on Contamination Tests / 224

   4.13.1. Standard Correction for Air Density and Humidity / 224

   4.13.2. Pressure Corrections for Contamination Flashovers / 227

4.14. Temperature Effects on Pollution Flashover / 229

   4.14.1. Temperatures Above Freezing / 229

   4.14.2. Temperatures Below Freezing / 231

4.15. Précis / 233

   References / 234

5. **CONTAMINATION FLASHOVER MODELS**   **241**

5.1. General Classification of Partial Discharges / 242

   5.1.1. Discharges in the Presence of an Insulating Surface / 243

5.2. Dry-Band Arcing on Contaminated Surfaces / 246

   5.2.1. Wetted Layer Thickness and Electrical Properties / 246

   5.2.2. Surface Impedance Effects / 247

   5.2.3. Temperature Effects Leading to Dry-Band Formation / 247

   5.2.4. Dry-Band Formation / 248

   5.2.5. Arcing and Enlargement of Dry Bands / 249

   5.2.6. Nuisance Factors from Discharges on Wetted Pollution Layers / 250

   5.2.7. Stabilization or Evolution to Flashover / 255

5.3. Electrical Arcing on Wet, Contaminated Surfaces / 255

5.3.1.   Discharge Initiation and Development / 256

5.3.2.   Arc V–I Characteristics in Free Air / 256

5.3.3.   Arc V–I Characteristics on Water or Ice
Surfaces / 259

5.3.4.   Dynamics of Arc Propagation / 261

5.4.   Residual Resistance of Polluted Layer / 262

5.4.1.   Observations of Series Resistance of Pollution
Layer / 262

5.4.2.   Mathematical Functions for Series Resistance of
Pollution Layer / 265

5.4.3.   Resistance of Arc Root on Conducting Layer / 268

5.5.   dc Pollution Flashover Modeling / 271

5.5.1.   Analytical Solution: Uniform Pollution Layer / 271

5.5.2.   Analytical Solution Using Insulator Form
Factor / 273

5.5.3.   Numerical Solution: Nonuniform Pollution
Layer / 274

5.5.4.   Comparison of Different Models for Pollution
Layer / 274

5.5.5.   Introduction of Multiple Arcs in Series / 276

5.5.6.   dc Arc Parameter Changes with Pressure and
Temperature / 277

5.6.   ac Pollution Flashover Modeling / 278

5.6.1.   ac Arc Reignition / 278

5.6.2.   ac Reignition Conditions Versus Ambient
Temperature / 281

5.6.3.   Mathematical Model for Reignition Condition / 282

5.6.4.   Comparison of dc and ac Flashover Models / 283

5.7.   Theoretical Modeling for Cold-Fog Flashover / 284

5.8.   Future Directions for Pollution Flashover Modeling / 285

5.9.   Précis / 286

References / 287

6. **MITIGATION OPTIONS FOR IMPROVED PERFORMANCE IN
POLLUTION CONDITIONS**   **291**

6.1.   Monitoring for Maintenance / 292

6.1.1.   Insulator Pollution Monitoring / 292

6.1.2.   Condition Monitoring Using Leakage Current / 296

6.1.3.   Condition Monitoring Using Corona Detection
Equipment / 301

6.1.4.   Condition Monitoring Using Remote Thermal Monitoring / 303

6.2.   Cleaning of Insulators / 305

6.2.1.   Doing "Nothing" / 305

6.2.2.   Insulator Washing: Selecting an Interval / 306

6.2.3.   Insulator Washing: Methods and Conditions / 307

6.2.4.   Case Study: Southern California Edison, 1965–1976 / 309

6.2.5.   Insulator Washing Using Industry Standard Practices / 310

6.2.6.   Insulator Washing: Semiconducting Glaze / 313

6.2.7.   Insulator Washing: Polymer Types and RTV Coatings / 314

6.2.8.   Insulator Washing: Procedures in Freezing Weather / 316

6.2.9.   Insulator Cleaning: Dry Media / 316

6.3.   Coating of Insulators / 319

6.3.1.   Oil-Filled Insulators / 319

6.3.2.   Greases / 319

6.3.3.   Silicone Coatings / 321

6.4.   Adding Accessories / 324

6.4.1.   Booster Sheds / 324

6.4.2.   Creepage Extenders / 325

6.4.3.   Animal, Bird or "Guano" Guards / 327

6.4.4.   Corona Rings / 329

6.4.5.   Arcing Horns / 331

6.5.   Adding More Insulators / 332

6.6.   Changing to Improved Designs / 334

6.6.1.   Anti-Fog Disk Profile with Standard Spacing and Diameter / 334

6.6.2.   Aerodynamic Disk Profile / 335

6.6.3.   Alternating Diameter Profiles / 337

6.6.4.   Bell Profile with Larger Diameter and Spacing / 338

6.6.5.   Anti-Fog Disk Profiles with Larger Diameter and Spacing / 340

6.6.6.   Station Post and Bushing Profiles / 342

6.7.   Changing to Semiconducting Glaze / 343

6.7.1.   Semiconducting Glaze Technology / 343

6.7.2.   Heat Balance: Clean Semiconducting Glaze Insulators / 345

6.7.3. Heat Balance: Contaminated Semiconducting Glaze Insulators / 348

6.7.4. On-Line Monitoring with Semiconducting Glaze Insulators / 349

6.7.5. Role of Power Dissipation in Fog and Cold-Fog Accretion / 350

6.7.6. Considerations for Semiconducting Insulators in Close Proximity / 351

6.7.7. Application Experience / 352

6.8. Changing to Polymer Insulators / 352

6.8.1. Short-Term Experience in Contaminated Conditions / 353

6.8.2. Long-Term Performance in Contaminated Conditions / 354

6.8.3. Interchangeability with Ceramic Insulators / 355

6.8.4. Case Study: Desert Environment / 357

6.9. Précis / 357

References / 358

7. ICING FLASHOVERS   363

7.1. Terminology for Ice / 364

7.2. Ice Morphology / 365

7.2.1. Crystal Structure / 365

7.2.2. Supercooling / 366

7.2.3. Lattice Defects from Pollution / 367

7.3. Electrical Characteristics of Ice / 367

7.3.1. Conductivity of Bulk Ice / 368

7.3.2. Conductivity of Ice Surface / 368

7.3.3. High-Frequency Behavior of Ice / 371

7.4. Ice Flashover Experience / 373

7.4.1. Very Light Icing / 374

7.4.2. Light Icing / 377

7.4.3. Moderate Icing / 380

7.4.4. Heavy Icing / 381

7.5. Ice Flashover Processes / 384

7.5.1. Icing Flashover Process for Very Light and Light Ice Accretion / 385

7.5.2. Icing Flashover for Moderate Ice Accretion / 386

7.5.3. Icing Flashover Process for Heavy Ice Accretion / 387

7.6. Icing Test Methods / 388

    7.6.1. Standard Electrical Tests of Insulators / 389

    7.6.2. Standard Mechanical Ice Tests for Disconnect Switches / 390

    7.6.3. Natural Icing Tests in Outdoor Test Stations / 390

    7.6.4. History of Laboratory Ice Testing / 391

    7.6.5. Recommended Icing Test Method / 398

    7.6.6. Recommended Cold-Fog Test Method / 402

7.7. Ice Flashover Test Results / 403

    7.7.1. Outdoor Text Results / 403

    7.7.2. Laboratory Tests with Very Light Icing / 403

    7.7.3. Insulators with Light Ice Accretion / 406

    7.7.4. Insulators with Moderate Ice Accretion / 408

    7.7.5. Insulators Fully Bridged with Ice / 415

    7.7.6. Arresters Under Heavy Icing Conditions / 422

    7.7.7. Ice Flashover Under Switching and Lightning Surge / 422

    7.7.8. Effect of Diameter on ac Flashover for Heavy Icing / 425

    7.7.9. dc Flashover Results for Heavy Icing / 427

7.8. Empirical Models for Icing Flashovers / 431

    7.8.1. The Icing Stress Product for ac Flashover Across Leakage Distance / 432

    7.8.2. The Icing Stress Product for ac Flashover across Dry Arc Distance / 434

    7.8.3. Implementation of ISP Model for dc Flashover Under Heavy Ice Conditions / 439

    7.8.4. Comparison of Ice Flashover to Wet Flashover / 440

7.9. Mathematical Modeling of Flashover Process on Ice-Covered Insulators / 441

    7.9.1. dc Flashover Modeling of Ice-Covered Insulators / 442

    7.9.2. Influence of Insulator Precontamination on dc Flashover of Ice-Covered Insulators / 448

    7.9.3. ac Flashover Modeling of Ice-Covered Insulators / 450

    7.9.4. Application Details: Flashover Under Very Light Icing / 457

    7.9.5. Application Details: Flashover Under Light Icing Conditions / 459

7.9.6. Application Details: Flashover Under Moderate Icing Conditions / 461

7.9.7. Application Details: Flashover Under Heavy Icing Conditions / 463

7.10. Environmental Corrections for Ice Surfaces / 465

7.10.1. Pressure Correction for Heavy Ice Tests / 465

7.10.2. Arc Parameter Variation with Temperature and Pressure / 467

7.10.3. Heat Transfer and Ice Temperature / 467

7.11. Future Directions for Icing Flashover Modeling / 469

7.11.1. Streamer Initiation and Propagation on Ice Surfaces / 470

7.11.2. Dynamics of Arc Motion on Ice Surfaces / 470

7.11.3. Dynamic Model for Ice Temperature / 471

7.12. Précis / 472

References / 472

8. SNOW FLASHOVERS   **481**

8.1. Terminology for Snow / 481

8.2. Snow Morphology / 482

8.3. Snow Electrical Characteristics / 484

8.3.1. Electrical Conduction in Snow, dc to 100 Hz / 487

8.3.2. Dielectric Behavior of Snow, 100 Hz to 5 MHz / 489

8.3.3. Products of Electrical Discharge Activity / 490

8.4. Snow Flashover Experience / 490

8.5. Snow Flashover Process and Test Methods / 493

8.5.1. Snow Flashover Process / 494

8.5.2. Snow Test Methods / 495

8.5.3. General Arrangements for Snow Tests / 496

8.5.4. Snow Deposit Methods / 497

8.5.5. Evaluation of Flashover Voltage for Snow Tests / 498

8.6. Snow Flashover Test Results / 500

8.6.1. Outdoor Tests Using Natural Snow Accretion / 500

8.6.2. Outdoor Tests Using Artificial Snow Deposit / 503

8.6.3. Indoor Tests Using Natural Snow Deposits / 503

8.6.4. Snow Flashover Results for dc / 505

8.6.5. Snow Flashover Under Switching Surge / 505

8.6.6. Snow Flashover Results for Long Insulator Strings / 508

8.7   Empirical Model for Snow Flashover / 508

     8.7.1.   Conversion of Test Results to Snow Stress Product / 509

     8.7.2.   Comparison of Snow Flashover to Ice and Cold Fog / 511

     8.7.3.   Comparison of Snow Flashover to Normal Service Voltage / 512

8.8.   Mathematical Modeling of Flashover Process on Snow-Covered Insulators / 513

     8.8.1.   Voltage–Current Characteristics in Snow / 514

     8.8.2.   dc Flashover Voltage / 517

     8.8.3.   ac Reignition Condition and Flashover Voltage / 518

     8.8.4.   Switching and Lightning Surge Flashover / 519

8.9.   Environmental Corrections for Snow Flashover / 520

     8.9.1.   Pressure / 520

     8.9.2.   Temperature / 520

8.10.   Case Studies of Snow Flashover / 520

     8.10.1.   In-Cloud Rime Accretion: Keele Valley, Ontario / 520

     8.10.2.   Temporary Overvoltage Problem: 420-kV Breaker in Norway / 522

     8.10.3.   Snow Accretion on Surge Arresters / 524

8.11.   Précis / 525

     References / 525

9. MITIGATION OPTIONS FOR IMPROVED PERFORMANCE IN ICE AND SNOW CONDITIONS   **529**

9.1.   Options for Mitigating Very Light and Light Icing / 530

     9.1.1.   Semiconducting Glaze / 532

     9.1.2.   Increased Leakage Distance / 535

     9.1.3.   Coating of Insulators with RTV Silicone / 539

     9.1.4.   Change for Polymer / 543

     9.1.5.   Insulator Pollution Monitoring and Washing / 544

     9.1.6.   Case Study: SMART Washing / 548

9.2.   Options for Mitigating Moderate Icing / 550

     9.2.1.   Use of Profiles with Greater Shed-to-Shed Distance / 552

     9.2.2.   Increased Dry Arc Distance / 553

     9.2.3.   Insulator Orientation / 555

     9.2.4.   Semiconducting Glaze / 555

9.2.5. Polymer Insulators / 559

9.2.6. Corona Rings / 561

9.2.7. Condition Monitoring Using Remote Thermal Measurements / 561

9.2.8. Silicone Coatings / 562

9.3. Options for Mitigating Heavy Icing / 564

9.3.1. Increasing the Dry Arc Distance / 564

9.3.2 Changing to Semiconducting Glaze / 566

9.3.3. Adding Booster Sheds / 568

9.3.4. Changing to Polymer Insulators / 572

9.3.5. Ice Monitoring Using Leakage Current / 573

9.3.6. Ice Stripping in Freezing Weather / 575

9.3.7. Corona Rings and Other Hardware / 576

9.3.8. Increasing Shed-to-Shed Distance / 577

9.3.9. Coating of Insulators with RTV Silicone / 577

9.4. Options for Mitigating Snow and Rime / 580

9.4.1. Increased Dry Arc Distance / 580

9.4.2. Insulator Profile / 580

9.4.3. Insulators in Parallel / 581

9.4.4. Polymer Insulators / 582

9.4.5. Surface Coatings / 583

9.4.6. Use of Semiconducting Glaze / 583

9.4.7. Use of Accessories / 583

9.5. Alternatives for Mitigating Any Icing / 584

9.5.1. Doing "Nothing" / 584

9.5.2. Voltage Reduction / 585

9.5.3. Post-Event Inspection Using Corona Detection Equipment / 585

9.6. Précis / 585

References / 586

10. INSULATION COORDINATION FOR ICING AND POLLUTED ENVIRONMENTS   **591**

10.1. The Insulation Coordination Process / 592

10.1.1. Classification of Overvoltage Stresses on Transmission Lines / 592

10.1.2. High-Voltage Insulator Parameters / 594

10.1.3. Extra-High-Voltage Insulator Parameters / 595

10.1.4.   Design for an Acceptable Component Failure
Rate / 596

10.1.5.   Design for an Acceptable Network Failure
Rate / 598

10.2.   Deterministic and Probabilistic Methods / 599

10.3.   IEEE 1313.2 Design Approach for Contamination / 604

10.4.   IEC 60815 Design Approach for Contamination / 606

10.5.   CIGRE Design Approach for Contamination / 607

10.6.   Characteristics of Winter Pollution / 611

10.6.1.   Days Without Rain in Winter / 612

10.6.2.   Rate of Increase of ESDD / 612

10.6.3.   Effect of Road Salt / 615

10.7.   Winter Fog Events / 617

10.8.   Freezing Rain and Freezing Drizzle Events / 618

10.8.1.   Measurement Units / 618

10.8.2.   Frequency of Occurrence / 619

10.8.3.   Time of Day and Time of Year of Freezing
Precipitation Occurrence / 622

10.8.4.   Severity of Freezing Rain Occurrence / 622

10.8.5.   Electrical Conductivity of Freezing
Rainwater / 623

10.9.   Snow Climatology / 625

10.9.1.   Standard Methods for Snow Measurements / 625

10.9.2   Snow Accumulation and Persistence / 625

10.9.3.   Snow Melting / 628

10.10.   Deterministic Coordination for Leakage Distance / 629

10.11.   Probabilistic Coordination for Leakage Distance / 630

10.12.   Deterministic Coordination for Dry Arc Distance / 631

10.12.1.   Dry Arc Distance Requirements for Icing
Conditions / 631

10.12.2.   Dry Arc Requirements for Snow
Conditions / 632

10.13.   Probabilistic Coordination for Dry Arc Distance / 634

10.14.   Case Studies / 635

10.14.1.   Ontario 500 kV / 635

10.14.2.   Ontario 230 kV / 637

10.14.3.   Newfoundland and Labrador Hydro / 640

10.15.   Précis / 641

References / 642

**APPENDIX A: MEASUREMENT OF INSULATOR
CONTAMINATION LEVEL** 645

**APPENDIX B: STANDARD CORRECTIONS FOR HUMIDITY,
TEMPERATURE, AND PRESSURE** 651

APPENDIX C: TERMS RELATED TO ELECTRICAL IMPULSES 659

INDEX 661

PREFACE

There are many causes of power system interruption. Breakdown of self-restoring air insulation is one of the most frequent. In a former analog world, the tolerable frequency and duration of supply interruptions was much greater: our grandparents were amazed that a power system worked at all, and our parents never took it for granted the way our children do.

Many causes of power system interruption have an obvious relation to adverse weather. Lightning flashes, either to overhead groundwire protection systems or to phase conductors, are the most frequent cause of transmission system insulation breakdown at many utilities. Ice loading on power lines can lead in extreme cases to excess overturning moment and tower failures, as was shown in the 1998 ice storm that devastated the province of Québec. Conductors and insulators buzz when there is heavy rain or fog, producing higher levels of electromagnetic interference and audible noise. In some cases, partial discharges in these conditions can develop into full electrical breakdown across insulator surfaces.

Electric power systems are carefully designed to withstand mechanical forces associated with wind and ice loads. The mechanical design specification, fully developed in standards, gives good results in climates with a wide range of icing risks. There is no equivalent quantitative, standards-based design process for electrical insulation performance in the same conditions. This book allows utility engineers, consultants, researchers, and students to evaluate the risks of electrical flashover in cold weather. The authors hope that this treatment will also serve as a suitable introduction to the insulation coordination process for other adverse-weather conditions and for other risk factors.

There were few problems with insulator choices in winter conditions at high-voltage transmission levels up to 230 kV, with the exception of pole fires on distribution systems. The insulation problems were recognized more quickly as extra-high-voltage (EHV) lines moved out of remote areas with limited development and low pollution. Problems with switching overvoltages were anticipated on EHV systems but improved switchgear and surge arresters mitigated the problems efficiently in the 1960s and 1970s. This led to the use of reduced insulation levels that exposed some utilities to low levels of reliability in a specific set of weather conditions, usually including ice accreting in thin or thick layers, melting temperature at 0 °C (32 °F), and some form of environmental pollution, either in precipitation or in surface deposits that accumulate over time. With the widespread use of EHV equipment in urban and suburban winter environments, the special case of winter flashovers at the melting point has now become an important design constraint in several areas. The public is less tolerant of this kind of fault because it is rare and relatively well tolerated by other systems, such as transportation. In many cases, as will be shown, the utility is caught in a trap because winter deicing treatments that ensure ice-free roads in the winter are a root cause of their electrical flashovers.

The duration of exposure to pollution buildup in winter can be similar to that found in typical desert exposure. Symptoms of this problem also show up when drift overspray from cooling towers condenses onto cold insulators or when unfavorable wind direction blows road salt spray directly onto insulators. The study of electrical flashover on iced or polluted and frosted insulators is also an interesting and accessible introduction to more difficult problems of insulation coordination for the electrical utility engineer or student in engineering or risk management. Much of the work described here also describes aspects of the general insulation coordination process. The specific reliability evaluations set out here also serve as working models for a coordinated risk management plan to treat other weather-related reliability issues in electrical power system installations.

After reading the appropriate parts of this book, technicians and environmental specialists will be able to carry out appropriate insulator contamination measurements, understand how these readings change with time and weather, and work out how the readings compare with the upper limits set by insulator dimensions in their existing stations. Design engineers will be able to assess the likely maximum pollution and icing limits at a substation or along an overhead line, and then select insulators that have appropriate withstand margins. Regulators will understand why modest ice accretion at a moderate 0 °C temperature on one occasion can qualify as a major reliability event day, while many similar days pass without power system problems. Educators will understand why the ice surface flashover is well behaved compared to the conventional pollution flashover, making it much more suitable for demonstrations, modeling, and analysis by students.

MASOUD FARZANEH
WILLIAM A. CHISHOLM

Chicoutimi, Québec, Canada
Toronto, Ontario, Canada
July 2009

ACKNOWLEDGMENTS

This book, with a provisional title "Fire in the Ice," was originally scoped on the back of a napkin at Ù Flekù after a seminal meeting of the CIGRE Task Force, *Influence of Ice and Snow on the Flashover Performance of Outdoor Insulators*, hosted by Vaclav Sklenicka in Prague in 1999. At that time, there were too many unknowns to complete important sections of the outline. Members of the CIGRE Task Force and the corresponding IEEE Task Forces on *Icing Test Methods* and *Selection of Insulators for Reliable Winter Performance* have supported and contributed to the free sharing and consolidation of experiences that now allows a cohesive presentation.

The authors have been surprised and delighted by the success of the CIGELE program at the Université du Québec à Chicoutimi (UQAC). This chair has been sponsored by the governments of Canada (NSERC and CRC) and Québec, by electrical utilities in Canada, the United States, and France, and by manufacturers in the United States and Canada. This program, now in its third five-year mandate, has engaged more than 40 graduate students in various aspects of electrical flashover modeling. Each student has found something of value and interest in the electrical performance of cold insulators and in return has left new findings and insight that expand our understanding exponentially.

During the writing process, we have received encouragement and positive feedback from anonymous reviewers for IEEE Press and also from the editors, Steve Welch, Mo El-Hawary, Lisa Van Horn, and Jeanne Audino. We were also encouraged by the helpful and detailed reviews of chapters and the entire text by several colleagues, notably Tony Baker, Jeff Burnham, Tony Carreira, Ed Cherney, Hiroya Homma, Ray Lings, Bill Meier, Farouk A. M. Rizk, Andy Schwalm, Mike Southwood, Jerry Stewart, and Jianhui Zhang.

Permission to use previously published material is gratefully acknowledged as noted in the references. We especially note the staff of Kinectrics, the former Research Division of Ontario Hydro, and of Hydro-Québec, for their collaboration starting from basic field studies and culminating in the release of many new test results presented here.

Finally, the authors acknowledge the help and support of their wives and families.

M. F.
W. A. C.

CHAPTER 1

INTRODUCTION

1.1. SCOPE AND OBJECTIVES

Electric power systems are carefully designed to withstand mechanical forces associated with wind and ice loads. The mechanical design specification, fully developed in standards, gives good results in a wide range of climates. There is no equivalent quantitative, standards-based design process for electrical insulation in winter conditions.

There were occasional problems with insulators operating in winter conditions at high-voltage transmission levels up to 230 kV, along with hazards of pole fires on distribution systems that made use of wood insulation. The insulation problems were recognized more quickly as extra-high-voltage (EHV) lines moved out of remote areas with limited development and low pollution. With the widespread use of EHV equipment in urban and suburban winter environments, and the challenges of ultra-high-voltage (UHV) engineering, the special case of winter flashovers at air temperature close to the melting point has now become an important design constraint in several areas.

This book is mainly about the electrical performance of power system insulators when coated with various forms of ice or snow. The roles of insulator precontamination and pollution in natural precipitation are important, so

Insulators for Icing and Polluted Environments. By Masoud Farzaneh and William A. Chisholm
Copyright © 2009 the Institute of Electrical and Electronics Engineers, Inc.

there is some overlap with the many studies of electrical performance of insulators in salt-fog and clean-fog conditions. The authors recognize that when conductors are weighted down by ice and snow, there are extreme loads on the insulators, towers, and poles, and that the low temperatures also affect the mechanical strength of the components. However, electrical problems with flashovers resulting from modest ice accretion on insulators are far more frequent than line collapse under heavy ice loads.

1.1.1. Problem Areas

The number of single-phase faults in transmission networks in winter increases substantially during and after accretion of cold precipitation, followed by a rise in air temperature above 0 °C. Farzaneh and Kiernicki [1995] reviewed problems from 11 countries. A detailed review of the North American experience, leading to significant event reports to the National Electric Reliability Council (NERC), was consolidated by Chisholm [1997]. Eighteen countries reported ice and snow electrical flashover problems in a 2005 CIGRE survey [Yoshida and Naito, 2005]. The flashover problems plagued 35 utilities with 400- to 735-kV transmission systems on both line and station insulators.

Worldwide, ice and snow test methods have been developed to a high degree of sophistication in Canada, Sweden, Japan, and China, with movement toward standardization in IEEE PAR 1783 [2008]. The topic has also been covered in panel sessions and papers at general conferences on insulation and high voltage such as:

- IEEE Power and Energy Society (*PES*).
- IEEE Dielectrics and Electrical Insulation Society (*DEIS*).
- Conseil Internationale des Grands Resaux Electriques (*CIGRE*).
- International Symposium on High Voltage Engineering (*ISH*).

Specialized conferences on icing such as the International Workshop on Atmospheric Icing of Structures (*IWAIS*) have also focused sessions on electrical flashover problems.

Geographically, the Köppen–Geiger World Climate Classification of cold regions, based on vegetation, can provide a good initial indication of where in the world there will be problems with electrical flashovers on insulators. The group D (continental) climates are the ones having an average temperature above 10 °C in summer and below −3 °C in winter. There are few areas in the Southern Hemisphere, for example, highland New Zealand, that have this classification, so icing flashover problems are mainly the domain of the northern regions, shown in Figure 1-1.

There are two other aspects to the classification in Figure 1-1. The second letter in the **Dxy** code has the following meanings:

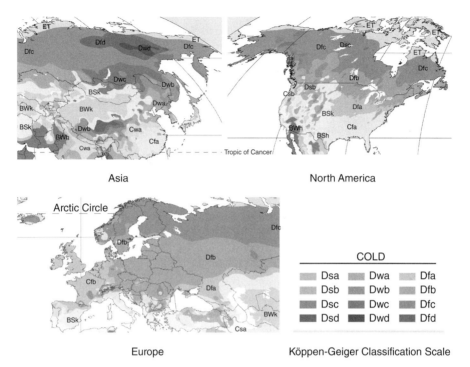

Figure 1-1: Köppen–Geiger classification of cold regions (from Norgord [2008]).

- Code s indicates an area with dry summer, with less than 30 mm precipitation and one-third of the amount in the wettest winter month.
- Code w indicates an area with dry winters, where the average winter month precipitation is less than 10% of an average summer month.
- Code f indicates significant precipitation in all seasons.

Generally, areas with relatively dry winters (code w) may be more prone to long-duration buildup of contamination, while those with heavy winter precipitation (codes f or s) will be more susceptible to problems from direct, short-term accretion of ice or snow.

The third letter in the Köppen–Geiger classification relates to the warmest monthly average temperature in summer temperature: a (above 22 °C), b (between 10 °C and 22 °C), and c (three or fewer months above 10 °C). This code is not relevant to icing performance except for areas with code d, indicating three or fewer months with mean temperatures above 10 °C in Siberia.

While it is formally classified as Cfb, the United Kingdom was the first country to report problems with flashovers in winter fogs at 0 °C on 132-kV and 275-kV systems. The code C reflects a coldest-month average temperature of somewhere between –3 °C and 18 °C.

1.1.2. Problem Characteristics

Many electrical flashover events in winter conditions have the following common features:

- A period of accumulation of pollution on insulator surfaces prior to the event, lasting from hours to months.
- Enough ice or snow accretion to fill in some of the space between insulator sheds or disks.
- Leakage currents that flow across insulators under normal operating voltage when ice, snow, or dew point conditions exist leading to ignition of pole fires if wood insulation is used in series with ceramic or nonceramic insulation.
- Flashovers that evolve slowly or quickly from leakage currents under normal operating voltage near 0 °C (32 °F), rather than in response to overvoltages from switching operations or winter lightning.
- Flashovers that occur from line to ground.
- Flashovers that occur seconds or minutes apart on parallel insulators that are exposed to the same conditions.
- Flashovers that occur mainly in areas near the sources of pollution.

These common factors lead to more problems at large transformer stations, with hundreds of insulators in parallel, rather than on lines where typical towers hold only three or six insulator strings in parallel.

The build up of pollution on insulator surfaces and the electrical characteristics of the ice, snow, or frost accumulation are dominant factors in winter flashover problems. In regions of major interest, the longest duration of exposure to pollution without cleansing rain occurs in the winter. Paraphrasing Brunnel [1972], "the winter is a desert with a perfect disguise." Also, in many of these countries, salt is used to ensure road safety under icing conditions. The road salt exposure is similar in most respects to ocean salt conditions. This brings forward an important task in this book: consolidation, use, and extension of existing methods used to evaluate insulator performance in contaminated regions.

Symptoms of icing problems also show up when drift overspray from cooling towers condenses onto cold insulators or when unfavorable wind direction blows sea salt spray or salt-laden wet snow directly onto insulators. The occurrence of dew point temperatures below freezing plays a role in these events, rather than freezing ambient temperatures.

1.1.3. Intended Audience

Reliability and power quality specialists for high-voltage networks, and their regulators, often question why outdoor insulators can fail some, but not all, of the time in winter conditions. Modest ice accretion at a moderate 0 °C winter temperature on one occasion can qualify as a major reliability event day, while many similar days pass each winter without power system problems. This book consolidates the authors' combined industrial and academic experiences with this problem, supplemented with a rich set of resources that include trouble reports from around the world originating in the 1930s, laboratory test results starting in the 1960s, and mathematical models for the flashover process that evolved in the 1990s.

Electrical utility design standards departments in many countries may refer to this book to evaluate why they have occasional unexplained reliability problems, often at dawn after a cool evening, and may not have recognized the critical role of melting temperature in their insulator flashover problems. Design engineers will also be able to assess the likely maximum pollution and icing severity at a substation or along an overhead line, and then select insulators that have appropriate withstand margins. Case studies and design tools in this book are useful guides for selecting and specifying the most appropriate insulators, bushings, and maintenance plans for the local winter conditions.

Electrical utility maintenance engineers and power system operators may need to explore root causes and quantify the cost of various mitigation options prior to making an informed recommendation for specific problem areas. The extensive coverage of mitigation options for four levels of icing severity is provided to help with this decision process. After studying the appropriate parts of this book, engineers and environmental specialists will also be able to carry out representative insulator contamination measurements, understand how these readings change with time and weather, and work out how the readings coordinate with the insulator dimensions in their existing networks.

The contents of this book lead toward a final chapter that demonstrates specific insulation coordination processes for the icing and polluted environment. Universities with programs in high-voltage engineering will find this to be a working example of how to introduce the uncertainties of real-life electrical engineering into quantitative evaluations of reliability. The process can be adapted to evaluating the risks of reliability issues with other types of adverse weather.

For engineering programs with some power system emphasis, the book may serve well as a graduate-level textbook for special topics in high-voltage electrical engineering, electrochemistry and environmental science for high-voltage engineers, insulation coordination, high-voltage transmission system design, or other courses in power engineering related to power transmission and distribution. Educators will also understand why the ice surface flashover is well behaved compared to conventional pollution flashover, making it much more suitable for demonstrations, modeling, and analysis by researchers and graduate students working in this area. The study of electrical flashover on iced or polluted and frosted insulators is also an interesting and accessible introduction to more difficult problems of insulation coordination in conventional pollution conditions for the electrical utility engineer or student in engineering or risk management.

1.2. POWER SYSTEM RELIABILITY

The modern electric power system is a complex organization of large machines, highly integrated with transmission and distribution systems for delivering continuous and real-time service to customers. At present, the ability to store electrical energy is limited mainly to the rotating inertia of the generators. This means that a fault in any of the equipment functional zones—generation, transmission, and distribution—will lead to customer disturbances. These disturbances may be minor—a voltage dip of less than 10% for a fault far away—or can be significant, if it leads to misoperation or damage of customer equipment.

The degree of customer satisfaction with the output of the electrical system is a strong function of the perceived reliability, which is the ability to supply the demand at any moment, under any circumstance, including adverse weather. Utilities and their regulators have always been concerned with measuring reliability by examining line and station fault rates, generator availability statistics, and other related data.

The occupation of the power system planner, especially for EHV systems, is to develop a design that offers the best trade-off between cost and reliability. While investment costs in lines and generating stations are relatively easy to calculate, the trend has been to include societal costs, such as visual impact, that are much more uncertain in this evaluation. The evaluation of reliability has always accommodated uncertainty, as there are constant interactions among power system components, and also large numbers of states that can satisfy the demand in a satisfactory way through the general use of fault tolerance.

1.2.1. Measures of Power System Reliability

In the United Kingdom, problems with winter-related insulation failures date from about 1935. Coincidentally, the UK Electricity Supply Regulations of 1937 specified four objectives that have evolved as follows [Argent and Hadfield, 1986]:

- To develop and maintain an efficient, coordinated, and economical system of bulk electricity supply.
- To provide a constant supply *except in case of emergency.*
- To maintain the supply frequency within ±1% of 50 Hz.
- To maintain the customer's supply voltage within limits of ±6% of the declared value.

The first, economic objective is balanced against the other three measures of reliability and security of supply. A "case of emergency" at the Central Electricity Generating Board (CEGB) was judged to be an event interrupting more than 1500 MW of customer load. In the traditional deterministic approach to power system planning, three transmission circuits would thus be needed to connect two large generators to the system. This configuration would provide full output following the loss of any two of the transmission circuits. Clearly, a common set of faults at the electrical substation supplying these three lines would have more severe consequences. This has led to development of probabilistic methods that consider the variation in probabilities and consequences of faults [Argent and Hadfield, 1986] at CEGB and many other large, regulated utilities.

In North America, standards for reporting the reliability of generation facilities are also well established, using component availability statistics. Of particular interest are historic forced outage rates, which are used to extrapolate the probability that the unit will be unavailable at some time in the future, and also the mean time to repair a unit after a typical or unexpected fault. Generators may also have partial failure modes, as they may be derated for some periods of time to respect temperature or emission limits.

Transmission and distribution systems tend to have "clean" failure modes and fast times to repair in most cases. The restoration time could be the days needed to replace a transformer with a spare component, or the milliseconds needed to perform switching actions to isolate and clear a fault. Generally, transmission system forced outages are classified in units of *outage per 100 km per year*. Momentary outages include those that are detected by relays, cleared by protective equipment, and successfully reclosed in the case of transmission systems. Sustained outages are classed as those with duration of more than 1 minute [Billington, 2006] by the CEA Equipment Reliability Information System [CEA, 2002, 2004]. Also, outages are segregated by voltage class, as many adverse weather hazards such as lightning tend to diminish as insulation levels increase.

There is a strong trend in Figure 1-2 toward reduced rates of sustained outages with increasing voltage level, while terminal problems are more severe for EHV systems. This reflects the fact that electric field stresses across

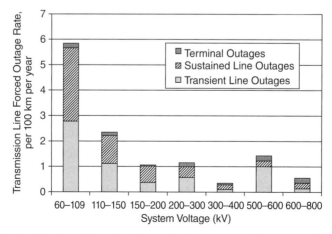

Figure 1-2: Sustained and transient forced outage rate performance of transmission lines (data from Billington [2006] and CEA [2002]).

insulators are 30–50% higher on EHV equipment than stress on typical HV systems.

There are two main measures of power system reliability that form the basis for industry comparisons. With their wide acceptance, these indices are being used as well by electric utility industry regulators.

- SAIFI is the System Average Interruption Frequency Index, given by the number of delivery point interruptions, divided by the number of delivery points monitored. It is usually averaged over a year. Areas of high lightning incidence may have a high level of SAIFI, caused by frequent but brief interruptions. To get around this, most utilities disregard outage durations of less than 1 or 3 minutes in their data collection. As a general guide, IEEE Standard 1366 [2003] reported a median value for North American utilities of 1.1 interruptions per customer with this adjustment.
- SAIDI is the System Average Interruption Duration Index, given by the total duration of all interruptions, divided by the number of delivery points. It is also reported as an annual value and sometimes converted to "system minutes." A median value of 90 minutes per year in IEEE Standard 1366 [2003] has been steadily improving at most North American utilities in response to regulatory incentives.

With the sensitivity of industrial process equipment to momentary outages, caused by the operation of protective equipment, there is increasing interest in the measure of MAIFI, or Momentary Average Interruption Frequency Index. In a Japanese survey [CRIEPI, 2004], 41% of customers' concerns were related to instantaneous voltage dips of less than 0.1 s, and another 13% were concerned with voltage fluctuations, compared to only 15% who worried about power outages of long duration. The MAIFI is given by the number of momentary outages, typically with a voltage dropout of 50% or more, divided by the number of delivery points.

The MAIFI has proved to be highly sensitive to the local level of lightning activity. Since, in many regions, there are significant variations in the ground flash density from year to year, it has become common to monitor the number of cloud-to-ground flashes in a region and to normalize the MAIFI values against this annual value. This is a desirable practice in weather normalization that is feasible because there are wide-area systems for monitoring lightning density. An example of the strong correlation for a specific area of a distribution network is given in Gunther and Mehta [1995]:

$$[\text{Sags/Month}] = 2.09\, N_g + 0.9, \quad r = 0.991 \tag{1-1}$$

where

[Sags/Month] is the number of voltage dips, typically less than 10 ac cycles in duration and larger than 10% of nominal voltage,

N_g is the lightning ground flash density (flashes/km^2/yr), and
r is the Pearson correlation coefficient.

For other adverse weather risks that lead to significant outages, the task of weather normalization is of great interest but is hampered to some extent by the limited measures of occurrence. One innovation in IEEE Standard 1366 [2003] for distribution system reliability measurement is the definition of a *major event day*. This recognizes that there are typically two modes of power system operation:

- A day-to-day mode, with reliability problems that a utility has come to expect and with planned response, putting the right staff and materiel in place as part of normal operation.
- A major event day that typically calls for emergency response procedures, including staff overtime, maintenance, or replacement procedures that, for example, have a place in the annual budget but no clear expectation of where or when the resources will be required.

The segmentation of reliability data into these two distinct sets allows utilities to analyze and report the activities that occur on major event days, and to assess their day-to-day performance against regulation targets or previous achievements on a more consistent basis.

The quantitative measure of whether an adverse weather event is a "major event day" uses a simple metric—when the SAIDI for the day is more than 2.5 standard deviations away from the daily mean SAIDI. This is known as the "2.5-Beta" method in IEEE Standard 1366 [2003]. This provides a good indication of the days on which an electrical network is experiencing stresses, such as severe weather, that are beyond those normally used.

There can be some odd results when the 2.5-Beta method is used in areas where specific reliability risks are low or high. Taking, for example, portions of the west coast of the United States, the lightning ground flash density and number of days with thunderstorms are low. Any day with lightning is classed as a "major event day." In contrast, in the southeast United States, the lightning flash density is up to 20 times greater, and dealing with the lightning activity that peaks every summer afternoon at 4 PM is a part of normal utility operation, not a major event day, using the same 2.5-Beta criterion.

Mechanical failures of transmission and distribution systems under severe ice, snow, or wind conditions typically qualify as major event days. One of the goals of this book is to indicate that there may be days with moderate ice accretion and other weather conditions, notably melting temperatures, which also qualify as major events when they lead to repeated flashovers of station and line insulators on the most critical HV and EHV networks.

1.2.2. Achieving Reliability with Redundant Components

The transmission line forced outage rates reported by Billington [2006] in Figure 1-2 are based on Canadian data and introduce an interesting anomaly for 500-kV systems. One reporting utility accepts a very high rate of 500-kV transient line outages [Mousa and Srivastava, 1989] as a consequence of the economies it achieved by eliminating overhead groundwires. Every lightning flash to the unprotected phases causes a flashover and circuit breaker operation, whether from the initial surge or from one of the subsequent strokes that follows the same flash. This is an example where a low level of component reliability, in this case transmission lines with lightning tripout rates of six to nine outages per 100 km per year, can still provide high system reliability, because resources not allocated to overhead groundwires were diverted to construction of redundant 500-kV lines along main point-to-point power transfer corridors.

In many places, the EHV networks are configured as grids. This provides a number of redundant paths for power delivery and can be particularly effective in icing conditions. A plan for operating the power system with the complete loss of a single substation may be effective for these areas. This would move the risk of repeated icing flashovers from a Class D to a Class C category in the NERC Definitions of Normal and Emergency Electric Power System Condition (given later in Table 1-2).

Long-range transport of energy using ac or dc EHV or UHV transmission lines may not have the luxury of redundant stations or parallel paths. The design of the series components of these systems to withstand the anticipated icing and contamination environments is thus more critical to the delivery of reliable and continuous electrical service.

1.2.3. Achieving Reliability with Maintenance

Icing problems causing electrical flashovers of insulators are a good example of a hazard to power systems that can be addressed with reliability-centered maintenance. A reliability-centered maintenance program is a series of orderly steps [IEEE Standard 100, 2000] for:

- Identifying system and subsystem functions, functional failures, and dominant failure modes,
- Putting the failure modes in priority order, and
- Selecting applicable and effective preventive maintenance tasks to address the classified failure modes.

The dominant failure modes—insulator line-to-ground flashovers under melting conditions—are identifiable in the case of icing problems. The risk of icing failures can also be established based on experience and modeling, to place the hazards into their correct priority compared to other power system risks. Finally, there are a fairly wide range of preventive maintenance options that can be used to prevent future failures.

1.2.4. Cost of Sustained Outages

Utilities are often careful to tally the number of customers affected, and to total up the costs of restoring service, after severe outage events. As an example, ampacity problems with low wind conditions on August 14, 2003 led to a widespread blackout that affected 50 million customers and led to financial losses estimated at US\$6 billion. This represents a loss of \$120 per person per day. The severe 1-week ice storm in eastern Canada and the United States in January 1998 also cost the economy about US\$6 billion. For the 700,000 people without power for up to 3 weeks, the specific power system restoration costs were about US\$2 billion, working out to \$140 per affected person per day.

1.2.5. Cost of Momentary Outages

There are two ways to estimate the cost of a momentary voltage dip or dropout. One way is to evaluate what a utility typically and specifically spends to eliminate common sources of line-to-ground faults. The other way is to survey customer perception and behavior, for example, in the costs associated with battery backup or redundant supply to critical loads, with lost production or other factors.

Chisholm and Anderson noted in EPRI [2005] that the cost of lightning protection on overhead transmission lines can be broken out of the overall line cost and then used to establish utility spending to avoid customer momentary dips. The additional cost of overhead groundwire protection includes three components:

- The capital cost of the wires and the heavier, taller towers to support them.
- The energy cost of induced currents in the overhead wires at average load.
- The cost of providing additional capacity to meet the induced currents at peak load.

These costs varied considerably with system voltage and line configuration, with single-circuit horizontal EHV lines having the greatest losses at peak loads. Double-circuit lines with low-reactance phasing (ABC/CBA) had peak losses that were eight times lower than lines with superbundle (ABC/ABC) phasing.

The benefits of lightning protection can be calculated from the outage rate of the line, with and without overhead groundwires. After expressing the power transfer in terms of affected customers for each phase-to-ground fault, it was possible to establish a median cost per avoided customer momentary outage of about US\$0.08 for single-circuit lines and \$0.02 for double-circuit lines with low-reactance phasing.

The benefit of lightning protection scales linearly with the lightning ground flash density. For areas with a ground flash density of 10 rather than 1 flash/km², the utility spending to avoid each customer momentary dip will be a factor of 10 lower.

It is far more effective for electrical utilities to mitigate momentary lightning outages than for the customers to undertake this protection. As an example, a customer worried about momentary disturbances from lightning flashes to a 230-kV line affecting a computer may purchase a battery-operated laptop or an uninterruptible power supply system. The customer options tend to have costs of $0.30 to $70 per avoided momentary dip [Chisholm, 2007] compared to utility options such as improving the grounding of towers ($0.03 per avoided dip), fitting some transmission line surge arresters (about $0.08 per avoided dip), or replacing 2-m insulator strings with 3-m strings ($0.15 per avoided dip).

There is also a wide gulf between what utilities spend to avoid customer momentary dips, and what the customers perceived and reported as their costs for each dip. Industrial customers advise that the cost of a momentary disturbance is on the order of $4 to $16 per kW of load [IEEE Standard 493, 2007]. An EPRI survey [CEIDS, 2001] went further, surveying all sectors of the industrial and digital economy, and estimated the annual cost of power system outages for all business sectors. These data are sorted in Table 1-1 according to the relative incidence of winter icing and fog conditions, measured by the number of days with snow cover.

While there are some regions, such as California, that do not face the direct effects of winter conditions, much of the energy supply to this region must traverse areas to the north where there are at least 84 days of snow cover per year. Even ignoring this fact, and using the median area values for snow cover in Table 1-1, the weighted sum of all electrical outage problems that occur in

TABLE 1-1: Cost of Electric Power System Outages by Region, Along with Average Number of Days with Snow Depth Greater than 25 mm

Region	Median Number of Days with Snow Depth >25 mm	Estimated Annual Cost of Outages, US$$_{2001}$ (Min–Max)
New England	84	6.1–9.6
Middle Atlantic	56	15.4–24.3
Mountain	56	6.1–9.5
West North Central	42	8.1–12.6
East North Central	35	17.3–27.0
South Atlantic	7	18.3–20.7
East South Central	7	6.3–9.9
West South Central	3	11.2–17.6
Pacific	0	16.2–25.2

Source: CEIDS [2001] and NOAA.

winter conditions falls between US$8 and US$12 billion per year for the United States.

An interesting point was raised in the CEIDS [2001] study, related to the effect of multiple reclosing sequences on customer loads. It is normal for utilities to protect against transmission system insulation flashovers of all sorts with automatic circuit breakers. Customers reported that the average cost of equipment damage after such a recloser operation was more than $2000, while the median value for any other outage duration was only $550. This was attributed to the strain of stopping and restarting machines and processes so quickly. The occurrence of several icing flashovers in a short period of time, each leading to a recloser operation, will exacerbate this problem for some customers.

1.2.6. Who is Responsible for Reliable Electrical Systems?

Electric power system reliability has many stakeholders, including:

1. Reliability coordinators.
2. Regional reliability organizations.
3. Transmission operators.
4. Balancing authorities.
5. Generator operators.
6. Transmission service providers.
7. Load-serving entities.
8. Purchasing–selling entities.

In the United States, the National Electric Reliability Council (NERC) gives reliability coordinators in their specific areas the mandate to redispatch generation, reconfigure transmission, or reduce load to mitigate critical conditions to return the electric power system to a reliable state. Reliability coordinators can delegate tasks to other stakeholders, but this delegation does not change their responsibility to comply with NERC and regional standards and to act fairly.

The design of a reliable transmission system places higher costs on transmission service providers. An alternative, which is to operate around problems by allowing selected failures, shifts these costs to the transmission operators and may also affect the ability of generator operators to deliver power into the grid.

Operation of the transmission system also has important rules and regulations that may be relevant, including identification of severe weather and emergency operating procedures in effect. NERC Standard TPL-003 classifies the normal and abnormal system operation as shown in Table 1-2.

NERC Standard TPL-003-00 [2005] also mandates that power systems should be designed to limit cascading outages for Category A, B, and C

TABLE 1-2: NERC Definitions of Normal and Emergency Electric Power System Condition

Category	Initiating Events and Contingencies	System Limits or Impacts
A: No contingencies		No loss of demand or curtailed firm transfers
B: Event resulting from the loss of a single element	Loss of an element without a fault. Single line-to-ground (SLG) or three-phase (3φ) fault and normal clearing: 1. Generator 2. Transmission circuit 3. Transformer Single-pole block and normal clearing: 4. Single pole (dc) line	System stable; thermal and voltage limits within ratings
C: Event(s) resulting in the loss of two or more (multiple) events	SLG fault with normal clearing: 1. Bus section 2. Breaker (failure or internal fault) 3. Type B fault, manual System adjustments, followed by another Type B fault: 3. Bipolar (dc) line fault with normal clearing 4. Fault on any two circuits of multiple circuit tower line with normal clearing SLG fault with delayed clearing (stuck breaker or protection system failure) on: 5. Generator 6. Transformer 7. Transmission circuit 8. Bus section	Loss of demand or curtailed firm transfers

14

D: Extreme event resulting from two or more (multiple) elements removed or cascading out of service

3φ Fault with delayed clearing on:

1. Generator
2. Transformer
3. Transmission circuit
4. Bus section
5. 3φ Fault with normal clearing of breaker failure
6. Loss of tower line with three or more circuits
7. All transmission lines on a common right-of-way
8. Loss of a substation (one voltage level plus transformers)
9. Loss of a switching station (one voltage level plus transformers)
10. Loss of all generating units at a station
11. Loss of a large load or major load center
12. Failure of a fully redundant special protection system (or remedial action scheme) to operate when required
13. Operation, partial operation, or misoperation of a fully redundant special protection system (or remedial action scheme) in response to an event or abnormal system condition for which it was not intended to operate
14. Impact of severe power swings or oscillations from disturbances in another regional reliability organization.

Source: NERC Standard TPL-003-00 [2005].

conditions. However, widespread ice storms or contamination problems with multiple line-to-ground faults in stations and lines can often lead to loss of a substation or switching station, leading to a Category D condition. For these events, system performance is evaluated for risks and consequences of:

- Substantial loss of customer demand and generation in a wide area, and
- Inability of parts or all of the interconnected systems to achieve new, stable operating points.

Evaluation of these Category D events may require joint studies with neighboring systems.

1.2.7. Regulation of Power System Reliability

Voluntary standards for electric power quality have been negotiated among industry groups such as Information Technology Industry Council [ITIC, 2005] and electric utilities. This is expressed as an envelope of tolerance around the nominal ac voltage at the customer premises. Household (120/240V) supply voltage that stays inside this envelope can be tolerated with no interruption in function by most information technology equipment. While the compliant equipment would not be damaged by a short-duration voltage reduction, a drop of line voltage to 70% of nominal for more than 20 ms would lead to an interruption in function—for example, a reset of a computer or machine control system. This means that, even if automatic protective relaying and successful reclose operations occur after a power system fault, many pieces of customer equipment will still be affected, since even the best high-speed breakers take more than 20 ms to complete their open-and-reclose function.

Some industries, such as semiconductor processing, have opted to improve the tolerance of specialized process equipment to short-duration dropouts. However, many more are relying increasingly on regulatory pressure. This has evolved into modern electric power system reliability standards, for example, the US Energy Policy Act of 2005, that set out responsibilities and reporting requirements. In order to ensure compliance, regulatory agencies have discretion to apply sanctions whenever a violation is verified. The prior attitude within an organization toward regulatory reporting and its commitment to meeting responsibilities is an important factor in the level of sanction that is levied. By recognizing insulator faults under icing and polluted conditions, utilities can be more confidently prepared for regulatory audits of their reliability compliance program.

1.3. THE INSULATION COORDINATION PROCESS: WHAT IS INVOLVED?

There are two main aspects in the process of insulation coordination. These are the design of insulation systems to withstand overvoltages, and the design to withstand adverse weather.

The most common example of an overvoltage related to adverse weather is the lightning surge. In general, the lightning performance of a transmission line is calculated with statistical methods that have a high state of development. Provided that the main input data, such as the distribution of soil resistivity or footing resistance and the local ground flash density, are known with sufficient detail and accuracy, the quality of prediction of lightning outage rates is very good. As an example, there is a strong linear relation between annual ground flash density and the number of lightning outages on the Japanese high-voltage transmission networks [Ishii et al. 2002].

There have been fewer reports of the performance of transmission line failures under constant line voltage and adverse contamination and wetting conditions. One recent summary was provided by Gutman et al. [2006] and important details are summarized in Table 1-3.

To date, there has been no organized presentation of the typical outage rates of power systems under ice and snow conditions that cause insulator flashovers because the incidence rates are low, even compared to the number of pollution events in Table 1-3 for Norway. Instead, experience has been consolidated in papers such as those of CIGRE [TF 33.04.09, 1999; TF 33.04.09, 2000; Yoshida and Naito, 2005], IEEE [Farzaneh et al. 2003, 2005, 2007; IEEE PAR 1783, 2008], and CEATI [Farzaneh and Chisholm, 2006].

1.4. ORGANIZATION OF THE BOOK

The remainder of the book is organized as follows.

Chapter 2 describes the general nature of electrical insulation, with a focus on HV and EHV systems. The chapter identifies terms used to describe the various electrical stresses that occur in service and introduces the standard tests used to quantify electrical strength.

Chapter 3 introduces the methods and terminology used to describe the effects of the natural environment on the insulation systems. The role of environmental pollution in the electrical conductivity of wetted surface deposits and of the precipitation itself is built up from the background measurements. Then, for the important and specific pollution sources such as the ocean, road salting, or cooling tower effluent, suitable pollution accumulation models are consolidated from literature around the world. Enough practical electrochemistry is provided for users to carry out their own evaluations of the electrical conductivity of local pollution and precipitation.

Chapter 4 describes the various test methods that have evolved to reproduce the natural processes of contamination build up and wetting. Representative test results from salt-fog, clean-fog, and cold-fog tests are shown. Empirical models are developed to interpolate some important results to give initial insight about how the physical processes of contamination flashover vary with contamination level, voltage stress, leakage distance, and other insulator characteristics. This chapter also suggests that the cold-fog flashover

TABLE 1-3: Comparison of Pollution Flashover Rates of Transmission Lines in Norway, Russia, and South Africa

Country	Line Voltage (kV)	Line Length (km)	Altitude (km)	Number of Pollution Events per Year	ESDD and NSDD (mg/cm^2)	Contamination Flashover Rate per 100km-yr
Russia	117	148	200	200	0.13^a/1.0	0.43
	119	87	200	150	0.04^a/0.5	1.43
South Africa	400	163	100	72	0.1/0.5	0.6
	400	60	1600	40	0.1/0.5	1.0
	400	74	1500	40	0.1/0.5	0.17
	400	336	1200	15	0.06/0.5	0
Norway	420	106	200	10	0.1/0.1	0
	420	134	1200	10	0.05/0.1	0
	420	118	650	8	0.1/0.1	0
	300	48	200	5	0.1/0.1	0

^{a}Using conversion of layer conductivity (μS) = 71.2 (ESDD)$^{0.786}$.

Source: Adapted from Gutman et al. [2006].

of polluted insulator surfaces is a relatively simple extension of the clean-fog flashover mechanism.

Chapter 5 on modeling of the contamination flashover builds on the knowledge gained from Chapters 3 and 4, using the likely resistance of a polluted and wetted insulator surface, and the observed effects that this has on electrical strength. These data are inputs to mathematical modeling that was developed starting in the 1980s to predict the ac voltage stresses that will lead initially to arcing and, at higher stresses, to flashover.

Chapter 6 reviews the wide variety of methods and options for improving the leakage distance flashover performance of insulators in polluted environments. Examples are drawn from both laboratory and field exposure. Results are compared with the empirical and theoretical models of flashover strength.

Chapter 7 on icing flashovers takes advantage of the ability to study flashover in controlled conditions to build a series of mathematical models for very light, light, moderate, and heavy icing severity. These four levels of severity are also used to organize the available test results in a wide range of conditions.

Chapter 8 completes the treatment of winter precipitation by analyzing the effects of dry and wet snow on the insulators. As in Chapter 6, a combination of visual observations, leakage current measurements, and results from several different physical scales are presented in an organized sequence.

Chapter 9 reviews the wide variety of options for improving flashover performance of insulators in icing environments. Case studies describe utility experiences with a number of different methods to deal with recurring winter flashover problems. These include ways to keep ice from forming (heat lamps or booster sheds), ways to deal with build up of ice pollution (SMART washing or steam cleaning) and ways to make insulators perform better (modified insulator surfaces or accessories).

Chapter 10 closes two loops by the insulation coordination process selection of dry arc and leakage distances. The existing process for selecting an appropriate leakage distance for contaminated environments is adapted for freezing conditions. Statistical modeling is introduced to set this contamination flashover problem into the overall context of substation and transmission line reliability evaluation. This can be done because the numerical models developed for pollution flashover actually work even better in cold-fog conditions, once arc parameters are corrected. Provision of adequate leakage distance is usually not onerous: the electrical industry has been successfully adapting leakage distance to local environments for more than 160 years. Cold-fog conditions in winter are an additional factor—the "winter desert with a perfect disguise"—that must be considered as a severe problem for EHV systems constructed near expressways subject to road salting.

Chapter 10 also treats the selection of an appropriate insulator dry arc distance. Unlike the selection of leakage distance, where there is a strong set of precedents and standards, the advice in this chapter is relatively new and, while well supported by test data, remains to some extent based on analytical and empirical models of the flashover process. Since the provision of adequate dry arc distance for all conditions can be very costly and affects insulation coordination of costly components such as transformers and circuit breakers, it is very important to place the icing problem within the overall context of substation and transmission line reliability evaluation.

1.5. PRÉCIS

Power systems achieve excellent overall levels of reliability, as measured by the frequency and duration of interruptions. This reliability is achieved through a combination of redundancy—such as a grid of parallel paths—and adequate performance of individual components such as transmission lines and stations.

Adverse weather of all types has a strong effect on reliability of power system components. Risks from lightning in summer storms are well understood, and protection measures are adapted to the local climate. The costs of this protection are balanced against regulatory requirements using SAIDI, SAIFI, and MAIFI metrics. The same process is not well established for risks from icing and contamination flashovers in winter. This book addresses the rational selection of station and line insulators in areas where these problems recur occasionally or frequently.

REFERENCES

References and bibliography for further study are provided at the end of each chapter. Also, the figures for this book and electronic links to relevant papers in the IEEE Xplore system and the public domain will be available at the following IEEE/Wiley web site, using the access code provided with this volume: www.ieee.org.

Argent, S. J. and P. G. Hadfield. 1986. "Probabilistic Transmission Planning Procedures Within the CEGB," in *Probabilistic Methods Applied to Electric Power Systems*, (S. Krishnasamy, ed.) Oxford, UK: Pergamon, pp. 215–222.

Billington, R. 2006. "Reliability Data Requirements, Practices and Recommendations," in *IEEE Tutorial Course: Electric Delivery System Reliability Evaluation*, Publication 05TP175. May.

Brunnel, D. 1972. "A Horse with No Name." Los Angeles, CA: Warner Brothers.

CEA. 2002. "Forced Outage Performance of Transmission Equipment—1997–2001." Canadian Electricity Association, Montreal, Québec.

CEA. 2004. "Bulk Electricity System: Delivery Point Interruptions and Significant Power Interruptions—1999–2003 Report." Canadian Electricity Association, Montreal, Québec.

CEIDS (Consortium for Electric Infrastructure to Support a Digital Society). 2001. "The Cost of Power Disturbances to Industrial & Digital Economy Companies." Report by Primen for EPRI. June.

CIGRE TF 33.04.09. 1999. "Influence of Ice and Snow on the Flashover Performance of Outdoor Insulators, Part I: Effects of Ice," *Electra*, No. 187 (Dec.), pp. 91–111.

CIGRE TF 33.04.09. 2000. "Influence of Ice and Snow on the Flashover Performance of Outdoor Insulators, Part II: Effects of Snow," *Electra*, No. 188 (Jan.), pp. 55–69.

CRIEPI. 2004. "Development of an Analysis Tool for Countermeasures Against Voltage Dips," *CRIEPI News*, No. 400 (July).

Chisholm, W. A. 1997. "North American Operating Experience: Insulator Flashovers in Cold Conditions." CIGRE SC33 Colloquium, Toronto, Canada.

Chisholm, W. A. 2007. "Placing a Price on Momentary Customer Dips," *Insulator News and Market Report* (INMR), Vol. 15, No. 1, pp. 38–47.

EPRI. 2005. *EPRI AC Transmission Line Reference Book—200 kV and Above*, 3rd edition. Palo Alto, CA: EPRI.

Farzaneh, M. and J. Kiernicki. 1995. "Flashover Problems Caused by Ice Build-Up on Insulators," *IEEE Electrical Insulation Magazine*, Vol. 11, No. 2 (Mar.), pp. 5–17.

Farzaneh, M., T. Baker, A. Bernstorf, K. Brown, W. A. Chisholm, C. de Tourreil, J. F. Drapeau, S. Fikke, J. M. George, E. Gnandt, T. Grisham, I. Gutman, R. Hartings, R. Kremer, G. Powell, L. Rolfseng, T. Rozek, D. L. Ruff, D. Shaffner, V. Sklenicka, R. Sundararajan, and J. Yu. 2003. "Insulator Icing Test Methods and Procedures: A Position Paper Prepared by the IEEE Task Force on Insulator Icing Test Methods," *IEEE Transactions on Power Delivery*, Vol. 18, No. 3 (Oct.), pp. 1503–1515.

Farzaneh, M., T. Baker, A. Bernstorf, J. T. Burnham, T. Carreira, E. Cherney, W. A. Chisholm, R. Christman, R. Cole, J. Cortinas, C. de Tourreil, J. F. Drapeau, J. Farzaneh-Dehkordi, S. Fikke, R. Gorur, T. Grisham, I. Gutman, J. Kuffel, A. Phillips, G. Powell, L. Rolfseng, M. Roy, T. Rozek, D. L. Ruff, A. Schwalm, V. Sklenicka, G. Stewart, R. Sundararajan, M. Szeto, R. Tay, and J. Zhang. 2005. "Selection of Station Insulators with Respect to Ice and Snow—Part I: Technical Context and Environmental Exposure," *IEEE Transactions on Power Delivery*, Vol. 20, No. 1 (Jan.), pp. 264–270.

Farzaneh, M. (Chair), A. C. Baker, R. A. Bernstorf, J. T. Burnhan, E. A. Cherney, W. A. Chisholm, R. S. Gorur, T. Grisham, I. Gutman, L. Rolfseng, and G. A. Stewart. 2007. "Selection of Line Insulators with Respect to Ice and Snow—Part I: Context and Stresses, A Position Paper Prepared by the IEEE Task Force on Icing Performance of Line Insulators," *IEEE Transactions on Power Delivery*, Vol. 22, No. 4 (Oct.), pp. 2289–2296.

Farzaneh, M. and W. A. Chisholm. 2006. "Guide to Define Design Criteria for Outdoor Insulators Taking into Account Pollution and Icing," *CEATI Wind and Ice Storm Mitigation* (WISMIG) Report T043700-3326. April.

Gunther, E. W. and H. Mehta. 1995. "A Survey of Distribution System Power Quality—Preliminary Results," *IEEE Transactions on Power Delivery*, Vol. 10, No. 1 (Jan.), pp. 322–329.

Gutman, I., J. Lundquist, K. Halsan, L. Wallin, E. Solomonik, and W. L. Vosloo. 2006. "Line Performance Estimator Software: Calculations of Lightning, Pollution and Ice Failure Rates Compared with Service Records," in *Proceedings of 2006 CIGRE Conference*, Paris, Paper B2-205, pp. 1–11.

IEEE Standard 100. 2000. *The Authoritative Dictionary of IEEE Standard Terms*, 7th edition. Piscataway, NJ: IEEE Press.

IEEE Standard 1366. 2003. *IEEE Guide for Electric Power Distribution Reliability Indices*. Piscataway, NJ: IEEE Press.

IEEE Standard 493. 2007. *IEEE Recommended Practice for the Design of Reliable Industrial and Commercial Power Systems*. Piscataway, NJ: IEEE Press.

IEEE PAR 1783. 2008. "IEEE Guide for Test Methods and Procedures to Evaluate the Electrical Performance of Insulators in Freezing Conditions." Joint DEIS/PES Task Force on Insulator Icing, PAR 1783. January 14.

Ishii, M., T. Shindo, T. Aoyama, N. Honma, S. Okabe, and M. Shimizu. 2002. "Lightning Location Systems in Japan and Their Applications to Improvement of Lightning Performance of Transmission Lines," in *Proceedings of CIGRE Session 2002*, Paper 33–201.

Information Technology Industry Council (ITIC). 2005. "ITI (CBEMA) Curve Application Note." Available at http://www.itic.org.

Mousa, A. M. and K. D. Srivastava. 1989. "The Lightning Performance of Unshielded Steel-Structure Transmission Lines," *IEEE Transactions on Power Delivery*, Vol. 4, No. 1 (Jan.), pp. 437–445.

North American Electric Reliability Council (NERC). 2005. "Standard TPL-003-00— System Performance Following Loss of Two or More BES Elements." Available at ftp://www.nerc.com/pub/sys/all_updl/standards/rs/.

Norgord, D. 2008. "Köppen–Geiger World Climate Classification," *Geographic Techniques—Geographic Analysis & Map Design*. Available at www.scribd.com/doc/2164459/KoppenGeiger-World-Climate-Classification-Map.

Yoshida, S. and K. Naito. 2005. "Survey of Electrical and Mechanical Failures of Insulators Caused by Ice and/or Snow," CIGRE WG B2.03 Report, *Electra*, No. 222 (Oct.), pp. 22–26.

CHAPTER 2

INSULATORS FOR ELECTRIC POWER SYSTEMS

This chapter describes the general nature of electrical insulation, with a focus on *high-voltage* (HV) and *extra-high-voltage* (EHV) systems. This chapter identifies terms used to describe the insulator characteristics, the various stresses encountered in service, and the standard tests used to quantify electrical strength. In the contaminated or icing environment, the insulator serves merely as an insulating substrate on which a pollution layer may accumulate. It is mainly the shape, size, and surface finish of the insulator rather than the material that affects the accumulation process. The response of the insulator to wetting is influenced by the insulator surface material, with a distinction between hydrophilic (ceramic) materials that sheet water, and hydrophobic (nonceramic or polymer) materials that bead water. North American practice commonly uses "polymer" insulators, while the interchangeable term "nonceramic" is used internationally. Since several references already provide details on the mechanical construction of insulators [Looms, 1988; Gorur et al., 1999; EPRI, 2005], this will not be covered in detail.

2.1. TERMINOLOGY FOR INSULATORS

An efficient electrical system makes more use of high voltage rather than high current. For example, the ratio between household current (15 A) and transmission line current (1500 A) is 100:1, while the ratio of voltages (115 V to

Insulators for Icing and Polluted Environments. By Masoud Farzaneh and William A. Chisholm
Copyright © 2009 the Institute of Electrical and Electronics Engineers, Inc.

115 kV) is 1000:1. It is much more effective to provide adequate high-voltage insulation than to increase conductor cross section.

2.1.1. Electrical Flashover

The dielectric insulation between metal fittings offers a high resistance to the passage of current and also resists disruptive discharge (flashover). *Flashover* is defined as a disruptive discharge through air around or over the surface of a solid or liquid insulation, between parts of different potential or polarity, produced by the application of voltage wherein the breakdown path becomes sufficiently ionized to maintain an electric arc [IEEE Standard 100, 2004]. The *disruptive discharge* completely bridges the insulation under test, reducing the voltage between the electrodes to practically zero.

Insulation is *self-restoring* if it completely recovers its insulating properties after a disruptive discharge; insulation of this kind is generally, but not necessarily, gas or liquid rather than solid insulation. Once a disruptive discharge has been interrupted, the air around a solid insulator will be self-restoring within a fraction of a second.

The electrical performance of power system insulators is normally tested using procedures such as those found in IEEE Standard 4 [1995] and IEC Standard 60060-1 [1989]. These standards specify the voltage waveshapes and other test details to be used in the evaluation of insulator performance. The *critical flashover level* is defined as the amplitude of voltage of a given waveshape that, under specified conditions, causes flashover through the surrounding medium on 50% of the voltage *applications*.

2.1.2. Mechanical Support

High voltage *insulators* are devices intended to give flexible or rigid support to electrical conductors or equipment and to insulate the conductors or equipment from ground or from other conductors or equipment [IEEE Standard 100, 2004]. The mechanical ratings and the methods for verifying insulator compliance can be found in standards such as those published by the International Electrotechnical Commission (IEC) and many national standards writing bodies. For example, the American National Standards Institute (ANSI) in the United States develops nation-specific standards such as the C29 series for insulators. Industry groups such as the Canadian Electricity Association (CEA) Light-Weight Insulator Working Group (LWIWG) have also developed common purchase specifications for polymer insulators that may be adopted by participating utilities or by the Canadian Standards Association (CSA).

2.1.3. Insulator Dimensions

The flashover performance of a self-restoring insulator is established by the dimensions of the insulator, the shape of the sheds or convolutions on its surface, the environmental conditions, and the nature of the surface material.

- *Dry arc distance* is the shortest distance in air between the line (high voltage) and the ground electrodes. The dry arc distance of an insulator determines the BIL (basic lightning impulse insulation level) of the system. This BIL in turn is used to establish safety clearances in national electric safety codes [IEEE/ANSI Standard C2, 2007] and is also an important aspect in guides to insulation coordination [IEC Standard 60071-2, 1996].
- *Shaft diameter* is the measure of the insulator size at its most narrow part. For ceramic post and long-rod insulators, shaft diameter normally determines the material cross section and the contact area to, and size of, metal end fittings. These areas, to a large extent, establish the mechanical breaking strength of the insulator. Some ceramic insulators have tapered profiles, with increasing shaft diameter toward the base of the post to provide additional strength where the mechanical stresses are greatest. Polymer insulators have a central core member with a rubber sheath. The shaft diameter and mechanical breaking strength for a polymer insulator are usually defined by the core member diameter, which is normally a fiberglass rod. The rubber sheath does not usually add much strength or thickness, and the shaft diameter is constant along the entire length.

A solid cylinder of insulating material with adequate dry arc distance would perform well in dry conditions, but would be susceptible to electrical flashovers whenever the surface became wet. The sheds break up columns of water that might otherwise run down as continuous vertical channels along the insulator's entire length. The sheds work as multiple umbrellas to keep some parts of the surface dry. Properly designed sheds increase the wet flashover strength of insulators in heavy rain conditions.

- *Shed diameter* is the measure of the insulator size at its widest part. This dimension is usually of more interest than shaft diameter in conditions of heavy ice or snow accretion. A crescent of ice can accumulate on the windward side, spanning the full outer diameter of the insulator.
- *Shed spacing* is the distance between the centers of the tips of each shed.

Insulator sheds have another benefit in outdoor conditions. The electric field stress along the surface of the complete length of the insulator is reduced. This means that the insulators tolerate a higher degree of pollution accumulation before partial discharge activity is initiated. The electrical stress across the surface of the complete length of the insulator is given by the normal power system operating voltage, divided by the leakage distance.

- *Leakage distance* is the shortest distance, or sum of the shortest distances, along the insulating surface between the line and ground electrodes [IEEE Standard 4, 1995; CSA Standard C411.1, 1989]. The distance across any cement or conducting jointing material is not counted, but distance across any high-resistance coating is included. Surfaces coated with semi-conducting glaze are considered as effective leakage surfaces.

Leakage distance is typically selected on the basis of the maximum contamination levels expected in the installed field location. For chains of standard porcelain or glass disks, the shed spacing and diameter are fixed. However, leakage distance can be increased by incorporating several ribs on the undersurface of the disk or by adding additional disks to the chain. Deep undersurface ribs may have diminishing returns, as numerous deep ribs will tend to load up with contamination over time and impede natural or high-pressure washing. If there is sufficient clearance, insulators with larger diameters or alternating diameters can also be used. For porcelain post insulators, many, relatively shallow ribs or "petticoats" are cut into solid cylinders of clay in a lathe before glazing and firing.

At average voltage stresses above 25 kVrms per meter of leakage distance, for high levels of surface pollution level, discharges will be initiated, propagate, and extinguish in a process called "dry-band arcing." The arc current is usually not more than 1 A but is still sufficient to maintain arc plasma with a temperature of about 4000 K. At this temperature, the arc is electrically conductive and can short out portions of the leakage distance. If an arc grows to a sufficient length along the insulator, the remaining air gap will become too weak to withstand the line voltage.

- *Shed-to-shed separation* is measured from the bottom of one shed surface to the top of another. For station post insulators with thick ribs, the shed-to-shed separation can be half of the shed spacing. The ratio of shed-to-shed separation to *shed depth* measured from the outer diameter is of some interest in contamination performance. The air space in the shed-to-shed separation distance can break down if there is too much voltage drop along the portion of leakage distance between the sheds. The shed-to-shed separation also influences how much ice accretion is needed to bridge the dry arc distance with a continuous ice layer.

- *Form factor* of an insulator is one measure used to quantify the relation between the shed-to-shed separation and shed depth (shed overhang). It is defined as the integral of incremental leakage distance dl divided at each measurement point by the insulator circumference $2\pi r$. The form factor of a cylinder of length 1 m and radius 10 mm is 15.9; if the radius is increased to 100 mm, the form factor reduces to 1.59. Form factor is used in electrical testing to ensure that the desired precontamination and wetting conditions have been achieved. Form factor may also be used when comparing the performance of insulators with the same length, diameter, and leakage distance but different profiles. Disks with lower form factor tend to make full use of their leakage distance, but they also tend to have lower leakage distances than disks with high form factors. Figures 2-1 to 2-3 show that there can be a very wide range in the form factor for typical power system insulators, ranging from about 1.0 for individual disks of a multidisk chain of ceramic insulators to 50 or more for long-rod single-piece polymer insulators.

• *Wall thickness* of hollow insulators, used to enclose high-voltage equipment (surge arrester blocks, instrumentation) or to transfer potentials into the interior of metal housings, establishes both the electrical and mechani-

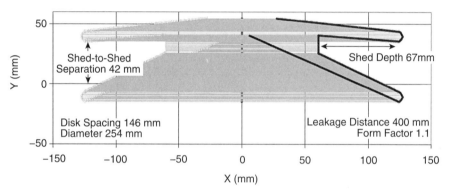

Figure 2-1: Typical dimensions of exterior-rib porcelain disk insulator for high-contamination areas.

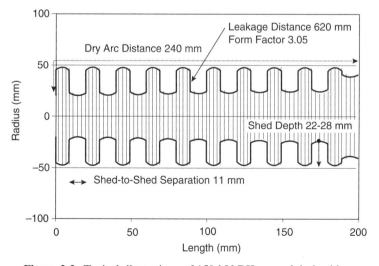

Figure 2-2: Typical dimensions of 150-kV BIL porcelain bushing.

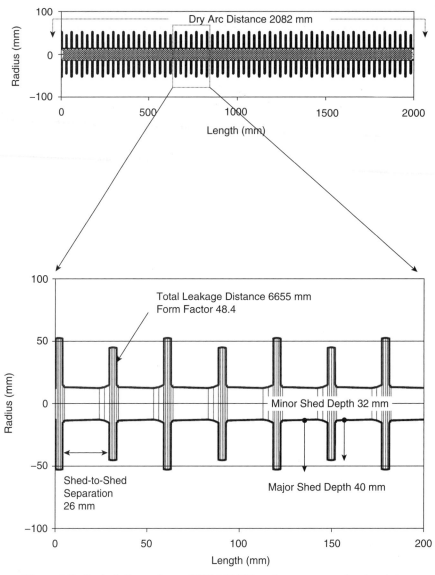

Figure 2-3: Typical dimensions of 900-kV BIL polymer suspension insulator.

càl strength. The thickness is selected so that an external flashover will occur in air across the outer surface of the insulator, rather than a *puncture*, a disruptive discharge through the solid insulation or an *internal flashover* along the inside surface. Puncture can also occur through the cement and shell of porcelain insulators, or through the individual sheds of polymer insulators. Flashover through the insulating parts of an insulator, such as between the rubber cover and fiberglass core of a polymer insulator, has recently been labeled as a *flashunder* phenomenon.

2.1.4. Interpretation of Terminology for Winter Conditions

For winter conditions, dry arc and leakage distance dimensions must provide reliable electrical service for the anticipated precipitation. However, to a very important degree, these solid accumulations change both the shape and the surface properties of the insulators.

Figure 2-4 shows how the accumulation of snow or ice on insulators can reduce the leakage distance to a value as low as the dry arc distance. For design purposes, it is thus common to consider the dry arc distance, bridged by contaminated ice or snow, as the most critical dimension and situation.

This ice and snow accumulate in a crescent shape on the sides of insulators that face into the wind (the windward side). Accumulation usually extends across the full insulator shed diameter. The small shed spacing and shallow shed depth of typical station post insulators in Figure 2-4 are easily filled with ice or snow. In contrast, it takes a heavy ice deposit to bridge the shed-to-shed separation of a string of line insulators, whether in vertical suspension positions or when mounted in a horizontal orientation, like those in Figure 2-5.

Snow or ice tends to accumulate on the top half of insulators with a horizontal orientation. Snow also tends to be trapped in the spaces between double or quadruple chains of insulators, also as shown in Figure 2-5.

In Figures 2-4 and 2-5, snow or ice accretion fills in most of the space in the convoluted insulator surfaces. In the accepted technical terms given earlier,

Figure 2-4: Accretion of snow and ice on 735-kV ceramic station post insulators. (Courtesy of TransEnergie.)

Figure 2-5: Accumulation of snow on horizontal EHV transmission line disk insulators. (Courtesy of TransEnergie.)

the leakage distance is bridged by the accretion. In worst case conditions, as shown in Figure 2-4, this leaves an ice or snow path across the entire dry arc distance.

Some of the large-diameter post insulators in Figure 2-4 are hollow. These are distinguished mainly by the size of the metal fittings at each end. Hollow core insulators enclose measurement equipment for line potentials and currents, switchgear, or connections from the line into transformers that have internal insulating oil or gas-insulated switchgear. Accretion of snow and ice on the exterior insulator surfaces affects the electrical performance of all of these porcelain components.

The highest level of insulator pollution may occur in the winter, where there is no natural rain to wash away soluble deposits. A fraction of this precontamination will dissolve into ice or snow layers, increasing electrical conductivity. The polluted top-surface area of insulators is the substrate that will have the greatest influence on the ice conductivity in these conditions, because the bottom surface area of the insulator will not be directly exposed to ice accumulation.

2.2 CLASSIFICATION OF INSULATORS

Generally, insulators are assembled from metal fittings and insulating materials. There are many ways to classify them. Three main parameters are normally considered first:

- Insulator material, addressed in Section 2.1
     Ceramic (glass and porcelain)
     Polymeric (nonceramic)

- Application, addressed in Section 2.2

  Stations

  Transmission or distribution lines

- Nature of mechanical load, addressed in Section 2.3

  Tension mainly in typical distribution and transmission line suspension and dead-end applications

  Combination of compression, torsion, and cantilever in both station and line applications

2.2.1. Classification by Ceramic or Polymeric Material

There are two basic materials for power system insulators—ceramic (porcelain and glass) and polymeric (nonceramic).

Ceramic materials rely on the strong polar chemical bonds between silicon and oxygen to give materials that are stable at high temperature, strong, and highly resistant to chemical attack. These advantages are offset by a high value of surface free energy, which causes ceramic surfaces to be easily wetted and, as a consequence, easily polluted in outdoor environments.

Polymer materials have large molecules that are weakly bonded by van der Waals forces. Successful polymers for insulators have been based on materials with carbon–carbon bond backbones (such as ethylene-propylene-diene monomer or EPDM) or silicon–oxygen bonds. They can decompose at moderate 300–500°C temperatures to an electrically conductive track across the insulator surface. The disadvantages of weak bonding and low decomposition temperature are offset by a low initial value of surface free energy that is *hydrophobic* (beads water), as illustrated in Figure 2-6.

Some hydrophobic materials also tend to resist accumulation of contamination.

For the best polymer insulator materials such as silicone rubber, hydrophobicity can persist for a long exposure time. Figure 2-7 shows that the contact angle of water drops to the surface, a good measure of hydrophobicity, can recover after temporary loss from contamination accumulation, abrasion, or electrical arcing activity.

Once surface oils from manufacturing processes are lost to the environment, EPDM materials no longer recover hydrophobicity with time and must rely instead on the superior tracking resistance to provide long service life.

The viscosity of light molecular weight (LMW) silicone oil molecules that provides the desirable hydrophobicity tends to increase as the temperature goes down. The recovery process shown in Figure 2-8 takes a longer time in freezing conditions, on the order of weeks at −4°C rather than days at 28°C [Chang and Gorur, 1994].

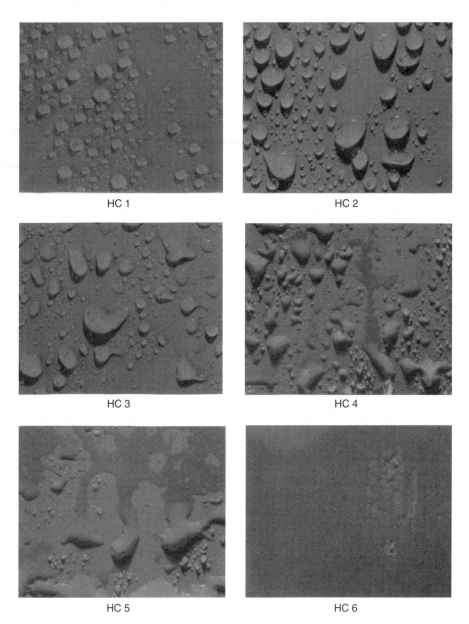

HC 1   HC 2

HC 3   HC 4

HC 5   HC 6

Figure 2-6: Classification chart for surface hydrophobicity of polymer materials (from STRI [1992]; courtesy of STRI).

2.2.2. Classification by Station or Line Application

Station insulators are used to support bus work near power apparatus such as switchgear and transformers. Hollow bushings that transfer high-voltage potential to the interior of power apparatus are also considered to be a type

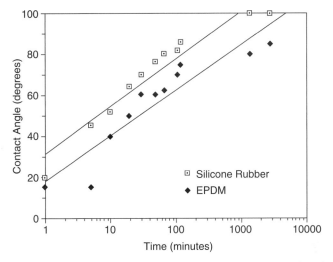

Figure 2-7: Recovery of hydrophobicity after exposure in fog chamber (from Gorur et al. [1990]).

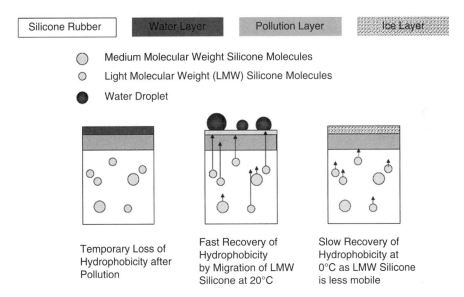

Figure 2-8: Hydrophobicity loss and recovery for silicone materials.

of station insulator. These bushings may be mounted directly on the metal enclosures or walls, may enclose surge protective devices or instrumentation, or may provide an interface to a high-voltage buried cable.

Station insulators are selected with particular care as part of a system, including protective levels of surge arresters and withstand capability of transformers, breakers, cables, and instrumentation. The level of electrical insulation within a substation varies in a controlled or "coordinated" way. The electrical strength of a transformer will be selected to be higher than the strength of the switchgear, which in turn will be stronger than the line-to-ground insulation strength. This insulation coordination process is normally supplemented with surge protective devices, which clip the peaks of overvoltages to levels that remain below the line-to-ground insulation strength.

Short-circuit currents in phase to ground faults on overhead lines are limited mainly by the series impedance of the line back to the source. The shorter the line length is from a substation, the greater the fault current. Station faults are of greater consequence than faults on overhead lines because the higher fault currents are more likely to damage insulators or equipment. Forces on the flanges of feed-through bushings for extra-high-voltage transformers, switchgear, and walls are particularly difficult to manage because of the cantilever length and static moment.

Flashover damage to ceramic insulators may lead to important hazards of flying or falling fragments of porcelain and glass sheds. With polymer insulators, fragmentation is not a safety hazard, making them much more suitable for housing metal oxide surge arresters.

Line insulators are used to suspend or support overhead conductors from towers. These insulators face a much wider range of environmental conditions.

Line and station outdoor insulators both provide mechanical support and insulate energized conductors and/or buses for all the applied mechanical and electrical stresses. Stations are confined to small areas compared to lines. Lines may traverse several, or even hundreds, of miles between terminal points. Environmental conditions, including the possible exposure to icing and snow, may be fairly constant or vary significantly along the length of a line.

For example, line insulators are exposed to direct overvoltages from lightning surges, where station insulators are normally located inside a protection zone. Lines may traverse areas of high contamination, high soil resistivity, high wind load, heavy ice accretion, or high level of large-bird activity. In response to these design uncertainties, electrical utilities may tolerate a certain number of phase-to-ground faults on overhead transmission and distribution lines.

In cold climates, it has often been the disappointing performance of station insulators relative to line insulators, and the relatively high vulnerability of EHV stations that should have been the most reliable, that have led to investigation of the effects of ice and snow on flashover performance.

2.2.3. Classification by Nature of Mechanical Load

Line suspension insulators are used to suspend conductors from line structures such as towers and poles. *Dead-end* insulators at strain structures are used at line terminations and changes in line direction. Dead-end insulators are used on both distribution and transmission lines while suspension insulators are more commonly used on transmission lines and less so on distribution lines.

Both suspension and dead-end types of insulators are subjected to tension loads and possibly some limited torsion loads. Suspension and dead-end (tension) insulators are unrestrained in two axes and do not support mechanical compression loads along the insulator axis.

In most transmission line applications the conductor is suspended from a tower crossarm. However, transmission lines may also be constructed using stand-alone horizontal posts or braced post designs.

Stand-alone line posts will typically see combined mechanical loads. These loads can expose the line post to compression, cantilever, and even torsional forces. The braced post design uses a typical line post mounted to the structure and supported at the conductor attachment point by a suspension type insulator. In this case, the line post is typically under compression load while the suspension insulator is under tension load. These line post designs can be used for improved visual appearance or restraint of conductor movement.

Distribution line conductors and the large-diameter bus bars in substations are supported with *distribution line posts* and *station posts*, respectively. *Post* insulators in general are those that support conductors from beneath or from the side, normally with a combination of compression and cantilever loads. Under wind loads, the "suspension" insulator on a transmission line will swing horizontally and the load along the axis of the insulator will remain a pure tension load. In contrast, the wind load on a line post or station post insulator will be transferred to the support as a cantilever force on the insulator. The bus or line conductor weight also exerts compressive or cantilever load on these insulators. These forces contrast with those pulling the insulator in tension in the normal line suspension and dead-end applications on overhead lines.

For each application, a string of disks or a rigid, hollow or solid core insulator will be selected according to the nature of the mechanical loads.

2.3. INSULATOR CONSTRUCTION

2.3.1. Ceramic Materials

Even before power systems were practical, electrical insulators with excellent performance and durability for outdoor service were being manufactured from electrical porcelain or electrical glass materials. The native materials for porcelain are clay, feldspar, and quartz and for glass are silica, soda ash, dolomite, limestone, feldspar, and sodium sulfate [Looms, 1988].

The earliest telegraph systems, for example, the 21-km line between London Paddington and West Drayton rail stations in 1839, had 60-m span lengths [STE, 1878]. Each point of contact from wire to wood pole led to signal loss, calling for the use of many repeaters. It became evident that leakage at each wire support could be minimized with the use of porcelain insulators finished with a leadless glaze that gave high insulation resistance. Wet glazed surfaces gave a comparatively low resistance, an important consideration in climates like the United Kingdom, and their telegraph insulators incorporated large sheltered surface areas as shown in Figure 2-9.

Insulators for telegraph systems were threaded at the base, with an exterior groove for a single wire. Porcelain was normally glazed with white, beige, or chocolate brown coating. The London & North Western Railway started using insulators with the LNW casting mark in the late 1870s, and this style in Figure 2-9 is similar to the standard pattern telegraph insulator of the UK Post Office [Johnson, 2007]. Glass insulators, such as the CD 162 pattern shown in Figure 2-10 [Meier, 2007] had similar rib structures for the same purpose.

The tradition of casting a company logo and date of patent or manufacture continued when insulator manufacturers started to produce designs for the

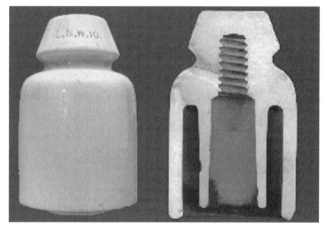

Figure 2-9: Typical porcelain telegraph system insulator showing sheltered creepage distance (from Johnson [2007]; courtesy of Teleramic).

Figure 2-10: Glass telegraph system deep groove double petticoat insulator showing sheltered creepage distance (from Meier [2007], courtesy of B. Meier).

electric power industry starting in the 1890s [Grayson, 1979]. Also, the successes of sheltered creepage distance on telegraph and telephone lines led to power system insulator designs that use similar ribs on the bottom surface of standard disk insulators, with typical profiles shown in Figure 2-11.

In service, the areas of sheltered leakage distance under fog type insulators can have some problems. These rib shapes generate vortexes that can deposit high levels of pollution in areas that are not easily washed [Looms, 1988]. A faster rate of pollution accumulation and higher rate of retention can degrade or reverse the performance that was originally gained with the extra leakage distance of the fog profile design. As an alternative, the open or aerodynamic profile as shown in Figure 2-12 has less leakage distance, but may also accumulate significantly less pollution in some conditions.

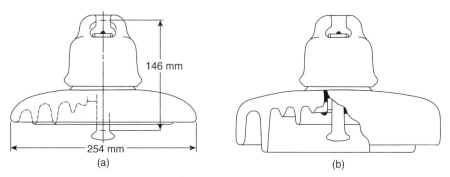

Figure 2-11: Porcelain disk insulators for power systems: (a) standard and (b) fog profiles.

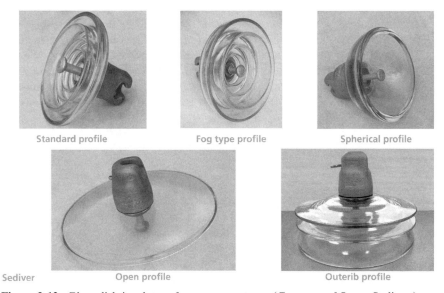

Figure 2-12: Glass disk insulators for power systems. (Courtesy of Seves-Sediver.)

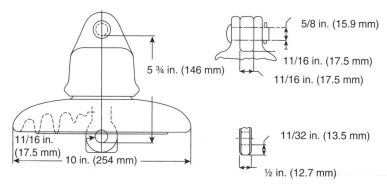

Figure 2-13: Dimensions of typical "clevis" fitting for ANSI Class 52-6 insulator with 30-kip M&E rating (adapted from NGK/Locke).

Traditional transmission line insulation has in the past settled on chains of individual ceramic disks, typically with 146-mm cap-to-pin separation and 254-mm (10 in.) diameter. These are designated as "ball-and-socket" or "cap-and-pin" types. The term cap-and-pin will be used in this book. Insulators are also produced with "clevis" and tongue mechanical fittings as shown in Figure 2-13. Insulators with this attachment method weigh more than cap-and-pin insulators with the same mechanical rating. Clevis attachments use a clevis pin and cotter key, while cap-and-pin units have a split cotter key, to guard against accidental decoupling from vibration or handling.

The highly refined clay of the porcelain body is pressed into molds that give the desired shape, then coated with glaze and fired in a well-regulated kiln. Ribs are pressed into the undersurface, with the exception of large-diameter open profile designs like the one shown in Table 2-1, which are offered in both porcelain and glass materials. The ribs extend the sheltered leakage distance of the insulator to reduce the effective voltage gradient under line voltage. Ribs with deep ratios of draw depth to thickness become fragile and are more easily damaged during firing and handling. The metal fittings are cemented onto a porcelain body with bands of sand to provide increased grip surface for the cement grout. The metal pins are secured into the adjacent caps with 50-mm cotter keys that are installed or released with a mallet and large screwdriver or a specialized tool.

Cap-and-pin suspension disks are most common on transmission lines. Table 2-1 shows that these are offered with a moderate range of variation among manufacturers, with all having some common features, such as a 146-mm cap-to-pin spacing and a 254-mm diameter. These are all used in tension applications, mounted either vertically (to allow a conductor to be suspended from a tower) or horizontally (at a strain tower, in line with the conductor axis). The long-rod porcelain line insulators are listed in this table because their main function is also to support tension loads.

TABLE 2-1: Typical Ceramic Insulators[a] for Tension Loads

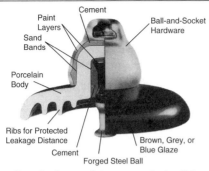

Standard porcelain cap-and-pin disk
(Lapp)

Glass cap-and-pin aerodynamic disk
(Seves)

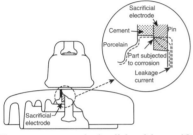

Fog type cap-and-pin disk with sacrificial electrode for high-corrosion areas
(NGK)

380-kV Cap-and-pin suspension disk
string (Seves)

Ceramic disk insulator profiles with bottom-surface ribs and outer ribs (Dalian)

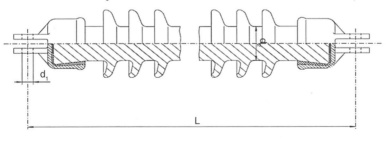

Long-rod porcelain suspension insulator (PPC)

[a]Manufacturer given in parentheses.

39

A main distinction is made between line insulators for tension, and station or post insulators that may support tension, compression, or cantilever loads. Since glass is relatively difficult to manage for this complex set of loads, ceramic post insulators tend to be made of porcelain. Table 2-2 shows that there are hollow core as well as large and small solid core diameter porcelain post insulators. In general, the spacing of sheds on these insulators is much smaller than the spacing between the disks of a typical suspension string.

TABLE 2.2: Typical Porcelain Insulators[a] for Cantilever or Compression Loads

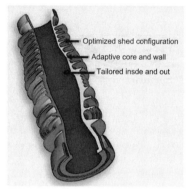

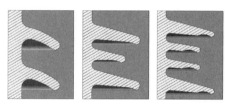

Typical profiles of porcelain station post insulators (Lapp)

Interior view of typical hollow porcleian bushing (PPC)

230-kV Two-section ceramic station post insulators with resistive grading glaze (Lapp)

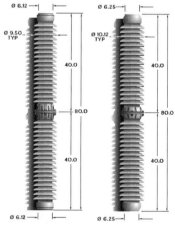

230-kV Two-section ceramic station post insulators: (left) standard and (right) high strength (Lapp)

TABLE 2-2: *Continued*

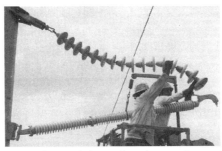

Multidisk suspension string and three-piece long-rod porcelain strut (Lapp)

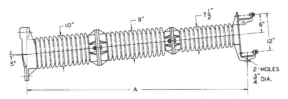

Porcelain horizontal line post (Lapp)

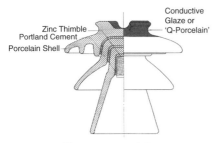

Pin type porcelain

Pedestal post cap-and-pin type

*a*Manufacturer given in parentheses.

Long-rod porcelain insulators are also common for transmission line applications in Europe. Glass disk insulators remain competitive with porcelain disks for cap-and-pin suspension disks. Due to limitations of the glass material, glazed porcelain posts and bushings are more common in stations where the insulators must withstand considerable cantilever and torque loads under short-circuit conditions.

Looms [1988] noted that the production of electrical porcelain was more art than science, and the same is somewhat true today, even with many high-tech improvements. Raw materials are crushed and filtered through sieves to obtain the desired particle sizes with very little processing. Porcelain manufacturers filter out impurities and adjust the mix to minimize bloat and other defects. Shrinkage between the shaping stage and the fired stage is a variable that is difficult to control and results in general tolerance of about 3% in the dimensions after the insulators are fired.

Porcelain insulators that are correctly fired have uniform internal pore size and few if any microcracks that degrade electrical and mechanical performance. For older insulators, prior to the 1930s, the porcelain body was relatively porous and a cover glaze was needed to prevent water penetration. Now, correctly fired wet-process porcelain is impervious to moisture, as proved by high-pressure dye penetration tests. However, the cover glaze is still used to achieve a smoother surface than is possible on a raw porcelain surface.

Glaze can lead to a remarkable improvement in the surface impedance. Moisture in any form—rain, dew, fog, melting ice, and snow—lowers surface insulation resistance significantly from the dry state. Surface resistance is measured in units of ohms per square, using square electrode geometry. Figure 2-14 shows that the measured surface resistance of glazed porcelain falls by five orders of magnitude as relative humidity at room temperature changes from 40% to 100%.

For relative humidity greater than 80%, the surface resistivity of glazed porcelain ($10^9 \Omega$) is seen to be about 50 times higher than surface resistivity of unglazed porcelain ($2 \times 10^7 \Omega$).

The glaze on porcelain insulators has an additional advantage. Its rate of linear expansion is smaller than the body of the insulator. When a fired

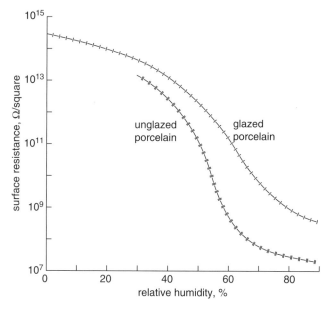

Figure 2-14: Relation of porcelain surface impedance to humidity (from Looms [1988]; courtesy of IET).

insulator cools, the glaze is "frozen." This puts a contracting envelope around the entire unit. Since the insulator body is strong in compression, the overall effect is an increase in flexural, tensile, and compressive strengths of about 30% for siliceous porcelain and 15% [Looms, 1988] to 25% for the stronger aluminous porcelain.

Typical standards for the quality of glaze on porcelain disk insulators [CSA C411.1, 1989] require that the diameter of any glaze defect be less than about 8 mm, and that any single inclusion (glaze over sand) should not protrude more than 2 mm from the finished surface. A 5-mm limit is applied for the diameter of any bubbles that form in the manufacture of toughened glass insulators, and surface defects are not acceptable for this material.

Suspension disk insulators are normally produced to either IEC or ANSI standard dimensions. This simplifies interchange since all units will have the same shed diameter and cap-to-pin spacing. However, there are still some significant differences in leakage distance and form factor from manufacturer to manufacturer. Standardization of dimensions allows for the development of application guidelines, for applying porcelain and glass insulator shapes. For example, the insulator leakage distance is commonly increased to a *unified specific creepage distance* (USCD) of 55 mm per kV of line-to-ground ac rms voltage near the sea coast or heavy industry, while insulators with only 28 mm/ kV are recommended for areas of low pollution.

The mechanical ratings of ceramic (glass or porcelain) insulators are traditonally kilopounds (kip) for the United States, United Kingdom, and Canada, and metric units of force in kiloNewtons (kN) for the rest of the world. It is important to remember, when making conversions, that "a Newton weighs about one apple," in other words a rating of $15 \text{ kip} \times 0.454 \text{ kg/lb} = 6800 \text{ kg} \times 9.8 \text{ m/} s^2 = 67 \text{ kN}$. Polymer insulators use metric ratings in Canada.

Mechanical strength of ceramic insulators is evaluated with several different tests. The ultimate breaking strength of the insulator is an important factor under heavy ice loads. However, some porcelain insulators may fail electrically with an internal crack between pin and cap under high mechanical load, before they fail mechanically. These are said to be *punctured* or *partially punctured*, depending on the breakdown voltage of the remaining porcelain. Glass insulators and high-quality ceramic insulators with low porosity tend to be more resistant to electrical puncture, and they usually fail electrically at the same time that they break under rated load.

In some national standards, such as in CSA C411.1 [1989] Clause 3.6, the expansion of Portland cement used to assemble the insulators is limited to a maximum of 0.12% in a standard autoclave test. This ensures that the compressive stress caused by cement expansion does not shorten the expected service life of the ceramic insulators. This is especially important in tangent applications at dead-end structures (strain towers), where insulator load requirements are higher and where the cement around the pin is wetted more often by rain.

The use of a steep voltage rate-of-rise specification, such as 2500 kV/μs in CSA C411.1 [1989], simulates the typical *dV/dt* associated with subsequent lightning flashes (median 40 kA/µs) and relatively high tower and ground impedance. Tests with this steep rate of voltage rise cause punctures in poor-quality porcelain insulators with inadequate design. Thus this is an important test especially for unshielded lines, lines in areas of high resistivity, and lines with tall double-circuit towers.

A combined mechanical and electrical (M&E) test is used to establish the mechanical load at which electrical puncture occurs. The combined M&E test for wet-process porcelain insulators, for example, as specified in CSA C411.1 [1989] Clause 6.6.1, applies 75% of the dry flashover voltage to the disk under test. The disk is then pulled to destruction at a controlled rate of increase of mechanical load. Failure load is established by the load at electrical failure or the breaking load, whichever is lower.

2.3.2. Polymeric Materials

The evolution of polymeric materials suitable for outdoor electrical insulation has been well described in several books and documents [CIGRE Working Group 33.04.07, 1999; Gorur et al., 1999]. Starting in the 1970s, the use of polymers as an outer weather-shed to protect a central fiberglass core proved to be a successful alternative to glass and porcelain. Metal hardware is fitted to each end of the fiberglass rod and sealed against moisture ingress. The "long-rod" and "line-post" insulators in Figure 2-15 have been especially successful for lower-voltage distribution and transmission line insulators. At this time, the polymer insulators and similar polymer-housed distribution surge arresters have a service defect rate of about 0.5 per 10,000 per year, which is actually lower than the average defect rates of 0.5–1.0 per 10,000 per year that have been reported for ceramic units [INMR, 2002].

When they are new, polymer materials such as ethylene-propylene-diene monomer (EPDM) and polydimethylsiloxane (PDMS) silicone rubbers offer much higher surface impedance than ceramic or glass. The ratio can reach 10^6, comparing silicone to porcelain under saturated and contaminated conditions. The best materials also bead water, improving wet performance, and retain this attribute for many years in most service conditions.

Some polymer materials do not have an inherently high resistance to electrical tracking. It is thus common to use fillers such as alumina trihydrate (ATH) to provide this property. Rather than releasing water of hydration, the filler cools the arc by increasing thermal conductivity [Meyer et al., 2004].

The resistance of the soft polymer surfaces to mechanical abrasion is not as high as ceramic materials. This can be a problem when applying them in desert areas. Abrasion damage will tend to increase the surface roughness, which is initially quite smooth except at mold lines and interfaces from the manufacturing process. A rough polymer surface will retain pollution more readily and surface damage to the sheath over the fiberglass core may lead to internal degradation and failure. However, some service experience over 25 years where sandblasting removed glaze from porcelain shows that the silicone materials may provide longer service life, in spite of their soft surfaces.

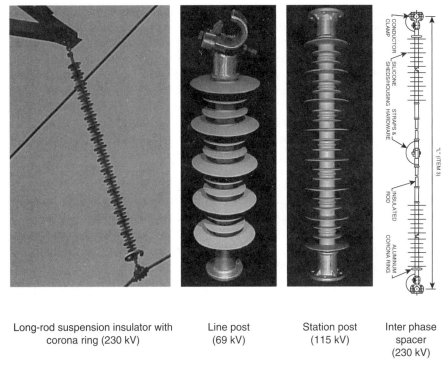

| Long-rod suspension insulator with corona ring (230 kV) | Line post (69 kV) | Station post (115 kV) | Inter phase spacer (230 kV) |

Figure 2-15: Types of polymeric insulators for transmission and distribution systems. (Courtesy of K-Line Insulators.)

Electrical puncture strength of the polymer materials is a relatively important factor in icing and contaminated conditions. Some polymer materials have twice the resistance to electrical puncture of porcelain [Looms, 1988], but the individual sheds of polymer insulators tend to be five to ten times thinner than typical ceramic ribs, sheds, or disks. Under arcing activity, it is common to have a large fraction of line voltage stress imposed on a small fraction of the insulator surface. If the polymer sheds are too thin, the applied voltage will puncture through them. This usually occurs at the area of highest electrical stress concentration—near the shaft of the insulator, at a mold line.

The polymer and composite construction is also used for hollow-core post insulators, for bushings, and for housings of surge arresters. The electrical puncture strength in the radial direction, from the outside of the insulator to the interior, is an important design consideration in contamination conditions. In addition, any capacitive coupling of current through the insulator can overheat surge arresters or electric stress grading materials.

When they were applied in areas of high contamination, the first generation of polymer insulators did not perform as well as anticipated from laboratory tests. In severe environments, dry-band arcing damaged the weather-shed material and allowed moisture to attack the fiberglass core. This could lead to either flashunder (between the polymer and core) or mechanical separation of the insulator.

The response of the polymer surfaces to long-term accumulation of soluble and insoluble pollution has been the focus of considerable research. Like various types of soil, the best polymer materials have a finite ability to encapsulate pollution, called a "critical loading" in environmental science.

Unfortunately, polymer surfaces lose some of their advantages over ceramic materials at the freezing point. The surface impedance of a wetted and polluted polymer insulator above the freezing point is about 10^6 times higher than the surface impedance of a porcelain insulator in the same conditions, but this advantage falls to only a 10:1 ratio below 0°C [Chisholm et al., 1994]. The critical loading level for encapsulating pollution is greatly reduced at low temperature. Also, while the polymer materials may be hydrophobic, they are not ice-phobic. The general advantage of a smaller diameter for polymer insulators, compared to ceramic units of the same rating, means that less ice will build up in heavy icing conditions. However, in many polymer designs with long leakage distance, the close spacing of sheds means that the amount of ice for full bridging may be lower than for strings of disk insulators, making the polymer insulators more vulnerable to flashover under moderate icing conditions.

2.3.3. End Fittings

The ferrous fittings for ceramic disk insulators are generally made from:

- Malleable cast iron, per ASTM Standard A47 or A220.
- Ductile cast iron, per ASTM Standard A536, Grades 60-40-18 or 65-45-12.
- Steel forging, per ASTM A668.

These materials give good mechanical performance down to −50°C. The fittings are galvanized with zinc to a typical density of 455 g/m^2.

End fittings for polymer insulators are normally made from steel, iron, or aluminum and then machined to a close tolerance to fit over the fiberglass core. Some manufacturers offer insulator hardware that is tested for fracture energy in a Charpy impact method to infer the fracture toughness that will ensure good performance at extremely low temperatures. Most manufacturers now compress the end fittings to secure them, and then seal them against moisture ingress during the remaining manufacturing steps. In previous generations of polymer insulators, fittings may have been attached with epoxy adhesives or mechanical wedges, but these did not prove to be as reliable as swaged or crimped fittings. Moisture diffusion through the polymer materials into internal air spaces of end fittings can lead to long-term problems, both with end fittings and with the internal parts of polymer housings for bushings and surge arresters. The most reliable polymer construction methods strive to eliminate internal air spaces by design, production quality, and displacement with more suitable insulating materials. Bonding of the sheath rubber to the fiberglass rod also helps eliminate this potential problem.

The thermal shock on insulator end fittings under flashover conditions can lead to mechanical damage. Normally, insulators are qualified for this possibility with a power-arc test at a specified level of fault current.

2.3.4. Other Materials in Series or Parallel with Insulators

In order to reduce the number of electrical flashovers from lightning that hits the ground near overhead distribution lines, many utilities take advantage of the relatively good electrical strength of wood crossarms and fiberglass components. These are inserted in series with the porcelain or polymer insulators, so that the base of the insulator does not have a metallic connection to ground. While these materials have limited ac strength under wet conditions, they can successfully resist impulse voltage gradients of up to 200 kV per meter.

One trade-off for improved reliability in the summer is that, in winter conditions, leakage currents across insulators may cause charring damage or ignition of the wood at areas of high current density.

Concrete, metal, or composite fiberglass poles are often substituted for wood poles in areas of heavy icing. Concrete and metal poles provide a good electrical contact to the base of insulators and do not contribute any additional insulation strength. Fiberglass poles do provide the possibility of adding considerable insulation strength in icing conditions with certain configurations.

Steel lattice transmission towers may serve as a source of rust pollution on insulators, once zinc galvanizing has eroded. Erosion of galvanizing on tower steel also serves as a cumulative indicator of local pollution problems and can be a guide to locations where insulators with better contamination performance may be appropriate.

In freezing conditions, accumulation of ice or snow will change the shape of the insulator, filling in spaces and shorting out some or all of the leakage distance. Thus ice and snow can be classed as a different type of insulator material, one with a particularly high dielectric constant and good electrical strength when the temperature is low. There can be an increased risk of flashover that will persist for as long as the ice deposit remains on the surface, especially when the ice or snow is melting. The retention time of the ice or snow deposit is mainly a function of the surface temperature, the insulator surface material, and its roughness.

Ice adhesion strength is measured in a number of ways, using centrifuges or vibrating beams. The surface finish of the material can have a strong influence. For example, ice adhesion to aluminum finished with 180-grit sandpaper has a strength of 2.8 MPa, compared to 1.5 MPa when smoothed with 400-grit sandpaper [Javan-Mashmool, 2005]. In contrast, adhesion to Plexiglas had ice adhesion strength of about 0.05 MPa, corresponding to a specific adhesion strength of about 15 kg/cm^2 in Figure 2-16 [Raraty and Tabor, 1958].

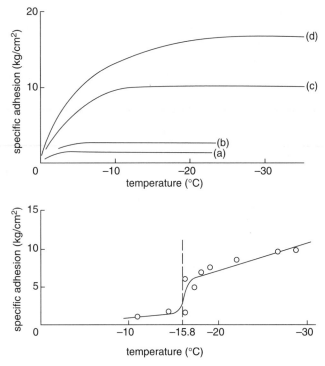

Figure 2-16: (Upper graph) Ice adhesion as a function of temperature to (a) polytetrafluoroethylene, (b) solid stearic acid, (c) polystyrene, and (d) Plexiglas. (Lower graph) Ammonium chloride solution on stainless steel, showing transition in adhesion at eutectic temperature of −15.8 °C (from Raraty and Taylor, [1958]; courtesy Royal Society of London).

The adhesion strength of ice can be influenced by the presence of electrically conductive pollution. For example, in Figure 2-16 there is a transition in ice adhesion strength to stainless steel around −16 °C when ammonium chloride is applied as a deicing agent. Similar effects limit the use of sodium chloride for road deicing to temperatures warmer than −20 °C.

2.4. ELECTRICAL STRESSES ON INSULATORS

Insulators have to perform under a wide range of service conditions. Many stresses act on line insulators in various combinations. Some of these stresses change drastically under ice and snow conditions, and these are highlighted in this section. Insulators will work reliably in service for many years only if their design and application has considered all of the relevant stresses. This section defines the relevant measures of the electrical stress on insulators.

2.4.1 Power Frequency Electrical Stresses

Each component of the power system is designed to withstand a continuous high voltage with high reliability. The voltage stress on insulation is defined by a gradient, typically in kilovolts per meter (kV/m). The distance over which this voltage gradient is measured can vary from a few millimeters for microgap corona discharge, up to ten meters for the space between phases of an EHV transmission line.

For ac systems, it is most common to consider the line-to-ground root-mean-square (rms) voltage across the insulator terminals. The peak voltage is related to the rms voltage by a factor of $\sqrt{2}$. The line-to-ground voltage is related to the system voltage by a factor of $\sqrt{3}$ in three-phase systems. For distribution systems, with a typical designation of 27.6/16 kV, the second value indicates the line-to-ground voltage on single-phase lateral circuits.

For dc systems with ripple, the peak voltage should also be used to establish insulator stress.

The *voltage stress per meter of dry arc distance* is given by dividing the relevant line-to-ground voltage by the dry arc distance. Depending on utility practice, an overvoltage factor of 5–10% may be added to this value. Also, for long transmission lines and modest loads, the Ferranti effect can cause the receiving-end voltage to be higher than the sending-end voltage. This should also be included in analyses.

The *voltage stress per disk* is given for short insulator strings of less than about ten standard cap-and-pin disks by the system voltage divided by the number of disks. For longer insulator strings, the voltage stress per disk is not constant. Instead, it is higher at the line end and ground end insulators and lower on the insulators in the middle of the chain, as a consequence of electrostatic potentials that are modified somewhat by the internal capacitances of the disks.

The same considerations apply to the normal calculation of *voltage stress per meter of leakage distance*. In IEC practice, the stress is often inverted, to give units of *unified specific creepage distance* (USCD) in units of mm/kV line-to-ground voltage. An older classification system [IEC 60815, 1986] used *specific creepage distance* (SCD) in units of mm/kV of line-to-line voltage, which led to some confusion in application.

There are two other definitions that have been applied to calculations of electric stress across leakage distance. Baker and Kawai 1973 defined *effective leakage distance* as the distance between the 10% and 90% potentials on an insulator, similar to the rise time. Also, some authors consider that the stress across *sheltered leakage* or *protected creepage distance*, facing downward, is a better predictor of contamination performance. For a typical standard disk insulator, about two-thirds of the leakage distance on the bottom surface is sheltered.

Radial electric stress for bushings and arresters is defined by the local potential difference from the outside to the inside of the hollow insulator in the radial direction, divided by the wall thickness. One particular problem with ice accretion is that it tends to bring ground potential on the exterior of the hollow insulator very close to the high-voltage end, giving a significant increase in voltage gradient across the wall.

2.4.2. Impulse Electrical Stresses

Overvoltage impulses generated by lightning or switching impose a higher, albeit momentary, stress on the insulator. For both lightning and switching impulses, electrical flashover voltage is mainly dependent on the dry arc distance, that is, the shortest air distance between the conductor and the grounded part of the insulator.

A review of the electrical strength of rod-to-rod gaps in IEEE Standard 4 (Table 11) gives a good indication of the expected strength of air around the insulator. For standard conditions of 20 °C, 101.3 kPa (760 mm Hg), and absolute humidity of 11 g/m^3, the rod-to-rod sparkover voltage, expressed as the peak of 60-Hz voltage, is

$$V_{ac\,pk} = 22 \text{ kV} + (500 \text{ kV/m}) \cdot L \qquad (2\text{-}1)$$

for a gap spacing L (m) with 0.02 m > L > 2.2 m.

Generally, it is the peak of the ac voltage wave that will initiate discharges. There is thus a reasonably strong relation between the flashover strength of air at the peak of ac voltage and the flashover strength under fast lightning impulse conditions, with a gradient of 540 kV/m for the *basic lightning impulse insulation level* (BIL). This is why, for dc systems with ripple, the peak voltage should also be used to establish insulator stress.

Voltage surges produced by lightning and/or switching operations have usually been the major factors considered in the electrical design of transmission line insulators intended for clean and very light polluted areas. The flashover values of these surges are defined by the BIL and the *basic switching impulse insulation level* (BSL). Insulators are normally selected to withstand most switching-surge overvoltages, but not to withstand all lightning impulse overvoltages since these have a very wide statistical distribution of magnitudes.

Insulation coordination for HV systems with voltages less than 230 kV is dominated by lightning performance in insulation coordination guides [IEC Standard 60071-2, 1996; Hileman, 1999; IEC Standard 60071-1, 2006]. Insulation coordination of higher voltage levels, including EHV systems with higher reliability requirements, is dominated by switching-surge control using protective devices such as arresters or closing resistors. However, regardless of the line voltage, insulator selection for polluted areas takes careful note of the contamination performance to provide adequate leakage distance. This evaluation process is even more important when selecting insulators for areas with ice and snow. Even distribution line insulators are affected by pole fire considerations and should be selected with care in areas where pollution can accumulate over an entire winter.

Under switching-surge conditions, the transient voltage wave has rise time and time-to-half-value of 250 μs and 2500 μs, respectively, rather than 1.2 μs and 50 μs for lightning impulse voltages. This means that predischarge activity is important for switching surges. There is normally a lower switching-impulse flashover gradient for rod-to-rod gaps than either ac peak or lightning gradient.

The effects of metal structures near the air gap for switching surges are relatively strong and lead to the use of "gap factors" that are multiplied by the *basic switching impulse insulation level* (BSL). This subject is covered in excellent detail in Hileman [1999]. The BSL for the rod-to-rod gap is used without gap factors to establish the suitable electrical clearances around high-voltage tests in IEEE Standard 4 [1995]. Normally, switching-surge stresses are expressed in a per-unit system based on the crest value of the line-to-ground voltage [Hileman, 1999].

2.4.3. Major Electrical Factors in Freezing Conditions

In cold-fog conditions, the full leakage distance of the polluted insulator is covered with a thin layer of ice. The *electric stress per meter of leakage distance* establishes whether a flashover will occur in these conditions.

As ice accretion continues, shed-to-shed intervals may be partially or fully bridged by ice or snow accumulation. Each time this happens, the contribution of the shed depth to the overall insulator leakage distance is reduced or eliminated. Under partial bridging, the leakage path shifts off the insulator surface, over to the ice caps and icicles. In the limit, the insulator leakage distance is fully bridged by ice or snow. The new leakage path along the ice surface can be about a third of the insulator leakage distance. This means that icing performance is normally evaluated using the *electrical stress per meter of dry arc distance.*

Especially in cold-fog conditions but generally for all icing conditions, the *onset of arcing activity* is one of the important signatures of upcoming problems with electrical performance. Partial discharges occur in the air gaps between areas of ice accretion, especially at areas of high local electric stress, such as the tips of icicles. Heat generated by partial arc activity can have positive and negative effects on icing performance at the same time:

- Arcing activity can melt the ice sufficiently to loosen its adhesion to the insulator surface, leading to ice shedding.
- Arcing can burn back icicle tips, restoring some of the original shed-to-shed separation.
- Arcing can warm the ice sufficiently to change its electrical conductivity in the range of −2 °C to 0 °C.
- Arcing can promote the formation of a conductive water film on the ice surface.

Arcing onset is more important in freezing conditions than arcing initiation associated with wet-weather corona or insulator performance in heavy rain conditions.

2.5. ENVIRONMENTAL STRESSES ON INSULATORS

2.5.1. Major Environmental Factors in Temperate Conditions

A heavy rain test is performed on every type of insulator and a wet flashover result is reported in its specifications. The environmental stress related to heavy rain is described by σ_{20}, the *electrical conductivity* of the rainwater, corrected to 20 °C. The electrical conductivity of most ionic solutions, including rain, increases with temperature.

Air density affects the electrical strength of the air around the insulators. Air density is a function of both *ambient temperature* and *ambient pressure*. In high-altitude areas, the average air density is lower and the electrical strength is reduced. *Humidity* has the opposite effect. Water in the air improves insulation strength of the air.

While the leakage distance is given in the insulator dimensions, there is no corresponding unique contamination flashover result. In part, this is because there is no single suitable test for the rate of deposit of pollution on the top and bottom surfaces of insulators. It is left up to the power system designer to select a suitable electric stress for the anticipated environmental conditions, using guidance such as IEC Standard 60815 [2008].

Salt-fog pollution conditions are reproduced in laboratory tests using spray nozzles with a controlled electrical conductivity and *flow rate*. This leads to the desired *fog density*.

The self-wetting rate of a polluted surface depends simply on the local *relative humidity* and the pollution chemistry. In conditions of fog, the environmental wetting rate is a function of the difference between *dew point temperature* and ambient, as well as the *wind speed* and the fog density.

The characterization of insulator pollution relevant to clean-fog conditions is relatively complex. Chapter 3 provides some insight about regional and local variations. However, some terms and measures are well established.

The rate of pollution accumulation near strong sources will be a function of the wind speed, the prevailing *wind direction*, and the *distance* from source to insulator.

The electrical conductivity of a surface deposit is normally expressed as an *equivalent salt deposit density* (ESDD). The use of the weight of sodium chloride per cm^2 of surface area that gives the same electrical conductivity as the wetted deposit simplifies testing and performance calculations.

A dry pollution layer poses little danger to insulation until it is wetted in some way—by scavenging moisture from the air at high relative humidity, by condensation, or by direct impingement of fog or natural or artificial precipitation. The electrical conductivity of a wetted pollution layer can be expressed as *layer conductivity*. This value is obtained by dividing the form factor of an insulator by its measured resistance. There is a nonlinear relation between layer conductivity and ESDD.

The layer conductivity can only be measured if the entire surface area is wet. One environmental stress that ensures full wetting is the presence of a nonsoluble deposit. The *nonsoluble deposit density* (NSDD) is measured in the same units as ESDD, weight per unit surface area.

Ceramic insulators are not affected by ultraviolet (UV) light, but some polymer materials are. Long-term UV radiation from sunlight can degrade the surfaces of polymer insulators. This type of damage is simulated in some long-term aging tests, either directly from sunlamps providing the same *UV intensity* or with the use of UV radiation from local arcing activity in tracking-wheel tests.

Weather-related outages such as lightning are the most common root cause of short-circuit currents in many power systems. Station insulators, in particular, are designed to meet the anticipated cantilever forces associated with heavy station fault currents. Mechanical forces related to *vandalism* from rocks or gunshots are also a form of environmental stress, as are consequences of *bird streamers* and *animal* or *bird contacts* across the insulator dry arc distance.

2.5.2. Major Environmental factors in Freezing Conditions

Electrical conductivity of ice or snow accretion plays the same fundamental role as conductivity of rain, fog, or wetted surface layers in insulator performance. The value of σ_{20}, electrical conductivity at 20 °C, serves as a common measure for both conditions. However, it is important to realize that the actual conductivity of the ice or snow is a strong function of temperature. The conductivity decreases by a factor of 1000 at –20 °C, often making the ice deposit at this temperature a relatively good insulator with high dielectric constant, providing superior voltage grading. Close to the melting point, the electrical conductivity of snow and ice is very sensitive to temperature in the range of –2 °C to 0 °C. In natural conditions, the snow or ice temperature is influenced by the *ice point temperature*, which is the dew point temperature below 0 °C, and by the evaporation and sublimation rates of water off the surface.

Any surface layer of water that forms on melting ice will tend to have a conductivity that is enhanced by temperature, by rejection of impurities from the original freezing of ice, and also by action of corona activity [Farzaneh and Melo, 1990; Farzaneh, 2000].

Ice, snow, and cold fog play a secondary role in stabilizing preexisting surface pollution. In this sense, the frozen ice can be considered to be a form of inert, heavy deposit similar to the NSDD. However, as heavy accretion melts and slides off surfaces, it tends to leave them in a relatively clean condition, with ESDD levels similar to those observed after heavy rain.

Wind speed affects the shape of ice and snow accumulation on insulators. The angle of icicles from vertical, and the proportion of ice in ice caps versus icicles, both increase with increasing wind speed [Farzaneh and Kiernicki, 1995].

The altitude has a significant effect on the nature of ice and snow accretion. Rime ice and snow are more common at high altitude, with glaze ice more common at altitudes near sea level. The difference in the electrical properties of these accretions relates to the *density* of the deposit.

The combined effect of accretion conductivity and density on the resistance of the resulting deposit can be described with a single *icing stress product* (ISP) value of $\sigma_{20} \times$ density.

Air pressure has proved to have about the same effect on the flashover strength in freezing conditions as it has on contamination flashovers. Reduced pressure leads to reduced flashover strength.

Viscosity of liquids tends to increase as the temperature decreases. This slows the ability of polymer insulators to encapsulate surface pollution.

2.6. MECHANICAL STRESSES

2.6.1. Important Factors in Temperate Conditions

In areas with little or no icing, as specified, for example, in IEEE/ANSI Standard 2 [2007], normally the highest mechanical design loads for insulators will be those associated with the maximum design wind speed. In overhead transmission lines, wind will exert a combined horizontal and vertical downward force on the conductor catenary. The vertical load will add to the static weight of the conductor itself. The wind direction to the span will influence the overall force.

In summer, the combination of *solar radiation* and air temperature can combine to cause insulator temperatures of up to 70 °C in low wind. With maximum conductor operating temperatures of 150 °C being used now at some utilities and thoughts of using high-temperature conductors at 300 °C, there is also a possibility that metal hardware may be even warmer in the future. Most insulators tend to lose mechanical strength at high temperatures, but these high temperatures are not achieved in the high-wind conditions that are associated with maximum design loads. High-temperature operation can also cause torques in conductors that can place unexpected stresses on midphase spacers.

Perhaps the most dangerous time for an insulator is during the line construction. Shock loadings from transport, handling, tension stringing, and other activities can lead to hidden internal damage and premature failure.

2.6.2. Important Factors in Freezing Conditions

In areas of moderate or heavy icing defined in IEEE/ANSI Standard 2 [2007], there are two design constraints: (1) heavy ice load and (2) combined wind-on-ice load. These static loads are usually limiting constraints in a line design for several reasons:

- Wind pressure on conductor diameter is increased by cross-sectional area of snow or ice and drag coefficient is increased by rough surfaces.
- Ice or snow adds to static weight of the conductor, increasing the load on both suspension and dead-end insulators.
- Conductor tension at low temperature is higher, increasing the load on dead-end insulators.
- Torques associated with conductor rotation during heavy ice or snow accretion can be transferred to insulators and spacers.

Dynamic loads under winter conditions are also more severe. These are associated with the following:

- Aeolian vibration is a high-frequency, low-amplitude conductor motion with most energy in the range of 0 to 25 Hz, that has more energy input at higher conductor tension at low temperature, along with constant wind speed.
- Galloping is a low-frequency, large-amplitude conductor motion that occurs for asymmetrical ice or snow accretion.
- Ice or snow shedding, whether natural or forced (deicing), will cause dynamic transient loads that are transferred to structures through the insulators.

In some cases, large-amplitude conductor motions can be controlled with the use of interphase spacers. Carreira and Gnandt [2004] provide details of latent system damage that can occur after dynamic loads in freezing conditions.

2.7. PRÉCIS

Insulators for electric power systems first took advantage of the materials and pin-type shapes that gave good performance on telegraph systems. The evolution of cap-and-pin disk insulators for transmission systems allowed increased system voltage and modularity. Fiberglass and polymer insulating materials have also proved to be highly suitable for certain types of power system insulators.

The electrical stress across an insulator under normal service conditions is applied across a convoluted surface. The "leakage distance" across this surface is between two and four times longer than the "dry arc" distance measured from metal-to-metal terminals. The insulator performance and lifetime in contamination conditions are improved when the electric stress across the leakage distance is reduced. When ice or snow accumulates on an insulator, it fills in the spaces of the insulator and shorts out substantial fractions of the leakage distance. Thus the dry arc distance of an insulator plays a much more significant role than the leakage distance in freezing conditions.

REFERENCES

Baker, A. C. and M. Kawai. 1973. "A Study on Dynamic Voltage Distribution on Contaminated Insulator Surface," *IEEE Transactions on Power Apparatus and Systems*, Vol. PAS-92, No. 5 (Sept.), pp. 1517–1524.

Carreira, A. J. and E. P. Gnandt. 2004. "Inspection Techniques for Detecting Latent Damage to Existing Overhead Transmission Lines from Previous Ice and Wind Storms." CEA Technologies, CEATI Report No. TO23700-3308. December.

Chang, J. W. and R. S. Gorur. 1994. "Surface Recovery of Silicone Rubber Used for HV Outdoor Insulation," *IEEE Transactions on Dielectrics and Electrical Insulation*, Vol. 1, No. 6 (Dec.), pp. 1039–1046.

Chisholm, W. A., K. G. Ringler, C. C. Erven, M. A. Green, O. Melo, Y. Tam, O. Nigol, J. Kuffel, A. Boyer, I. K. Pavasars, F. X. Macedo, J. K. Sabiston, and R. B. Caputo. 1996. "The Cold Fog Test," *IEEE Transactions on Power Delivery*, Vol. 11, pp. 1874–1880.

CIGRE Working Group 33.04.07. 1999. "Natural and Artifical Ageing and Pollution Testing of Polymeric Insulators." CIGRE Technical Brochure No. 142. June.

CSA Standard C411.1. 1989. *AC Suspension Insulators*. Toronto, Canada: CSA.

EPRI. 2005. *EPRI AC Transmission Line Reference Book—200 kV and Above*, 3rd edition. Palo Alto, CA: EPRI. *AC Suspension Insulators*. Toronto, Canada: CSA.

Farzaneh, M. 2000. "Ice Accretions on High-Voltage Conductors and Insulators and Related Phenomena," *Philosophical Transactions of the Royal Society*, Vol. 358, No. 1776 (Nov.), pp. 2971–3005.

Farzaneh, M. and O. Melo. 1990. "Properties and Effect of Freezing Rain and Winter Fog on Outline Insulators," *Cold Regions Science and Technology*, Vol. 19, pp. 33–46.

Farzaneh, M. and J. Kiernicki. 1995. "Flashover Problems Caused by Ice Build-Up on Insulators," *IEEE Electrical Insulation Magazine*, Vol. 11, No. 2 (Mar./Apr.), pp. 5–17.

Gorur, R. S., J. W. Chang, and O. G. Amburgey. 1990. "Surface Hydrophobicity of Polymers Used for Outdoor Insulation," *IEEE Transactions on Power Delivery*, Vol. 5, No. 4 (Oct.), pp. 1923–1933.

Gorur, R. S., E. A. Cherney, and J. T. Burnham. 1999. *Outdoor Insulators*, Phoenix, AZ: Ravi S. Gorur Inc. Available at www.insulators.net.

Grayson, M. 1979. "Stutzen-Isolatoren: Manufacturers of Pintype Insulators in Germany." Available at http://cjow.com/archive/article.php?month=6&a=06Stutzen%20Isolatoren%20Insulators%20of%20Germany.htm&year=1979.

Hileman, A. R. 1999. *Insulation Coordination for Power Systems*. Boca Raton, FL: CRC Press, Taylor & Francis Group.

IEC Standard 60060-1. 1989. *High-Voltage Test Techniques Part 1: General Definitions and Test Requirements*. Geneva, Switzerland: IEC.

IEC Standard 60071-2. 2006. *Edition 8.0, Insulation Co-ordination—Part 1: Definitions, Principles and Rules*. Geneva, Switzerland: IEC.

IEC Standard 60815. 2008. *Selection and Dimensioning of High-Voltage Insulators Intended for Use in Polluted Conditions*. Geneva, Switzerland: IEC.

IEC Standard 60071-2. 1996. *Third Edition, Insulation Co-ordination-Part 2: Application Guide*. Geneva, Switzerland: IEC.

IEC Standard 60815. 1986. *Guide for the Selection of Insulators in respect of Polluted Conditions*. Geneva, Switzerland: IEC.

IEEE Standard 4. 1995. *IEEE Standard Techniques for High Voltage Testing*. Piscataway, NJ: IEEE Press.

IEEE Standard 100. 2004. *IEEE Dictionary of Standard Terms*. Piscataway, NJ: IEEE Press.

IEEE/ANSI Standard C2. 2007. *National Electrical Safety Code*. Piscataway, NJ: IEEE Press.

INMR. 2002. "Arresters: Market Forces, Current Technologies and Future Directions," *INMR Quarterly Review*, Vol. 10, No. 6 (Nov.).

Javan-Mashmool, M. 2005. "Theoretical and Experimental Investigations for Measuring Interfacial Bonding Strength between Ice and Substrate." M. Inginerie Mémoire, Université du Québec à Chicoutimi. November.

Johnson, G. 2007. "Cordeaux Country." Available at www.teleramics.com/type/ telegraph/ standard.html.

Looms, J. S. T. 1988. *Insulators for High Voltages*. London: Peter Peregrinus Ltd.

Meier, B. 2007. "New CD 152.2 Assignment." Available at http://www.insulators.com/ news/, http://www.insulators.info/news/ (Mar. 17).

Meyer, L. H., E. A. Cherney and S. H. Jayaram. 2004. "The Role of Inorganic Fillers in Silicone Rubber for Outdoor Insulation: Alumina Tri-Hydrate or Silica," *IEEE Electrical Insulation Magazine*, Vol. 20, No. 4 (July), pp. 13–21.

Raraty, L. E. and D. Tabor. 1958. "The Adhesion and Strength Properties of Ice," *Proceedings of the Royal Society of London, Series A, Mathematical and Physical Sciences*, Vol. 245, No. 1241 (June), pp. 184–201.

Sediver. 2007. Now Seves catalog. Available at http://www.seves.com/catalogo.php.

STE (Society of Telegraph Engineers). 1878. "Insulators for Aerial Telegraph Lines."

STRI. 1992. *Guide 1, 92/1: Hydrophobicity Classification Guide*. Swedish Transmission Research Institute. Available at www.stri.com.

CHAPTER 3

ENVIRONMENTAL EXPOSURE OF INSULATORS

This chapter introduces the methods and terminology used to describe the effects of the natural environment on insulation systems. The role of environmental pollution in the electrical conductivity of wetted surface deposits and of the precipitation itself is built up from the background levels across the continents. Then, for the important and specific pollution sources such as the ocean, road salting, or cooling tower effluent, suitable pollution accumulation models are consolidated from the literature. Enough practical electrochemistry is provided for users to carry out their own evaluations of the electrical conductivity of local pollution and precipitation.

3.1. POLLUTION: WHAT IT IS

Pollution is the introduction of substances or energy into the environment that can endanger human health, harm living resources and ecosystems, or impair assets. While much focus is placed on pollution sources from human activities such as electric power generation, these normally occur in the context of global cycles that also include natural pollution sources. A simple example would be the elevated corrosion rates of metals exposed to ocean salt. Lightning releases about 5Tg/year of nitrogen as NO_x [Meijer et al., 2001], mostly over land. A more complicated example can be found in the global sulfur cycle prior to electric power generation pollution controls [Friend, 1973] in Table 3-1.

Insulators for Icing and Polluted Environments. By Masoud Farzaneh and William A. Chisholm
Copyright © 2009 the Institute of Electrical and Electronics Engineers, Inc.

TABLE 3-1: Sources, Transport Paths, and Sinks in the Global Sulfur Cycle

Source Description	Strength (Tg/yr)	Transport Path	Strength (Tg/yr)	Sink Description	Strength (Tg/yr)
Volcanoes	+7	Rainfall, land	−86		
Organic decay, ocean	+48	Dry deposition, land	−20	Ocean bed Sediment	−100
Organic decay, land	+58	River runoff	−136	Dead organic matter, land	−15
Rock weathering	+42	Ocean/land (net)	4		
		Ocean spray	+44		
Coal combustion	+65	Rainfall, ocean	−71		
Coal residue (as fertilizer)	+26	Ocean surface absorption	−25		

Source: Friend [1973].

Air pollution is the release of chemicals and particulates into the atmosphere. Common examples include carbon monoxide (CO), sulfur dioxide (SO_2^-), and nitrogen oxides (NO_x) produced by industry and motor vehicles. Photochemical smog and ground-level ozone (O_3) are formed when nitrogen oxides and hydrocarbons react to sunlight. Ozone is also formed by electrical arcing and corona activity on high-voltage transmission lines in wet conditions. Chemicals that attack the ozone layer, such as chlorofluorocarbons (CFCs) or the electrically insulating sulfur hexafluoride (SF_6) gas, are also treated as pollutants.

Water pollution occurs through processes of surface runoff and leaching to groundwater. In the sulfur cycle in Table 3-1, coal combustion, mainly for electric power, mobilizes 91 Tg/yr. A total of 94 Tg/yr finds its way from river runoff to the oceans, driven mostly by 86-Tg/yr deposit from rainfall over land. Generally, water pollution increases the electrical conductivity of the runoff water. The increased electrical conductivity and decreased pH of precipitation are also factors in the reliability of outdoor insulators and the service life of metal components.

Soil contamination occurs when chemicals released by spill or leakage are retained for long periods of time. Examples of soil contaminants with long persistence are long-chain and chlorinated hydrocarbons, heavy metals, herbicides, and pesticides. Electric utilities have been aggressive users of herbicides to ensure safe substations and reliable transmission rights-of-way, free from vegetation.

Radioactive contamination, often in combination with heavy metal contamination, became prevalent during the development phases of atomic physics. Air, water, and soil pollution with radioactivity have all been problems, leading to some resolutions such as the censure of atmospheric nuclear testing. The ongoing use of nuclear power to produce electricity is accompanied by nonzero risk of radioactive release. The reality of continuous radioactive release from burning of coal or wood is seldom acknowledged with the same fervor. For some types of wood, fireplace ash with cesium and strontium [Farber and Hodgdon, 1991] had 100 times more radioactivity than the 1 picocurie per kilogram limit used to classify low-grade nuclear waste at nuclear power plants.

Noise pollution includes roadway noise, aircraft noise, and industrial noise from steam management, cooling tower fans, and coal conveyors. Substation transformer hum is another source of annoyance from power systems. Audible noise from high-voltage transmission lines also occurs near many homes, both from conductor corona in rain and from arcing activity on insulators in fog.

Visual pollution refers to the presence of overhead power lines, advertising billboards, scarred landforms (from strip mining or highway construction), or open storage of trash. It is interesting that bridges with construction similar to overhead power lines, fast food restaurants that are essentially billboards enclosing an interior volume, and the Grand Canyon are not considered visual pollution. A complex psychology related to the origin and purpose of the feature and the possibility for interaction plays a role in these perceptions. Arcing on contaminated insulators draws visual attention to power lines at night, and single-disk flashovers on dc lines are also a form of visual pollution.

Thermal pollution is the temperature change in natural bodies of water caused by human activity. Locally, cooling water is used at most thermal plants to maximize thermal efficiency. With superheated steam working between 310 and 800 K, the theoretical efficiency of the Carnot cycle is $(800 - 310)/800 = 61\%$. Practical coal-fired power plants can achieve 38%, leaving about 62% of the total heat energy in the local environment. Cooling ponds provide a large surface area to allow for radiative and evaporative cooling prior to returning water to nearby lakes or rivers. In wet cooling towers, hot water is sprayed into a cool air stream, with heat of vaporization providing highly efficient heat transfer to the atmosphere. Both of these tend to produce microclimate effects of fog and rain downwind. All of the energy produced by electric power plants will eventually appear in the environment as heat after use. Globally, warming from increased CO_2 in the atmosphere is another form of indirect thermal pollution.

While electrical power utilities and countries work with varying degrees of diligence to minimize their various forms of pollution, this is not the primary focus of our work here. Air and water pollution degrades the reliability of critical power system amenities in the long term through accelerated corrosion rates of metal and insulating polymer components. Pollution also impairs the electrical performance of critical power system components in the short term, typically under winter conditions, through electrical flashovers and faults on insulators.

3.2. POLLUTION DEPOSITS ON POWER SYSTEM INSULATORS

Power systems must deliver reliable service under a wide range of outdoor conditions. The insulators used in outdoor substations and overhead transmission and distribution lines must withstand normal service voltage, without flashover failures and their related network disturbance. These insulators must also withstand some overvoltages—power system switching transients being the most relevant in the winter environment but lightning transients also considered in the polluted operating environment.

Electrical station equipment and line equipment located near generating stations have had a wide range of associated problems, including accelerated corrosion damage, high rate of buildup of surface pollution, and heavy localized wetting from cooling tower plumes.

3.2.1. Typical Sources

In CIGRE Task Force 33.04.01 [2000] some examples of typical pollution environments are defined as follows.

- *Marine environment*, where the presence of the sea puts Na, Cl, Mg, K, and other marine salts into the atmosphere. Temperatures near the seacoast are normally moderated by the large body of water but can occasionally reach −18 °C (0 °F). Wind speed [Taniguchi et al., 1979] plays a large role in the deposit rate of sea salt in Japan.
- *Industrial environment* includes sources of soluble pollution such as electric generating stations burning coal or oil, steel mills, refineries, or sources of inert dust such as quarries and cement factories.
- *Agricultural environment* includes sources of highly soluble fertilizers and excrement as well as relatively insoluble dust and chaff.
- *Desert environment* includes inert content from sand as well as a high degree of salt in some areas [Rizk et al., 1975; Akbar and Zedan, 1991]. Typical inland desert areas are dry, dusty, windy, and hot, with cold temperatures overnight.

These environments may appear alone or in some combination. The *winter environment*, especially concerning the period with snow cover (low absolute humidity, dusty, windy, and cold) and exposure to sand and salt (from road salting) is most similar to the desert environment. This means that some seemingly illogical combinations, such as agricultural and desert conditions, do occur in winter in some areas.

The degree of pollution can be quantized using the rate of change of deposit on exposed surfaces. Since every area on earth has a measurable deposit rate, the *clean* areas are those that have a combination of pollution flux and duration of exposure that leads to very light pollution deposits of less than

$0.01\,mg/cm^2$ of soluble deposit or $0.1\,mg/cm^2$ of insoluble deposit as set out in international standards [IEC Standard 60815, 2008]. A rate of increase of less than $0.001\,mg/cm^2$ $(1\,\mu g/cm^2)$ per day of soluble deposit, or $0.01\,mg/cm^2$ per day of insoluble deposit, in an area with rain every week would generally be considered "clean" since the insulator surfaces would usually accumulate very light contamination level. If this area had the same pollution deposit rate but only had rain every six months, insulators would have a heavy pollution level.

Areas of *local* pollution have a high rate of change of deposit that decays rapidly with distance from the source. For point sources such as chimneys, the pollution flux is usually reported to pollution control agencies. For line sources, such as road salt flux from highways, the decay with distance is not as rapid.

Wide-area *regional* sources of pollution, such as the sea, heavily industrialized or coal-burning districts, or forest fires, have a high rate of change of pollution deposit that cannot be traced to a single point or line. Long-range transport of pollution from the central United States to Canada, and from the United Kingdom to Norway are examples of man-made regional sources.

3.2.2. Deposit Processes

Two different deposition processes lead to the accumulation of pollution on insulator surfaces.

- *Dry deposition* buildup on insulating surfaces of the ions that are electrically conductive when in solution with water. Dry deposition mass flux due to turbulence affects both top and bottom insulator surfaces. Mass flux due to gravitational settling affects mainly the upward-facing surfaces, leading to more rapid rates of increase in dry conditions. Accumulation of sodium chloride, sulfate, nitrate, and ammonium ions are most important although a range of other salts and metals can accumulate.

- *Occult deposition* of precipitation, such as fog, rain, ice, or snow. Depending on conditions, polluted ice, snow, or fog can be an important additional source of electrically conductive ions as well as a stable source of water to dissolve existing surface pollution.

3.2.3. Monitoring Methods for Site Pollution Severity

Around the world, there have been a number of different methods used to monitor the pollution levels at specific sites and on insulators [Lambeth et al., 1972; CIGRE, 1979; Looms, 1988; CIGRE, 1994]. As shown in Table 3-2, these methods range from the extremely simple—station staff need to clean the insulators when they can't see out of the office windows—to highly quantified, such as on-line monitoring systems for leakage current.

TABLE 3-2: Electrical Utility Practices for Assessing Pollution Severity

Method Used	Measured Parameters (Typical Units)
On-line monitoring of insulator leakage current	Current (mA)
On-line monitoring of air quality	Particulate concentration (e.g., $PM_{2.5}$)
On-line contamination monitoring on specialized sensors (such as liquid water sensor)	Resistance or capacitance versus humidity
Daily contamination monitoring of chilled insulator samples	Insulator leakage resistance (Ω) and surface resistance (Ω)
Daily samples on porcelain coupons	Electrical conductivity of wash water (μS/cm) and chemical composition (moles/L)
Daily visual inspection of insulators or windows nearby	Light transmission level, color
Systematic visual inspection of ability to bead water when wet (hydrophobicity)	State of insulator surface, Visual Guide (HC0-HC6)
Weekly or monthly samples on exposed sample insulators	ESDD (equivalent salt deposit density), NSDD (nonsoluble deposit density) (mg/cm^2 or $\mu g/cm^2$), and rate of increase
Systematic leakage current pulse monitoring	Weekly or monthly number of pulses (#) above threshold (mA)
Per-event or monthly samples of precipitation chemistry	pH and chemical composition (moles/L), electrical conductivity (μS/cm)
Monthly samples of dustfall	Soluble and insoluble deposit rates (g/m^2 per month)
Long-term withstand stress	Number of fuses remaining intact in tests on strings of disk insulators after exposure period (kV rms line-to-ground per unit)
Annual or long-term flashover frequency	Flashovers per 100 km of line length per year
Long-term climate statistics	Days with fog per year; mean and maximum periods without rain
Long-term corrosion rates	Rate of change of thickness of zinc, steel, copper, aluminum, or paint (μm/yr)

The methods in Table 3-2 are used:

- To assess site pollution severity (SPS), giving the input data to determine adequate insulator designs, including dimensions and materials, for reliable long-term electrical performance.

- To trigger maintenance actions such as insulator washing, cleaning, or renewal of grease or coatings.
- To quantify long-term outdoor exposure tests that compare the performance of different insulators, thereby allowing reproduction in controlled testing conditions.

Monitoring methods to trigger maintenance actions are discussed in Chapter 6.

3.2.4. Short-Term Changes in Pollution Levels

At any instant in time, the pollution on an insulator surface represents a balance between accumulation processes and cleaning processes.

The most important change in the pollution on insulators occurs during dry periods. Natural rain of moderate intensity and amount is reasonably effective at removing contamination, as detailed in the next section. The peak levels of insulator contamination are reached just before rain washing events. In winter or desert conditions, this period can extend to months or years, respectively.

Pollution accumulation on insulators is governed mostly by:

- The concentration (g/m^3) of pollutants in the air.
- The mass flow rate (m/s) past the insulator surface.
- The orientation of the surface relative to the wind and the air flow patterns around the disk profile.
- The capture efficiency of the surface material and condition.

There are both direct and indirect measurements of the concentration of pollution. For example, the $PM_{2.5}$ measurements are expressed as concentration. However, the samplers, filters, and analysis needed for accurate measurements are expensive. Normal measurements of pollution, dry and wet deposition rates, are carried out on plastic materials in a sheltered enclosure. The units of accumulation are g/m^2 on an upward-facing surface. These provide an integrated measurement of the product of pollution concentration and mass flow rate. The dry deposition values are representative for pollution accumulation rates on the upward-facing surfaces of insulators but less useful for establishing the values on the downward-facing surfaces.

The wind speed can exert a strong influence on the rate of accumulation. For areas of constant pollution flux in every direction, the influence should be linear. In cases where there is a strong directional source of ions, such as salt from the ocean, the wind direction starts to play a role. After studying a number of models, Taniguchi et al. [1979] found the best relation between salt deposit density and wind speed off the Sea of Japan was obtained using the cube of the hourly wind speed. Their observed insulator contamination values are shown in Figure 3-1 and their fitted expression is given in Equation 3-1.

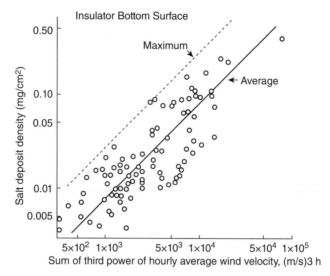

Figure 3-1: Observed salt deposit density on top and bottom surfaces of insulators near Sea of Japan (from Taniguchi et al. [1979]).

$$ESDD = C\sum_i \left(V_i^3 \cdot t_i\right) \tag{3-1}$$

where:

ESDD is the salt deposit density (mg/cm^2) on the bottom surface of the insulator surface,

C is a constant between 5 and 8×10^{-6}, depending on the location from sea and type of insulator, and

V_i is the average wind speed (m/s) for the time interval t_i (hours).

Surfaces at right angles to the wind flow will accumulate pollution more rapidly than those that face downwind. Measurements show that the pollution capture efficiency of aerodynamic disks with completely open profiles is 30–60% lower than standard disk insulators with convoluted bottom surfaces in the same exposure [Akbar and Zedan, 1991].

Hall and Mauldin [1981] carried out a series of wind-tunnel studies of the accumulation of salt pollution on energized insulators. They compared the three forces acting on a salt particle: F_g from gravitation, F_a from viscous force, and F_e a dielectrophoresis force from field-induced dipoles in the dielectric particle under ac electric field. The F_e force directs the particle toward areas of maximum electric field.

$$F_g = \frac{4\pi}{3}\rho r^3 g$$

$$F_a = 6\pi\eta r v \qquad\qquad (3\text{-}2)$$

$$F_e = 2\pi r^3 \varepsilon_0 \left(\frac{k-1}{k+2}\right)|\nabla E^2|$$

where

r is the particle radius (m),
ρ is the particle density (kg/m^3),
g is the force of gravity (9.8 m/s^2),
η is the viscosity of air (kg/(m-s)),
v is the particle velocity (m/s),
ε_0 is the permittivity of free space (8.854 × 10^{-12} F/m),
k is the relative permittivity of the particle (usually 81 if it is hydrated),
E is the electric field (V/m), and
∇ is the vector differential operator.

For NaCl particles of $r = 64\,\mu$m with a density of $\rho = 2.1$ kg/m^3 and an air viscosity of $\eta = 1.74 \times 10^{-5}$ kg/m-s, Hall and Mauldin calculated at a wind speed of 0.8 m/s that $F_g = F_a = 2 \times 10^{-8}$ kg-m/s^2. The dielectrophoresis force, typically acting against gravity and toward the direction of the insulator at the location shown in Figure 3-2, was $F_e = 2 \times 10^{-12}$ kg-m/s^2.

As the salt particles move near the insulator surface, or into a zone of still air between sheds on the bottom surface, F_e becomes the dominant force. This leads to a characteristically nonuniform deposit, with the greatest density being measured near the pin of the insulator as illustrated in Figure 3-3.

The electric field is highest near the pin of the insulator. It would seem that this also leads to the high density of salt particles in Figure 3-3.

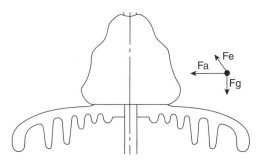

Figure 3-2: Direction of gravitational, viscous, and dielectrophoresis forces on salt particle near insulator (from Hall and Mauldin [1981]).

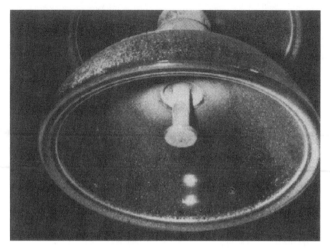

Figure 3-3: Accumulation of salt near pin of smooth shell-type insulator, energized in wind tunnel (from Hall and Mauldin [1981]).

The capture efficiency of pollution on energized insulators can also be increased if there is heating from leakage current activity. This tends to dry out occult deposition before it can drip away. This process was found to be stronger than ac dielectrophoresis [IEEE Working Group on Insulator Contamination, 1979].

Hall and Mauldin measured the accumulation of salt on the bottom surface for a 146-mm by 254-mm fog-type disk insulator with 445 mm of leakage distance. With no voltage, the ESDD was measured to be 0.019 mg/cm^2. This increased considerably to 0.049 mg/cm^2 with 7 kV ac per disk and to 0.064 mg/cm^2 with 12 kV ac per disk for the same wind-tunnel exposure. These are upper bounds for the increase in ESDD associated with ac fields.

The electrophoresis process for dc fields is stronger than the dielectrophoresis force for ac fields. This leads to higher accumulation rates on the positive poles of HVDC insulator systems. The difference in accumulation diminishes to less than 20% in areas where the wind speed is high and there is only natural pollution, but may reach a factor of 2 or 3 in areas of low industrial contamination [Wu and Su Zhiyi, 1997].

The diameter of station post insulators affects the wind flow and exposed surface area. When posts of different diameters were exposed under deenergized conditions near the sea [Matsuoka et al., 1991], this is one factor that could be an influence but differences in washing efficiency are more important.

Soft insulator surfaces such as EPDM and silicone may attract somewhat more pollution than hard ceramic surfaces. For EPDM, surface roughness is a factor while for silicone insulators, the presence of light molecular weight oil serves to capture and encapsulate pollution. These lead to significant differences in the washing efficiency among polymer materials.

3.2.5. Cleaning Processes and Rates

There are two main processes that naturally remove pollution from insulator surfaces—wind and rain. In addition, winter precipitation such as snow or ice can also dissolve surface ions and release them into drip water during melting and/or shedding.

Wind was also identified as a dominant factor in the accumulation of pollution, especially in wind tunnels and near the sea. Looms [1988] suggests that the time to reach equilibrium between accumulation and cleaning by wind action varies from days to years, and possibly follows the relation $M = A \log(t) + B$, where M is the mass of the deposit, t is time, and A and B are constants that vary with insulator shape, size, material, and orientation. In the desert [Akbar and Zedan, 1991] the rate of change of pollution on the bottom surfaces of insulators was initially linear with exposure duration, but after about 18 months this tended to level off to a constant value. This long time to reach equilibrium was also found in measurements of leakage current on disks with extensive sheltered creepage distance [Forrest, 1936]. The values are initially low but then rise to high levels over a 2-year exposure time.

Rain is very effective for cleaning the upward-facing surfaces of insulators [Beauséjour, 1981] but the efficiency for the bottom surface varies with intensity and quantity [Kimoto et al., 1971]. For upper surfaces and cylindrical posts, 10 mm or more of rain removes 90% of the salt deposit. Figure 3-4 shows some typical reductions observed for salt pollution, near the sea and near urban expressways in winter [Beattie, 2007].

Beattie found that relatively clean insulators with an ESDD less than $15 \mu g/cm^2$ did not have much reduction in ESDD with heavy rain, compared to insulators with higher pollution levels. The rain itself leaves a residual ESDD when it dries that is usually less than $5 \mu g/cm^2$.

Insulator profile has a strong influence on the effectiveness of natural rain washing. The sheltered leakage distance that protects insulators from pollution and some forms of icing flashover can only be washed by natural rain of high

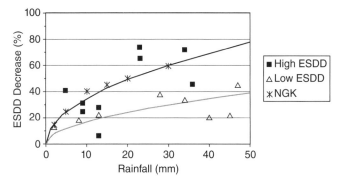

Figure 3-4: Reduction in ESDD versus rainfall amount: average of top and bottom insulator surfaces [Beattie, 2007] and bottom insulator surfaces [NGK, 1991].

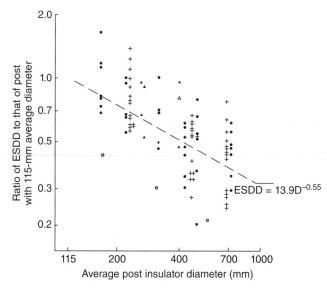

Figure 3-5: Ratio of ESDD as function of station post diameter (adapted from Matsuoka et al. [1991]). Reference: 1.0 at 115-mm diameter.

intensity under high wind speed. Insulators mounted with the cylindrical axis horizontal [Lambeth et al., 1973] or even at a small angle [Looms, 1988] rather than vertical tend to be washed more effectively and perform better. Station post insulators with large diameter and shallow shed profiles stay cleaner than those with smaller diameter with deep sheds. For example, Figure 3-5 shows that 450-mm posts had half of the ESDD of 115-mm posts, exposed off-potential together near the sea.

Soft insulator surfaces such as EPDM and silicone retain more pollution than hard ceramic surfaces. Measurements near the sea and inland suggest that ESDD values on silicone surfaces are twice those found on porcelain disks [CIGRE Task Force 33.04.01, 2000]. There are significant differences in the way water rolls off hydrophobic surfaces. Also, the silicone oil that promotes hydrophobicity also forms a barrier between the water and the pollution layer. The active nature of the silicone surface and its "target loading," or ability to absorb pollution, is discussed in more detail in Chapter 6.

3.2.6. Long-Term Changes in Pollution Levels

Over the time scale of years, pollution will tend to accumulate on downward-facing insulator surfaces. This can be simulated in the short term with wind-tunnel exposure at high salt density levels, as shown in Figure 3-3. It has also been observed, starting in the 1930s, that insulators with deep skirts perform well for the first few years in polluted conditions, but eventually start to cause more problems than insulators with uniform or no ribs on a flat bottom surface.

On a scale of decades, the pollution conditions around power system components have changed over time, and will continue to evolve. Industrial plants are often developed near electrical substations. New technologies for agriculture and transportation, such as manurigation or road salting of expressways, were not anticipated at the time when many insulators were specified. Some of these changes, such as pollution controls on coal-fired generating stations and automobiles, have actually led to a decrease in pollution level from maxima in the 1970s and this also needs to be considered.

Porcelain insulators removed from transmission lines placed in service 100 years ago may still have some excellent electrical properties. They usually reach the end of service life because the zinc protection on metal caps and pins erodes, leaving the underlying metal to rust.

While the evidence for global warming is strong, the effects of climate change are slow and it is unlikely that winter flashover problems will go away any time soon. However, another aspect of global warming is the appearance of erratic swings in weather patterns and this is of more interest. Effects of climate change on the duration of dry periods may be the most important aspect to consider in the insulation coordination process.

3.2.7. Other Factors in Pollution Problems

The presence of pollution alone does not necessarily lead to electrical flashover problems. Additional factors in the process, as detailed in Chapter 4, include the following:

- *Wetting* of the polluted surface, whether through hygroscopic scavenging from the atmosphere, accumulation or melting of natural precipitation, local fog from cooling towers, or other means. Surface deposits of insoluble materials play an important role in stabilizing the wetting process. The ability of a surface to repel water (its hydrophobicity) affects the continuous or discrete-drop nature of the wetted area.

- *Heating* of the electrically conductive pollution layer by resistive heating from leakage current. The electrical resistance of the pollution decreases with increasing temperature. Also, the power dissipation causes local evaporation and formation of *dry bands* that interrupt the flow of current. The high-voltage distribution along the insulator from line to ground end changes, with most electric field stress across the dry bands.

- High local electric stress across dry bands causes initiation and propagation of *arcing*. Arcing produces local ozone (O_3) that reacts quickly to NO_x, and also produces distinctive audible noise pollution. If arc plasma bridges all the way across an insulator, it will short-circuit the power system, causing current flow to local grounding.

The additional steps in the pollution flashover process after wetting add additional ions to those that accumulate in advance by dry and occult deposition.

3.3. NONSOLUBLE ELECTRICALLY INERT DEPOSITS

Each region may have a mixture of electrically conductive (soluble) and relatively nonconductive (insoluble) deposits on insulator surfaces. The main focus in electrical flashover is on the resistance of the wetted pollution layer, determined from the specific conductance approach (Section 3.4). Any nonsoluble deposit plays an indirect role in the resistance of the wetted pollution layer. Dense layers of insoluble deposit do not contribute to conductivity but instead change the insulator surface roughness, with effects on:

- The run-off rate of soluble material.
- The hydrophobicity of the insulator surface.
- The evaporation rate of the wetted layer.
- The local electric field strength.

3.3.1. Sources and Nature of Nonsoluble Deposits

According to a consolidation of global trends [Jickells et al., 2005], dried-out lake systems such as the Bodele Depression in the Sahara Desert are particularly important sources of dust. Dust is formed from the action of wind on erodible material, which has the effect of drying out the remaining land, destroying the vegetation cover, and leading to further desertification. Sandblasting and saltation occur when wind speeds are high enough. Several laboratory wind-tunnel studies [e.g., Duce, 1995] show that dust production is proportional to the cube of wind speed. The desert dust has a mean particle size of $2\,\mu m$, small enough to be transported in the atmosphere for long distances. The dust production and transport depend on rainfall, wind speed, surface roughness, temperature, topography, and vegetation cover.

At a certain critical velocity v_c, Cheng and Yeh [1979] observed and computed that dust particles of 0.5–$10\,\mu m$ bounce off dry plates rather than adhere. Their low, median, and maximum values of v_c are shown in Figure 3-6.

In the case of insulators exposed to natural winds of at least $1\,m/s$, Figure 3-6 suggests that the nonsoluble deposit that accumulates on insulator surfaces will typically result from accumulation of particles with aerodynamic diameters less than $10\,\mu m$. Since the wind speed tangent to the surface is lower, the nonsoluble deposit may accumulate faster on insulator surfaces that are nearly horizontal than on those facing into the wind.

3.3.2. Direct Measurement Method for NSDD

It has become common to evaluate the insoluble fraction of the pollution layer that accumulates on insulators. This is reported as a nonsoluble deposit density (NSDD), in units of mg/cm^2 or sometimes $\mu g/cm^2$ for low density. The details for obtaining insulator wash water are fully described in Annex C of IEC Standard 60815 [2008]. The NSDD is obtained by weighing a dried filter paper, pouring

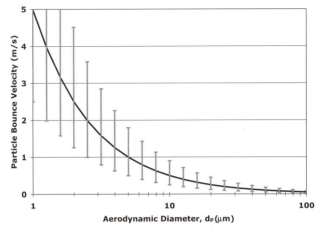

Figure 3-6: Critical velocity for onset of particle bounce (data from Cheng and Yeh [1979]).

a sample of insulator wash water through it, and then redrying and reweighing the filter. The change in weight (in mg or µg) divided by the surface area wiped (in cm^2) gives the NSDD.

3.3.3. Indirect Measurement Methods for NSDD

Some utilities simply monitor the NSDD from its appearance on insulators. It is also possible to adapt optical measurements for this purpose [CIGRE Task Force 33.04.03, 1994], or to compare the deposit thickness against thin feeler gauges.

Many countries around the world have federal and local agencies that monitor and regulate the quality of air in their jurisdictions. Examples were given of acid-rain monitoring programs in the United States and Canada [Canada–U.S. Subcommittee on Scientific Co-operation, 2004] and of dry and wet deposition, as shown in Figures 3-15 and 3-16.

Standard, low-cost passive dust samplers are often used for local pollution measurements. Although these environmental survey tools require no special operator skill or electrical power at the measurement sites, they still give quantitative measurements of all types of pollution. Active instruments that measure dry and wet deposition are described in the following section.

The simplest *dust samplers* use a single, open-top polyethylene bucket, fitted with a fresh 3-mil plastic bag. The bucket is about 450 mm tall and has a diameter of 150 mm, located 3.6 m above ground. Every month, the plastic bag is removed and sealed for analysis. At the laboratory, distilled water is added to make a specific volume and the electrical conductivity is measured. The sample is filtered and the weight change of the redried filter paper gives a measure of insoluble deposit. The soluble and insoluble deposits are both expressed in g/m^2 per 30 days.

According to Potvin [2006], dustfall samples contain the large and visible particulate matter in the 25–100-μm size range. Since these particles settle out quickly by gravity, dustfall is a good indicator of local pollution sources. Natural dustfall sources include vegetation fibers, biological material, and particles from soil erosion. Man-made sources can include particles of coal, coke, ash, wood fibers, wood char, paint chips, grain dust, and road dust. As a guide, regulations in the Canadian province of Ontario call for limits of 7 g/m² in any given month and an annual average of less than 4.6 g/m²/30 days. Perfect rank correlation was found between total dustfall and insulator contamination level at several sites in Ontario [Chisholm et al., 1993]. Empirical relations have also been developed between soluble dustfall and insulator contamination for other regions.

Directional dust deposit gauges (DDDG), consisting of four tall, thin, slotted cylinders, pointing North, South, East, and West, give an additional indication of the source location [Lambeth et al., 1972]. In IEC Standard 60815 [2008], the cylinders of a DDDG are specified to be 500 mm long, with 75-mm outside diameter. The long axis is oriented vertically. The slots are 40 mm wide with 20-mm radius at each end. The center-to-center distance is 351 mm. The bottom of the slot is about 3 m off the ground. As for nondirectional gauges, the four sample jars are removed every 30 days. For IEC practice, 500 mL of demineralized water is added, and twigs and leaves are removed. The conductivity of each sample is measured at 20 °C and the four results are averaged to give a Pollution Index (PI) in units of μS/cm. The nonsoluble deposit is measured the same way, by filtering the dust gauge contents through a dry and preweighed filter paper, which is then dried and reweighed.

When a series of DDDG measurements have been completed, the results are used to adjust site pollution severity classification according to a five-by-three level matrix, given in Table 3-3.

The IEC recommends that the PI value of conductivity be modified by multiplying by a climate correction factor C_f:

$$C_f = \sqrt{\frac{F_{\text{days}}}{40} + \frac{D_{\text{months}}}{6}} \qquad (3\text{-}3)$$

where F_{days} is the number of days with visibility less than 1 km per year and D_{months} is the number of months with total precipitation less than 20 mm per year. This formula tries to address the largest uncertainty in relating inexpensive dust deposit measurements to insulator contamination levels using exposure duration.

As an alternative to ground-based measurements, satellite measurements of iron flux have been processed to evaluate, on a large scale, the regions where dust originates and also to observe the seasonal and climate variations. Jickells et al. [2005] validated and consolidated three independent modeling studies of the annual dust deposition in order to estimate that the oceans of the world receive about 450 Tg/yr. This was developed from a global map of dust deposition in Figure 3-8 in the case studies presented later.

TABLE 3-3: Site Pollution Severity Assessment Using Directional Dust Gauge Results According to the IEC

Insoluble Deposit (g/month)		Conductivity of Soluble Dust Sample in 500 mL of Water (µS/cm)					
Annual Average	Maximum in Any Month	<25	25 to 75	76 to 200	201 to 350	>350	Annual Average
		<50	50 to 175	176 to 500	501 to 850	>850	Maximum in Any Month
<0.5	<1.5	a	b	c	d	e	
0.5–1	1.5–2.5	b	c	d	e	e	
>1	>2.5	b-c	c-d	d-e	e	e	

a: Very Light; **b**: Light; **c**: Medium; **d**: Heavy; **e**: Very Heavy

3.3.4. Role of NSDD in Insulator Surface Resistance

The role of nonsoluble deposit in obtaining consistent laboratory results is recognized by including kaolin (clay) in the mixture used for precontamination. Normally, Roger's kaolin (quartz kaolinite) of about 46% SiO_2, 37% Al_2O_3, and 1% Fe_2O_3 is used in North America and Europe; tonoko (quartz muscovite) is common in Japan [Matsuoka et al., 1995], and it has less ignition loss and 6% Fe_2O_3. While other clays have similar composition, Roger's kaolin gives a surface resistance that is a factor of 4 lower than other kinds, as shown in Figure 3-7.

In freezing conditions, the accumulation of natural fog in thin layers on polluted insulator surfaces acts as a form of NSDD [Chisholm, 2007]. A typical 50-µm layer of frozen fog with density of 0.8 g/cm³ would have a NSDD of 4 mg/cm². This would exert a strong effect in the IEC classification system (see Figure 3-9).

3.3.5. Case Studies: NSDD Measurements

Global Source Map Figure 3-8 shows the global dust deposit rate, derived from the flux of iron observed from satellite studies and consolidated by Jickells et al. [2005].

Some interesting aspects are noted in the global picture. Dust deposit in Europe is two to four times higher than levels in North America, with the exception of California, where most utilities have specialized practices

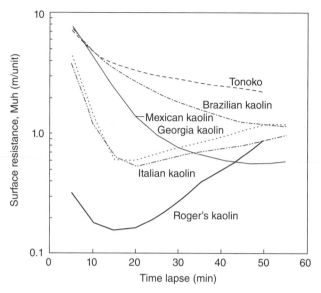

Figure 3-7: Time variation of surface resistance during clean-fog (controlled wetting) tests (from Matsuoka et al. [1995]).

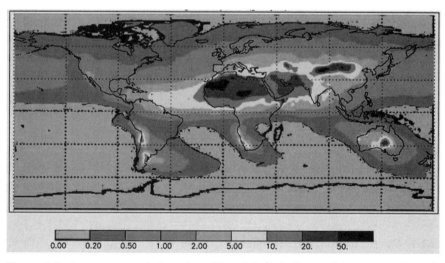

Figure 3-8: Average annual dust deposition (g/m²/yr) (from Jickells et al. [2005]], reprinted with permission from AAAS).

for contamination performance. Dust deposit levels in parts of Africa, the Middle East, and Asia are significantly higher than in Europe or North America.

West Coast of Persian Gulf, Iran Detailed measurements of nonsoluble deposit density (NSDD) and equivalent salt deposit density (ESDD) were carried out on exposed insulators and directional dust gauges along the northeast cost of the Persian Gulf in Iran by Shariati et al. [2005]. Their results in Figure 3-9 confirm that this region has a severe combination of salt (decaying with distance from the shore) and dust. These results are plotted on a classification system, also shown in Figure 3-9, that was adopted in IEC Standard 60815 [2008].

The Persian Gulf is indicated to have a high annual rate of dust deposit of 10–20 g/m^2 per year in the global map of Figure 3-8.

1000-km Transmission Line Route, Russia Figure 3-10 shows the variations in nonsoluble deposit density (NSDD) recorded along a 1000-km line route in Russia [Farzaneh et al., 2007]. In this case, the surface conductivity, associated with soluble contamination, was relatively constant at 0.3–0.4 μS. The NSDD varied considerably from 0.02 to 0.14 mg/cm^2. Figure 3-10 also shows a drop in insulator flashover voltage from 31 kV to 24 kV, associated mainly with the peak values of NSDD.

Russia is indicated to have a modest annual rate of dust deposit (1–2 g/m^2 per year) in the global map of Figure 3-8.

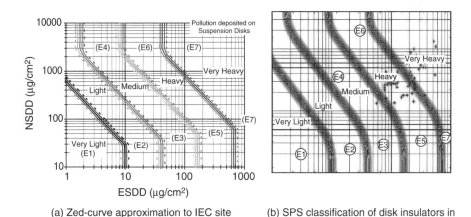

(a) Zed-curve approximation to IEC site pollution severity (SPS) guidelines

(b) SPS classification of disk insulators in Iran [Sharati et al., 2005]

Figure 3-9: Measurements of soluble (ESDD) and nonsoluble (NSDD) insulator pollution in region northeast of Persian Gulf [Shariati et al., 2005], plotted over IEC Standard 60815 [2008] guidelines for disk insulators.

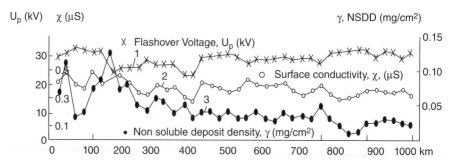

Figure 3-10: Observed pollution parameters along 1000-km transmission line (from Farzaneh et al., [2007]).

3.4. SOLUBLE ELECTRICALLY CONDUCTIVE POLLUTION

It is important to distinguish among measures of air quality (levels of CO or ozone) from those of air pollution, which are normally expressed as particle density and rates of mass deposit on surfaces. A precise knowledge of the chemical content of the air cannot, by itself, give a reliable indication of the pollution levels on an insulator. Additional factors are needed to establish this correlation, such as:

- The wind speed (mass flow past the insulator cross section).
- The duration of the exposure (days without rain).
- The state of the insulator surface during the air flow (wet or dry).
- The orientation of the insulator surface (at right angles or parallel to the wind flow, facing up or facing down).

Air pollution measurements of the dry and wet deposition on the upward-facing surfaces of samplers include most these factors in well-controlled ways. This is one reason why these measurements can give a satisfactory correlation with the ESDD levels on insulator surfaces.

3.4.1. Electrical Utility Sources

Electrical utilities that burn coal contribute significantly to the flux of electrically conductive pollution that accumulates on insulators. This led to increasing regulation, for example, the National Environmental Policy Act of 1969 [NEPA, 1969] in the United States, taking a federal rather than state responsibility for oversight and enforcement of limits. Since 1971, limits on three air pollution components have been set:

Sulfur dioxide (SO_2), $2620\,\mu g/m^3$,
Nitrogen dioxide (NO_2), $3750\,\mu g/m^3$, and
Particulate matter, $1\,mg/m^3$.

Operational experience showed that flue gas desulfurization (scrubbers) using lime or limestone was effective in removing about 80% of the SO_2 from coal-fired power plants [US EPA, 1977]. The concept of emissions trading began in 1992 [US EPA, 1990] with the parameters given in Table 3-4. (See Figure 3-11.)

The Tennessee Valley Authority [Thomas et al., 1969] reported the parameters given in Table 3-5 prior to pollution controls.

The time-averaged concentration of pollution at the ground level downwind of an isolated, elevated chimney source is

$$\bar{s}(x, y, z = 0) = \frac{Q}{\pi \bar{u} \sigma_y(x) \sigma_z(x)} \exp\left(\frac{-y^2}{2\sigma_y^2}\right) \exp\left(\frac{-h^2}{2\sigma_z^2}\right)$$

where

Q is the source intensity (kg/s),

$\bar{s}$ is the time-averaged concentration (kg/m^3),

$\bar{u}$ is the mean wind velocity (m/s),

σ_y and σ_z are the standard deviations in the horizontal and vertical directions (m),

TABLE 3-4: EPA Regulation Limits for Electrical Utility Generation Plants

Year	Number of Regulated Plants	Regulation Limit (tonnes SO$_2$)
1995	110	5,700,000
2000	810	14,600,000

Figure 3-11: Electrical plants affected by Phase I of the US EPA Acid Rain Program (from US EPA [1998b]).

TABLE 3-5: Thermal Plant Design and Operation Data at the Tennessee Valley Authority

Parameter	Paradise	Gallatin	Shawnee	Johnsonville	Colbert	Widows Creek
Number of units	2	4	10	4	4	1
MW/unit	704	314	175	173	296	575
Number of stacks	2	2	10	2	4	1
Height (m)	193	152	76	122	81	152
Diameter (m)	7.9	7.6	4.3	4.3	5.9	6.3
Design flue gas temperature (K)	413	410	413	425	444	414
Stack gas velocity			7.7–29.2 m/s			
Volume emission rate			136–663 m^3/s			
Heat emission rate			22–103 MJ/s, or about 8.3 MJ/s per 100 MW			
Sulfate emission rate[a]			0.13 m^3/s per 100 MW for 1.6% sulfur coal at 23 MJ/kg			
Wind speed			1.0–16.8 m/s			
T_a vertical gradient			−5.3 to +37.4 K/km			

[a]Moore in Thomas et al. [1969].

Source: Thomas et al. [1969].

y is the direction downwind (m), and

h is the relative height of the source above ground (m).

The increasing regulation has made significant reductions in the flux of SO_2 and NO_2 as a fraction of total air pollution. Aherne et al. [2005] suggest in Figure 3-12 that there has been nearly a 50% reduction in sulfate deposition from the peak levels in 1975. Hopefully, this trend will continue to decline with improving pollution control technology and satisfactory operating experience.

Other countries have made similar or greater progress in reducing sulfur emissions, as shown in Figure 3-13.

The report by Mikhailov [2002] is particularly interesting as it focuses on the effects of air pollution on metal corrosion rates. This was important in a country that historically lost 10–12% of its annual metal production to corrosion each year. After 1945, when air pollution monitoring was initiated, SO_2

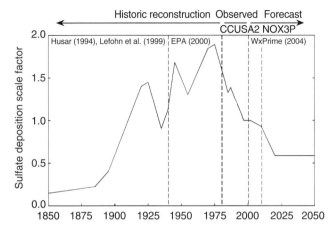

Figure 3-12: Estimated historic and future sulfate deposition (scaled to 2000 = 1.0) for the period 1850–2050. (from Aherne et al. [2005]).

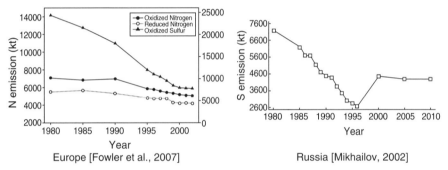

Figure 3-13: Emissions of sulfur and nitrogen in Europe and Russia, 1980–2005. (Courtesy of Springer Science+Business Media.)

levels of 200–400 µg/m³ were common in the urban and industrial regions of the former Soviet Union, with an average level of 250 µg/m³ for Moscow in 1950. With substitution of coal with gas, the SO_2 levels declined continuously to a minimum 16 µg/m³ in 1994–1995, and have since recovered. There was also a factor-of-4 reduction in the corrosion rates of steel, zinc, and copper over the same observation periods.

The Canadian concept of "target loading" for soil was proposed and accepted during transboundary pollution control negotiations of sulfates and other acid pollutants with the United States in the early 1980s. The target loading hypothesis was originally used to manage deposition of sulfates to aquatic systems. Each lake has an ability to buffer some acid deposition, described by its *critical load*. The excess of current deposition over the critical load is called *exceedence*. At the worst period in 1975, more than 30% of the lakes modeled in Canada [Aherne et al., 2005] had no remaining ability to neutralize acid, due to the exceedence of sulfate deposits.

Some nonceramic insulator materials have an ability to absorb pollution. In particular, silicone materials provide light molecular weight polydimethyl-siloxane (PDMS) molecules with low viscosity that float on top of the polluted surface. However, this ability can be overwhelmed by heavy deposit density and the rate also depends on ambient temperature. In this context, insulators also have target loading levels that can be managed with the same process used for soil and lake pollution.

Critical loading is generally defined as a quantitative estimate of exposure to one or more pollutants below which significant harmful effects on specified elements of the environment do not occur, according to present knowledge [Nilsson and Grennfelt, 1988]. Critical loads for air pollution components, based on lake and soil chemistry, have been widely accepted in Europe for the Convention on Long-range Transboundary Air Pollution [e.g., Hettelingh et al., 1991; Posch et al., 2005] as well as North America.

The levels of critical loading for soil vary widely, as shown, for example, in values for forest soils in Table 3-6.

The concept of critical loading can readily be adapted to insulator surfaces. The electrically conductive deposit on insulators is expressed as an equivalent salt deposit density (ESDD) in units of mg or µg of sodium chloride per cm^2 of surface area. Values of $10 \mu g/cm^2$ ESDD are considered light, while levels of $400 \mu g/cm^2$ are classed as very heavy [IEC Standard 60815, 1986]. The total (wet and dry) sulfate deposition rate of $32 kg/(ha\text{-}yr)$ in Table 3-6 converts to $320 \mu g/cm^2$ per year. Sulfate and chloride have the same equivalent conductance. Adding in the contribution of sodium anions to ESDD, the Class-5 critical load works out to a background ESDD accumulation rate of $200 \mu g/cm^2$ per year.

Aherne et al. [2005] calculated that pollution controls have been remarkably effective in bringing sulfate emissions down below critical loading levels. A deposition and runoff model was calibrated for dominant anions $\left(H^+, Ca^{2+}, Mg^{2+}, Na^+, K^+, NH_4^+\right)$, cations $\left(SO_4^{2-}, Cl^-, NO_3^-\right)$, and dissolved organic carbon. The ability to neutralize acid is expressed by the difference

TABLE 3-6: Classification of Critical Loads for 50-cm Soil Depth in Europe with Conversion to Equivalent Salt Deposit Density (ESDD) for Electrical Insulators

Class	Total Acidity (kmol H^+) / (km²-yr)	Equivalent Amount of Sulfur (kg/(ha-yr))	Equivalent Amount of NaCl ((µg/cm²)/yr)
1	<20	<3	<20
2	20–50	3–8	20–50
3	50–100	8–16	50–100
4	100–200	16–32	100–200
5	>200	>32	>200

Source: Nilsson and Grennfelt [1988].

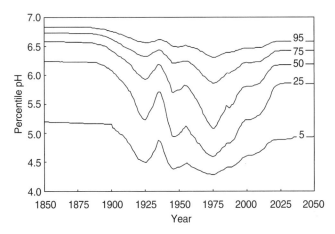

Figure 3-14: The 5th, 25th, median, 75th, and 95th percentile pH in Canadian lakes, 1850–2050, based on soil and lake water chemistry (from Aherne et al. [2005]).

between these anions and cations. Based on the sulfate loading in Figure 3-12, the median pH in 398 lakes was reported to fall from an original value of 6.1 to a minimum of 5.1 in 1975. Figure 3-14 shows the median pH recovering to the present level of 5.8 after the Clean Air Act came into effect.

The history and future of air pollution control has two important effects on the selection of insulators for power systems.

1. Insulators that have given good performance through the mid-1970s, when air pollution peaked, are likely to give better performance now and in the future as the deposition of electrically conductive ions such as sulfate continues to decline.

2. With advances in pollution control, less sulfate is released to the atmosphere, meaning more is retained near generating stations. Local emissions of small fractions of this sulfate when washing the pollution control equipment during maintenance have proved to be problems for electrical equipment at some coal-fired generation plants.

3.4.2. Other Fixed Sources

To obtain an appropriate balance between electrical utility and other industrial emissions, a wide range of inventory and reporting programs for release of chemicals have been implemented in many countries. Many of these focus on particulate matter (PM) and wet deposition as well as dry deposition rates for electrically conductive ions such as hydrogen, sulfate, nitrate, ammonium, sodium, and chloride.

Wet deposition is particularly easy to study because it involves ionic analysis of precipitation samples. As an example, the US EPA reports wet deposition rates for nine ion sources as shown in Figures 3-15 and 3-16. The ion sources are expressed in kg per hectare, with 10^8 cm^2/ha. The contribution of each ion to electrical conductivity of precipitation depends both on the ion molecular weight and on the specific conductance as shown in Table 3-7.

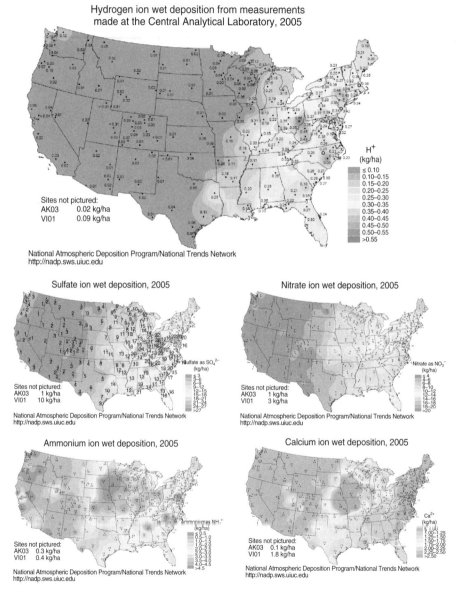

Figure 3-15: Wet deposition of electrically conductive ions in the United States: inland sources (from NADP [2007]).

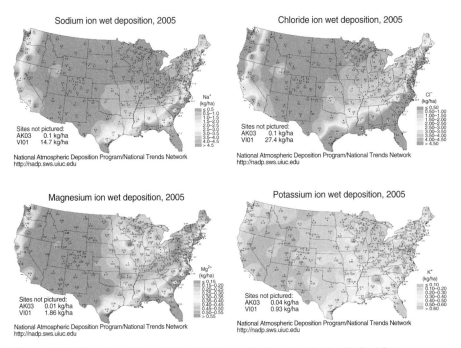

Figure 3-16: Wet deposition of electrically conductive ions in the United States: ocean sources (from NADP [2007]).

TABLE 3-7: Contribution of Ion Sources to Electrical Conductivity of 10 × 10-cm (Exposed) × 100-cm² (Deep) Annual Precipitation Sample

Source	Equivalent Ionic Conductance $\lambda°$ (S-cm²/mEq)	Maximum Level in USA (kg/ha)	Molecular Weight (g/mEq)	Contribution, Conductivity at 0 °C (µS/cm)	Contribution, Conductivity at 20 °C (µS/cm)
H^+	350	0.5	1.0	12.0	16.3
$\frac{1}{2}SO_4^{2-}$	80	27	48.0	2.3	4.1
NO_3^-	71.5	20	62.0	1.3	2.1
NH_4^+	73.5	4.5	18.0	1.0	1.7
$\frac{1}{2}Ca^{2+}$	59.5	2.5	20.0	0.4	0.7
Na^+	50.1	30	23.0	3.4	5.9
Cl^-	76.4	50	35.4	5.8	9.7
$\frac{1}{2}Mg^{2+}$	53.0	3.5	12.15	0.8	1.4
K^+	73.5	1	39.1	0.1	0.2
Total				*27 µS/cm*	*42 µS/cm*

The concentrations of ions are small, allowing the use of the infinite dilution values for equivalent ionic conductance in Table 3-7. Generally, the contributions of hydrogen and ocean salt (Na^+ and Cl^-) form more than three-quarters of the total in an annual sample of 100 cm of precipitation that combines the worst wet deposition rate of every ion. The values of conductivity are given for both 20 °C, the normal reference temperature, and at 0 °C, the temperature relevant to flashovers during the melting phase in winter conditions.

The pH of the precipitation (negative logarithm of $[H^+]$) is established by an acid–base balance, as sulfuric and nitric acids will be neutralized by basic species such as ammonia and calcium carbonate. Typically, two-thirds of the acid input to the atmosphere is neutralized. Carbonate is sometimes assumed to make up any deficit when carrying out an anion–cation balance for a typical precipitation sample when the pH has been established.

Figure 3-15 shows long-range transport of sulfates downwind from the Ohio Valley into the northeastern United States. In Fikke et al. [1993], a similar long-range transport of pollution from the United Kingdom to Norway was found in ice samples.

It is clear in Figure 3-16 that prevailing winds carry salt pollution from the ocean a long distance inland. This was also quantified as the monthly rate of increase of insulator salt deposit density by Kimoto et al. [1971] for distances from 1 to 100 km from the ocean in Figure 3-17.

A detailed analysis of precipitation and dry deposition chemistry by Hara et al. [1995] used wet and dry samplers at 29 sites throughout Japan. The relative concentrations of Na^+, Cl^-, and Mg^{2+} indicated that half of the ions in precipitation in Japan originated from sea salt (Table 3-8).

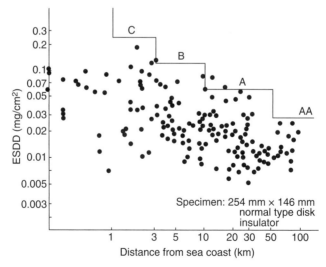

Figure 3-17: Monthly rate of increase of insulator salt deposit density versus distance from the ocean (from Kimoto et al. [1971]).

TABLE 3-8: Concentrations and Deposition of Major Ions in Japan

Quantity	Concentration (µEq/L)			Deposition (mEq/(m²-yr))		
	Minimum	Mean (Standard Deviation)	Maximum	Minimum	Mean (Standard Deviation)	Maximum
pH	4.5	4.8 (0.7)	5.8	—	—	—
H	1.5	17.3 (9.7)	31.6	2.2	24.2 (15.1)	60.9
Na	6.4	49.1 (70.2)	275	8.2	86.9 (90.9)	365
Cl	13.7	63.5 (80.4)	322	15.4	109 (105)	429
Mg	1.8	13.6 (15.9)	60.9	2.7	22.9 (21.2)	80.7
Total Ca	4.9	16.0 (8.4)	37.2	7.4	21.7 (13.4)	59.0
Non-sea-salt Ca	2.0	14.2 (9.1)	34.5	3.0	18.0 (13.6)	54.7
K	0.6	3.1 (2.8)	13.6	0.7	4.9 (3.0)	11
NO_3	1.8	14.1 (4.1)	25.0	3.1	19.4 (7.4)	40.8
SO_4	20.5	44.4 (12.5)	63.6	22.5	62.5 (22.8)	105
Non-sea-salt SO_4	5.2	38.6 (12.5)	58.9	9.4	52.2 (22.6)	99.5
NH_4	0.6	18.3 (6.7)	29.8	1.1	25.9 (12.3)	55.4
Rainfall (mm/yr)	590			1403 (319)		2041

Source: Hara et al. [1995].

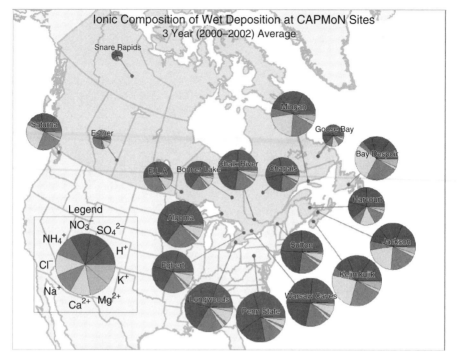

Figure 3-18: Ion composition of wet deposition. Slices represent fraction of total wet deposition (eq) contributed by that ion. Areas of circles directly proportional to total wet deposition at each site (from Environment Canada [2004]).

Ion species with a high relative standard deviation (Na, Cl, and Mg) originate from sea salt and vary from site to site. Species with low relative standard deviation $(SO_4^{2-}, NO_3^-, \text{and } NH_4^+)$ are transported from distant sources on a national scale.

In Canada [Environment Canada, 2004] cations and anions contribute to the total electrical conductivity, in the fractions shown in Figure 3-18.

The CAPMoN program was established to monitor transboundary pollution arriving from the coal-burning power pants in the United States. The measurements suggest that the areas of high pollution are in the east, but, in fact, the highest levels of sulfur and nitrogen emissions and deposits are downwind of coal-fired generation in the province of Alberta.

While the chemistry of precipitation and wet deposition is fundamental to the performance of insulators in heavy icing, the dry deposition onto insulator surfaces is the most important aspect for cold-fog conditions. Total acid deposition is determined using both wet and dry deposition measurements.

- Wet deposition is the portion of acid deposit that is dissolved in cloud droplets and then deposited during precipitation. As illustrated previously, the usual chemical components of wet deposition include sulfate, nitrate, and ammonium with elevated sodium and chloride near the ocean.

· Dry deposition is the portion deposited on dry surfaces during periods of no precipitation, generally including winter if surfaces are sheltered from direct snowfall. The dry deposition may be solid particulate matter such as sulfate (SO_4) and nitrate (NO_3), or it may be in the form of gaseous nitric acid (HNO_3), sulfur dioxide (SO_2), and ammonium (NH_4).

Generally, wet deposition of SO_4 and dry deposition of SO_2 gas are the main, and roughly equal, sources of sulfate. Dry deposition of HNO_3 particulate has the same magnitude as wet deposition of NH_4 and NO_3, with the three fractions forming the majority of the total nitrate deposition.

In Environment Canada [2004] the 5-year average values of dry and wet deposition were established at 11 sites.

Figure 3-19 shows that the relation between total sulfur and nitrogen deposition is rather strong, suggesting that in cases where only the sulfur values are monitored, a simple linear correction factor may be sufficient to establish the total deposit of all important electrically conductive ions. The relationship between wet deposition, shown in Figures 3-15 and 3-16, and the dry deposition that would accumulate on insulator surfaces is still statistically significant although the correlation is lower.

In the United States, the EPA CASTNET program monitors dry deposition, with results expressed in annual kg/ha as is done for wet deposition [US EPA, 2000; NADP, 2007].

Figure 3-20 shows high levels of dry sulfate deposition in the same regions with high wet deposit rates of sulfate in Figure 3-15.

The influence of distance from the Persian Gulf, a saltwater body, has also been quantified using 45 insulator and dust deposit measurement sites in Iran

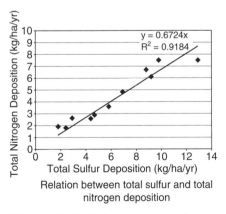
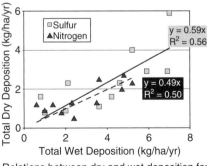

Figure 3-19: Relations among total and wet and dry sulfur and nitrogen deposition in Canada (data from Environment Canada [2004]).

Figure 3-20: Dry deposition of sulfur in the United States in 1998 (from US EPA [2000]).

[Shariati et al., 2005]. They classified their results using the IEC site pollution severity classes, as defined in IEC Standard 60815 [2008]. For purposes of standardization, the values are defined from (a) very light, (b) light, (c) medium, (d) heavy, and (e) very heavy, and further classified on the basis of soluble (ESDD) and nonsoluble (NSDD) content. The densities are both expressed in mg/cm^2. Mixed pollution, with both ESDD and NSDD, is referred to as "Type A." "Type B" occurs when liquid electrolytes (such as salt water from the ocean) are deposited on insulator surfaces with little or no NSDD. While locations close to the sea suggest that the region of Iran in Figure 3-21 should have Type B pollution, the studies demonstrated that this region also had high dust levels (transported from the Sahara Desert) bringing it into Type A.

Many European countries participate in COST programs [COST, 2007a, b] that support cooperation among scientists and researchers across Europe with open access to results on the Internet. Examples of COST programs that address mapping of fixed pollution sources include:

COST Action 715 [COST, 2005] to apply meteorology to urban air pollution problems.

COST Action 728 to enhance mesoscale meteorological modeling capabilities for air pollution and dispersion applications.

COST Action 729 to assess and manage nitrogen fluxes in the atmosphere–biosphere system in Europe.

COST Action 735 to develop methods and results for estimating global air–sea fluxes of compounds relevant to climate and air pollution. The compounds include CO_2, sulfates, dimethyl sulfate (DMS) from phytoplankton in oceans, halogenated hydrocarbons, nitrogen oxides, trace metals, and nutrients.

Figure 3-21: Pollution regions on northeast coast of Persian Gulf (from Shariati et al. [2005]).

COST Action 722 (short-range forecasting methods of fog, visibility, and low clouds) and Action 727 (measuring and forecasting atmospheric icing structures) provide additional resource data that can also be used when carrying out insulation coordination for winter conditions in Europe.

3.4.3. Conductance of Electrolytes

Electrolytes conduct an electric current when they are melted to a liquid state or alternately dissolved in a solvent such as water. Electrolytes are composed of positively charged species, called cations, and negatively charged species, called anions. For example, sodium chloride (NaCl) is an electrolyte composed of sodium cations (Na$^+$) and chlorine anions (Cl$^-$). An electrically neutral electrolyte will have a fixed ratio of cations to anions.

Anions and cations from electrolytes can be monoatomic, like NaCl, or composed of polyatomic ions, such as ammonium nitrate (NH$_4$NO$_3$). Hydration of molecules into anions and cations is described by reaction expressions such as $NaCl \rightarrow Na^+ + Cl^-$ or $NH_4NO_3 \rightarrow NH_4^+ + NO_3^-$.

The hydrogen cation, H^+, has a special role in electrochemistry. When H^+ ions are produced in water solution along with a monatomic anion (such as Cl^-) or polyatomic anion (such as NO_3^-), the electrolyte is an acid—hydrochloric acid (HCl) or nitric acid (HNO_3). The acidity (or alkalinity) of a solution is given approximately by the negative logarithm (to base 10) of the concentration of H^+ ions (pH) as follows:

$$pH \approx -\log_{10} \frac{[H^+]}{1\,mol/L} \tag{3-4}$$

Water naturally disassociates to equal concentrations of $[H^+] = [OH^-] = 1 \times 10^{-7}\,mol/L$ at $25\,°C$, giving a pH of 7.0.

For a constant applied voltage to two metal electrodes immersed in an electrolyte, the amount of electric current is proportional to the number of ions dissolved. Pure water with pH = 7.0 is a good insulator and the current flow will be low. The resistance will decrease as the amount of electrolyte in the solution is increased.

Electrical conductance of ionic solutions is the reciprocal of ionic resistance. To make the measurement of conductance practical, correction for the shape and size of the sample leads to the measurement of conductivity σ. This is defined by

$$\sigma = \frac{L}{AR} \tag{3-5}$$

where

σ is the conductivity of the medium, in $\mu S/cm$ ($\Omega^{-1}cm^{-1}$),
L is the length of the sample in cm,
A is the cross-sectional area of the sample in cm^2, and
R is the measured resistance across the sample length L.

Electrolytes that yield a hydroxyl (OH^-) anion in solution, such as sodium hydroxide (NaOH), are called bases. Ammonia (NH_3) is a polyatomic cation that forms basic solutions, with the reaction $NH_3 + H_2O \rightarrow NH_4^+ + OH^-$.

Electrolytic substances that dissociate completely into component ions in solution are called strong electrolytes, acids or bases, depending on how they change the concentration of H^+ ions. The chemical reactions shown above would proceed completely to the right of the arrow, leaving dissolved ions with no associated, electrically neutral molecules. For example, when sodium chloride is dissolved in water, all of the dissolved material is present as Na^+ and Cl^- ions, with no dissolved NaCl molecules.

Weak electrolytes dissolve in water but do not dissociate completely. For example, acetic acid ($HC_2H_3O_2$) is dissolved in water because it has polar O—H bonds. In solution, some but not all molecules will dissociate to form H^+ and $C_2H_3O_2^-$ ions, making it a weak acid. Similarly, the reaction of ammonia with water, $NH_3 + H_2O \rightarrow NH_4^+ + OH^-$, proceeds to only a small extent and some molecules remain as NH_3. Ammonia is thus classified as a weak base.

Ethanol (CH_3CH_2OH) is an example of a nonelectrolyte molecule that dissolves in water but does not produce any ions, making it a useful additive when washing insulators in cold weather.

The conductivity of a solution of a strong electrolyte such as sodium chloride (NaCl) will decrease as the concentration of salt increases. For dilute solutions, the change with concentration occurs because the number of charge carriers per unit volume tracks the ion concentrations closely. For these cases, it is practical to define the *equivalent conductance* λ to factor out the effect of concentration c.

Concentration c can be expressed in moles per cubic meter (mol/m^3), molarity (mol/dm^3), molality (mol/kg solvent), or normality. In the case of water, which is the only solvent we are concerned about, molarity and molality are equal. As a reminder about molar weights and terminology: to make a 1-molal aqueous solution of salt water, measure 1 kg of water and add 1 mole of the solute, NaCl. The atomic weight of chlorine is 35.4 and the atomic weight of sodium is 23, so the formula weight of NaCl is 58.4. This means that 58 grams of NaCl dissolved in 1 kg of water would give a 1-molal solution of NaCl. The process is nearly the same for producing a 1-molar solution: 58.4 g of salt is placed in a flask, and enough water is added to bring the volume up to 1 liter.

The use of equivalent weight factors out the valance of the ion species in conductivity calculations. The *equivalent weight* is obtained by dividing the molar mass by the number of H^+ ions per mole, corresponding to the number of protons exchanged in a reaction. For sodium and chlorine, the equivalent weight equals the molar mass because these are univalent ions.

The equivalent conductance in cgs units $\lambda = \sigma/c$, where σ is conductivity in μS/cm and c is the concentration in equivalents per cm^3. Kohlrausch's law suggests that the equivalent conductance will vary with the square root of concentration as follows:

$$\lambda = \lambda^\circ - B\sqrt{c} \tag{3-6}$$

The empirical constant B is fixed for each solute and solvent. However, for dilute solutions of strong electrolytes, it is helpful to focus in Equation 3-6 on the fundamental value of λ°, the molar conductivity extrapolated to infinite dilution. For the near-infinite dilution associated with natural rain and wetting of polluted insulator surfaces, each species of ion contributes a fixed amount to the total ionic conductivity, independent of the nature or presence of other ions. This means that the sum of the n individual ionic conductivities gives a good estimate of the total conductivity as *Kohlrausch's axiom*:

$$\sigma = \sum_{i=1}^{n} \lambda_i c_i \approx \sum_{i=1}^{n} \lambda_i^\circ c_i \tag{3-7}$$

TABLE 3-9: Equivalent Conductance of Ions

Cation	Atomic Weight	λ° (S-cm²/eq)	Anion	Atomic Weight	λ° (S-cm²/eq)
H^+	1 g	350	OH^-	20 g	198.4
K^+	39.1 g	73.5	Cl^-	35.4 g	76.4
Na^+	23 g	50.1	Br^-	79.9 g	78.2
Li^+	6.9 g	38.7	I^-	126.9 g	76.9
$\frac{1}{2}Ba^{2+}$	137.3 g/2	63.7	NO_3^-	62 g	71.5
$\frac{1}{2}Ca^{2+}$	40.1 g/2	59.5	HCO_3^-	61 g	44.5
$\frac{1}{2}Mg^{2+}$	24.3 g/2	53.1	CH_3COO^-	59 g	40.9
$\frac{1}{2}Pb^{2+}$	207.2 g/2	69.5	$C_6H_5COO^-$	115 g	32.4
$\frac{1}{2}Cu^{2+}$	63.6 g/2	53.6	$\frac{1}{2}SO_4^-$	96.1 g/2	80.0
Ag^+	107.9 g	61.9	F^-	19 g	55.4
NH_4^+	18.0 g	73.6			

Source: CRC [1989] and Plambeck [1982].

Table 3-7 gave values of equivalent ionic conductance for some of the most important pollution ions. A more complete set of values appears in Table 3-9.

3.5. EFFECTS OF TEMPERATURE ON ELECTRICAL CONDUCTIVITY

Environmental contamination greatly increases the electrical conductivity of precipitation compared to pure water. This effect has relatively strong temperature dependence, often stated to be about 2.2% per centigrade degree (C°) of temperature change, referenced to a 20°C value (σ_{20}). The root cause of this dependence is actually related to changes in the water viscosity rather than to any changes in the ions themselves. Remarkable changes in the electrical conductivity also occur when ice forms, as a result of freeze–thaw concentration effects as well as fundamental changes in the organization of the water molecules.

3.5.1. Equivalent Conductance of Ions

The molar conductivity of any salt in dilute conditions, $\lambda^\circ_{\text{ion}}$, can be obtained using the data from Table 3-9 and some additional considerations [Plambeck, 1982] including the Stokes–Einstein equation, leading to

$$\lambda^{\circ}_{\text{ion}} = \frac{zFe}{6\pi\eta}\frac{z_{\text{ion}}}{r_{\text{ion}}}$$

$$\Lambda^{\circ} = \frac{zFe}{6\pi\eta}\left[\frac{z_{\text{ion}}(+)}{r_{\text{ion}}(+)} + \frac{z_{\text{ion}}(-)}{r_{\text{ion}}(-)}\right] \tag{3-8}$$

where

η is the viscosity (CP) of the medium,

z_{ion} is the valence (ionic charge),

e is the electronic charge (F/N),

Λ° is the molar conductivity of the salt, and

r_{ion} is the particle radius.

According to Walden's rule, the product of $\Lambda^{\circ}\eta$ from Equation 3-8 is constant. This means that changes in viscosity will affect the electrical conductivity of the solution.

3.5.2. Effect of Temperature on Liquid Water Conductivity

The viscosity of water has strong temperature dependence, above and below the freezing point. Hallett [1963] measured viscosity of supercooled water with a capillary flow technique; the results are plotted along with values for 0 °C to 25 °C in Figure 3-22.

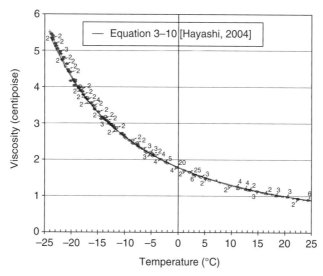

Figure 3-22: Viscosity of water from −23.8 °C to +25 °C. Plotted points with figures (n) are the mean values of n points [Hallett, 1963; Hayashi, 2004].

As a result of the viscosity–temperature relation, the electrical conductivity of water and aqueous solutions of ions also varies with temperature [Robinson and Stokes, 1965]. As temperature increases, the water becomes less viscous and ions can move more easily.

Normally, a linear relation is used to correct the electrical conductivity of a solution to a value at a reference temperature. For example, in IEEE Standard 4 [1995] a correction to 20 °C is given by

$$\sigma_{20} = \sigma_\theta [1 - b(\theta - 20)] \tag{3-9}$$

where

θ is the solution temperature (°C),

σ_θ is the volume conductivity (S/cm) of the solution at temperature θ,

σ_{20} is the conductivity (S/m) corrected to the reference temperature of 20 °C, and

b is a factor that depends on ambient temperature.

The adequacy of a linear correction to solution conductivity was evaluated by Hayashi [2004] for seawater and a range of fresh lake and ground waters at dilution ratios from zero (full strength) to 1:49 (2%). The constant factor b in Equation 3-9 was in fact found to vary by about ±5% with ion concentration in the range of total dissolved solids from 1 to 100 g/L. A more precise approach to temperature correction of conductivity makes use of the temperature dependence of viscosity as follows:

$$\log \frac{\eta_\theta}{\eta_{25}} = \frac{A(25 - \theta) - B(25 - \theta)^2}{\theta + C} \tag{3-10}$$

where

η_θ is the viscosity of the solution (N-s/m²) at the temperature θ (°C),

η_{25} is the viscosity of the solution at 25 °C, equal to 0.0008903 N-s/m² or 0.8903 cP,

A is a constant, equal to 1.1278,

B is a constant, equal to 0.001895 °C⁻¹, and

C is a constant, equal to 88.93 °C.

The correction for the electrical conductivity σ then becomes

$$\sigma_\theta = \sigma_{25} \left(\frac{\eta_\theta}{\eta_{25}} \right)^{-b} \tag{3-11}$$

where

σ_θ is the conductivity of the NaCl solution at temperature θ,

θ is the solution temperature (°C),

κ_θ is the conductivity of the NaCl solution corrected to 25 °C, and

η_{25} is a dimensionless constant ranging between 0.806 and 0.933 with an average of 0.877.

In typical freezing rain conditions, the water can be supercooled to about −5 °C. Water viscosity at this temperature is about 2.15 cP, as observed in Figure 3-22, compared to 0.89 cP at 25 °C. Using Equation 3-11, the conductivity of supercooled water at −5 °C is about 46% of its value at 25 °C and is 51% of σ_{20}.

Appendix A, dealing with details of the measurement of ESDD, gives a nonlinear equation for temperature correction of conductivity between 5 °C and 30 °C [IEC Standard 60507, 1991]:

$$\sigma_{20} = \sigma_\theta[3.2 \times 10^{-8} \cdot \theta^4 - 1.096 \times 10^{-5} \cdot \theta^2$$
$$+ 1.0336 \times 10^{-3} \cdot \theta^2 - 5.1984 \times 10^{-2} \cdot \theta + 1.7088] \qquad (3\text{-}12)$$

This gives a good fit to the measurements of saline solution conductivity given by Hewitt [1960] and provides a correction to the preferred temperature of 20 °C, but the empirical equation does not give any physical insight about why the change occurs, and it is not intended for use below 5 °C.

3.5.3. Effect of Temperature on Ice Conductivity

Under freezing conditions, ice forms in a two-phase system of solid and liquid as shown in Figure 3-23. If a glass of water is placed in a freezer and left for a while, the ice that forms will have significantly lower conductivity than the unfrozen water that remains. If the ice is not melted before the measurement, it will appear to have very low conductivity, verging on being a good insulator.

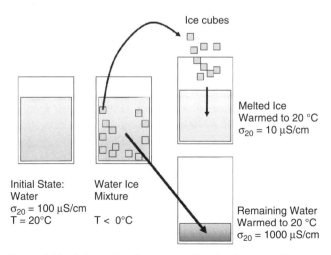

Figure 3-23: Schematic of water purification by crystallization.

When both fractions are warmed back up to room temperature, a significant difference will still persist as a result of freeze–thaw purification: the water from the melted ice will be much lower in conductivity.

The effect of "purification by crystallization" rejects ions from the solid fraction into the remaining liquid fraction. It is normal to measure a factor-of-10 decrease in the conductivity of ice removed from an ice–water mix during the freezing process, compared to the initial value of σ_{20}. It is also normal that the conductivity of the remaining water is elevated to conserve the total ion content in the sample.

Freeze crystallization has been used to separate a wide variety of contaminants from water, such as dissolved minerals, organic chemicals, and particulates. Freezing methods have also been used to desalinate seawater and to purify silicon for semiconductor chips. In one study using snow melt and municipal water, Conlon [1992] found that sodium, chloride, and sulfate concentrations were reduced to 4–5% of their initial values after freezing. (See Table 3-10.)

The same process occurs during the solidification of supercooled water droplets in a wet icing regime. This leads to elevated drip-water conductivity and high conductivity of icicles compared to ice caps in ice accretion on insulators [Farzaneh and Melo, 1990].

The bulk conductivity of ice at $-4\,°C$ is roughly a factor of 1000 less than the initial conductivity of the water, σ_{20}. Figure 3-24 shows that the conductivity increases by about a factor of 10 in the ice temperature range from $-4\,°C$ to $0\,°C$.

Farzaneh et al. [1994] established the overall effect of temperature on 280-mm triangular ice samples, oriented vertically in a test chamber (Figure 3-25). The measurements of resistance on the ice sample can be expressed as an equivalent thickness of the freezing water conductivity σ_{20}. Table 3-11 shows that the derived water film thickness for the data in Figure 3-25 was constant for a wide range of freezing water conductivity.

TABLE 3-10: Effect of Batch Crystallization on Separation of Minerals from Water

Contaminant	Initial Concentration	Concentration After Freezing	Rejection
Bicarbonate	120 ppm	6 ppm	95.00%
Calcium	42 ppm	4 ppm	90.50%
Chloride	26 ppm	0.95 ppm	96.30%
Magnesium	15 ppm	2.7 ppm	82.00%
Sodium	17 ppm	0.65 ppm	96.20%
Sulfate	13 ppm	0.62 ppm	95.20%
Nitrate	190 ppm	0.66 ppm	99.65%
Lead	190 ppm	0.88 ppm	99.54%

Source: Conlon [1992].

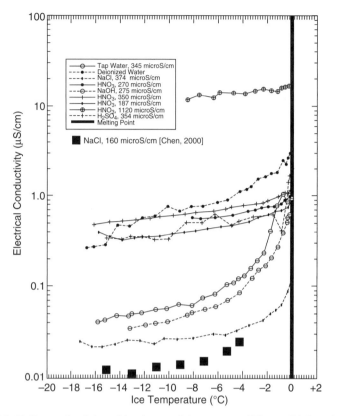

Figure 3-24: Bulk conductivity of ice in coaxial geometry [Vlaar, 1991] and flat-plate geometry [Chen, 2000].

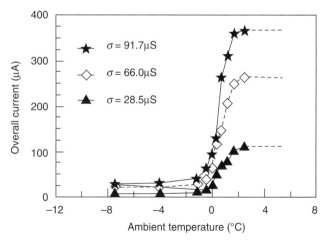

Figure 3-25: Variation of overall current in 280-mm triangular ice sample at constant dc voltage (from Farzaneh et al. [1994]).

TABLE 3-11: Maximum Thickness of Water Film h_{max} Corresponding to the Surface Conductivity of Ice for Different Values of Freezing Water Conductivity σ_{20}

Freezing water conductivity σ_{20} (μS/cm)	300	91.7	66	28.5
Ice surface conductivity σ_S (μS)	5.14	1.52	1.09	0.48
Derived water film thickness on vertical ice surface, h_{max} (μm)	171	166	165	167

3.6. CONVERSION TO EQUIVALENT SALT DEPOSIT DENSITY

The composition of ions in a surface deposit will be a mixture of various ions, with highly variable exposure illustrated in Figures 3-15 and 3-16 leading to wide variation in composition, as shown later in Figures 3-26 and 3-27.

It is standard practice in the electrical utility industry to refer the electrical conductivity of this complex ionic solution to that of a sodium chloride solution.

The traditional unit of *equivalent salt deposit density* (ESDD) is mg/cm^2. This normally gives values that are less than unity, and the centimeter is not a preferred unit in the SI either, but tradition remains strong among insulation specialists. Units of μg/cm^2 are very helpful to electrical utility staff because the values are similar to speed limits, and this reduces the errors during staff training and reporting. Where appropriate, and especially when dealing with ESDD for EHV insulation [Chisholm et al., 1994], both units will be used.

ESDD is normally measured with a rag-wipe method, as described in detail in Appendix A. A clean cloth is rinsed several times in deionized water. Insulator surfaces are wiped, the cloth is rinsed again, and the change in the electrical conductivity of the wash water is measured. The ESDD is calculated from this change, correcting for the surface area, wash water volume, and temperature.

Normally, top and bottom surfaces of individual suspension disks are wiped independently. In summer rain conditions, the sheltered bottom surfaces tend to have higher levels, because top surfaces are washed more frequently. In dry or winter conditions this is reversed, as the accumulation rate on upward facing surfaces is also higher.

For long-rod or station post insulators, it can be practical to take a series of conductivity measurements as the top and bottom surfaces of each shed are wiped. These results give the distribution of pollution along the insulator string or post, which can be nonuniform if the shed profile is not uniform or if booster sheds or creepage extenders are used.

The ESDD level on the uppermost surface of the top insulator may be different from the levels found on other upward-facing surfaces. Rust stains from hardware, bird streamers, wind flow, exposure to rain, and many other factors are different.

3.6.1. Insulator Case Study: Mexico

Chemical analysis is often carried out on the contamination in wash water after establishing ESDD. As a set of examples, Ramos et al. [1993] provided the ionic composition of ESDD from seven locations around Mexico, shown in Figures 3-26 and 3-27. In half the cases where they were measured sepa-

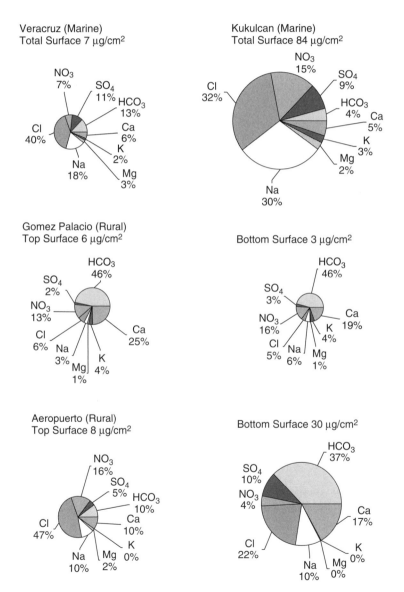

Figure 3-26: Ionic composition of insulator contamination at rural and marine locations in Mexico (data from Ramos et al. [1993]).

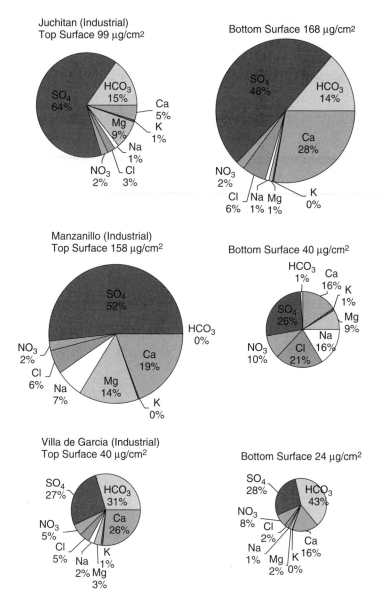

Figure 3-27: Ionic composition of insulator pollution at industrial locations in Mexico (data from Ramos et al. [1993]).

rately, ESDD on the top surface of the insulator was greater than that on the bottom surface.

The concentration of sulfate (SO_4^{2-}) and carbonate (HCO_3) on insulator surfaces was found to be high in industrial areas. As anticipated from dry and wet deposition work, sodium and chloride are important for marine exposure.

The majority of pollutants in Figures 3-26 and 3-27 are strongly ionic and soluble in water, and thus satisfy the basic principle of the ESDD approach.

3.6.2. Insulator Case Study: Algeria

Winds off the desert (sirocco) in Algeria [El-A Slama et al., 2000] caused buildup of contamination on 220-kV insulators that led to disruptions in September 1998. Prompt measurements of the contamination levels on 146-mm × 280-mm fog-type cap-and-pin disks with 445 mm of leakage distance, 881 cm² top surface area, and 1646 cm² bottom surface area gave the results in Table 3-12.

The conclusions were that the dominant desert pollution salts on the insulators were $CaCl_2$, $MgCl_2$, and $CaSO_4$ rather than NaCl, and that the top and bottom surface pollution levels were nearly equal for the fog-type disk at the time of flashover.

3.6.3. Insulator Case Study: Japan

Overall, there is good agreement between the ionic composition of precipitation [Hara et al., 1995], presented in Table 3-8, and the ionic composition of contamination on exposed insulators [Takasu et al., 1988], shown in Figure 3-28. Insulators were sampled in summer and winter conditions at two inland test sites, Yonezawa, in the center of Japan about 70 km from each coast, and Takeyama, about 600 m from the seaside. In most cases, chemical analysis showed that the anion–cation balance remained within 10%. The levels of contamination were about the same on energized (dc-) and unenergized strings.

Over the 18-month exposure, peak levels of ESDD reached 50 µg/cm² on bottom surfaces in winter near the seacoast site at Takeyama, but remained in the range of 7–16 µg/cm² inland at Yonezawa.

3.6.4. Surface Resistance of Insulator

The surface resistance of the insulator is the ratio of the power-frequency voltage on an insulator to the resistive part of the leakage current that flows on its surface. There is also a capacitive component to the leakage current that is fixed for constant applied voltage. This component is negligible for highly polluted insulators but should be considered in measurements on clean insulators.

The applied voltage is held on the insulator for a sufficient time to obtain a reading [IEC Standard 60507, 1991; IEEE Standard 4, 1995] but not so long as to cause heating. Also, the voltage level must not provoke electrical discharges on the surface. The result of a surface resistance measurement can reflect the entire state of a naturally exposed insulator—its ESDD, and also the degree of wetting as affected by any NSDD. However, for monitoring purposes, an insulator can be artificially wetted in steam or clean fog.

TABLE 5-12: ESDD, NSDD, and Ion Chemistry of Pollution Deposit Causing 220-kV Flashovers

Surface	ESDD (mg/cm^2)	NSDD (mg/cm^2)	Cl^- (mg)	SO_4^{2-} (mg)	Ca^{2+} (mg)	Mg^{2+} (mg)	Na^+ (mg)	NO_3^- (mg)	K^+ (mg)	Cu^{2+} (mg)
Top	0.41	0.040	346	51	34	10	3.0	2.6	1.2	0.0
Bottom	0.42	0.114	878	104	88	36	5.6	4.8	1.8	2.8
Ratio[a]	0.98	0.35	0.74	0.92	0.72	0.52	1.00	1.01	1.25	0.00

[a] Area weighted ratio for mg of ions.

Source: E–A Slama et al. [2000].

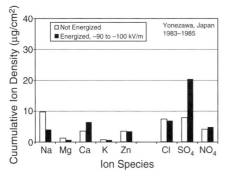 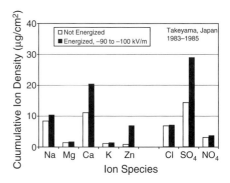

Figure 3-28: Chemical analysis results of salt contamination on bottom surface of insulators (data from Takasu et al. [1988]).

Normally, calculations of insulator resistance go to infinity for ac voltage after each zero crossing of current. It is more convenient to plot insulator conductivity. This value is then converted to surface conductivity (μS) using a "form factor" F as follows:

$$\sigma_{inst} = F \cdot \left(\frac{i_{inst}}{v_{inst}} \right)_{pk}$$

$$F = \int_{0}^{L} \frac{dl}{\pi \cdot D(l)}$$

where

i_{inst} is the instantaneous leakage current (mA),

v_{inst} is the instantaneous supply voltage (kV),

σ_{inst} is the peak value of conductivity (μS) over the half-cycle of ac,

F is the dimensionless form factor,

l is the distance (mm) along the leakage path of the insulator, and

$D(l)$ is the insulator diameter (mm) at length l along the leakage path.

There are many advantages and disadvantages to measuring surface resistance rather than ESDD and NSDD. For example, if the pollution deposit is not uniform, this effect will automatically be included in the surface resistivity result. The method finds wide use in artificial laboratory tests [IEC Standard 60507, 1991; IEEE Standard 4, 1995], because it is a convenient way to establish that the desired wetting and precontamination levels have been achieved prior to initiating a flashover test. However, measuring small leakage currents accurately in the high-voltage environment is difficult and the source supply is costly.

3.6.5. Insulator Leakage Current: Case Studies

For insulators exposed at constant line-voltage stress, dry-band formation and partial discharge activity will normally take place under natural wetting conditions. Investigations show that large peak currents of up to 200 mA can flow during these conditions.

Beauséjour [1981] carried out field studies of the accumulation of ESDD on distribution insulators, with exposure sites at various distances from pollution sources including the sea, a fertilizer plant, and a paper mill. At two sites, leakage currents were measured on pin-type distribution insulators, energized at 14.4 kV. At a location 30 m from the sea, ANSI Class 56-2 insulators with 404 mm of leakage distance and a form factor of 1.05 were exposed, and ESDD levels on these insulators increased to winter-peak levels of 200–500 mg/cm^2. Leakage current of more than 200 mA was noted for at least 6 hours per year. At another location, ANSI Class 55-5 insulators with 322 mm of leakage distance and a form factor of 1.09 were installed about 300 m from a fertilizer plant. The annual maximum ESDD levels for these insulators also peaked in the month of March after a long duration of winter exposure without rain. The accumulated total ESDD levels of 0.03 mg/cm^2 led to leakage currents that exceeded 10 mA for 3.5 hours per year [Beauséjour, 1981].

The outdoor exposure tests were supplemented with measurements of the highest peak voltage and current that did not cause flashover in artificial fog tests. The results for the insulators exposed in the field are shown in Figure 3-29. There is a general trend toward higher withstand voltage for lower peak levels of leakage current, but there is also a 10:1 scatter in the currents associated with each voltage. This is one of the aspects that make leakage current monitoring, on its own, an uncertain predictor of flashover performance.

With suitable processing of leakage current data, Amarh [2001] suggested that it could be possible to predict the flashover of single-disk insulators with reasonable reliability. The extension of this concept to establishing the flashover or withstand of insulator strings is of great commercial and academic interest, as discussed in Chapter 6. However, in spite of more than 70 years of work with this approach, the accuracy of predictions and insulator rankings from leakage current still do not tally with observed in-service flashover results. Instead, leakage current monitoring serves utilities as warning systems that insulators need to be cleaned to reduce noise complaints or other ancillary problems.

3.6.6. Estimating ESDD from Environmental Measures for Corrosion

A visual review of the fraction of ions in Figures 3-26 and 3-27 suggests that sulfate forms a significant fraction of the pollution that has been measured on insulator surfaces. The fraction of chloride is high for marine exposure but falls in the range of 3–22% for industrial exposure.

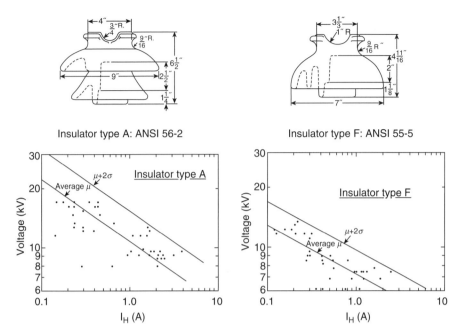

Figure 3-29: Peak of voltage that did not cause flashover versus peak leakage current in artificial fog tests at three levels of ESDD (from Beauséjour [1981]; courtesy of CEATI International Inc.).

International standards for the corrosion of metals and other materials, such as ISO 9223 [1992], consider three factors when classifying sites:

- The daily rate of dry deposit of sulfur dioxide, expressed as SO_2 in $(mg/m^2)/day$.
- The air salinity, given by rate of deposit of chloride, expressed as Cl in $(mg/m^2)/day$.
- The time of wetness, *Tow* (hours/year).

In cases where the sulfate deposit rate is not known, it is estimated from the SO_2 concentration in air using the relation $(SO_2 (mg/m^2)/day = 0.8 SO_2 (\mu g/m^3))$.

Regions are divided into four classifications (P_0 to P_3) with regard to SO_2 and another four (S_0 to S_3) with regard to salinity as shown in Table 3-13.

ISO 9223 uses units of mg per m^2 per year that are in accord with standard scientific practice. For reference, the accumulation rate of $60 mg/m^2$ per day corresponds to $6 \mu g/cm^2/day$ or $0.006 mg/cm^2/day$. This level of pollution buildup has been observed on post insulators near urban roads in winter conditions.

The ISO standard classifies regions according to their P and S values, and the annual number of wet hours, to obtain an overall corrosion rate index ($C_{1.5}$). This index is then used to look up the rate of corrosion of carbon steel, zinc, copper, and aluminum components. This is an approach remarkably

TABLE 3-13: Classification of Site Pollution Severity for Corrosion of Metals

Category	P_0	P_1	P_2	P_3
SO_2 ($\mu g/m^3$)	≤12	12–40	40–90	90–250
SO_2 ((mg/m^2)/day)	≤10	10–35	35–80	80–200
Category	S_0	S_1	S_2	S_3
Cl^- ((mg/m^2)/day)	≤3	3–60	60–300	300–1500

Source: ISO 9223 [1992].

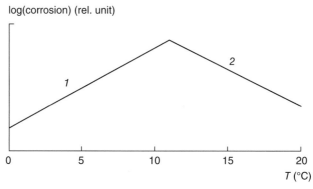

Figure 3-30: Schematic of observed dependence of corrosion rate with annual mean temperature (from Mikhailov et al., [2004], courtesy Springer Science+Business Media).

similar to that adopted by IEC Technical Committee 36 for establishing site pollution severity for insulators, as described in Table 3-3.

According to Mikhailov et al. [2004], the corrosion rates of carbon steel, zinc, copper, and aluminum are well predicted when the annual average temperature and relative humidity are used rather than the time of wetness. This improvement is important when considering corrosion over a wide range of observed annual average values.

The corrosion rates of metals typically have a break point above and below a mean annual temperature of 9–11 °C. Below this temperature, as shown in the schematic in Figure 3-30, slope 1 shows a corrosion rate that increases with increasing temperature in a range of low temperatures, while slope 2 shows a deceleration of corrosion with an increase in temperature above the break point.

The reduction in corrosion rate with annual mean temperature of more than 10 °C is held to be a consequence of the inverse correlation between relative humidity and temperature, coupled with improved heating of surfaces by sunshine to accelerate the evaporation of rain and dew from wetted surfaces (Figure 3-31). Incorporation of these factors leads to a pair of expressions, above and below 10 °C, for the corrosion rate of zinc from galvanized components on insulators:

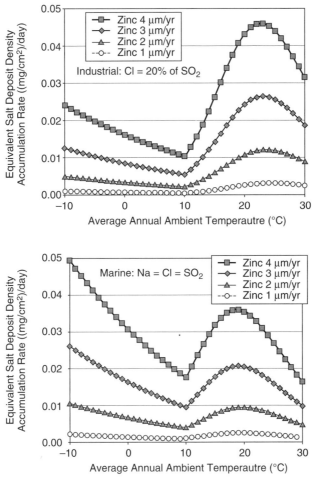

Figure 3-31: Estimated daily rate of accumulation of ESDD versus average annual temperature, based on annual corrosion rate of zinc.

$$C = \begin{cases} T \le 10^\circ C & \begin{aligned} & 0.0129[SO_2]^{0.52}\exp(0.046RH)\exp(0.038(T-10^\circ)) \\ & +0.0175[Cl]^{0.57}\exp(0.008RH+0.085T) \end{aligned} \\ T > 10^\circ C & \begin{aligned} & 0.0129[SO_2]^{0.52}\exp(0.046RH)\exp(0.071(10^\circ -T)) \\ & +0.0175[Cl]^{0.57}\exp(0.008RH+0.085T) \end{aligned} \end{cases}$$

where

C is the annual rate of loss (μm/yr) of zinc from an exposed surface,

T is the average annual temperature, from $-17\,^\circ C$ to $+28.7\,^\circ C$,

RH is the average relative humidity, from 34% to 93%,

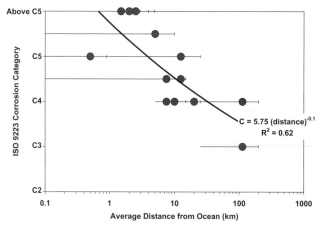

Figure 3-32: 2006 Estimate of corrosion levels in South Africa using ISO 9223 index (data from Leitch, [2006]).

SO_2 is the sulfate deposit level, from 0.7 to 150 (mg/m^2)/day, and

Cl is the chloride deposit level, from 0.4 to 700 (mg/m^2)/day.

Specialists in many countries, such as Leitch [2006], have mapped out coastal regions where high levels of corrosion have been found on exposed metal coils. The zones of high corrosion in Figure 3-32 extend up to 15 km inland, except for 2 areas, where effects are measured up to 25 km away from the seacoast. As a point of evaluation, Durban on the Indian Ocean has an annual average temperature of 21 °C with average RH of 77%. A high rate of zinc loss exceeding 4 µg/yr suggests a corresponding rate of increase of ESDD on insulators of 0.033 mg/cm^2 per day.

Etching of glass was also studied by Mikhailov [2002] for the European part of Russia. Figure 3-33 shows a pattern of high glass leach rates of 4–8 µm/yr in the heavily industrialized central areas.

The relation of the glass etching rate to the pollution level has not yet been established, but the fact that the pollution levels may be sufficiently high to etch glass insulators may be significant in its own right.

3.6.7. Statistical Distribution of the Conductivity of Natural Precipitation

The electrical conductivity of precipitation samples is not constant. There is a tendency over the course of individual rain, snow, or ice storms for elevated conductivity in the first few samples, followed by decay to a background level. There are also significant differences from storm to storm and changes over the course of the year. One simplified model can be used to establish an

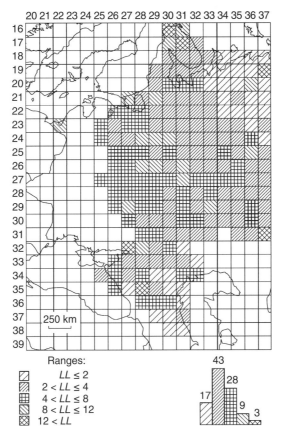

Figure 3-33: Depth (*LL*, in μm) of leached glass layer from outdoor exposure to corrosive environment in Russia (from Mikhailov [2002]); courtesy of Springer Science+Business Media).

appropriate value of melt-water conductivity, approximating the annual variations as a log-normal distribution. The probability of exceeding precipitation conductivity, σ, can be approximated with

$$P(\sigma) = \frac{1}{1 + \left(\dfrac{\sigma}{\sigma_a}\right)^{2.3}} \tag{3-13}$$

where σ_a is the median of precipitation conductivity. The exponent value of 2.3 can be adjusted to match the observed log standard deviation of other local measurements. This equation has the value of being easily inverted to give the appropriate precipitation conductivity for any desired probability level.

For many sites described in the US national network of precipitation measurements, σ_a is typically 25 μS/cm and the log standard deviation of the distribution is 0.8. Statistical distributions of precipitation conductivity can be obtained from a number of sources, such as the National Atmospheric Deposition Program on-line resources in the United States.

The form of Equation 3-13 can also be used to grade the distribution of snow conductivity. For example, the conductivity of snow at the Ishiushi test site in Japan [Jiang et al., 2005] is reasonably represented when the median conductivity, σ_a, is updated to $14\,\mu S/cm$. Also, melted ice and snow in the winter were shown to have nearly the same conductivity as precipitation in the summer in four locations in Ontario [Tam, 1989].

3.6.8. Mobile Sources

Vehicles are a known source of pollutants themselves. However, the most important influence of vehicle traffic near electric power networks is that associated with road salt. At distances of 50–220 m from an overhead expressway, Figure 3-34 shows that the dustfall in summer months (July and August) was relatively constant and remained well below the $9\,g/m^2/month$ guideline. In the winter months where data were available (January and March), the dustfall was nearly double the limit at 50 m from the expressway—as a consequence of the heavy application of salt and sand mixtures to ensure safe travel.

The speed of vehicle travel and the number of lanes have an effect on the rate of change of pollution accumulation. Figure 3-35 shows that the accumulation has reached the local median value about 20 m from an arterial road with four lanes of traffic moving at 60 km/h, while the traffic moving at 100 km/h throws pollution more than six times further.

A utility in Ontario carried out field studies to establish the rate of increase of insulator contamination levels from road salting. These included important and problematic stations such as Hamilton Beach, Strachan in downtown

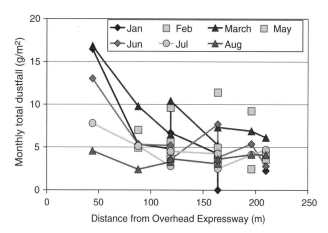

Figure 3-34: Monthly dustfall versus distance from expressway (adapted from Chisholm et al. [1993]).

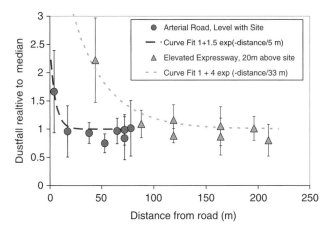

Figure 3-35: Observations of normalized dustfall (g/m²/month) relative to distance from road (adapted from Chisholm et al. [1993]).

Toronto, and Richview near a major expressway. As examples of the wide range of ESDD values observed near expressways, Figures 3-36 and 3-37 show the levels measured at a total of five locations along urban transmission corridors during winter seasons.

Proximity to expressways thus leads to high levels of ESDD in the winter. The locations of these stations relative to local expressways are shown in Figures 3-38 to 3-41. The closest arterial road to the Cherrywood site in Figure 3-38 is 1000 m to the south and there are no expressways within 2000 m. This station has the lowest values of ESDD in Figure 3-37. The peak values at Claireville and Milton are moderate. Figure 3-39 shows a divided highway 300 m east of Claireville, and Milton in Figure 3-40 is 900 m south of an expressway. At the Richview station in Figure 3-41, 60 m east of the nearest highway and with the busiest expressway in Canada on its eastern border, pollution levels reach a peak ESDD of 100 μg/cm² every year. This is a station where a Smart Washing program has been highly effective.

Detailed measurements of ion content of precipitation and insulator surface pollution were carried out at the Richview station, with the location shown in Figure 3-41. Not surprisingly, the heavy deposits of salt, exceeding 2000 kg/km per week in Figure 3-42 led to an elevation of salt levels in both wet deposition samples and insulator surface deposits in Figure 3-43.

The contribution of NaCl to wet deposition and insulator ESDD peaked at 86% and 75%, respectively, in the week of January 16–22, a week when more than 2000 kg of salt per km of expressway were deposited. A month later, the NaCl contribution to conductivity of the wet deposition fell to 41%.

Recent work in Sweden has been evaluating the spread of road salt into sensitive aquifers and other areas. Lundmark and Olofsson [2007] took measurements upwind from a six-lane expressway, with 90,000 vehicles per day

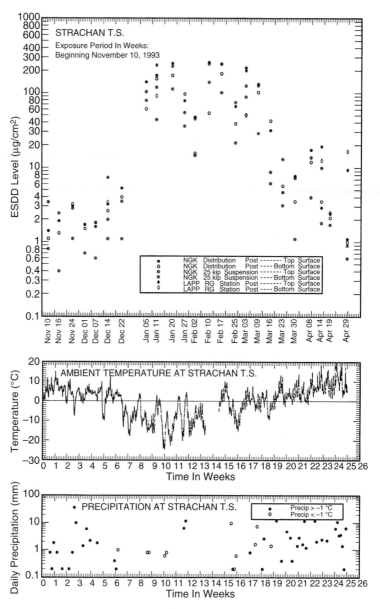

Figure 3-36: Insulator contamination levels at the Strachan 115-kV transformer station near an urban expressway [Chisholm, 1995]. (Courtesy of Kinectrics.)

and a 2003–2004 season salt deposit rate of 8 tonnes (8 Mg) per km. Figure 3-44 from the Swedish study shows the total chloride deposit rate on the ground as a function of distance from the edge of their southbound lanes, located downwind from the measurement sites.

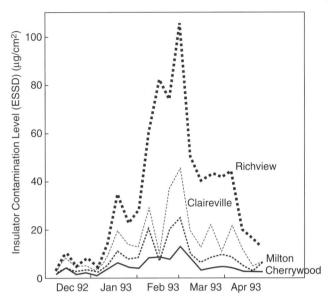

Figure 3-37: Observed ESDD at four locations within 40 km (from Farzaneh et al. [2007]).

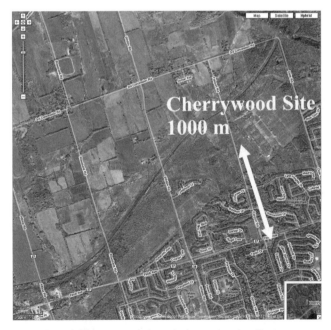

Figure 3-38: Location of Cherrywood site relative to local pollution sources. (Courtesy of Google Maps © 2006 Google Imagery; © DigitalGlobe; Map data © 2006 NAVTEQ™.)

Figure 3-39: Location of Claireville site relative to local pollution sources. (Courtesy of Google Maps © 2006 Google Imagery; © DigitalGlobe; Map data © 2006 NAVTEQ™.)

Figure 3-40: Location of Milton test site relative to local pollution sources. (Courtesy of Google Maps © 2006 Google Imagery; © DigitalGlobe; Map data © 2006 NAVTEQ™.)

Figure 3-41: Location of Richview site relative to local pollution source. (Courtesy of Google Maps © 2006 Google Imagery; © DigitalGlobe; Map data © 2006 NAVTEQ™.)

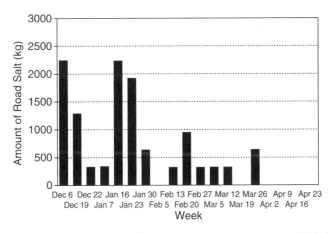

Figure 3-42: Road salt deposit per kilometer of expressway near Richview Station (Figure 3-41) in 1991–1992 study period (from Chisholm et al. [1993]; courtesy of Kinectrics).

Lundquist and Olofsson also fitted the total daily deposition $D(x)$ of chloride at a distance from the edge of the road x with a pair of exponential functions, one for splash and another for the spray transport mechanism, along with a constant background level. The fitted function was

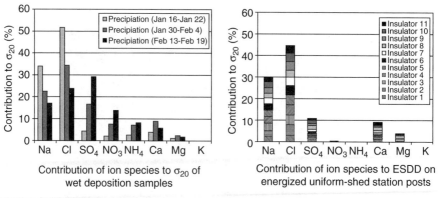

Figure 3-43: Effects of road salt from Figure 3-42 on ion content of wet deposition conductivity (σ_{20}) and insulator surface contamination (ESDD) [Chisholm et al. 1993].

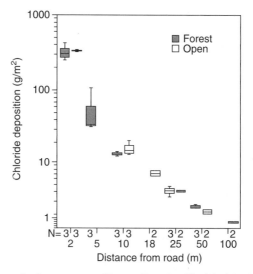

Figure 3-44: Measured winter season (December–April) chloride deposit rate versus distance from edge of six-lane 110-km/h expressway in Sweden (from Lundmark and Olofsson [2007]); courtesy Springer Science+Business Media).

$$D(x) = a_{\text{splash}} \cdot e^{-x/2m} + a_{\text{spray}} \cdot e^{-x/20m} + a_{\text{background}} \qquad (3\text{-}14)$$

where

$D(x)$ is the chloride deposit density (g/m^2 per day),

$a_{\text{splash}} = 8\,\text{g/m}^2/\text{day}$,

$a_{spray} = 0.12\,g/m^2/day$, and

$a_{background} = 0.003\,g/m^2/day$.

The chloride deposit density in Figure 3-44 fell by an order of magnitude in the distance range from 5 to 25 m from the expressway. In contrast, Higashiyama et al. [1999] noted a salt flux of 0.02–0.4 mg/m^2/s, corresponding to 1.7–350 g/m^2/day, at a location 30 m from the ocean beach in winter conditions, depending on wind speed and wave height. This level fell by a factor of 10 for each wind condition at a distance of 550 m from the beach.

The numerical values for Equation 3-14 can be converted to ESDD escalation rate in μg/cm^2/day by changing the units to $(1\,g/m^2) \cdot (10^6\,\mu g/g) \cdot (1\,m^2/10^4\,cm^2)$ and by multiplying the chloride weight by a factor (23 + 35.5)/35.5 to correct for the stoichiometric ratio of NaCl to Cl$^-$. This gives values of $a_{splash} = 1320$, $a_{spray} = 20$, and $a_{background} = 0.5\,\mu g/cm^2$ per day upwind from a six-lane expressway. The density values can be scaled by the number of traffic lanes (Figure 3-45).

The deposit rate $D(x)$ will accumulate on upward-facing surfaces of a post insulator, which have been shown to increase and decrease dramatically with winter road salting and rain washing in Figure 3-36.

Splash can be ignored for stations that are more than 20 m from the edge of an expressway. A distance constant of 33 m for measurements of dustfall next to an Ontario expressway is similar to the 20-m value found for the "spray" term of salt deposition in Sweden (Table 3-14).

Most of the pollution from vehicles is electrically conductive, but some is insoluble in water. This is illustrated in the observations of chemical composition of Toronto and Vancouver. Particulate matter less than 2.5-μm diameter (PM$_{2.5}$) was measured daily for a year [Lee et al., 2003]. Sources were classified

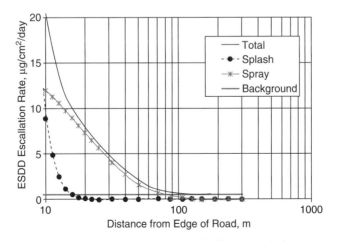

Figure 3-45: Components of road salt accumulation.

TABLE 3-14: Calculated and Observed (Figure 3-37) Rates of Increase of ESDD

Site	Distance from Edge of Highway	Number of Lanes	ESDD from Splash ($\mu g/cm^2/$ day)	ESDD from Spray ($\mu g/cm^2/$ day)	ESDD from Background ($\mu g/cm^2/$ day)	Predicted Δ ESDD ($\mu g/cm^2/$ day)	Observed Increase ($\mu g/cm^2/$ day)
Richview a	37 m	4	0.00	2.07	0.33	2.40	2–3
Richview b	108 m	14	0.00	0.21	1.15	1.36	2.6
Richview c	127 m	14	0.00	0.08	1.15	1.23	1.4
Claireville	206 m	4	0.00	0.00	0.33	0.33	1–1.5
Milton	600 m	6	0.00	0.00	0.49	0.49	1
Cherrywood	>1000 m	4	0.00	0.00	0.33	0.33	0.5

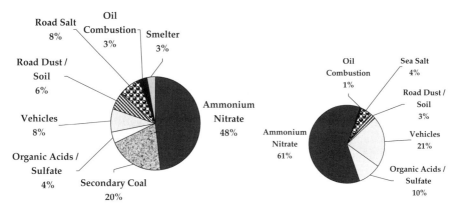

Figure 3-46: Contribution of pollutants to particulate matter ($PM_{2.5}$) in winter for (left) Toronto, Ontario, and (right) Vancouver, British Columbia (data from Lee et al. [2003]).

into eight and six categories in Toronto and Vancouver, respectively, as shown in Figure 3-46 for the winter season. The $PM_{2.5}$ concentration in Toronto was 78% higher than in Vancouver. Wind trajectory analysis suggests that coal sources in Ohio added 20% to the Toronto $PM_{2.5}$ values. The overall influence of local vehicle-related sources was estimated to be 36% in Toronto and 51% in Vancouver.

There are seasonal variations in $PM_{2.5}$ values. For example, Figure 3-47 Nejedly et al., [2003]; Meteorological Services of Canada, [2004] shows that

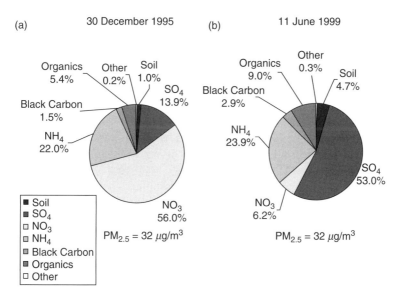

Figure 3-47: Composition of particulate matter ($PM_{2.5}$) in winter and summer pollution episodes at Egbert, Ontario (from Nejedly et al. [2003]).

NO_3 nearly disappears in the summer but is replaced by SO_4 as the overall concentration remains constant.

3.7. SELF-WETTING OF CONTAMINATED SURFACES

Many common insulator pollutants absorb water readily from the environment and are thus said to be "hygroscopic." Sulfuric acid, sodium hydroxide (NaOH), calcium chloride, and magnesium chloride are highly hygroscopic and sodium chloride has this property to some extent.

The hygroscopic nature of a material can be described by its water activity a_w, given approximately by Raoult's law as the ratio of the vapor pressure of the saturated solution to the vapor pressure of pure water at the same temperature. Hygroscopicity of a solid is evaluated experimentally by plotting the water content as a function of relative humidity at a certain temperature, giving the sorption isotherm. The shape of the sorption isotherm depends on how water is bound in the solid–air interface [Wolf et al., 1972]:

- For water activity up to about 0.3, polar sites of relatively high energy hold the water in a *monolayer*.
- For water activity between 0.3 and 0.7, *multilayer water* holds several more layers over the initial monolayer by hydrogen bonds.
- For water activity over 0.7, the flow moisture point, *condensed water* is relatively free of bonds and behaves like an aqueous solution.

Each salt normally has a sorption curve with two sections. In the anhydrous form, from dry conditions up to its critical water activity level (a_w), there is very little water content. Above the flow moisture point in the sorption isotherm, the material rapidly absorbs large amounts of water vapor. As an extreme case, calcium chloride is so hygroscopic that it will deliquesce from solid to liquid solution if exposed to the atmosphere. Table 3-15 shows the water activity levels of several common pollutants.

Any insulating surface, completely coated with a uniform film of these hygroscopic materials, can be wetted at relative humidity in the range of 8–97%, where normally full surface wetting would occur only in fog at 100% RH.

Sulfuric acid is a compound with a more complicated water activity relation because its equilibrium reaction $HSO_4^- \rightleftharpoons (H^+ + SO_4^{2-})$ ions in solution shifts toward $(H^+ + SO_4^{2-})$ with decreasing temperature. Concentrated sulfuric acid will absorb enough water to reduce its concentration to less than 40% when exposed to air with relative humidity of more than 55%. Heavy deposits of sulfuric acid, at equivalent salt deposit density above $300 \mu g/cm^2$, have been observed to be wet at ambient temperatures as low as $-17\,^\circ C$ [Chisholm, 1998].

Self-wetting of insulators shows up in most measurements of insulator leakage resistance, or leakage current at fixed supply voltage. In the case of

TABLE 3-15: Water Activity of Saturated Salt Solutions

Material	Water Activity at 5 °C, $a_w = p / p_0$	Water Activity at 25 °C, $a_w = p / p_0$
Sodium hydroxide	—	0.082
Lithium chloride	0.113	0.113
Calcium chloride	0.23	0.18
Magnesium chloride	0.336	0.328
Potassium carbonate	0.431	0.432
Magnesium nitrate	0.589	0.529
Sodium nitrate	0.786	0.709
Sodium chloride	0.757	0.753
Ammonium sulfate	0.824	0.810
Potassium chloride	0.877	0.843
Potassium Nitrate	0.963	0.936
Potassium sulfate	0.985	0.973

Source: Sahin and Sumnu [2006].

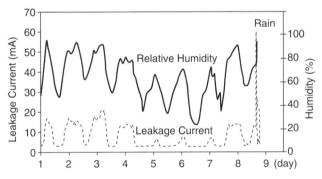

Figure 3-48: Leakage current on precontaminated distribution insulators, demonstrating self-wetting at 50–60% RH (adapted from Richards et al. [2003]).

Figure 3-48, insulators were precontaminated with NaCl to an ESDD of 0.37 mg/cm^2 [Richards et al., 2003] before exposure to the natural environment.

Whenever the relative humidity rose above 60%, leakage current levels on the precontaminated insulators exceeded the background level of 3 mA. At the water activity level of 75% for NaCl in Table 3-15, leakage currents stabilized at about 15 mA. This serves as a good practical illustration of the self-wetting phenomenon and also suggests a direction toward practical pollution monitoring systems.

3.8. SURFACE WETTING BY FOG ACCRETION

3.8.1. Fog Measurement Methods

Fog density is normally measured at airports by evaluating contrast targets at known distances from the observer. The relation between visibility and fog density is a function of the parameters of the fog, including the volume mean diameter and the density of the drops.

Fog density can be measured independently of visibility using aspirators to draw a calibrated volume of air over a set of fine threads that extract the drops (Figure 3-49). Typical natural fogs have a liquid water content (LWC) in the range of 0.01–$0.3\,g/m^3$, while artificial fog made in insulator test chambers can reach values of 1–$3\,g/m^3$.

The distribution of fog droplet size is evaluated using a specialized *forward-scattering spectrometer probe* (FSSP). This is an optical particle counter that detects individual drops and measures their size using the intensity of the light that is scattered as the drop passes through a laser beam. If there is no drop present in the light path, the beam stays focused and is blocked from the optical receiver by a spot mask. Particles that move past the laser beam scatter light in all directions. Some of this light will pass around the mask into optical receivers that have a typical collection angle of $4°$ to $12°$. In addition, the optical focus only views particles in a small area, to allow calculation of particle concentration from the air flow rate. The size of the particle is established by the intensity of scattered light, using Mie scattering theory and the known refractive index of water. The size of each measured particle is sorted into

Figure 3-49: Optical sensor, aspirator, and filter string pack to measure fog density and ionic composition.

Figure 3-50: Forward-scattering spectrometer probe (FSSP) for measuring volume mean diameter of fog droplets.

multiple channels and stored. Normally, the FSSP is mounted on an aircraft wing but a ground-based installation is shown in Figure 3-50.

3.8.2. Typical Observations of Fog Parameters

Typical results from an FSSP show a distribution of fog droplet sizes ranging from 0.5 to 20 μm with few or no drops having diameters at the upper measurement limit of 50–90 μm. Figure 3-51 shows that the particle size distribution peaks at a radius of 1.5 μm with particles of 10–20 μm observed only for dense fog with visibility of 54 m.

Gultepe et al. [2006] provide some comparisons of results from FSSPs designed to record particle sizes in a standard (2.1–48.4 μm, FSSP-96) and extended size range (4.6–88.7 μm, FSSP-124). Their plot of visibility versus total droplet (number) concentration N_d in Figure 3-52 shows a steep decline in visibility as droplet counts increases from 10 to 100 per cm³.

Figure 3-52 shows that there is not much difference in the results between the two instruments in spite of their differing sensitivity to droplet size, compared with the scatter in the individual measurements. The FSSPs are also used to calculate liquid water content (LWC) or ice water content (IWC) of fogs or clouds, considering both the particle count and size. Figure 3-53 shows the overall relation between visibility of fog and LWC or IWC values. Fitted to these data is a power-law relation between visibility *Vis* (m) and LWC (g/m³):

$$Vis = \frac{22}{(LWC)^{0.96}} \qquad (3\text{-}15)$$

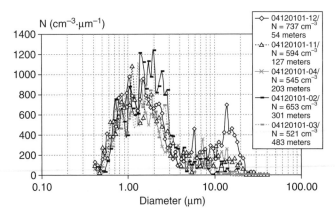

Figure 3-51: Fog droplet distribution for visibility in the range of 50–500 m [from COST Action 722, 2007b].

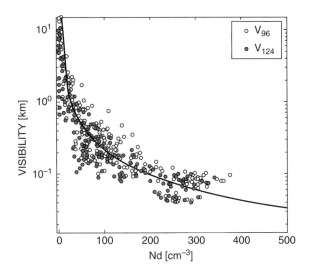

Figure 3-52: Visibility versus water droplet number concentration (N_d) of fog using two different forward-scattering spectrometer probes (FSSPs) (from Gultepe et al. [2006]).

The combined effect of LWC and water droplet number concentration N_d (#/cm^3) on visibility (m) is adapted from Gultepe et al. [2006] in Equation 3-16:

$$Vis = \frac{0.305}{(LWC \cdot N_d)^{0.6473}} \tag{3-16}$$

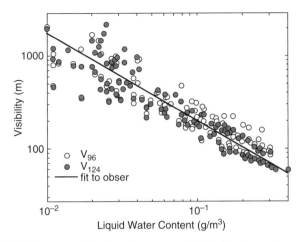

Figure 3-53: Visibility versus liquid water content (LWC) or ice water content (IWC) of fog using two different forward-scattering spectrometer probes (FSSPs) (adapted from Gultepe et al. [2006]).

3.8.3. Fog Climatology

Fog is not a form of precipitation, but it is initiated from processes related to precipitation systems, along with advection, radiation, and condensation of vapor. Normally, fog will occur at a relative humidity that exceeds 97% with a wind speed less than 3 m/s.

Fog droplets have a small particle size. The drops have a relatively slow fall speed, as their aerodynamic drag nearly balances the gravitational force. This complicates prediction of the lifetime of fog layers after they form. This small size also ensures that the fog droplets will rapidly reach thermal equilibrium with the dew point temperature.

Forecasting fog is difficult because it only takes a small difference in temperature or humidity to determine whether or not condensation will occur. This is also a problem with determining the occurrence of fog from infrared satellite images. The evaluation is performed using data from multiple channels of short-wave infrared, with 0.6-, 1.6-, 4-, and 11-μm wavelengths being common. Estimates of cloud-top temperatures are compared with values established for a 2-m height above ground. When these are in close agreement, and other conditions are satisfied, then an observation pixel can be classified as fog. Studies by the National Oceanic and Atmospheric Administration (NOAA) in the United States and Cooperation Scientifique et Technique (COST) in Europe continue to make progress in validating this observation approach.

In freezing conditions, fog often occurs as a result of rain falling on a snow surface. This releases considerable latent heat, leading to snow evaporation that can easily saturate the air near the ground and thereby fulfill a necessary

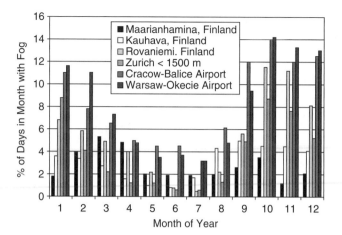

Figure 3-54: Monthly frequency of days with fog at locations in Finland, Poland, and Switzerland (data from COST [2007b]).

condition for fog occurrence. Gultepe and Colomb [COST, 2007b] noted an increase in fog formation in the winter of 2005–2006 in southern Ontario relative to the 30-year climatology, which illustrated this effect clearly. Other studies in this report for Finland, Poland, and Switzerland also show a greater incidence of fog in the winter compared to spring and summer, as shown in Figure 3-54.

There is a peak in radiation fog activity in many regions in the fall. Meyer and Layla [1990] suggest that, for the Albany, New York area, this is the consequence of adequate cooling periods overnight, along with sufficient moisture in the air and ground. The time needed to saturate the air is a function of the relative humidity at sunset and the overnight cooling rate. These factors also led to a diurnal pattern in fog occurrence in a ±2-h window centered at sunrise. For the Albany, New York region, the months of August, September, and October were the only ones that had sufficient overnight hours to reach saturation. A similar pattern of high occurrence of fog in the September to December period is noted in Figure 3-54 for locations in Europe.

The combined diurnal and annual variations in fog occurrence have been tabulated for most first-order weather stations in Canada [Hansen et al., 2007]. Four sample plots in Figure 3-55 show that the probability of fog varies widely, from a peak of 25% in St. John's, Newfoundland, at sunrise in April to about 1.5% at sunrise in January in Edmonton, Alberta.

3.8.4. Fog Deposition on Insulators

The rate of impact of fog on an insulator of diameter D (m), exposed to fog of density δ (g/m^3) with a wind speed v (m/s) will be $D \cdot \delta \cdot v$ g per meter of dry arc distance. For a typical station post insulator with $D = 0.33$ m, a typical

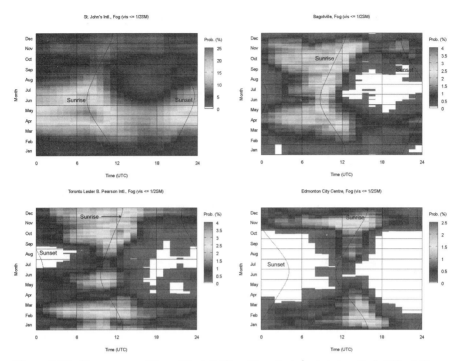

Figure 3-55: Occurrence of fog with visibility of less that $\frac{1}{2}$ statute mile ($\cong$805 m) (from Hansen et al. [2007]; courtesy of the Government of Canada).

wind speed in fog of v = 1 m/s (3.6 km/h), and a typical winter fog density of 0.1 g/m^3, the accumulation rate will be about 0.033 g/s per meter of dry arc, or 100 mg/(m^2-s), in units used later to calculate evaporation rate. In a test labora-tory, these values will be higher, with v = 3 m/s and a density of δ = 1 g/m^3 being typical and giving an accumulation rate of 1 g/s. For natural exposure, the rate of increase of fog layer thickness $d\tau/dt$ is more easily computed in centimeters, with $d\tau/dt$ = 0.033 g/s / (33 cm × 100 cm) or 1×10^{-5} cm/s, and then converted to 0.1 μm/s.

For reference, Salama et al. [1998] suggest that a water film thickness of between 3 and 8 μm forms in tracking wheel tests on insulators. They suggest that film thickness decreases from 8 to 3 μm as water solution conductivity increases from 100 to 500 μS/cm. They also found a stronger decrease in water film thickness from 8 to 0.5 μm as electric field stress varied from 100 to 150 kV/m. This evaluation can also be interpreted for fog accretion. Thirty to eighty seconds of natural fog exposure should thus be sufficient to re-wet a dry area on an insulator surface normal to the wind direction to an extent that will support tracking and arcing activity before formation of another dry band area.

3.8.5. Heat Balance Between Fog Accretion and Evaporation

The rate of evaporation from a water surface, m_c, is proportional to the difference in vapor pressures between the water surface temperature and the dew point temperature. The vapor pressure at the dew point is given by the relative humidity multiplied by the saturated vapor pressure at ambient P_a. A common expression [Carrier, 1918; ASHRAE, 1991] is of the form

$$m_c = (A + Bv)\frac{P_w - RH \cdot P_a}{h_{va}}$$ (3-17)

where

m_c is the rate of water evaporation, in kg/ (m²-s),

P_w and P_a are the saturated vapor pressures of water and air at their respective temperatures T_w and T_a, in N/m²,

h_{va} is the latent heat of vaporization of water, 2.45×10^6 J/kg,

v is the wind speed, in m/s,

RH is the relative humidity, expressed in this case as a fraction ($0 < RH < 1$), and

A and B are constants (0.0887 and 0.07815) for disturbed surfaces and (0.06741 and 0.0515) for calm surfaces.

The traditional equation for calculating vapor pressure over water [Goff and Gratch, 1946; Goff, 1957] is

$$\log_{10}(P_w) = 10.79574\left(1 - \frac{T_0}{T}\right) - 5.02800 \cdot \log_{10}\left(\frac{T}{T_0}\right) + 1.50475 \times 10^{-4}$$
$$\cdot (1 - 10^{[-8.2969(T/T_0 - 1)]}) + 4.2873 \times 10^{-4} \cdot (10^{[4.76955(1 - T_0/T)]} - 1)$$
$$+ 0.78614$$ (3-18)

where

T is the water temperature in kelvin units,

T_0 is 273.13 K,

and P_w is in hPa, with 100 hPa = 1 N/m².

This reference expression covers −50 °C to 102 °C (223–375 K).

The saturation vapor pressure in N/m² can be calculated more simply from temperature if T is in °C, using a series expansion centered at 42 °C:

$$P = 3226 \cdot \exp\left(-8 + 0.98\sqrt{T + 42°C}\right) \cdot \left(1 - 0.019 \cdot \exp\left(\frac{T}{42°C}\right)\right)$$ (3-19)

This expression is valid from 0 °C to 100 °C within ±0.8%. At 100 °C, the saturation pressure from Equation 3-18 is 10.14 hPa, while Equation 3-19 gives the vapor pressure of boiling water in desired units of 101420 N/m².

An experimental case for an insulating surface given by Textier and Kouadri [1986] can be used to cross-check these expressions for large, horizontal pools exposed to the natural environment. The still-water coefficients are used in Equation 3-17.

- At zero wind speed ($v = 0$), water and ambient temperature at 20°C, and 60% relative humidity, the calculated evaporation rate is 30 mg/(m²-s) and the observed is 28 mg/(m²-s).
- At an average water temperature of 55°C ($P = 15760$ N/m²) against a background level of 20°C / 60% RH ($P = 1400$ N/m²), the predicted evaporation rate is 460 mg/(m²-s) and the observed value is 530 mg/(m²-s).
- The average water temperature that gives the evaporation rate of 530 mg/(m²-s) is 58°C.

Allowing for a 2-C° uncertainty in the infrared measurement of water surface temperature, the agreement between model and measurement is satisfactory.

- If the wind speed had been 3 m/s (10.8 km/h) rather than still air, the evaporation rate would triple, to 89 mg/(m²-s) at $T_w = 20$°C and to 1590 mg/(m²-s) at $T_w = 58$°C.

Consider the laboratory case described earlier where fog accretion from a density of 1 g/m³ is occurring at the rate of 3000 mg/(m²-s) at a wind speed of 3 m/s, 100% relative humidity, and 20°C ambient temperature. The insulator temperature in this artificial fog test would have to reach 72°C in order to evaporate the fog at the same rate at which it accumulates. This equilibrium temperature is not sensitive to wind speed: at 1 m/s the value is 61°C and at 10 m/s it reaches 78°C. Figure 3-56 shows that only for the rare combination of dense fog (>2 g/m³) and high wind speed (10 m/s) will the insulator dry-band temperature have to exceed 100°C to achieve this balance.

This finding also is interesting in the context of Textier and Kouadri [1986], who measured peak dry-band temperatures of about 70°C in still air with an infrared camera. For natural fog density with levels typically below 0.3 g/m³, the surface temperature only needs to reach 20 C° above ambient to initiate the dry-band process.

3.8.6. Critical Wetting Conditions in Fog

The critical insulator wetting conditions [IEEE Working Group, 1979] occur in fog that condenses at a rate sufficiently high to wet the entire insulator surface, but not so high that the film of deposited water runs or drips off the insulator and carries away the surface pollution. Critical wetting leads to the

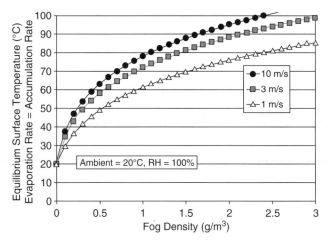

Figure 3-56: Equilibrium surface temperature for fog deposition and evaporation at 20 °C.

lowest electrical strength in standardized tests for both clean-fog and salt-fog environments.

Fog, dew, and cold-fog accretion on insulators does not usually lead to any indication on tipping-bucket rain gauges with 0.2-mm thresholds. Generally, an accumulation of 2 mm of rain on the top surface of an insulator is needed to reduce the contamination level by 50%. This level of rain will even reduce the bottom-surface contamination levels by 15% [Kimoto et al., 1971]. Liquid rain of 0.4 mm or more was sufficient to reduce the ESDD on station post insulators to background levels in winter conditions [Chisholm et al., 1993]. These precipitation amounts are thus well above the critical amount of water needed to produce a minimum flashover voltage in fog.

3.9. SURFACE WETTING BY NATURAL PRECIPITATION

Natural precipitation has a wide variety of types, divided into three categories in Table 3-16 along with their Météorologique Aviation Régulière (METAR) codes used to report routine avaiation weather.

Often, winter storms contain mixed precipitation. The standard practice is to report the dominant form first, so that a mix of rain with a bit of snow is RASN, while wet snow with a bit of rain will be indicated as SNRA.

In North American practice, the two-letter code ZR is sometimes used for freezing rain. Also, rate of accumulation can be indicated with the use of − and + signs, so that SN− is light snow and ZR+ is heavy freezing rain.

Electrical problems with precipitation on insulators are most often associated with forms that accumulate on the insulators in ways that wet the surfaces and/or shorten the leakage distance by bridging the shed spacing with a parallel water, ice, or snow path.

TABLE 3-16: Categories and Types of Precipitation and Their METAR Codes

Parameters	Liquid > 0 °C	Supercooled Liquid < 0 °C	Frozen or Solid
Size < 500 µm Precipitation rate ≤ 1 mm/day	Drizzle (DZ)	Freezing drizzle (FZDZ)	
Size ≥ 500 µm Precipitation rate > 1 mm/day	Rain (RA)	Freezing rain (FZRA, ZR)	Snow (SN) Snow pellets (SHGS) Snow grains (SG) Ice pellets (PL) Hail (SHGR) Graupel (GS) Ice crystals (IC)

Liquid precipitation tends to run off the surfaces, taking away some of the surface pollution in the process. If the rain accumulation is at a high rate, there is a possibility of water bridging of sheds that is simulated in heavy rain tests on insulators. In light rain or drizzle conditions, wetting of the downward-facing surfaces of polluted insulators can occur, leading to discharge activity similar to that observed in fog conditions.

Supercooled liquid precipitation can accumulate on insulator surfaces as ice. In the freezing condition, the typical runoff rate is slow. Soluble pollution may be retained near the surface of the accumulation until melting occurs, rather than being washed away in drip water. Multiple icicles with diameters of 3–10 mm can form, and with enough exposure, these can fully bridge the insulator sheds.

Most frozen or solid forms of precipitation such as snow pellets and ice grains are hard enough that they tend to bounce or slide off smooth parts of the insulators. Some forms such as dry and wet snow can accumulate on insulator surfaces at specific temperatures. Generally, electrical problems can occur during accumulation of heavy wet snow or during a melting period after a thick deposit of snow has been packed onto the insulator by wind or gravity.

3.9.1. Measurement Methods and Units

Precipitation is measured by noting the increase in volume in a graduated cylinder with an open top. The cylinder is normally filled with a small amount of low-density oil to prevent evaporation. Other guidance to meteorological observations is also given by the World Meteorological Organization [WMO, 1996]. In winter, there are two practices. The depth of accumulation of snow can be reported directly, or the water equivalent depth can be given by melting the accumulation. Automatic weather stations may use heated tipping-bucket rain gauges rather than graduated cylinders. Each tip of the bucket normally corresponds to about 0.2 mm of liquid precipitation, or 2 mm of light snow.

Atmospheric pollution stations distinguish between wet and dry deposition, using a form of rain gauge that has a wetness sensor (Figure 3-57). In this case, two cylinders, each fitted with plastic bags on the inside, are used. An automatic door exposes the "wet" bucket whenever the wetness sensor indicates that rain is falling.

Ice accumulation can be measured with passive or active systems. An example of a passive ice monitoring (PIM) instrument is shown in Figure 3-58.

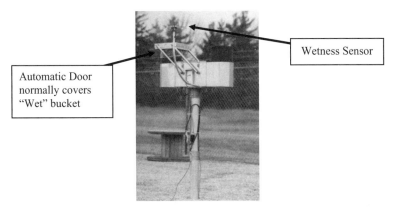

Figure 3-57: Automatic rain collection instrument for wet and dry deposition.

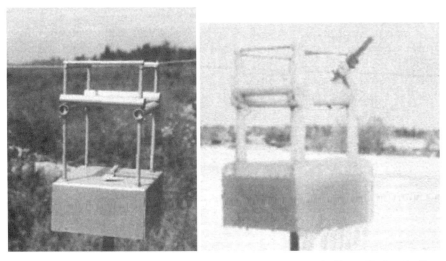

Figure 3-58: Passive ice meter used in Hydro-Québec Network (from Hydro-Québec [1993]).

The PIM collector was designed to collect ice deposits on a horizontal plate, four vertical plates at each point of the compass and two groups of four cylinders, the lower set with 25-mm diameter to simulate accumulation on phase conductors and the upper set with a smaller diameter to represent accumulation on overhead groundwires. The surface accumulations on the plates are used to establish ice loading on structures. Many icing measurements and standards are now calibrated against accumulation on the reference 25-mm cylinders.

With its special importance to aviation, real-time detection of icing conditions in the atmosphere forms an important supplement to weather forecasts. Automatic systems for measuring ice accretion have also been developed, as shown in Figure 3-59.

Typical icing rate meters detect and measure ice accumulation using the change in mechanical vibration frequency of a thin probe in contact with the air. With no accumulation, the probe's free vibration frequency can be as high as 40 kHz. The probe responds with a reduced frequency under accumulation of any type of atmospheric icing, whether caused by the passage of cold clouds, supercooled precipitation, or wet snow. Under heavy accumulation rates, it may be necessary to correct for the period of probe deicing. For example, in a test lab, the Rosemount detector in Figure 3-59 provided pulses indicating 0.51 mm (0.02 in.), with 377 s of accumulation and 96 s of heating. The correction for icing that should have occurred during heating cycles was (377 + 96)/377 or 1.25. With this correction, the Rosemount instrument indicated 10.4 mm of ice thickness at the same time that 20 mm accreted on a fixed monitoring cylinder.

Hydro-Québec SYGIVRE   Goodrich Rosemount

Figure 3-59: Icing rate meters.

Networks of icing rate meters provide information on the rate of accumulation and the extent of freezing rain and freezing drizzle events that are used to establish the need for airplane deicing and for defensive operation of electric power networks.

3.9.2. Droplet Size and Precipitation Conductivity

Rain and drizzle droplets initially form around the nuclei of airborne pollution. This means that the concentration of pollution in small droplets is relatively high. As additional pure water is scavenged from the atmosphere onto the droplet surface, the volume concentration of pollution will be reduced. The decrease in conductivity is a dynamic process for each droplet, eventually leading to low conductivity for large raindrops.

There is also a time dependence to precipitation conductivity over the typical duration of a storm. For storms of extended duration, the initial precipitation conductivity will be elevated, but as pollution is swept out of the air volume, the conductivity will reduce to a constant and lower level.

The effects of droplet size on winter precipitation were demonstrated in a series of field studies carried out in Ontario in the early 1990s at several important transformer stations. The major contributions to ionic conductivity at most sites were sulfates and nitrates as shown in Table 3-17.

Samples of winter fog had a conductivity that was nearly six times higher than the values of σ_{20} measured for snow and ice samples.

One site, located near major expressways, showed an elevated level of salt content in both precipitation and surface pollution.

TABLE 3-17: Measured Ionic Content of Winter Precipitation in Ontario, Canada

Parameter	Median Values (µmol/L) for Snow and Ice				Median (µEq) for Fog, Site T
	Site K	Site H	Site C	Site T	
H^+	63	63	63	63	204
SO_4^{2-}	22	23	22	47	386
NO_3^-	44	40	34	64	830
Cl^-	23	14	15	26	267
NH_4^+	26	21	14	30	389
Na^+	10	7	16	19	180
Ca^{2+}	7	6	8	8	262
Mg^{2+}	3	1	2	3	45
K^+	2	1	1	2	17
σ_{20} (µS/cm)	37	29	25	48	280

3.9.3. Effects of Washing on Surface Conductivity

The effect of rain washing on the surface pollution level was discussed in Section 3.2.5. Initially, the surface conductivity of insulators wetted by rain will increase, partly because the preexisting contamination layer migrates into solution or suspension. Depending on the rain rate and the hydrophobicity of the insulator surface, water will build up as a continuous film that eventually runs off, or will bead up right away and run off. Either way, conductive ions will be transported off the insulator surfaces to the ground.

The washing effects of precipitation leave ceramic surfaces with a low ESDD. Washing by natural rain of normal intensity, with typical conductivity of 30 μS/cm, leaves top surfaces of insulators with ESDD levels below a minimum detectable level (MDL) of 0.001 mg/cm^2 (1 μg/cm^2). If the rain is more conductive, the residual ion content on the insulator surfaces will also be higher. In a laboratory study by the authors, porcelain insulators were washed with 300-μS/cm water and allowed to dry. The resulting ESDD was measured to be 0.006 mg/cm^2 (6 μg/cm^2) after this process. Although this ESDD level is considered as very clean, some 500-kV insulators start to have flashover problems at ESDD levels of 0.02 mg/cm^2 (20 μg/cm^2) [Chisholm, 2007], a level only three times higher than the freshly washed insulators. As a consequence of this evaluation, it was recommended that EHV insulators be washed with deionized water in winter conditions.

3.9.4. Rain Climatology

One measure of the degree of wetness of a particular location is the annual number of hours of wetness (*Tow*) in hours per year. This measure is used, for example, in the ISO 9223 [1992] standard for corrosion of metals. This standard classifies the *Tow* values into five levels, as shown in Table 3-18.

Generally, it is the lack of rain, rather than its frequent occurrence, that leads to pollution flashover problems on insulator surfaces. Desert areas classified τ_1 with fewer than 10 hours of wetness per year are areas where pollution can build up on insulator surfaces for months or years. Areas classed τ_4 or τ_5 with more than 2500 h/yr may be associated with tropical climates or with direct exposure to ocean spray. There is enough rain in these climates to keep the insulator surfaces relatively free of contamination buildup. A basic problem

TABLE 3-18: Classification of Annual Time of Wetness (*Tow*) in ISO 9223 for Corrosion of Metals

Category	τ_1	τ_2	τ_3	τ_4	τ_5
Hours/yr	$\leq$10	10–250	250–2500	2500–5500	$\geq$5500

with the *Tow* index in ISO 9223 [1992] for areas classed as τ_2 or τ_3 is that it does not give the most important design factor for contamination, which is the duration of accumulation between rain periods.

In general, it is possible to obtain more than 20 years of daily observations for many weather stations located at airports in electronic format. It is relatively simple to interpret the records of rain observations, and to establish a typical distribution of "days without rain."

In climates that have a long winter period, the number of days without rain can be extracted more simply from the rain climatology. Figure 3-60 shows that Edmonton, Alberta, has a winter dry period that extends from the end of October to the start of April. Bagotville, Québec, has a shorter dry spell from early December to mid-March. The period in Toronto, Ontario, is short—about 2 weeks each side of the start of February. With its location near the Atlantic Ocean, St. John's, Newfoundland, has no winter dry spell at all, with a constant 5% chance of rain all winter.

Surface wetting by supercooled liquid or solid precipitation can occur in the winter. This process may increase the risk of electrical problems during the melting phase, but it is just as effective as rain in removing pollution from insulator surfaces.

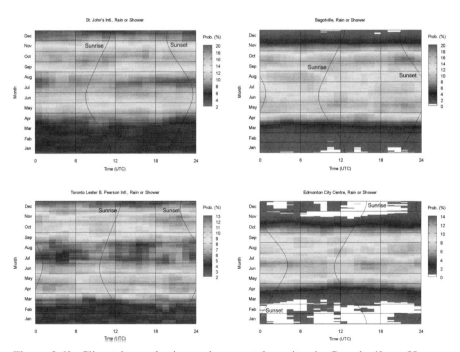

Figure 3-60: Climatology of rain or shower at four sites in Canada (from Hansen et al. [2007]; reproduced by permission of the Government of Canada).

3.10. SURFACE WETTING BY ARTIFICIAL PRECIPITATION

There are a wide range of artificial precipitation sources for transmission line exposure. These are described in order of increasing electrical conductivity.

3.10.1. Tower Paint

Transmission towers are painted by live-line utility workers or contractors who dip their entire hand into pails that may hold several gallons. Specialized mitts are used, similar to paint rollers, to retain large quantities of paint and speed up production. On occasion, thin streams of paint have dripped onto energized conductors, leading to flashovers. In general, the electrical conductivity of the zinc-rich paint is relatively low. Drips or streams of paint on insulator disk surfaces have proved to be less of a problem than the paint stream itself. Flashovers of the paint stream, observed at ac voltage gradients less than 50 kV/m, are related to volatile content.

3.10.2. Bird Streamers

While it could be classed as a natural precipitation, the viscous stream of excrement ejected into the air as large birds take off from perches above insulators has proved to be a significant overhead transmission line reliability problem. This form of pollution has proved to be difficult to diagnose directly. Burnham [1995] outlines the scope of this problem in Florida, with additional mitigation progress described in an IEEE Task Force report [Sundararajan et al., 2004].

Important background to this pollution problem was described by West et al. [1971]. These researchers gathered samples of stream specimens from osprey, eagles, and hawks (Figure 3-61). The observed resistivity was between 30 and 120 Ω-m (σ = 333–83 μS/cm). Insulator flashover tests used simulated bird streamers with raw scrambled eggs (130 Ω-m) with a bit of salt to give 83 μS/cm conductivity. The simulated streamers were ejected from pressurized balloons. Their test results showed that a discharge volume of 50–60 cm^3 was sufficient to produce an electrically conductive stream of 2.4–2.7 m with about 5-mm average diameter. This stream caused flashovers on 3.17-m insulator strings for 500-kV systems at 320-kV ac line-to-ground and also caused flashovers on 3.42-m insulators energized at 400 kV dc.

3.10.3. Dam Spray

Many electrical substations serving hydraulic generation are located tens or hundreds of meters downstream from dams. Water falling over the edge of a dam will hit the water and rocks below. The impact will form aerosol droplets. When water temperature is warmer than the air, these droplets mix with the air and warm it. The mixture expands and a plume of water will rise, sucking

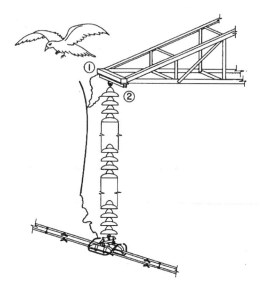

Figure 3-61: Bird streamer flashover mechanism (from West et al. [1971]).

in even more misty air [Bursik, 2005]. This plume can carry downstream to the location of substation equipment or lines feeding the plant output to the grid.

The height of the mist plume H_T is a function of the flow rate Q, the difference between air and water temperature ΔT, and the ambient temperature T_a:

$$H_T = C_1 \cdot \left(\frac{Qg\,\Delta T}{T_a} \right)^{1/4} \tag{3-20}$$

where

Q is the flow rate in m³/s, ranging from 1500 to 3000 m³/s for Niagara Falls, and

g is the gravitational constant, 9.8 m/s².

A large temperature difference between the air and the water leads to a higher mist plume of greater density. Plume height of 50–400 m is typical at Niagara Falls, with values as high as 1100 m for wind speeds less than 25 km/h. Bursik [2005] also established that the plume from Niagara Falls rose to heights of 50 m during the day when air and water temperatures were equal, but reached 300 m at night in the fall, when the water temperature of 15 °C was significantly higher than the air temperature of 0–5 °C.

The plume from waterfalls can blow downstream, wetting insulators on hydraulic generating substations in summer and coating them with ice in winter. Figures 3-62 and 3-63 illustrate the overspray from a 40-m waterfall over the Kinzua Dam in Warren, Pennsylvania, along with the thick coating

Figure 3-62: Overall situation of pumped storage dam and 230-kV substation.

Figure 3-63: Summer and winter views of overspray effects on 230-kV substation (from Dawson et al. [2001]; courtesy of First Energy).

of rime ice that accumulates on metal bus work for the 400-MW hydroelectric plant.

Generally, the conductivity of freshwater lakes such as Ontario, Huron, Michigan, and Erie is in the range of 80–120 μS/cm. River water conductivity varies more widely depending on the sources and local geology.

3.10.4. Irrigation with Recycled Water

The reuse of water for irrigation brings about the possibility that accidental spray will hit power system insulators. Todd et al. [1990] provide the information in Table 3-19 for water quality standards for irrigation use. (See also Todd and Mays [2005].)

Todd et al. [1990] define the threshold values in Table 3-19 as those where an irrigator might become concerned about water quality, and start to add

TABLE 3-19: Water Quality Standards for Irrigation Use

Quality Factor	Threshold Concentration	Limiting Concentration
Electrical conductivity	750 μS/cm	2250 μS/cm
Range of pH	7.0–8.5	6.0–9.0
Total dissolved solids	500 mg/L	1500 mg/L
Sodium absorption ratio[a]	6.0[c]	15
Residual sodium carbonate[b]	1.25[c]	2.5
Arsenic	1.0 mg/L	5 mg/L
Boron	0.5 mg/L	2 mg/L
Chloride	100 mg/L[c]	250 mg/L
Sulfate	200 mg/L[c]	100 mg/L
Copper	0.1 mg/L[c]	1 mg/L

[a]Defined as SAR = $Na/(Ca + Mg)^{0.5}$, where concentrations are expressed in mEq/L.
[b]Defined as sum of equivalents of normal carbonate and bicarbonate, minus the sum of equivalents of calcium and magnesium.
[c]Not to be exceeded more than 20% of the time.
Source: Todd et al. [1990]; Todd and Mays [2005].

additional water from other sources for leaching. The limiting values will have drastic effects on landscape or crop quality.

3.10.5. Cooling Pond Overspray: Freshwater Makeup

Cooling ponds can be an inexpensive alternative to cooling towers. These systems consist of multiple spray systems and large, shallow ponds to cool process water by evaporation and radiation. Figure 3-64 shows examples of cooling ponds located next to high-voltage electrical substations. Figure 3-65 illustrates a preferred practice, with the cooling pond located far (and downwind) from the electrical insulators.

The water in cooling ponds can have high electrical conductivity, especially compared to natural rain or river water. The evaporation process increases the concentration of ions. In one case, shown in Figure 3-66, the intake and outflow of water was managed to respect an upper limit of 2000 μS/cm. The water conductivity was managed to comply with local regulations, including pH and total dissolved solids, before release back into the environment.

For the pond water records in Figure 3-66, the conductivity ranged between 1000 and 2000 μS/cm, with peak values reaching 3500 μS/cm. The pH of the pond was managed in the range of 8.0–8.5. Total suspended solids, consisting in this case of iron oxide particles, was found to vary between 100 and 200 mg/L. Hexane extraction was used to test for oil and grease, with significant oil leaks giving peak readings of 600 mg/L against a background level 30–90 mg/L.

The pond water in this case is 10–20 times more conductive than the water used to qualify electrical insulators in the heavy rain tests of IEEE Standard 4 [1995] and IEC Standard 60060-1 [1989].

Figure 3-64: Examples of overspray from cooling ponds near electrical substations.

Figure 3-65: Example of preferred location of cooling pond, away from electrical substation and lines.

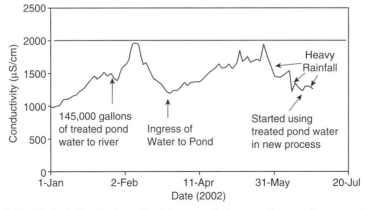

Figure 3-66: Typical electrical conductivity records for a cooling pond near steel mill.

3.10.6. Cooling Tower Drift Effluent

Santee Cooper [Johnson and Salley, 2004] reported details of flashover on a 230-kV structure, located near the Winyah Generating Plant Cooling Tower Units 3 and 4. The plant is located in Georgetown, South Carolina, about 18 km from the Atlantic Ocean and 56 km southwest of Myrtle Beach, South Carolina. There are three cooling towers, with two located about 300 m west of a large switchyard.

The contamination environment at Winyah is hostile, with frequent wetting and contamination from sea mist and vapor from the cooling towers. One 230-kV steel pole dead-end tower was stated to be within a few feet of the cooling towers. Records show that the conductivity of water from the cooling tower varied with time from 2000 to 8000 μS/cm, peaking at 12,000 μS/cm on some occasions.

From 1999 to 2003, flashovers on the Charity–Winyah 230-kV transmission line caused 35% of the duration of outages on the Santee Cooper 230-kV transmission system. Johnson and Salley [2004] also noted that the majority of the outages took place during the month of November. One specific event had the following parameters, measured at Myrtle Beach about 60 km to the northeast:

- Previous day, November 28, 2003:

    Wind from the south shifting to 30 km/h from the west at 6 PM

    Ambient temperature falling from 23 °C to 5 °C

    Dew point falling from +13 °C to −4 °C
- Day of flashover, November 29, 2003:

    Cooling tower effluent conductivity 4400 μS/cm and 6390 μS/cm

    Wind 16–32 km/h from the west, ambient temperature 2–9 °C, dew point −4 °C

The flashover on the nearby tower was stated to be caused by the shift in wind speed and direction and may also be related to the drop in dew point temperature below the freezing point. Fine drops of mist will quickly come to equilibrium with the dew point temperature, rather than ambient, through evaporation, and as the dew point temperature on both days was −4 °C, the drops are likely to have been supercooled, meaning they would freeze on impact.

Similar weather conditions existed on November 14, 2003. On November 29, 2003, the cooling tower water conductivity was measured to be 4400 μS/cm for Unit 3 and 6390 μS/cm for Unit 4. Johnson and Salley [2004] also used ultraviolet image intensification (DayCor) to recognize and quantify corona and partial discharge activity on tower 110 and other nearby equipment.

The problems with insulator flashover were solved with a number of improvements in 2003 including:

- Reduction of cooling tower effluent conductivity to 300 µS/cm by introduction of fresh water from a new supply line.
- Increase in leakage and dry arc distance to suspension insulators rated for 730 kV compared to a normal 13 or 14 disk for 230-kV lines.
- Application of a silicone–rubber surface coating to 230-kV substation insulators to repel water and encapsulate pollution.
- Corona rings at the high-voltage ends of 230-kV post insulators to make the electric field more uniform.
- A monthly maintenance program of hand washing the insulators.

Johnson and Salley [2004] reported that the corrective actions were effective, with problems only when the freshwater supply to the cooling towers is out of service for repairs.

3.10.7. Cooling Water Overspray: Brackish or Saltwater Makeup

Power plants located on rivers that are close to the ocean face additional problems with electrical insulators when cooling water is drawn from brackish water. The normal electrical conductivity of the ocean is about 40,000 µS/cm (40 mS/cm). For locations on the shore of the ocean, such as Galveston, Texas, Figure 3-67 shows that there can be daily fluctuations in conductivity of 20–50 mS/cm. At a location where the tide affects the mixing of salt and fresh water, such as Lewisetta, Virginia, there are much greater swings, with a sustained low value of 2000 µS/cm corresponding to a large outflow of fresh water from tropical storm Ernesto, passing on August 31, 2006. Further inland, for example, in Mobile, Alabama, the salt water remains dilute all the time but the conductivity still has a daily variation with the tides from 2000 to 3500 µS/cm.

Plants that draw water for cooling towers or ponds from brackish rivers should organize water intake schedules to take advantage of daily minima in the local conductivity. New cooling systems with seawater or brackish drift or overspray should be located well away from existing electrical insulators.

3.10.8. Manurigation

Depending on crop type and species, it has become common practice to store and reuse animal urine as crop fertilizer. In Minnesota, Martens et al. [1998] reported that 671 million gallons of liquid manure was pumped onto farmland by 17 commercial liquid manure service companies in one year. Liquid manure is sprayed on land controlled by the livestock operators who generate the manure. In contrast, solid applications are usually purchased by organic crop farmers who want to replace commercial fertilizer nutrients with manure.

Laboratory analysis by Shapiro et al. [1998] showed that swine effluent contains about 80 lb total nitrogen, 100 lb K_2O, and 10 lb P_2O_5 (phosphate)

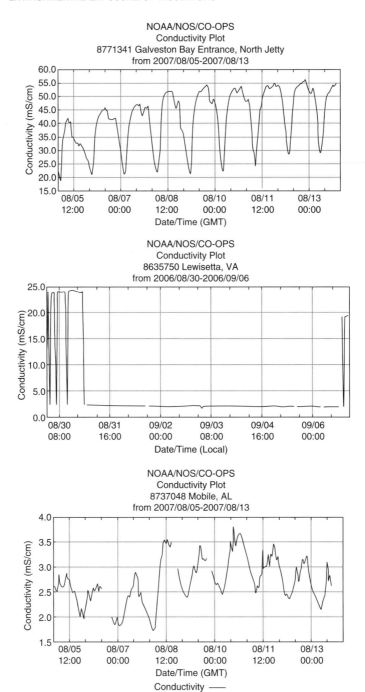

Figure 3-67: Variation in electrical conductivity of water at Galveston Bay, Texas; Lewisetta, Virginia; and Mobile, Alabama (from NOAA [2007]).

TABLE 3-20: Typical Chemical Composition of Pig Urine

Element	Weight (% or ppm)
Nitrogen	0.61%
Potassium	0.17%
Phosphorus	0.16%
Calcium	0.16%
Magnesium	600 ppm
Sodium	294 ppm
Iron	178 ppm
Sulfur	40 ppm
Zinc	39 ppm
Manganese	23 ppm
Copper	6 ppm

Source: Shapiro et al. [1998].

per acre-inch of water. Table 3-20 provides additional details of typical composition.

The high concentrations of ions lead to extremely high electrical conductivity, meaning that manurigation can create two hazards:

- Deposit of contamination on insulators downwind of operations.
- Direct risk of flashover should a manurigation stream come too close to an energized conductor.

3.11. PRÉCIS

A wet layer of electrically conductive ions reduces insulation strength. This chapter has shown that the conductive ions may be a result of dry deposition of the same ions that make up the local air pollution—for example, sodium chloride near the oceans, or sulfates downwind from coal-fired power plants. The conductive ions may also be present in the natural precipitation, but wet deposition also tends to wash insulator surfaces of any preexisting dry deposit.

Anything that stabilizes a layer of conductive ions on the insulator surface will increase the risk of electrical flashover. A rough insulator surface performs worse than a smooth surface because it is more likely to have extensive areas with continuous layers of water. Insoluble deposits such as desert sand will reduce flashover strength because they have a similar effect. Winter fog accretion can also stabilize a pollution deposit into a continuous layer of thin ice that becomes electrically dangerous at the melting point.

In addition to the ions in natural rain, power lines and stations may be exposed to a wide range of artificial wetting, such as overspray from road salting or nearby cooling towers. These local problems should be considered in the design stage but may need to be mitigated after-the-fact when problems occur.

REFERENCES

Aherne, J., T. A. Clair, I. F. Dennis, M. Gilliss, S. Couture, D. McNicol, R. Weeber, P. J. Dillon, W. Keller, D. S. Jeffries, S. Page, K. Timoffee, and B. J. Cosby. 2005. "Dynamic Modelling of Lakes in Eastern Canada," in *European Critical Loads and Dynamic Modelling* (M. Posch, J. Slootweg, and J.-P. Hettelingh, eds.). Report 259101016/2005, UN ECE CCE Status Report 2005.

Akbar, M. and F. Zedan. 1991. "Performance of HV Transmission Line Insulators in Desert Conditions, Part III, Pollution Measurements at a Coastal Site in the Eastern Region of Saudi Arabia," *IEEE Transactions on Power Delivery*, Vol. 6, No. 1 (Jan.), pp. 429–436.

Amarh, F. 2001. "Electric Transmission Line Flashover Prediction System: Ph. D. Thesis and Final Report." Power Systems Engineering Research Center Publication 01-16, Arizona State University, May.

ASHRAE. 1991 and 1999. *ASHRAE Handbook—HVAC Applications*. Atlanta, GA: American Society of Heating, Refrigerating and Air-conditioning Engineers.

Beauséjour, Y. 1981. "Insulator Contamination Study for Distribution Systems." CEA R&D Report for Project 75-05. October; available from CEAI International Inc.

Beattie, J. 2007. "Road Salt Contamination of Insulators." Sask Power Report 07-362, January 23, 2007.

Burnham, J. T. 1995. "Bird Streamer Flashovers on FPL Transmission Lines," *IEEE Transactions on Power Delivery*, Vol. 10, No. 2 (Apr.), pp. 970–977.

Bursik, M. 2005. "Field Trip Guide BGS05," in *Proceedings of 36th Binghamton Geomorphology Symposium*, University of Buffalo. October 7–9.

Canada–U. S. Subcommittee on Scientific Co-operation, in Support of the Canada–U.S. Air Quality Agreement. 2004. Canada-US Transboundary Particulate Matter Science Assessment. Catalogue No. En56-203/2004E.

Carrier, W. H. 1918. "The Temperature of Evaporation," *ASHVE Transactions*, Vol. 24, pp. 25–50.

Chen, X. 2000. "Modeling of Electrical Arc on Polluted Ice Surfaces." PhD Thesis, École Polytechnique de Montréal, Département de Génie Électrique. February.

Cheng, Y. S. and H. C. Yeh. 1979. "Particle Bounce in Cascade Impactors," *Environmental Science and Technology*, Vol. 13, pp. 1392–1396.

Chisholm, W. A. 1995. "Cold-Fog Test Results on 115-kV Bus Support Insulators from Hamilton Beach TS." Ontario Hydro Technologies Report A-G-95-77-P.

Chisholm, W. A. 1998. *Effects of Flue-Gas Contamination on Ceramic Insulator Performance in Freezing Conditions*. EPRI, Palo Alto, CA, and Tennessee Valley Authority, Chattanooga, TN: TR-110296.

Chisholm, W. A. 2007. "Insulator Leakage Distance Dimensioning in Areas of Winter Contamination Using Cold-Fog Test Results," *IEEE Transactions on Dielectrics and Electrical Insulation*, Vol. 14, No. 6 (Dec.), pp. 1455–1461.

Chisholm, W. A., Y. T. Tam, and T. Jarv. 1993. "Determination of Insulator Contamination Levels from Environmental Measurements." Ontario Hydro Research Division Report 92-264-K, April 16.

Chisholm, W. A., P. G. Buchan, and T. Jarv. 1994. "Accurate Measurement of Low Insulator Contamination Levels," *IEEE Transactions on Power Delivery*, Vol. 9, No. 3 (July), pp. 1552–1557.

CIGRE Working Group 33-04. 1979. "The Measurement of Site Pollution Severity and Its Application to Insulator Dimensioning for a. c. Systems," *Electra*, No. 64, pp. 101–116.

CIGRE Task Force 33.04.03. 1994. "Insulator Pollution Monitoring," *Electra*, No. 152, (Feb.), pp. 79–89.

CIGRE Task Force 33.04.01. 2000. "Polluted Insulators: A Review of Current Knowldege." CIGRE Brochure No. 158. June.

Conlon, W. M. 1992. "Recent Improvements to the Freeze Crystallization Method of Water Purification," in *Proceedings of IACT 92*, pp. 1–4.

COST. 2005. European Cooperation in the Field of Scientific and Technical Research. B. Fisher, S. Joffre, J. Kukkonen, M. Piringer, M. Rotach, and M. Schatzmann. *Meteorology Applied to Urban Air Pollution Problems—Final Report*. COST Action 715, Group B5.

COST. 2007a. European Cooperation in the Field of Scientific and Technical Research. *Earth System Science and Environmental Management (ESSEM) Domain Scientific Days Presentations 2007*. Available at http://www.cost.esf.org/.

COST. 2007b. European Cooperation in the Field of Scientific and Technical Research. *Short Range Forecasting Methods of Fog, Visiblity and Low Clouds*. Final Report for COST Action 722. May 29.

CRC. 1989. *Handbook of Chemistry and Physics*. Boca Raton, FL: CRC Press.

Dawson, C., B. Hogg, M. Wilkinson, and M. Clark. 2001. "Review of the Seneca Pumped Storage Plant for FirstEnergy Corp, Akron, Ohio." Kinectrics Report 9100-010-RA-0001-R00. July 10.

Duce, R. A. 1995. In *Atmospheric Forcing of Climate* (R. J. Charlson and J. Heinzenberger, eds.). Chichester, UK: Wiley, pp. 43–72.

Environment Canada. 2004. *2004 Canadian Acid Deposition Science Assessment*, Library and Archives Canada.

El-A Slama, S. Flazi, H. Hadi, T. Tchouar, and M. Belkadi. 2000. "Measurement of ESDD and Chemical Analysis of Pollution of 220 kV Transmission Line Insulator Exposed to a Severe Contamination," in *International Conference on Electrotechnics, ICEL 2000,* UST Oran, Algeria, pp. 567–572. November 13–15.

Farber, S. A. and A. D. Hodgdon. 1991. "Cesium-137 in Wood Ash: Results of a Nationwide Survey." Health Physics Society Annual Meeting, July 1991.

Farzaneh, M. and O. Melo. 1990. "Properties and Effect of Freezing Rain and Winter Fog on Outline Insulators," *Cold Regions Science and Technology*, Vol. 19, pp. 33–46.

Farzaneh, M., J. Zhang, and Y. Chen. 1994. "A Laboratory Study of Leakage Current and Surface Conductivity of Ice Samples," in *1994 Annual Report of IEEE Conference on Electrical Insulation and Dielectric Phenomena*, Arlington, TX, pp. 631–638.

Farzaneh, M., A. C. Baker, R. A. Bernstorf, J. T. Burnham, E. A. Cherney, W. A. Chisholm, R. S. Gorur, T. Grisham, I. Gutman, L. Rolfseng, and G. A. Stewart. 2007. "Selection of Line Insulators with Respect to Ice and Snow, Part I: Context and Stresses, A Position Paper Prepared by the IEEE Task Force on Icing Performance of Line Insulators," *IEEE Transactions on Power Delivery*, Vol. 22, No. 4, pp. 2289–2296.

Fikke, S. M., J. E. Hanssen, and L. Rolfseng. 1993. "Long-Range Transport of Pollutants and Conductivity of Atmospheric Ice on Insulators," *IEEE Transactions on Power Delivery*, Vol. 8, No. 3 (July), pp. 1311–1321.

Forrest, J. S. 1936. "The Electrical Characteristics of 132-kV Line Insulators Under Various Weather Conditions," *IEE Journal*, Vol. 79, pp. 401–423.

Fowler, D., R. Smith, J. Muller, J. N. Cape, M. Sutton, J. W. Erisman, and H. Fagerlo. 2007. "Long Term Trends in Sulphur and Nitrogen Deposition in Europe and the Cause of Non-linearities," *Water Air Soil Pollution*, Vol. 7, Feb. 16, pp. 41–47.

Friend, J. P. 1973. "The Global Sulfur Cycle," in *Chemistry of the Lower Atmosphere* (S. I. Rasool, ed.). New York: Plenum.

Goff, J. A. 1957. "Saturation Pressure of Water on the New Kelvin Temperature Scale," *Transactions of the American Society of Heating and Ventilating Engineers*, pp. 347–354.

Goff, J. A. and S. Gratch. 1946. "Low-Pressure Properties of Water from −160 to 212 °F," *Transactions of the American Society of Heating and Ventilating Engineers*, pp. 95–122.

Gultepe, I., J. Milbrandt, and S. Belair. 2006. "Visibility Parameterization from Microphysical Observations for Warm Fog Conditions and Its Application to the Canadian MC2 Model," in *Proceedings of January 2006 American Meteorological Society Meeting*, Atlanta, GA.

Hall, J. F. and F. P. Mauldin. 1981. "Wind Tunnel Studies of the Insulator Contamination Process," *IEEE Transactions on Electrical Insulation*, Vol. EI-16, No. 3 (June), pp. 180–188.

Hallett, J. 1963. "The Temperature Dependence of the Viscosity of Supercooled Water," *Proceedings of the Physics Society*, Vol. 82, pp. 1046–1050.

Hansen, B., I. Gultepe, P. King, G. Toth, and C. Mooney. 2007. "*Visualization of Seasonal-Diurnal Climatology of Visibility in Fog and Precipitation at Canadian Airports*," in *Proceedings of 16th Conference on Applied Climatology, 87th Annual Meeting of the American Meteorological Society*, San Antonio, TX. January 14–18.

Hara, H., M. Kitamura, A. Mori, I. Noguchi, T. Ohizumi, S. Seto, T. Takeuchi, and T. Deguchi. 1995. "Precipitation Chemistry in Japan 1989–1993," *Water, Air and Soil Pollution*, Vol. 85, pp. 2307–2312.

Hayashi, M. 2004. "Temperature–Electrical Conductivity Relation of Water for Environmental Monitoring and Geophysical Data Inversion," *Environmental Monitoring and Assessment*, Vol. 96, pp. 119–128.

Hettelingh, J.-P., R. J. Downing, and P. A. M. de Smet (eds.). 1991. Mapping Critical Loads for Europe. UN Economic Commission for Europe (ECE) Coordination Center for Effects (CCE), Netherlands Environmental Assessment Agency (RVIM) Report No. 259101001.

Hewitt, G. F. 1960. *Tables of the Resistivity of Aqueous Sodium Chloride Solutions*, U. K. Atomic Energy Research Establishment Report AERE-R3497. London: HM Stationary Office.

Higashiyama, Y., T. Sugimoto, S. Ohtsubo, F. Sato, and H. Homma. 1999. "Heavy Salt Deposition onto a Distribution Line by a Seasonal Wind in Winter," in *Proceedings of IEEE/PES Summer Meeting*, Edmonton, pp. 882–887.

Hydro-Québec. 1993. *Manuel d'observation du givre et du verglas—Programme d'observation glacimétrique*, 8th edition. Division équipement de lignes, Services etudes et normalization.

IEC Standard 60815. 1986. *Guide for the Selection of Insulators in Respect of Polluted Conditions.* Geneva, Switzerland: Bureau Central de la Commission Electrotechnique Internationale. January 1.

IEC Standard 60060-1. 1989. *High-Voltage Test Techniques Part 1: General Definitions and Test Requirements*. Geneva, Switzerland: Bureau Central de la Commission Electrotechnique Internationale.

IEC Standard 60507. 1991. *Artificial Pollution Tests on High-Voltage Insulators to Be Used on a.c. Systems*. Geneva, Switzerland: Bureau Central de la Commission Electrotechnique Internationale. April.

IEC Standard 60815. 2008. *Selection and Dimensioning of High-Voltage Insulators Intended for Use in Polluted Conditions*. Geneva, Switzerland: IEC.

IEEE Standard 4. 1995. *IEEE Standard Techniques for High Voltage Testing*. Piscataway, NJ: IEEE Press.

IEEE Working Group on Insulator Contamination. 1979. "Application of Insulators in a Contaminated Environment," *IEEE Transactions on Power Apparatus and Systems*, Vol. PAS-98, No. 5 (Sept./Oct.), pp. 1676–1695.

ISO 9223. 1992. "Corrosion of Metals and Alloys. Classification of Corrosivity of Atmospheres."

Jiang, X., S. Wang, Z. Zhang, S. Xie, and Y. Wang. 2005. "Study on ac Flashover Performance and Discharge Process of Polluted and Iced IEC Standard Suspension Insulators String," in *Proceedings of 14th International Symposium on High Voltage (ISH)*, Beijing, Paper D-70, pp. 1–8.

Jickells, T. D., Z. S. An, K. K. Andersen, A. R. Baker, G. Bergametti, N. Brooks, J. J. Cao, P. W. Boyd, R. A. Duce, K. A. Hunter, H. Kawahata, N. Kubilay, J. LaRoche, P. S. Liss, N. Mahowald, J. M. Prospero, A. J. Ridgwell, I. Tegen, and R. Torres. 2005. "Global Iron Connections Between Desert Dust, Ocean Biogeochemistry, and Climate," *Science*, Vol. 308, Apr. 1, pp. 67–71.

Johnson, W. and D. T. Salley. 2004. "Winyah Generating Plant, Santee Cooper Corona Imaging," in *Proceedings of Third Annual EPRI/OFIL UV Inspection Users Group Meeting*, Orlando, FL. February.

Kimoto, I., K. Kito, and T. Takatori. 1971. "Antipollution Design Criteria for Line and Station Insulators," *IEEE Transactions on Power Apparatus and Systems*, Paper 71 TP 649 PWR.

Lambeth, P. J., H. Auxel, and M. P. Verma. 1972. "Methods of Measuring the Severity of Natural Pollution as It Affects HV Insulator Performance," *Electra*, No. 20, pp. 37–52.

Lee, P. K. H., J. R. Brook, E. Dabek-Zlotorzynska, and S. A. Mabury. 2003. "Identification of the Major Sources contributing to $PM_{2.5}$ Observed in Toronto," *Environmental Science and Technology*, Vol. 37 No. 21, pp. 4831–4840.

Leitch, J. E. 2006. "The Corrosivity Regions of Southern Africa—Long Term Trends and Implications for Materials Selection," in *Proceedings of Southern African Institute of Mining and Metallurgy 8th International Corrosion Conference*, pp. 1–9.

Looms, J. S. T. 1988. *Insulators for High Voltages*. London: Peter Peregrinus Ltd.

Lundmark, A. and B. Olofsson. 2007. "Chloride Deposition and Distribution in Soils Along a Deiced Highway—Assessment Using Different Methods of Measurement," *Water Air Soil Pollution*, Vol. 182, Jan. 13, pp. 173–185.

Martens, G., R. Martens, D. Martens, J. Hoeft, M. Hoeft, R. Elkins, and S. McCorquodale. 1998. "Custom Manure Application in Minnesota," in *Proceedings of Manure Management Conference*, Ames, IA. February 10–12.

Matsuoka, R., S. Ito, K. Sakanishi, and K. Naito. 1991. "Flashover on Contaminated Insulators with Different Diameters," *IEEE Transactions on Electrical Insulation*, Vol. EI-26, No. 6 (Dec.), pp 1140–1146.

Matsuoka, R., O. Kaminogo, K. Kondo, K. Naito, Y. Mizuno, and H. Kusada. 1995. "Influence of Kind of Insoluble Contaminants on Flashover Voltages of Artificially Contaminated Insulators." 9th *International Symposium on High Voltage*, Graz, Austria, Paper 3210. August 28 to September 1.

Meijer, E., van Velthoven, P., Brunner, D., Huntrieser, H., and Kelder, H. 2001. "Improvement and Evaluation of the Parameterisation of Nitrogen Oxide Production by Lightning," *Physical Chemistry of the Earth*, Vol. 26, No. 8, pp. 557–583.

Meteorological Service of Canada. 2004. Canada-U.S. Subcommittee on Scientific Cooperation, "Canada – United States Transboundary PM Science Assessment," available at http://www.epa.gov/airmarkt/usca/index.html

Meyer, M. B. and G. G. Lala. 1990. "Climatological Aspects of Radiation Fog Occurrence at Albany, New York," *Journal of Climate*, Vol. 3, May, pp. 577–586.

Mikhailov, A. A. 2002. "Estimating and Mapping the Material Corrosion Losses in the European Part of Russia with Unified Dose–Response Functions," *Protection of Metals*, Vol. 38, No. 3, pp. 243–257.

Mikhailov, A. A., J. Tidblad, and V. Kucera. 2004. "The Classification System of ISO 9223 Standard and the Dose–Response Functions Assessing the Corrosivity of Outdoor Atmospheres," *Protection of Metals*, Vol. 40, No. 6, pp. 541–550.

National Atmospheric Deposition Program [NADP]. 2007. NRSP-3, NADP Program Office, Illinois State Water Survey, 2204 Griffith Drive, Champaign, IL 61820.

NEPA. 1969. "The National Environmental Policy Act of 1969, as amended (Pub. L. 91-190, 42 U.S. C. 4321-4347, January 1, 1970, as amended by Pub. L. 94-52, July 3, 1975, Pub. L. 94-83, August 9, 1975, and Pub. L. 97-258, § 4(b), Sept. 13, 1982), An Act to establish a national policy for the environment, to provide for the establishment of a Council on Environmental Quality, and for other purposes," Senate and House of Representatives of the United States of America in Congress.

NGK. 1991. "Technical Guide." NGK Nagoya, Catalog No. 91-R.

Nilsson, J. and P. Grennfelt (eds.). 1988. *Critical Loads for Sulphur and Nitrogen.* Miljørapport 1988:15. Copenhagen: Nordic Council of Ministers.

Nejedly, Z., J. L. Campbell, J. R. Brook, R. Vet, and R. Eldred. 2003. "Evaluation of Elemental and Black Carbon Measurements from the GAViM and IMPROVE Networks," *Aerosol Science and Technology*, Vol. 37, No. 1, pp. 96–108.

NOAA. 2007. Center for Operational Oceanographic Products and Services. Available at http://tidesandcurrents.noaa.gov/.

Plambeck, J. A. 1982. *Electroanalytical Chemistry.* Hoboken, NJ: Wiley.

Posch, M., J. Slootweg, and J.-P. Hettelingh (eds.). 2005. *European Critical Loads and Dynamic Modelling*, Report 259101016/2005. UN ECE CCE Status Report 2005.

Potvin Air Management Consulting. 2006. "Informal Consultation on Local Air Issues in Sault Ste. Marie, Ontario–Michigan Under the Canada–United States Air Quality Agreement: Technical Support Document on Air Quality 2001–2003," Summary 2001–2003 Report. November.

Ramos, G. N., M. T. Campillo, and K. Naito. 1993. "A Study on the Characteristics of Various Conductive Contaminants Accumulated on High-Voltage Insulators," *IEEE Transactions on Power Delivery*, Vol. 8 No. 4 (Oct.), pp. 1842–1850.

Richards, C. S., C. L. Benner, K. L. Butler-Purry, and B. D. Russell. 2003. "Electrical Behavior of Contaminated Distribution Insulators Exposed to Natural Wetting," *IEEE Transactions on Power Delivery*, Vol. 18, No. 2 (Apr.), pp. 551–558.

Rizk, F. A. M, A. El-Arabaty, and A. El-Sarky. 1975. "Laboratory and Field Experiences with EHV Transmission Line Insulators in the Desert," *IEEE Transactions on Power Apparatus and Systems*, Vol. PAS-94, No. 5 (Sept./Oct.), pp. 1770–1776.

Robinson, R. A. and R. H. Stokes. 1965. *Electrolyte Solutions.* London: Butterworths, p. 128.

Sahin, S. and S. G. Sumnu. 2006. *Physical Properties of Foods.* New York: Springer.

Salama, M. A., R. Bartnikas, and M. M. Sallam. 1998. "A Tentative Tracking Model Based on Energy Balance Criteria," *IEEE Transactions on Power Delivery*, Vol. 13, No. 3 (July), pp. 824–833.

Shapiro, C. A., B. Kranz, M. Brumm, and B. Anderson. 1998. "Determining the Environmental Impact of Irrigating with Swine Effluent," in *Proceedings of Manure Management Conference*, Ames, IA. February 10–12.

Shariati, M. R., A. R. Moradian, M. Rezaei, and S. J. A. Vaseai. 2005. "Providing the Pollution Map in South West Provinces of Iran Based on DDG Method," in *2005 IEEE/PES Transmission and Distribution Conference & Exhibition: Asia and Pacific*, Dalian, China.

Sundararajan, R., J. Burnham, R. Carlton, E. A. Cherney, G. Couret, K. T. Eldridge, M. Farzaneh, S. D. Frazier, R. S. Gorur, R. Harness, D. Shaffner, S. Siegel, and J. Varner. 2004. "Preventive Measures to Reduce Bird Related Power Outages—Part II: Streamers and Contamination," *IEEE Transactions on Power Delivery*, Vol. 19 No. 4 (Oct.), pp. 1848–1853.

Takasu, K., T. Shindo, and N. Arai. 1988. "Natural Contamination Test of Insulators with DC Energization at Inland Areas," *IEEE Transactions on Power Delivery*, Vol. 3, No. 4 (Oct.), pp. 1847–1853.

Tam, Y.T. 1989. "Study of Freezing Precipitation and Fog Events—Data Report IV: March & April 1989," Ontario Hydro Research Division Report C89-61-K.

Taniguchi, Y., N. Arai, and Y. Imano. 1979. "Natural Contamination Tests at Noto Testing Station near Japan Sea," *IEEE Transactions on Power Apparatus and Systems*, Vol. PAS-98 No. 1 (Jan./Feb.), pp. 239–245.

Textier, C. and B. Kouadri. 1986. "Model of the Formation of a Dry Band on an NaCl-Polluted Insulation," *IEE Proceedings Part A*, Vol. 133, No. 5, pp. 285–290.

Thomas, F. W., S. B. Carpenter, and W. C. Colbaugh. 1969. "Plume Rise Estimates for Electric Generating Stations," *Philosophical Transactions of the Royal Society of London A*, Vol. 265, No. 1161, pp. 221–243.

Todd, D. K. and L. Mays. 2005. *Groundwater Hydrology*, 3rd edition. Hoboken, NJ: Wiley.

Todd, D. K., F. van der Leeden, and F. L. Troise. 1990. *The Water Encyclopedia*, 2nd edition. Chelsea, MI: Lewis Publishers.

United States Environmental Protection Agency (US EPA). 1977. "Flue Gas Desulfurization in Power Plants, Status Report." April.

United States Environmental Protection Agency [US EPA]. 1990. "Clean Air Act Amendments of 1990."

United States Environmental Protection Agency [US EPA]. 1998a. *National Air Quality and Emissions Trends Report, 1996*, EPA-454/R-97-013. Research Triangle Park, NC: Office of Air Quality Planning and Standards.

United States Environmental Protection Agency [US EPA]. 1998b. *National Air Quality and Emissions Trends Report, 1998*. Research Triangle Park, NC: Office of Air Quality Planning and Standards.

United States Environmental Protection Agency [US EPA]. 2000. "Clean Air Status and Trends Network (CASTNET)." Available at http://www.epa.gov/acidrain/castnet/sites.html. March.

Vlaar, J. 1991. "Electrical and Thermal Properties of Icicles." University of Waterloo 2B Honours Physics Report, SN 88104434.

West, H. J., J. E. Brown, and A. L. Kinyon. 1971. Simulation of EHV Transmission Line Flashovers Initiated by Bird Excretion," *IEEE Transactions on Power Apparatus and Systems*, Vol. PAS-90, No. 4 (July), pp. 1627–1630.

World Meteorological Organization (WMO). 1996. Document 8, *Guide to Meteorological Instruments and Methods of Observation*, 6th edition; and Supplement No. 1, December 1997.

Wolf, M., J. E. Walker, and J. G. Kapsalis. 1972. "Water Vapor Sorption Hysteresis in Dehydrated Foods," *Journal of Agriculture and Food Chemistry*, Vol. 20, No. 5, pp. 1073–1077.

Wu, D. and S. Su Zhiyi. 1997. "The Correlation Factor Between dc and ac Pollution Levels: Review and Proposal," in *Proceedings of 10th International Symposium on High Voltage Engineering*, Montreal, August 25–29, Vol. 3, pp. 253–256.

CHAPTER 4

INSULATOR ELECTRICAL PERFORMANCE IN POLLUTION CONDITIONS

Generally, the most important aspect of the electrical performance of an insulator is characterized by its electrical flashover strength in its intended pollution and wetting environments.

A secondary characteristic in insulator electrical performance is its ability to inhibit leakage currents, again in the intended pollution and wetting environments. Nonceramic insulators exhibiting long-term surface hydrophobicity excel in this regard above $0\,°C$ but tend to lose this property below freezing.

Power system reliability may be reduced whenever the flashover strength across an insulator falls below the electrical strength of air gaps in its local application environment. The minimum air gap strength is about 400–600 kV/m in most practical outdoor applications for a wide range of overvoltage waveshapes. The presence of a clean, dry, or wet insulator does not reduce the power-frequency strength appreciably, but test results do show many cases where contaminated insulators considerably degrade the electrical performance under service voltage stresses.

4.1. TERMINOLOGY FOR ELECTRICAL PERFORMANCE IN POLLUTION CONDITIONS

4.1.1. Terms Related to Pollution and Its Characterization

Conductivity: (1) A factor such that the conduction current density is equal to the electric field intensity in the material multiplied by the conductivity.

Insulators for Icing and Polluted Environments. By Masoud Farzaneh and William A. Chisholm
Copyright © 2009 the Institute of Electrical and Electronics Engineers, Inc.

Note: In general, it is a complex tensor quantity. (2) A macroscopic material property that relates the conduction current density (J) to the electric field ($\vec{E}$) in the medium. *Note*: For a monochromatic wave in a linear medium, that relationship is described by the (phasor) equation:

$$\vec{J} = \sigma^{=} \cdot \vec{E}$$

where $\sigma^{=}$ is a tensor, generally frequency dependent, and $\vec{J}$ is in phase with $\vec{E}$. For an isotropic medium, the tensor conductivity reduces to a complex scalar conductivity σ, in which case $\vec{J} = \sigma\vec{E}$.

Equivalent Salt Deposit Density (ESDD): A measure of the electrically conductive and water-soluble material per unit area on an insulator surface, expressed as the equivalent amount of NaCl that gives the same surface conductivity as the original deposit.

Form Factor: A dimensionless constant F given by the integral of the inverse of the insulator circumference $p(l)$ at a partial leakage (creepage) distance l along the surface, taken as l varies from the origin to the total leakage (creepage) distance L.

$$F = \int_0^L \frac{dl}{p(l)}$$

Layer Conductivity: A characteristic of wet, polluted insulators given by multiplying the layer conductance measured on the unenergized insulator by the form factor of the insulator.

Leakage Current: A component of the measured current that flows along the surface of the tool or equipment, due to the properties of the tool or equipment surface, including any surface deposit, when the device is connected as intended to the energized power system at rated voltage.

Leakage Distance (Creepage Distance): The sum of the shortest distances measured along the insulating surfaces between the conductive parts, as arranged for dry flashover tests. *Note:* Surfaces with semiconducting glaze are considered as effective leakage surfaces, and leakage distances over such surfaces should be included in the leakage distance. Creepage distance is the term commonly used in Europe.

Nonsoluble Deposit Density (NSDD): A measure of the inert, electrically nonsoluble and insoluble, material per unit area on an insulator surface.

Protected Leakage (Creepage) Distance: The leakage distance on an insulator that is not wetted in a specified precipitation condition. The protected leakage distance of an insulator in any orientation may vary with precipitation type, intensity, wind speed, and direction.

Surface Conductivity: The form factor of an insulator, divided by its resistance under high-voltage test conditions that do not lead to arcing or dry-banding.

Unified Specific Creepage Distance (USCD): For pure dc application, the specific creepage distance is the creepage (leakage) distance divided by the rated voltage for the dc system where the insulator is intended to be used. For insulators for combined voltage application, the specific creepage distance is the creepage distance divided by $Z \cdot V_d$, where Z is the number of six pulse bridges in series, and V_d is the dc-rated voltage per valve bridge. For ac applications, the USCD is the creepage distance divided by the rms value of the line-to-ground voltage for the system where the insulator is intended to be used.

4.1.2. Terms Related to In-Service Environment

Cold Fog (Freezing Fog): A fog whose droplets freeze upon contact with exposed objects and form a coating of hoarfrost and/or glaze.

Drizzle: Light precipitation consisting of liquid water drops smaller than that of rain, and generally smaller than 0.5 mm (0.02 in.) in diameter. Precipitation rates due to drizzle are on the order of less than a millimeter per hour at the ground.

Fog: Visible aggregate of minute water droplets suspended near the earth's surface. According to international definition, fog reduces visibility below 1 km. Fog differs from clouds only in that the base of fog is at the earth's surface while clouds are above its surface. Fog is easily distinguished from haze by its appreciable dampness and gray color.

Glaze Ice (Clear Ice): Type of precipitation icing resulting in pure ice accretion of density 700–900 kg/m³, sometimes with the presence of icicles underneath the wires. It very strongly adheres to objects and is difficult to knock off.

Hoarfrost: A type of low-density (<0.1 g/cm³), low-adhesion precipitation, with interlocking ice crystals (hoar crystals) formed by direct sublimation of moisture from saturated air below the freezing point onto objects of small diameter.

Mist: Visible aggregate of water droplets suspended in air, intermediate between fog and haze.

Rain: Liquid precipitation with drop size generally larger than 0.5 mm (0.02 in.) in diameter. Precipitation rates due to rain are on the order of a millimeter per hour or more at the ground.

Smog: Mixture of fog and smoke (near industrial areas) or vehicle exhaust (near heavy traffic areas).

4.1.3. High-Voltage Measurement Terminology

Accuracy: The degree of agreement of the observed value with the conventionally true value of the quantity being measured.

Critical Flashover Voltage: The amplitude voltage of a given waveshape that, under specified conditions, causes flashover through the surrounding medium on 50% of the voltage applications.

Conventional Deviation of Flashover Voltage: The difference between 50% and 16% disruptive discharge voltage.

Discharge: The passage of electricity through gaseous, liquid, or solid insulation.

Disruptive Discharge (Flashover): A discharge that completely bridges the insulation under test, reducing the voltage between the electrodes to practically zero.

Disruptive Discharge Probability (Flashover Probability): The probability that one application of a prospective voltage of a given shape and type will cause a disruptive discharge.

Disruptive Discharge Voltage (Flashover Voltage): The voltage causing the disruptive discharge for tests with direct voltage, alternating voltage, and impulse voltage chopped at or after the peak; the voltage at the instant when the disruptive discharge occurs for impulses chopped on the front.

Error: Any discrepancy between a computed, observed, or measured quantity and the true, specified, or theoretically correct value or condition.

External Insulation: Insulation that is designed for use outside buildings and for exposure to the weather.

Fifty Percent Disruptive Discharge (Critical Flashover) Voltage: The voltage that has a 50% probability of producing a disruptive discharge. *Note*: The term mostly applies to impulse tests and has significance only in cases when the loss of electric strength resulting from a disruptive discharge is temporary.

Flashover: A disruptive discharge through air around or over the surface of solid or liquid insulation, between parts of different potential or polarity, produced by the application of voltage wherein the breakdown path becomes sufficiently ionized to maintain an electric arc.

Minimum Flashover Voltage: The crest value of the lowest voltage impulse of a given waveshape and polarity that causes flashover.

Nondisruptive Discharge (Partial Discharge): A discharge between intermediate electrodes or conductors in which the voltage across the terminal electrodes is not reduced to practically zero.

Nonsustained Disruptive Discharge: A momentary disruptive discharge.

Peak Value of Alternating Voltage: The maximum value, disregarding small high-frequency oscillations (greater than 10 kHz) such as those arising from partial discharges.

p-Percent Disruptive Discharge Voltage (Flashover Voltage): The prospective value of the test voltage that has a p-percent probability of producing a disruptive discharge.

Precision: The discrepancy among individual measurements.

Random Error: Errors that have unknown magnitudes and directions and that vary with each measurement.

Scale Factor: The factor by which the output indication is multiplied to determine the measured value of the input quantity or function.

Self-restoring Insulation: Insulation that completely recovers its insulating properties after a disruptive discharge.

Systematic Error: Errors where the magnitudes and directions are constant throughout the calibration process.

Transfer Function [$H(f)$]: The quantity $Y(f)$ divided by $X(f)$, where $Y(f)$ and $X(f)$ are the frequency domain representations of the output and input signals, respectively.

Uncertainty: An estimated limit based on an evaluation of the various sources of error.

value of Test Voltage for Alternating Voltage: The peak value divided by the square root of 2, or the rms value as defined by the appropriate apparatus standard.

Voltage Ratio of a Voltage Divider: The factor by which the output voltage is multiplied to determine the measured value of the input voltage.

Withstand Probability: The probability that one application of a prospective voltage of a given shape and type will not cause a disruptive discharge.

Withstand Voltage: The prospective value of the test voltage that the equipment is capable of withstanding when tested under specified conditions.

4.2. AIR GAP BREAKDOWN

4.2.1. Air Breakdown in Uniform Field

An essentially uniform electric field is established between two energized electrodes with large surface dimensions compared to their separation gap. The electrode configuration can be parallel plates, two spheres, or hemisphere-capped rods of large radius. For constant air temperature, air pressure, moisture content, and gap separation the disruptive discharge voltage of an air gap with a uniform field is relatively constant for either polarity of dc, for the peak of the ac sine-wave voltage up to 1 kHz, and for the negative lightning impulse [e.g., IEEE Standard 4, 1995].

The breakdown levels in air have low standard deviation, typically 1% [Kuffel et al., 2000], making the sphere–sphere gap geometry an important reference method for calibrating high-voltage dc, ac, and lightning impulse test equipment. Reference tests using sphere gaps are relatively specialized, as the electrodes need to be conditioned with ten or more sparks to burn off rough points. These and other details about the use of sphere gaps for high-voltage measurements and the breakdown process are found in reference texts such as Kuffel et al. [2000] and Meek and Craggs [1978]. Standard switching-surge overvoltages, with their rise to a peak of 250 μs, tend to have more inconsistent flashover levels across sphere gaps, with a standard deviation of 6% being typical [IEC 60071-2, 1996].

For uniform-field gap breakdowns in atmospheric air, it is traditional to report the ac, dc, or lightning impulse flashover voltage as a function of the product of the gap separation d (mm) and the pressure p (bar = atmospheres = 101.3 kPa). Paeschen's law states that the breakdown voltage in a uniform field is a linear function of the product of pd at constant temperature. Dakin et al. [1974] consolidated results covering the minimum voltage for breakdown (340 V) up to 1 MV (for 450 mm) over five orders of magnitude of pd in Figure 4-1.

Paeschen's law, a linear relation of 2.5 kV/(atm-mm), holds for a useful range of gap sizes between 3 and 1000 mm. Converting to meters at atmospheric pressure, the 2500-kV/m slope for dc voltage corresponds to an ac breakdown strength of 1770 kV rms per meter, obtained by equating the peak of the ac wave with the dc breakdown value. At the upper end of the curve, a deviation from the straight line can be seen in the overlaid reference values from IEEE Standard 4 [1995]. This occurs when the spacing between sphere gaps is increased to more than a third of the sphere radius, causing a nonuniform field distribution that has lower breakdown strength.

To obtain a specific disruptive discharge voltage in a test with uniform field, it is just as important to measure the air pressure as the air gap distance. The detailed process for calculating the relative air density for any combination of ambient temperature, dew point temperature, and pressure is given in Appendix B.

While the disruptive discharge levels in uniform fields are well defined, insensitive to voltage waveshape, and useful for calibration purposes in high-voltage test laboratories, sphere gap flashover values have little meaning when carrying out high-voltage engineering studies for power system components. Conductors, insulators, and bushing terminals have nonuniform fields with breakdown voltage gradients that are, taken as an average across the entire

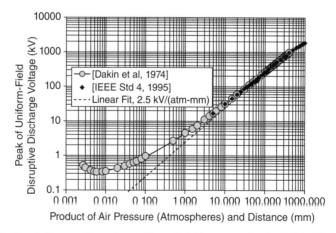

Figure 4-1: Breakdown voltage for uniform field in atmospheric air [Dakin et al., 1974; IEEE Standard 4, 1995].

gap, one-fifth to one-quarter of the 2500-kV/m gradient in a uniform field, depending on geometry and waveshape.

4.2.2. Air Breakdown in Nonuniform Field

The rod-to-plane geometry has been used in many investigations of discharge in high voltage. The electric field in this geometry is highly nonuniform, with higher field near the surface of the rod than in the center of the gap. The electric field has cylindrical symmetry but it is asymmetrical along the axis of the gap, with lower field near the ground plane. In these two ways, this geometry is most similar to the field around an insulator, attached to a small-diameter conductor or bus bar.

The asymmetrical field in a rod-to-plane gap leads to important differences in the physics of discharge that show up as differences in flashover levels as a function of polarity. Figure 4-2 shows a comparison of the 50% withstand values obtained for ac, dc, switching, and lightning impulse waves. The effect of polarity is strong for all waveshapes.

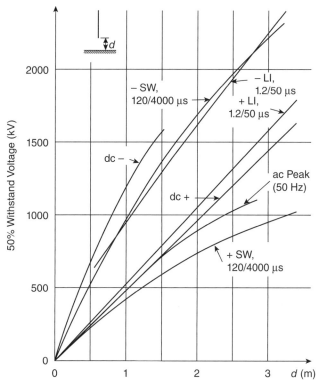

Figure 4-2: Comparison of 1-minute withstand values for dc and ac, with 50% withstand values for 120/4000 and 1.2/50 impulse flashover strength of rod-to-plane gaps (adapted from Knudsen and Iliceto [1970]).

For example, the negative dc withstand voltage is twice as high as positive dc withstand for a rod-to-plane gap of 1.5 m.

The tests of dc and ac withstand in Figure 4-2 are compared with the 50% withstand levels for switching impulse with a 120-μs rise time and 4000-μs time to half value, indicated by the standard 120/4000 notation, and with standard 1.2/50 lightning impulses. In each case, the negative polarity flashover of the rod–plane gap is higher than positive polarity.

4.2.3. Breakdown of Clean and Dry Insulators

When an insulator is attached to an energized conductor, the dielectric constants of the materials will alter the potentials around the conductor. In common with the rod-to-plane gap, however, breakdown will tend to be initiated at the area of highest electrical stress, which is the area near the energized end of the insulator where the field gradient in all directions is highest.

Figure 4-3 shows the increase in 50% flashover voltage with distance for a rod-to-plane gap with lightning impulse voltage of both polarities. In common

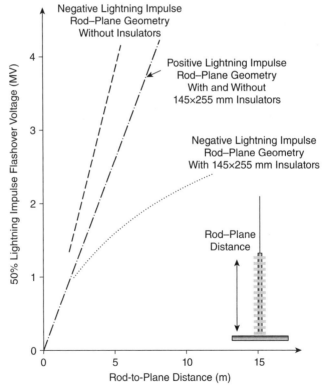

Figure 4-3: Comparison of lightning impulse flashover strength for rod-to-plane configuration, with and without cap-and-pin disk insulators between rod electrode and earth plane (adapted from Pigini et al. [1989]).

with Figure 4-2, the strength for a 3-m gap is about 1600 kV for positive polarity and 2200 kV for negative polarity.

Figure 4-3 also shows the changes in flashover level for positive and negative polarity lightning impulses when a string of cap-and-pin insulators is introduced into the air gap. The presence of the string of insulators has a negligible effect on the positive lightning impulse flashover levels of a rod-to-plane gap [e.g., Pigini et al., 1989]. In contrast, there is a very large reduction in the negative lightning impulse flashover voltage when the same string of insulators spans a gap of more than 2 m.

Under positive and negative switching impulse voltage, the effect of dry insulators in a conductor–crossarm gap is less than 3% and can be neglected [CIGRE Working Group 33.07, 1992].

Aleksandrov et al. [1962, 1965] noted that the ac flashover strength of insulator strings fell between their measurements on rod-to-plane and rod-to-rod gaps as shown in Figure 4-4.

Hileman [1999] noted that the Aleksandrov curve for ac flashover strength of insulators, with its stated standard deviation of 2%, was a very close match to the more recent expression for power-frequency critical flashover (V_c) of air gaps from IEC 60071-2 [1996]:

$$V_{c(rms)} = 750\left(1.35 \cdot k_g - 0.35 \cdot k_g^2\right)\ln\left(1 + 0.55 \cdot L^{1.2}\right) \tag{4-1}$$

where

L is the air gap distance (m), normally with $L \geq 2$ m,

k_g is the switching-surge gap factor for the particular tower window and is equal to $k_g = 1$ for the rod-to-plane gap, and

V_c is the critical flashover voltage (kV rms) at power frequency.

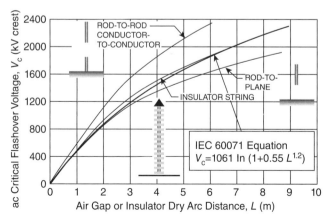

Figure 4-4: Power-frequency critical flashover of large air gaps and insulators (data adapted from Aleksandrov et al. [1962, 1965]).

Figure 4-4 plots the predicted value of ac rod-to-plane flashover, converted from rms to peak, against the Aleksandrov [1962] observations and there is a very close match when $k_g = 1$. This simplifies the estimate of the ac flashover strength of clean and dry insulators to

$$V_{c(rms)} = 750 \cdot \ln(1 + 0.55\, L^{1.2})$$
$$V_{c(peak)} = 1061 \cdot \ln(1 + 0.55\, L^{1.2})$$

$(4\text{-}2)$

where

L is the dry arc distance (m) of the insulator, and

$V_{c(rms)}$ and $V_{c(peak)}$ are, respectively, the rms and peak critical ac insulator flashover voltages (kV).

In general, the low capacitance of polymer insulators does not modify the electric fields as much as the disturbance introduced by strings of cap-and-pin ceramic disks. In many cases, polymer insulator breakdown levels are relatively close to those measured for the geometry without the insulator, but with any grading hardware such as corona rings that may reduce the overall dry arc distance.

4.2.4. Breakdown of Clean and Wet Insulators

The dc breakdown strength of insulators in wet conditions can be considerably lower than the minimum observed flashover gradient of 400 kV/m for either polarity of insulators in dry conditions [CIGRE Working Group 33.07, 1992]. Under rain conditions, the negative polarity flashover of 170–400 kV/m tends to be lower than positive polarity at 300–500 kV/m, depending on the insulator type and orientation, rain conductivity, intensity, and duration. Lambeth [1990] reported that nonuniform wetting on HVDC bushings gave extremely low flashover levels, sometimes dropping as low as 70 kV/m.

The influence of wet cap-and-pin insulators of various shapes on critical 60-Hz flashover level was reported by EPRI [2005] in the insulator length range from 0.3 to 6 m. Figure 4-5 shows that the shape can have a ±10% effect on the flashover voltage, with most of the results falling about 15% below the IEC estimate for the normal value of gap factor, $k_g = 1.2$.

The differences in the ANSI and IEC ratings relate to differences in test procedures, such as rain intensity, water conductivity, and test duration.

A factor of 0.87 ± 0.12 was suggested by the CIGRE Working Group 33.07 [1992] for the ratio between dry and wet ac flashover of insulator strings. This is in good agreement with the overall results in Figure 4-6. The value of 0.76 ± 0.06 was suggested for single post insulator columns, which have smaller shed-to-shed separations.

Figure 4-6 also shows that, for the ANSI wet test specification, the value varies from about 0.6 to 0.8 over the typical insulator string lengths of 1–3 m.

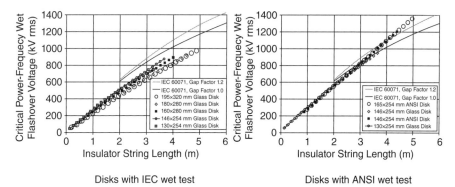

Disks with IEC wet test   Disks with ANSI wet test

Figure 4-5: Comparison of results from Equation 4-1 [IEC 60071-2, 1996] with catalog values of wet ac flashover strength of cap-and-pin insulators of various shapes.

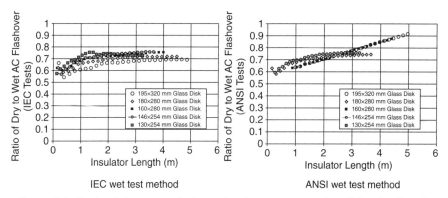

IEC wet test method   ANSI wet test method

Figure 4-6: Ratio of dry to wet flashover level versus suspension disk string length.

The IEC wet test gives a ratio of dry to wet flashover that is relatively constant at about 0.7 for longer string lengths.

4.3. BREAKDOWN OF POLLUTED INSULATORS

4.3.1. Breakdown Process on a Contaminated Hydrophilic Surface

Flashover on polluted surfaces of hydrophilic insulators (those that do not bead water) often involves a series of steps [Nasser, 1972; Le Roy et al., 1984], illustrated using Figure 4-7 on a simplified flat geometry. This geometry has been used in many series of experiments to establish the general nature and specific arc parameters for various wetted and polluted surfaces.

1. Deposit of a pollution layer on insulator surfaces from exposure to sea or road salt, insoluble dust, and various forms of industrial pollution

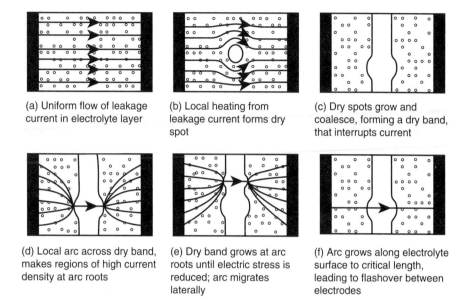

(a) Uniform flow of leakage current in electrolyte layer

(b) Local heating from leakage current forms dry spot

(c) Dry spots grow and coalesce, forming a dry band, that interrupts current

(d) Local arc across dry band, makes regions of high current density at arc roots

(e) Dry band grows at arc roots until electric stress is reduced; arc migrates laterally

(f) Arc grows along electrolyte surface to critical length, leading to flashover between electrodes

Figure 4-7: Process of arc formation on a polluted hydrophilic insulator surface (adapted from Le Roy et al. [1984]; courtesy of Eyrolles).

such as sulfates from coal-fired electric power plants, as described in Chapter 3.

2. Wetting of the pollution layer under atmospheric conditions that may include fog, drizzle, dew, rain, snow, or ice. A dry pollution layer has high resistivity, and consequently no significant effects on the electrical performance of insulators. It is only when a pollution layer is wet that the resulting electrolyte has sufficient conductivity to support the flow of measurable leakage current flowing on the insulator surface, shown in Figure 4-7a.

3. Small perturbations in the flow of leakage current lead to areas of higher current density and increased local heating effects, resulting in the formation of dry points, Figure 4-7b. Multiple dry points may spread and coalesce to form a single dry band, Figure 4-7c.

4. The conductivity of a dry band is much lower than that of the electrolyte layer. The voltage distribution, which may have originally been uniform, becomes nonuniform along the insulator surface. Almost all of the voltage applied to the insulator now appears across the dry band. If the electric field given by the voltage drop across the dry band, divided by the dry-band width, is high enough, a breakdown will occur, resulting in formation of a local arc across the dry band as shown in Figure 4-7d.

5. The dry band will grow from Joule and radiation heating near the arc roots. The local arc may move laterally to an area with higher electric

field stress, Figure 4-7e. The local arc may also move along the electrolyte surface, eventually causing a flashover as at a critical length shown in Figure 4-7f. The local arc may also extinguish, putting the process back at a state similar to that in Figure 4-7c.

The last step plays a very important role in the flashover on polluted surfaces and will be the focus of modeling in Chapter 5.

4.3.2. Breakdown Process on a Contaminated Hydrophobic Surface

Some nonceramic insulator materials are difficult to wet fully. The polluted surfaces of hydrophobic insulators have a different flashover mechanism, illustrated in Figure 4-8.

Environmental wetting produces water droplets (Figure 4-8a) rather than a continuous conductive layer of electrolyte, as found on hydrophilic surfaces. Pollution diffuses through the thin layer of surface oil and dissolves in the water droplets, which then become conductive.

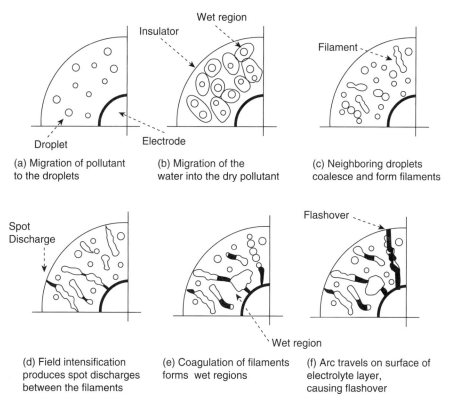

Figure 4-8: Process of arc formation on a polluted hydrophobic insulator surface (adapted from Karady et al. [1995]).

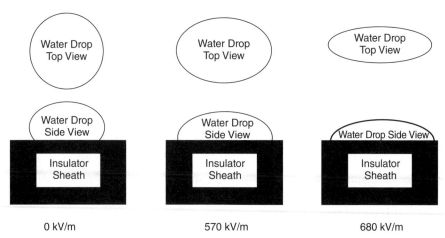

Figure 4-9: Deformation of water drop on sheath of NCI (from Phillips et al. [1999]).

The wet region around each water drop grows by diffusion, causing areas of increased conductivity as shown in Figure 4-8b. Diffusion continues to the point that wet areas coalesce and small amounts of leakage currents start to flow in several different paths. The Joule heat dries some wet areas and eventually an equilibrium is reached between evaporation and wetting. Low conductivity of the polymer surface persists between wet areas.

Interaction between the electric field and the droplets tends to elongate them into filaments as shown in Figure 4-8c and also flattens them out across the surface. Phillips et al. [1999] actually measured the electric field at which this flattening starts to occur and established a sharp threshold above 570–580 kV/m as shown in Figure 4-9.

Field intensification at the tips of each filament produces spot discharges that are randomly distributed along the insulator surface as shown in Figure 4-8d.

Surface discharges can erode the hydrophobicity, leading to irregular wetted areas. The electric fields continue to flatten the water drops, furthering this wetting process as shown in Figure 4-8e. Wetted regions remain surrounded by low-conductivity layers that are still covered by droplets.

Finally, the combination of filament growth and formation of wet areas may eventually short out the insulator with a conductive, electrolytic path. An arc can travel along this electrolyte path, causing flashover in Figure 4-8f.

4.3.3. Complications in the Process for Real Insulators

On real insulators, the process of pollution flashover may be much more complex than illustrated in Figure 4-7 or 4-8.

There are usually a large number of dry bands, rather than one. For example, because of the higher current density there, a dry band will often form near the pin of each cap-and-pin disk insulator in a chain. The effects of multiple arcs and dry bands in series should be considered in these cases. The multiarc model forms an important part of Chapter 6. The effect of multiple arcs on an electrical performance model can be summarized briefly: each arc in series will add to flashover voltage, making disk insulators perform somewhat better than uniform post insulators of the same leakage distance.

Insulators with a high ratio of leakage to dry arc distance may have sheds spaced close enough that the air space between sheds, rather than the dry bands, support local arcing. These sheds are said to be inefficient [Looms, 1988]. The effect on insulator performance can often be discerned in test results. Flashover voltages for various insulators will be reported in this chapter as a function of ESDD, with higher values of ESDD leading to lower flashover levels. Insulators with stable wetting (i.e., high NSDD or cold fog) and/or efficient sheds tend to have a flashover voltage that is proportional to $(ESDD)^{-1/3}$. Test results on insulators with uneven wetting or lower shed efficiency may show a relation closer to $(ESDD)^{-1/4}$ [Le Roy et al., 1984] or $(ESDD)^{-0.22}$ [CIGRE Task Force 33.04.01, 2000].

Before presenting test results for flashover levels along polluted insulator surfaces, it is also important to have a perspective of the air gap breakdown strength, and also an understanding about the ways in which a clean dielectric surface modifies this strength.

4.4. OUTDOOR EXPOSURE TEST METHODS

The first outdoor exposure test methods were simply to construct a power transmission line with a given insulator, and then to fit more insulators in problem areas. Utilities in the 1930s realized that this haphazard approach did not lead to optimal solutions that could be used to engineer new lines. A number of test stations were established, and many of the findings were documented in the technical literature of the period 1935–1975 [Forrest, 1936; Ely et al., 1971].

4.4.1. Field Observations of Leakage Current Activity

It has been common for the electrical utilities to construct exposure stations near the seacoast or known pollution sources. Examples include South Padre Island (Texas, USA), Moss Landing (California, USA), Brighton and Dungeness (UK), Noto, Akita, and Takeyama (Japan), Koeberg and Sasolburg (South Africa), Anneberg (Sweden), Glogow (Poland), Martigues (France), and St. Remy des Landes (France) and many others. These have proved to be excellent for establishing cumulative damage rates on existing and new insulation systems because the pollution exposure rates are relatively constant. Knowledge gained from test sites in England and France led directly to development of artificial test methods [e.g., Lambeth et al., 1970]. Also, while the unknown climate variables such as sunshine, wind, salt exposure rate, and temperature are uncontrolled, over the long term the exposure will be similar to that experienced by each local utility.

It is unfortunately difficult to relate exposure conditions at one site to those at other locations, even for moderate temperatures. In cold conditions, the relationships are even weaker. As the most important example, at most test sites, the bottom-surface contamination levels are considerably higher than top-surface levels. Frequent rain or sea spray tends to wet and wash contamination from upward-facing surfaces of the insulators, leaving the bottom surfaces more contaminated. In contrast, the winter exposure conditions generally involve long periods of pollution accumulation in cold conditions, leading to higher levels of contamination on upward-facing surfaces.

4.4.2. Field Observations of Flashover Performance

In some cases, an experiment setup to measure natural surface resistance or leakage current is plagued by the fact that, under some conditions, the insulator will flash over. This will tend to destroy any sensitive measurement equipment. An alternative exposure method takes better advantage of these flashovers. A string of insulators is mounted and placed at line potential. Independent ground connections to each insulator cap at the ground end are fused with explosive disconnect devices (Figure 4-10). Each time an insulator flashes over, its ground lead drops away, adding an additional insulator to the chain.

This approach gives the required insulation level directly at reasonable cost, although the high-voltage supply needs to have enough short-circuit current

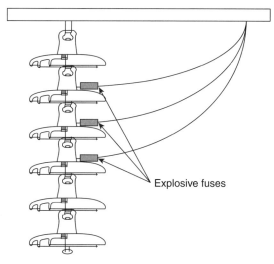

Explosive fuses

Figure 4-10: Use of explosive fuses to establish a suitable length of insulator strings.

to meet the requirements of [IEC 60507, 1991; IEEE Standard 4, 1995]. Unless video monitoring is carried out at the same time, however, the method does not give much information about the mechanism of flashover.

4.4.3. Field Observations of Other Variables

Originally, utilities established test stations for internal objectives, in order to rank performance of different insulators. However, test stations' roles have expanded, to provide exposure tests to external users such as manufacturers of insulators, arresters, cable terminations, or other apparatus.

Modern stations collect continuous readings of leakage current, cumulative charge, and meteorological variables to normalize exposure periods over time. For nonceramic insulators, the cumulative charge rather than the amplitude of leakage currents has proved to be a better indicator of surface degradation. Also, automatic systems may be used to measure the insulator contamination level every day. These systems usually immerse sample insulators into a steam environment, and then measure the leakage resistance at a low voltage that does not cause arcing activity.

4.4.4. Observations at Croydon, United Kingdom, 1934–1936

Problems with the performance of 132-kV line insulators under fog and cold-fog conditions were studied using leakage current measurements by Forrest [1936]. Twelve different insulator profiles were tested, including the traditional cap-and-pin disk of 254-mm diameter and 127-mm spacing with three ribs on the undersurface. Leakage currents were monitored on 30 insulator strings at the CEGB Croydon station, where there was a high incidence of fog, a power station with a cooling tower, a garbage incinerator, and a "gas works" that all provided high levels of atmospheric pollution.

Forrest noted that rain increased the leakage current on the insulator from about 0.3 to 1.1 mA, with an increase in power loss from 10 to 90 W at line voltage. He also set out the term "fog surging" to define the leakage current under fog conditions. At irregular intervals on the order of a minute, the leakage currents increased from their steady-state 1-mA condition to levels of 5–100 mA for fractions of a second. Forrest also demonstrated that the combination of fog and frost led to very severe surging, and he mentioned similar effects when melting snow was blown against contaminated insulators in high wind. Two flashovers occurred on a string of eleven, 254×127-m² disks under the following conditions:

- Snow and sleet, relative humidity 75–80%, temperature 2–3 °C.
- Fog and frost, relative humidity 90–100%, temperature 1 °C to –1 °C.

Using telephone relays with their internal springs retensioned to give 20-mA trip levels, Forrest measured 6000 surges on standard suspension insulator strings in a 1-week period of mist and fog in December 1935. This was compared to 2000 surges measured on anti-fog designs and less than 10 surges on standard tension strings, located horizontally.

Many other utilities have followed the monitoring principles set out by Forrest:

- Location near severe pollution sources.
- Use of fixed line-voltage stress.
- Wide dynamic range for leakage current, from 0.2 to 100 mA.
- Use of oscilloscopes to monitor leakage current as a function of time and also as a function of applied voltage, producing an elliptical Lissajous figure from the magnitude, phase angle, turn-on, and extinction voltages.
- Slow increase of applied voltage after flashover to avoid closing-in faults.
- Surge counters with 20-mA threshold to simplify data analysis.

Forrest's criterion, "a satisfactory insulator should not give rise to leakage-current surges having an amplitude of more than 10 mA at cleaning intervals of not less than 1 year," has proved to be rather restrictive. Even in 1936, the nine pages of discussions to Forrest's [1936] paper show that debates on insulator selection criteria using flashover incidence at line voltage, rather than leakage current magnitude, were well underway. For example, Ryle [1931] ranked the observed flashover performance of similar insulators in similar polluted conditions (Newcastle, UK) in a different order. Generally, high leakage currents in cold-fog conditions indicate that dry-banding is occurring at a rate that introduces considerable heating power, on the order of hundreds of watts per insulator disk. While intense radio and acoustic noise are consequences of this activity, insulators that sustain high and stable leakage current are less likely to flash over at line voltage.

4.4.5. Observations at Croydon, United Kingdom, 1942–1958

A summary of British practice in evaluating insulators under severe outdoor pollution conditions is given in Forrest et al. [1959]. This was developed at the Croydon Insulator Testing Station, where the solid deposition rate was measured to be 76–78 tons/mi^2 per month, equivalent to an accumulation rate of 27 g/m^2 per month. This is about three times the acceptable dustfall level now set by the Ministry of Environment in Canada. The solid deposit rate at a semirural area nearby, Leatherhead, was measured to be 3.2 g/m^2 per month. Sulfur concentrations were measured to be 70 and 20 parts per billion (ppb) at Croydon and Leatherhead, respectively.

The monitoring method used during this period employed electromechanical relay surge counters that operated at a leakage current of 25 mA, along with relay flag indicators for currents in excess of 150 mA. The station was established to test the susceptibility of insulator designs to flashover. Line-to-

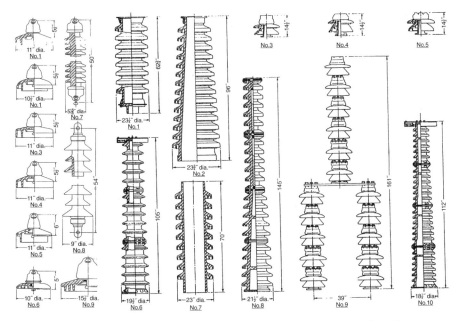

Figure 4-11: Insulator profiles tested at Croydon (from Forrest et al. [1959]; courtesy of IET).

ground test voltages were 85 kV ac, 175 kV ac, 231 kV ac, and 115 kV dc. The dc source had a filter capacitance of 0.19 μF. Figure 4-11 shows the disk insulators with ribs on the bottom surface, the long-rod insulators common in Europe at that time, and the station post and pedestal insulators that were tested in the 1950s.

Figure 4-12 plots the times to flashover observed by Forrest et al. [1959], covering exposure periods up to 5 years in duration, for the Croydon test site. Stress on the horizontal axis is expressed using the line-to-ground rms voltage, divided by the leakage distance of each type of insulator. The results were sorted into two main insulator types—ceramic suspension strings of standard or anti-fog profile, and long-rod porcelain or post insulators with a higher form factor.

There is a general downward trend in the time to flashover after installation with increasing voltage stress per meter of leakage distance. This led Forrest and co-workers to conclude that the leakage path length was the main factor in determining insulator performance in the polluted atmosphere. They recommended a ratio of leakage to dry arc distance of 3 for anti-fog suspension insulators.

The results of flashover and withstand performance on ceramic insulators of all types from the Croydon test site are compared in Figure 4-13 for 85- and 231-kV ac tests, compared to results with negative-polarity 115-kV dc excitation. Based on the short times to flashover, Forrest et al. [1959] concluded that 30% more leakage distance would be required for the dc application.

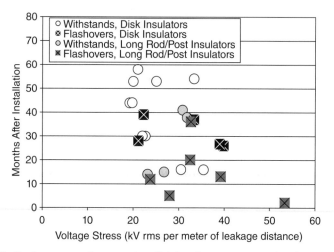

Figure 4-12: Performance of suspension disk insulators compared to long-rod and post insulators at Croydon pollution test station.

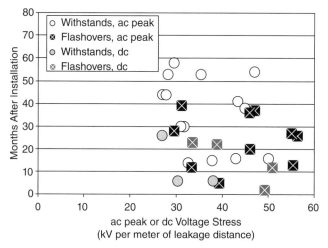

Figure 4-13: Comparison of months to flashover for ac and dc excitation of ceramic insulators at Croydon test site.

In the years since 1955, when they were tested at Croydon, some of the palliative measures suggested by Forrest et al. [1959], specifically surface treatments to increase resistivity (greasing or oiling) and treatments to decrease resistivity (semiconducting glaze), have been refined to be practical solutions to local contamination problems. These measures will be discussed in Chapter 6.

4.4.6. Observations at Brighton, United Kingdom

Many thermal and nuclear power plants draw their cooling water from the sea. Reliable operation of the substations and transmission lines that serve these plants thus requires insulators that will withstand marine exposure. Test stands at Brighton, United Kingdom [Ely et al., 1971] and Martigues, France [Lambeth et al., 1970] are good examples of utility commitment to insulator testing.

At Brighton, the main 20×20-m^2 test area is located about 65 m north of the high tide level. Insulators are normally mounted 2 m above ground and energized with a 900-kV, 30-A supply at normal 400-kV system operating voltage. Up to the 1970s, Ely et al. [1971] noted that the use of electromechanical counters with operating currents of 25 ± 1 mA rms and 150 ± 5 mA rms were used to monitor long-term activity. In the early 1970s, increasing use was made of leakage current monitoring using oscilloscopes, as reported, for example, in Figure 4-14.

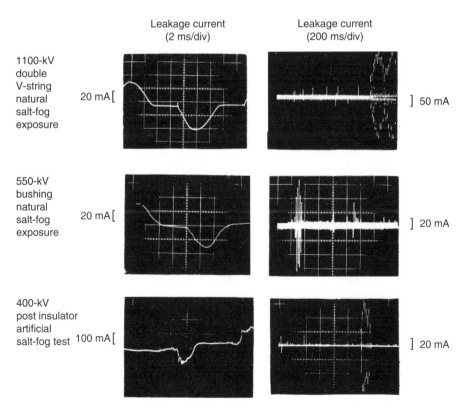

Figure 4-14: Discharge current waveforms in natural (Brighton) and artificial salt-fog exposure (adapted from Lambeth [1973]).

4.4.7. Observations at Martigues, France

The Martigues site was established by Electricité de France (EDF) in 1968 near Marseille in France. This test site has a combination of sea-salt exposure from the Mediterranean Sea and industrial pollution from a nearby power station and petrochemical industries. Starting in the early 1990s, a 24-kV test stand has been used to compare the performance of polymeric insulators [Spangenberg and Riquel, 1997]. For evaluation of material performance, insulator exposure is normalized, often using the stress levels specified in IEC 60815 [1986] as unified (phase-to-phase) creepage distance levels of 25, 20, and 16 mm/kV. In these types of exposure tests, visual observations of the extent and nature of hydrophobicity are also carried out and compared with results of artificial aging tests in the EDF laboratories.

Generally, test results from Martigues have shown that discharge activity on ceramic insulators can be minimized with the use of silicone coatings, and that semiconducting glaze insulators also offer improvements in performance. However, the goal of developing a single, comprehensive artificial aging method that accelerates the natural exposure of polymer insulators has, so far, not been achieved [CIGRE Task Force 33.04.07, 1999].

4.4.8. Observations at Enel Sites in Italy

For certain regions in Italy, strong winds carry salt pollution from the coast inland to a distance of about 50 km [Cortina et al., 1976]. This, compounded with industrial pollution in the north of the country, led to studies of insulator pollution performance at four locations: Sangione, near Torino; Brugherio, near Milan; Porto Marghera, near Venice; and Santa Caterina, near Cagliari. The first three sites had industrial pollution and the last two sites had marine pollution as well. Leakage current peaks and total charge were measured for six different shapes of porcelain and glass disks and four profiles of porcelain long-rod insulators.

One significant finding was the comparison of the layer conductivity that accumulated on three insulator profiles, called K (μS) in Figure 4-15. The 90th and 99th percentile levels of conductivity on the naturally exposed long-rod insulator were about four times higher than the levels on standard disks and eight times higher than the levels on the anti-fog disks. The high surface conductivity led directly to increased levels of leakage current activity on the long-rod units, and to lower levels of activity on the anti-fog strings, compared to strings of standard disks.

4.4.9. Observations at Noto, Akita, and Takeyama, Japan

The Central Research Institute of Electric Power Industry (CRIEPI) in Japan began investigation of accumulation of ocean salt pollution at their Takeyama test yard on the Pacific coast, south of Tokyo, at 500-kV ac system voltage

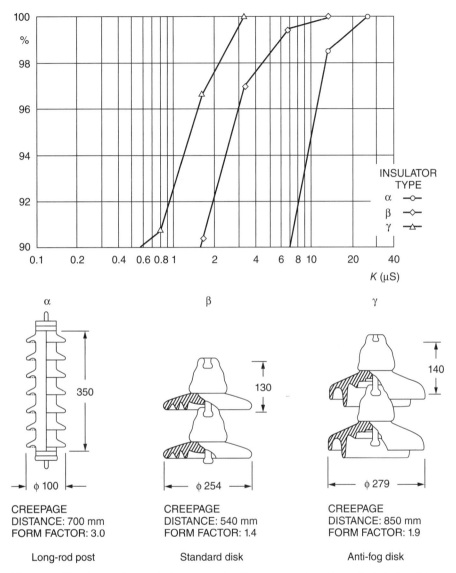

Figure 4-15: Comparison of layer conductivity on naturally polluted insulators [adapted from Cortina et al., 1976].

starting in 1967. The Noto and Akita stations are located on the Sea of Japan and testing began with 200 kV ac in 1971. Insulator ESDD levels were determined with automatic systems that wetted a five-standard-disk insulator chain each day and recorded the value of surface conductance of the middle three insulators. The sites were characterized by coefficients c in $(mg/cm^2/(m/s)^3$-h) that relate the sum of hourly average wind velocity to the ESDD on the

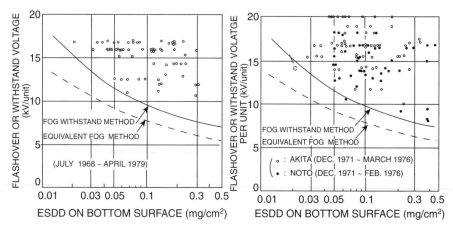

Figure 4-16: Flashover voltage characteristics of suspension disk insulators at CRIEPI Takeyama (left) and Noto and Akita (right) testing stations (from Arai [1982]).

bottom surfaces of standard disk insulators. The values of c were found to be 8×10^{-6} for the Noto site, 5×10^{-6} for Akita, and 1×10^{-6} for Takeyama.

Arai [1982] also remarked on the flashover of 12 disks at the Noto station under a stress of 16.7 kV per disk in snow conditions. The top-surface ESDD was found to be 0.016 mg/cm², with 0.076 mg/cm² on the bottom surface at a temperature of −2.5 °C with 5-m/s wind speed from the land side.

Typically, flashover voltages recorded outdoors at the CRIEPI test yards had wide dispersions, as shown in Figure 4-16 [Arai, 1982]. All results from natural sea-salt exposure were above performance curves established with an artificial fog test for wetting the bottom-surface pollution. This tendency continued, as shown in the consolidated results [Naito et al, 1990] for all stations in Figure 4-17.

The dispersion in natural flashover results did not average out with time. Naito et al. [1990] calculated a standard deviation of insulator flashover voltage of about 20%, compared with a value of 6% for withstand voltage in artificial tests.

4.5. INDOOR TEST METHODS FOR POLLUTION FLASHOVERS

Utilities with specific concerns about contamination problems, and sufficient resources, can invest in local insulator exposure test stations using their anticipated insulation systems at service voltage stress. For more general power system design, the experiences of test stations have been distilled into artificial test methods that reproduce marine and industrial pollution exposure and also the stress involved in heavy rain. The standard artificial tests are designed to be [CIGRE Task Force 33.04.07, 1999]:

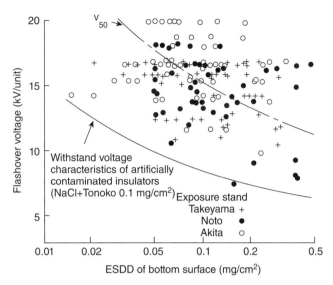

Figure 4-17: Flashover voltage characteristics of standard $146 \times 250\text{-mm}^2$ suspension disk insulators at CRIEPI Takeyama, Noto, and Akita testing stations (adapted from Naito et al. [1990]).

- *Representative:* The tests are carried out with the same range of salt conductivity, insulator precontamination, and rainwater conductivity as found in field exposure tests. The flashover test regime is typically within 30% of normal service voltage for contamination flashover tests.

- *Repeatable:* Expressed by the relative standard deviation of flashover results. For contamination tests, this is about 7% [Lambeth et al., 1987]. There are very few outdoor locations in the world that give consistent and repeatable natural wetting over a period of days or weeks [Kawai and Milone, 1969].

- *Reproducible:* The test methods have been reproduced at transmission-class voltages at many laboratories, with results summarized for a variety of materials and ac and dc voltage in Naito and Schneider [1995] and Baker et al. [1989].

- *Cost Effective:* Each test calls for withstand or, at most, two flashovers, which may destroy the pollution deposit [IEEE Standard 4, 1995] but typically do not damage the insulator under test. It takes at least five effective tests to establish a value of withstand voltage [IEC 60507, 1991]. The investment in high-voltage laboratory test time will typically be the same order, regardless of the voltage level, although it can be practical to test several small insulators in parallel. The business case for an insulator testing program can be especially successful for HV and EHV stations, where the cost of each insulator is relatively high and the consequences of flashover are most severe.

In order to normalize insulator exposure conditions, a number of international test standards have been developed. Though first developed for ceramic insulators, these are now intended to be used on all types of insulators including nonceramic ones. For example, IEEE Standard 4 and IEC 60507, standard electrical test methods for insulators, describe three types of contamination tests:

- *Salt-fog* tests, where salt aerosol with conductivity ranging from 4300 to 200,000 µS/cm is sprayed onto energized insulators.
- *Clean-fog* tests, where precontaminated insulators with salt deposit density of 25–400 µg/cm² are energized and then wetted with steam or water aerosol.
- *Heavy rain* tests, where water with controlled 100-µS/cm conductivity and rain rate is applied for a period of several minutes, then the insulator is energized.

Testing standards provide specific technical requirements for source impedance, pollution levels, and other factors that ensure repeatability from test lab to test lab. These requirements are found in Section 5.4.

The salt-fog and clean-fog test methods reproduce important aspects of contamination flashover performance on ceramic insulators and these are discussed in detail in Sections 5.2 and 5.3. With its strong relation to icing test methods, the heavy rain test is discussed in Chapter 8.

Flashover testing of nonceramic insulators calls for some changes to the traditional test methods, mainly to ensure that a uniform pollution layer is applied to hydrophobic surfaces. This will be covered in Section 5.5.

Some of the specialized tests for tracking, erosion, and accelerated aging of polymer materials have similar features to contamination flashover tests. This testing domain is important, but aging tests are not covered here. Readers are referred to CIGRE Task Force 33.04.07 [1999], IEC 61109 [1992], and IEC 62217 [2005] as an introduction to the literature and issues.

At present, three types of contamination tests for freezing conditions are derived from standard tests:

- The *cold-fog* test, derived from the clean-fog test, where a thin layer of ice wets and stabilizes the pollution on precontaminated insulators.
- The *icing* test, a variation of the heavy rain test, where the chamber temperature is reduced below freezing to produce a thin or thick ice layer.
- The *snow* test, also similar to the heavy rain test, where a thick layer of natural snow is packed systematically onto an insulator.

4.5.1. Comparison of Natural and Artificial Pollution Tests

Artificial pollution test methods can only be considered successful if they reproduce the arcing activity that would occur in the natural environment.

The arcing is characterized by several factors—the number of disks or sheds with activity, the number of arcing roots in parallel, the repetition rate, and the magnitude of leakage current peaks being the most important.

At present, artificial pollution tests for ceramic insulators have achieved this goal of reproducibility, as detailed in Table 4-1, but the same tests for polymer insulators have not succeeded in this way. Standards groups [IEC 62217, 2005] note that the results of pollution tests according to IEC 60507 or IEC 61245 on insulators made of polymeric materials do not correlate with experience obtained from service. This means that specific pollution tests for polymeric insulators are still under consideration.

In contrast, the heavy rain test and its related adaptations for ice and snow conditions would seem to be well developed and the heavy rain test has been accepted for both ceramic and polymer insulators.

4.5.2. Power Supply Characteristics for Pollution Tests

The technical requirements for source impedance, accuracy, and stability of the voltage supply pollution are set up to ensure that results are repeatable from test lab to test lab. These requirements are much more stringent than those needed to establish corona inception or radio influence voltage levels. The need for high-power sources was well illustrated by Garcia et al. [1991] in a comparison test of similarly contaminated insulators using two different sources, one rated at 150 kV/166 mA fed from a 25-kVA and the other at 200 kV/1000 mA continuous duty and fed from a 2550-kVA regulator (Figure 4-18).

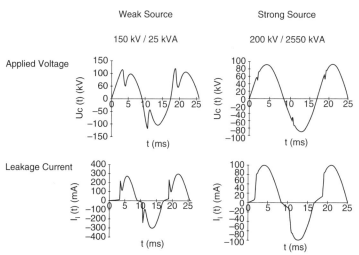

Figure 4-18: Distortion of supply voltage waveform during pollution testing (from Garcia et al. [1991]).

TABLE 4-1: Comparison of Processes in Natural and Artificial Contamination Tests on Ceramic Disk Insulators

Process	Natural Contamination	Salt-Fog Test	Clean-Fog Test	Cold-Fog Test
Clean insulator surface		Clean insulator surface	Clean insulator surface	Clean insulator surface
Deposit of contamination layer		None	Dried slurry of inert material and salt	Dried slurry of inert material and salt
Preferential cleaning of top surface		None	Not simulated (except DCM)	Does not occur ($T < 0\,°C$)
Wetting of contamination layer		Wetting by saline solution, 20 min	Wetting by condensation, 20–40 min	Accumulation of thin ice layer
Formation of dry bands		Typical current of 1–20 mA heats solution, leaves dry salt residue	Typical current of 1–20 mA, stabilized by nonsoluble deposit	Flash melting as thin ice layer warms up
Dry-band arcing		Typical current peaks: 20–200 mA	Typical current peaks: 20–200 mA	Flashover or withstand soon after arcing occurs
Growth or quenching of dry bands		Typical current peaks: 200 mA to 2 A	Typical current peaks: 200 mA to 2 A	Flashover or withstand soon after arcing occurs
Flashover of disk		Within 60 min if a current peak exceeds threshold	Within 15 min if a current peak exceeds threshold	Within 1 min if a current peak exceeds threshold

Source: Adapted from EPRI [2005].

As anticipated by the modeling work of Rizk and Nguyen [1984] and Rizk and Bourdages [1985], the interaction between the arcing activity on the contaminated insulator and the voltage is strong enough to compromise test results obtained when using a weak source.

In testing standards [IEEE Standard 4, 1995], the ac power supply minimum short-circuit current requirements for contamination testing are spelled out in Clause 15.3 in units of amps versus (mm per kV of line-to-ground voltage) rather than the IEC 60815 [1986] units of specific creepage distance (SCD). In a recent revision of the IEC application standard [IEC 60815, 2008], the use of unified specific creepage distance (USCD) and line-to-ground voltage is harmonized with IEEE practice.

- Every pollution test should be carried out with a supply with a short-circuit current $I_{SC} \geq 6$ A rms, a resistance/reactance ratio $R/X > 0.1$, and a capacitive current / short-circuit current ratio of $0.001 \leq I_c/I_{SC} \leq 0.1$.
- For insulators with levels of pollution in [IEC 60815, 1986] Site Pollution Severity (SPS) classes c and d (medium and heavy), $16 \leq SCD < 24.2$ mm per kV of line-to-line (L-L) voltage ($27.7 < USCD < 42$ mm per kV of line-to-ground (l-g) test voltage), the short-circuit current should exceed

$$I_{SC} \geq \frac{SCD \cdot 1\,\text{A}}{1\,\text{mm/kV}_{\text{line-line}}} - 10\,\text{A}$$

- For insulators with SCD > 24.2 mm/kV$_{l\text{-}l}$ (USCD > 42 mm/kV$_{l\text{-}g}$) and for supplies that do not comply with the requirements for SPS classes c and d, verification of a withstand level can still be carried out by measuring the peak value of the highest leakage current pulse, $I_{h\,max}$. If the short-circuit current $I_{SC} \geq 11$ ($I_{h\,max}$) then the test is valid. This means that, for a 6-A supply, no leakage current peak can exceed 545 mA.

Cascade-connected transformers have been widely used in many high-voltage laboratories to study EHV and UHV flashovers. These arrangements can have three problems—unequal voltage distribution, resonance at 60-Hz harmonics in the 500–800-Hz range, and compromised short-circuit capability. There are a number of analysis and control strategies that can identify and ameliorate some of these problems [Train and Vohl, 1976; Olivier et al., 1984].

Resonant high-voltage ac supplies rely on the capacitance of the test object as stable circuit element in a tuned circuit. Their main use is to excite long cables, sometimes at frequencies that fall outside the 45–65-Hz range permitted for contamination testing. More seriously, resonant test sets tend to fall out of tune with fast-changing leakage currents and cannot generally meet the voltage regulation requirements for contamination testing.

There is a wide degree of sophistication in HVDC test supplies, ranging from simple half-wave rectifiers with large capacitors to complex feedback-controlled SCR systems. Instead of a single prescription for dc contamination testing, the power supply requirements have the following performance criteria in Clause 15.4 of IEEE Standard 4 [1995]:

- Ripple factor of less than 3% with 100-mA load.
- Relative voltage drop of less than 10% from leakage current activity in any withstand test.
- Relative overshoot after extinction of electrical discharge of less than 10%.

With the sensitivity of high-voltage dc test sources to load release effects, flashovers that occur when the relative overshoot of the voltage is between 5% and 10% are not considered valid.

4.5.3. Electrical Clearances in Test Chamber

For indoor tests, as the dimensions of climate rooms are generally limited, the level of test voltage is normally restricted by the clearance of insulators from the climate room walls and from the floor. The electrical clearance should be at least 15% greater than the positive-polarity switching-impulse clearance for the highest test voltage. IEEE Standard 4 [1995] recommends the minimum clearance d as a function of the peak line-to-ground test voltage V_{Pk} of

$$d \geq \frac{8 \cdot V_{Pk}}{2890 \text{ kV} - V_{Pk}} \tag{4-3}$$

where d is in meters (m) and V_{Pk} is in kilovolts (kV).

Figure 4-19 shows that the required clearances have a nonlinear relation to peak voltage.

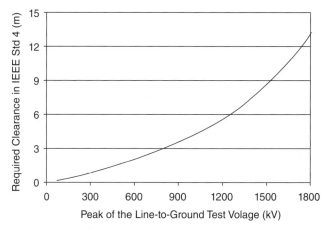

Figure 4-19: Recommended clearances from insulator to environmental chamber.

For peak voltages above 1800 kV, a more detailed calculation of the neces-sary clearance should be carried out using an appropriate gap factor in a modern switching-surge flashover model such as that in Rizk [1989].

4.6. SALT-FOG TEST

4.6.1. Description of Salt-Fog Test Method

The artificial salt-fog test was developed in Europe in the 1960s as a way of consolidating test stand marine exposure experience in the United Kingdom [e.g., Houlgate et al., 1990] and France [e.g., Holtzhausen, 1993]. The salt pol-lution layer does not contain insoluble material in marine exposure and it is applied wet.

The insulator is initially cleaned and mounted in its normal operating con-figuration in a sealed room. It is subjected to a salt aerosol with conductivity ranging from 4300 to 200,000 μS/cm. The construction of the atomizers, the air and saline solution flow rates, and the number and orientation of atomizers relative to the insulator are fully specified in IEC 60507 [1991] and IEEE Standard 4 [1995]. The flow rate to each nozzle should be 500 mL/min (0.5 dm^3/min ± 0.05 dm^3) as in IEC 60507 and CIGRE Task Force 33.04.07 [1999], which is 1000 times higher than the level of 0.5 mL/min (0.5 cm^3/min ± 0.05 cm^3) given in IEEE Standard 4 [1995]. This discrepancy may be a unit error (dm versus cm) that has not been noted because the salt-fog test is not widely used in North America.

In the salt-fog tests, the rated line-to-ground service voltage is normally applied for all tests. Salinity of the salt spray is increased or decreased by a factor of $\sqrt{2}$, called an "up-and-down" method, depending on whether the test result is a withstand or flashover, respectively. The test duration is a maximum of 1 hour before declaring the result to be a withstand. The instruc-tions for interpreting withstand salinity values are precisely detailed in the IEC and IEEE standards.

4.6.2. Validation of Salt-Fog Test Method

The salt-fog test method was found to be satisfactory for a number of reasons. Flashover levels were achieved at salinity levels similar to the 60-kg/m^3 (or 60-g/L) level of natural marine pollution measured at insulator test sites in England. Also, Figure 4-20 from Lambeth et al. [1973] shows that leakage currents in the salt-fog test reached the same equilibrium levels under normal service voltage as those observed in natural conditions.

The power supply requirements for the salt-fog test were evaluated as well. The peak leakage currents in Figure 4-21 were found to approach 800 mA prior to flashover for four insulator profiles with different withstand salinities.

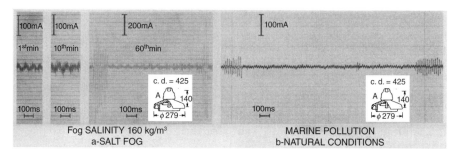

Figure 4-20: Comparison of leakage current activity in artificial salt-fog test and natural conditions for string of nine, $140 \times 279 \, \text{mm}^2$ cap-and-pin disk insulators (from Lambeth et al. [1973]).

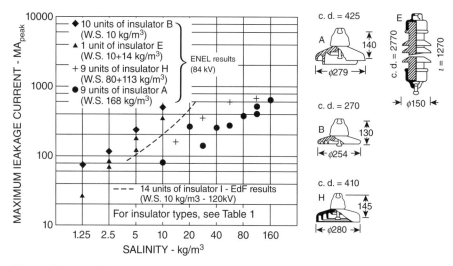

Figure 4-21: Maximum leakage current versus salinity in salt-fog tests of insulators (from Lambeth et al. [1973]).

With a good understanding of the test requirements, ac salt-fog tests from at least three laboratories were compared to ensure that the method was reproducible. This knowledge also guided the specification of adequate power supplies for contamination testing in IEEE Standard 4 [1995].

There is general agreement within the IEEE and IEC that the salt-fog test is suitable for ac voltage. However, round-robin tests [Naito and Schneider, 1995] led to some dispute about the applicability of the test method for dc insulators. This has not yet been resolved. As a consequence, IEC 61245 [1993] endorses the salt-fog test method for HVDC insulators while IEEE Standard 4 [1995] states explicitly that it is not suitable for standardization.

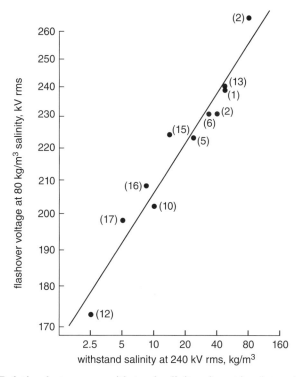

Figure 4-22: Relation between ac withstand salinity of post insulators in salt-fog test with flashover voltage at 80 g/L salinity in rapid flashover voltage tests (from Lambeth, 1988).

4.6.3. Quick Flashover Voltage Technique

Lambeth [1988] noted that an up-and-down voltage application method within the wetting phase of a salt-fog test could be used to reduce the number of tests required, by giving a critical flashover level (V_{50}) for every test. He evaluated the relation between the flashover voltage at constant salinity and the standard results of withstand salinity, obtained using the IEC 60507 test method (Figure 4-22).

The strongest criticism of the rapid flashover method is that the effect of multiple flashovers on some nonceramic insulator surfaces leads to a loss of hydrophobicity and a progressive decline in flashover voltage, rather than stabilization at a constant level as observed on ceramic insulators.

4.7. CLEAN-FOG TEST METHOD

The artificial clean-fog test originated in Japan [e.g., Fujitaka et al., 1968] as a way to simulate stable outdoor wetting environments [Kawai and Milone,

1969] that had proved to be very useful for insulator testing. The clean-fog test method was perfected by the IEEE and CIGRE Working Groups on Insulator Contamination [Cherney et al., 1983; Lambeth et al., 1987] in the 1970s as a way of consolidating test stand exposure experience in the United States for industrial environments, where dry deposition takes place over a period of time and then natural wetting by dew, fog, or rain occurs.

4.7.1. Precontamination Process for Ceramic Insulators

The pollution layer contains a substantial quantity of insoluble material. It is applied wet, allowed to completely dry, and is then rewetted with artificial fog that has a natural electrical conductivity.

The insulator is initially cleaned and then coated with slurry, consisting of a mix of 40 g/L kaolin and one of nine levels of salt concentration, giving a volume conductivity of the slurry ranging from 1 to 16 S/m in steps of $\sqrt{2}$. This leads to an ESDD on the insulator surface in the range of 0.025–0.400 mg/cm^2 (25–400 µg/cm^2), along with an NSDD of 0.05–0.07 mg/cm^2 (50–70 µg/cm^2) as shown in Figure 4-23.

The NSDD layer stabilizes the wetting of the pollution layer and leads to more severe test conditions.

In IEEE Standard 4 [1995], kaolin (a kind of clay readily available from pottery supply stores) is the recommended nonsoluble material permitted in the slurry. Tonoko is also permitted as it is in IEC 60507 [1991]. However, the effect of the type of clay on the resulting NSDD must be considered [Matsuoka et al., 1996b]. There is a 20–25% decline in dc flashover performance and about

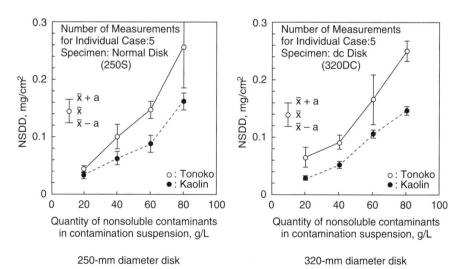

250-mm diameter disk   320-mm diameter disk

Figure 4-23: Relations between NSDD and quantity of kaolin or tonoko in precontamination slurry [from Matsuoka et al., 1996b].

a 10% decrease in ac flashover with kaolin compared to tonoko. As an alternative to these clay materials, in IEC 60507 [1991], a slurry with 100 g kieselguhr (diatomaceous earth, diatomite), and 10 g highly dispersed silicon dioxide, particle size 2–20 nm, per liter along with specified amounts of NaCl give about the same layer conductivities. Figures 4-24 and 4-25 show that the kieselguhr suspension gives much greater layer conductivity with lower levels of salt than the kaolin or tonoko recipes.

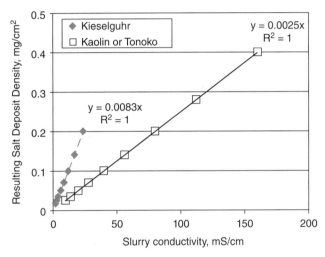

Figure 4-24: Relation of layer conductivity to insulator salt deposit density for three types of suspensions [IEC 60507, 1991, Clause 13].

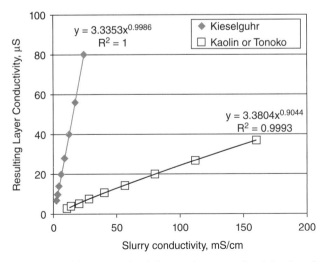

Figure 4-25: Relation of layer conductivity to slurry conductivity for three types of suspensions [IEC 60507, 1991, Clause 13].

The effect of the slurry recipe on salt deposit density is mainly a consequence of how a higher level of insoluble material (110 versus 40 g/L) can make a thicker layer on the insulator surfaces. The more significant effect on layer conductivity in Figure 4-24 is a better demonstration of how the increased NSDD can stabilize the wetting of the salt layer.

The contamination layer, applied by dipping or flow coating, is allowed to dry and then the insulator is mounted in its normal operating configuration in a sealed environmental chamber. The distance from the insulator to the walls of the chamber, its height above ground, and the other physical arrangements should simulate as closely as possible the electrical stress in service. As an example, if grading rings are needed for the full-scale insulators in service, they should be mounted in the same position in the test laboratory and the additional clearances should be provided.

Conductivity of the precontamination slurry is increased or decreased by a factor of $\sqrt{2}$, in an "up-and-down" method, depending on whether the test result is a withstand or flashover, respectively. The instructions for interpreting withstand ESDD values are precisely detailed in the IEC and IEEE standards.

4.7.2. Precontamination Process for Nonceramic Insulators

The flashover performance of polymer insulators when first tested in the laboratory is usually excellent, compared to that of ceramic insulators of the same dimensions. This performance may degrade with environmental exposure at rates that remain to be established. However, it has been recognized that some of the apparent advantage of polymers relates to laboratory test problems. When new, the insulators are difficult to contaminate uniformly and also difficult to wet in the clean-fog test.

Our focus is on test methods that reproduce nontracking flashover due to surface wetting of a precontamination layer, as tracking flashovers are more of an aging phenomenon. This insulation capability is one of the two acceptance criteria for polymer insulators, the other being acceptable tracking and erosion performance in natural or artificial aging tests such as IEC 62217 [2005].

The cleaning process in contamination test standards ensures that there are large wetted areas on ceramic insulators, and in salt-fog tests, conditioning is also controlled by the number of flashovers required. For polymer insulators, conditioning prior to short-term laboratory tests is intended to allow reproduction of the relatively uniform contamination layer that would accumulate after months of natural exposure. The conditioning is specifically intended to increase wettability, so that islands of highly conductive contamination slurry, each insulated from the next, do not form in the artificial wetting process.

Several methods have been evaluated [CIGRE Task Force 33.04.07, 1999] to make pollution application methods for nontracking flashover more representative, repeatable, and reproducible. These fall into two general categories:

1. Methods that condition the *surface*, temporarily destroying its hydrophobicity in a controlled way to allow uniform application of the standard precontamination slurry.
2. Methods that apply the pollution as a *dry powder* to the surface.

The surface conditioning methods include:

- Natural aging outdoors under service voltage electric stress.
- Artificial aging indoors with salt-fog exposure such as IEC 62217 [2005].
- Artificial aging with a cyclic method that may include dust, rain, fog, ultraviolet, and electric stresses.
- Scrubbing the surface with abrasive powder, detergent powder, or sandpaper.
- Sandblasting or bead blasting.
- Addition of alcohol to the kaolin/salt slurry for clean-fog tests.

The dry powder application methods include:

- Gentle application of a fine powder of kaolin, tonoko, or diatomaceous earth, which is then blown away from the surface.
- Application of a dry layer of salt using a piezoelectric sprayer.
- Application of a mix of ball-milled and sieved kaolin and salt with a dry sprayer apparatus.

The most representative conditioning methods are probably those that apply a nonsoluble layer first, followed by flow coating with standard salt-and-kaolin contamination slurry. De La O et al. [1994] proposed using a gentle application of dry kaolin for conditioning. Naito et al. [1996] instead used 1000 g/L of kaolin in slurry, rather than the normal 40 g/L. Gutman et al. [2001] noted that the larger particle size of diatomaceous earth (kieselguhr) in the dry application method of De La O and co-workers, and application of ultraviolet light, both led to more consistent and faster recovery time of hydrophobicity.

Most polymer surfaces have conditions that change, often continuously, during both the energized and deenergized phases of any particular test method. This means that all of the insulators in a comparison test series should receive the same time intervals between conditioning and testing.

At present, silicone is the only insulator material that has an ability to transfer its ability to bead water (hydrophobicity) to pollution layers. The hypothesis is that low-molecular-weight silicone fluid chains are split off from longer chains in the bulk. This fluid has a low viscosity, much less than water, for example, and tends to float above the pollution. In some specific cases, such as carbon black pollution, the resulting surface can be superhydrophobic with water contact angle in excess of $150°$. It takes some time for this process to occur, and, as well, the speed of recovery of hydrophobicity is sensitive to ambient temperature [Chang and Gorur, 1994].

Polymer insulator ranking based on the standard salt-fog contamination test in IEC 60507 [1991] did not match the performance observed at the Brighton exposure station in England. A series of retests, using half of the saline input rate to each nozzle (250 mL/min rather than 500 mL/min) gave the correct ranking, according to CIGRE Task Force 33.04.07 [1999]. Gorur et al. [1987] noted that the salt-fog conditions leading to the minimum electrical performance on nonceramic materials are not necessarily those with the highest level of conductivity.

Leakage current measurements on polymer insulators give some important indications about the test quality. Normally, the activity will proceed from a stable condition with low-level leakage current, through the inception of dry-band arcing to a quasi-stable level of dry-band arcing activity. Like the case for ceramic insulators, this third level of activity will normally reach a maximum and then fall off as pollution is washed away in the test process. For a well-prepared polymer insulator, however, this washing effect may be much lower because the pollution layer is properly stabilized.

Unlike porcelain insulators, there seems to be some initiation of surface damage for leakage currents of 3–5 mA on polymer insulators. This means that the leakage current monitoring systems need more accuracy and resolution than those normally used for ceramic insulators, where leakage current pulses of more than 1000 mA are common.

4.7.3. Artificial Wetting Processes

Once the insulator is mounted in place, there are then two test procedures, with the second one being in common use now:

1. The insulator is wetted after service voltage is applied.
2. The insulator is wetted before and during energization.

A steam generator is normally used to produce the fog, which is injected close to the test object at floor level. The steam tends to raise the chamber temperature, and this is a factor that needs to be monitored to ensure repeatability from lab to lab.

In test process 1, the service voltage is maintained for 100 minutes or until flashover, whichever comes first. It is traditional to monitor the leakage current and to record the maximum amplitude of the bursts of activity that will occur. If the discharge activity decays to 70% of the maximum value, it is considered that the insulator has lost enough of its pollution layer to terminate the test.

The greatest advantage of test process 2 is that it allows the measurement of the insulator resistance and pollution layer conductivity at a reduced voltage as a double check on the ESDD and NSDD measurements. As described in Chapter 3, the calculation of layer conductivity involves the insulator form factor as well as the ratio of applied voltage to resistive component of current.

There is some flexibility in the choice of method to generate fog in test process 2. The use of a steam generator is allowed, but chambers that make use of arrays of ultrasonic fog nozzles fed with warm or cold water can reduce stratification, build up fog faster, and give more repeatable results. With either source, the fog density is managed so that the maximum layer conductivity (maximum resistive current for constant voltage) is reached from 20 to 40 minutes after the start of the wetting period. At this point, the service voltage is applied and held for 15 minutes, or until flashover occurs.

With process 2, it is permissible to repeat the test with the same pollution layer, once, provided that the maximum layer conductivity in the second test is within 10% of the first test.

Lambeth [1988] suggested that considerable test time could be saved by allowing a voltage application process started at a high level, and then decreasing the applied voltage within a test to obtain a minimum value. This method has considerable merit—it cuts the testing time by a factor of at least 2—and is valid for insulation systems that can be repeatedly rewetted without surface damage. The method is probably suitable for clean-fog and cold-fog tests on the leakage distance of ceramic insulators, but is unsuitable for test on polymer insulators or on ice- and snow-covered insulators.

4.7.4 Validation of Clean-Fog Test Method

An indication of the representative and repeatable nature of the ac clean-fog test methods is obtained by comparing test results from Arai [1982] in Figure 4-26 with the results obtained from natural exposure of the same insulators, given in Figure 4-16. The test points for the "equivalent fog method," adapted to an NSDD of tonoko powder, follow a line with a power-law fit of about $5 \cdot (ESDD)^{-0.22}$ for the flashover voltage (kV rms) per insulator disk.

In the equivalent fog method, the insulator surface was contaminated and completely wetted before the voltage was applied and raised to the flashover level. Testing conditions in this method are more severe than the normal fog withstand method, where the insulators are wetted by fog under normal service voltage stress.

The artificial contamination on the insulators was optimally wetted with a steam fog density in the range of $3–7 \, g/m^3$, which is considerably denser than natural fogs, and with this steam input rate, the temperature in the test chamber also tends to rise. Both of these factors make the fog test less representative but are necessary to ensure the excellent repeatability demonstrated in Figure 4-26.

The clean-fog test method was evaluated in 11 different test laboratories to explore reproducibility [Cherney et al., 1983]. Two laboratories tested single disks with short-circuit currents of 2–4 A, rather than insulator strings of six to ten units with a median short-circuit current of 22 A. Two other

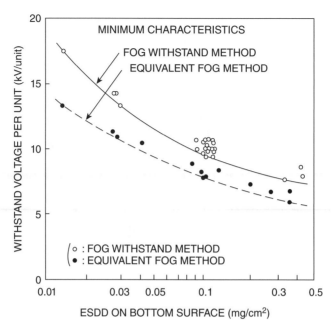

Figure 4-26: ac Withstand voltage per suspension insulator disk in artificial clean-fog test (from Arai [1982]).

TABLE 4-2: Summary of Multilaboratory Test Results on IEEE Disk Insulators

Number of Laboratories	Number of Insulators	Wetting Time (min)	Short-Circuit Current (A)	Flashover Voltage per Insulator Disk (Standard Deviation) (kV) ESDD (mg/cm²)		
				0.02	0.07	0.2
2	1	30	2–4	20.2 (2.8)	16.5 (2.3)	13.6 (0.1)
2	7–10	70–120	22–50	18.7 (5.9)	13.0 (3.0)	10.5 (0.5)
7	6–10	Median 37	Median 22	18.6 (0.7)	12.6 (1.0)	10.4 (1.0)

Source: Cherney et al. [1983].

laboratories had slow wetting rates, 70–120 minutes rather than the median of 37 minutes for all laboratories. When these results were excluded, the test results on IEEE standard disk insulators had good reproducibility, as shown in Table 4-2.

The IEEE results for the seven reference laboratories can be fitted with a flashover voltage per insulator disk of $6.7 \cdot (ESDD)^{-0.25}$.

TABLE 4-3: Characteristics of Suspension Disks in Clean-Fog Tests by Joint CIGRE/IEEE Task Force

Characteristic	IEEE Standard Disk	Anti-Fog Disk
Shed diameter (mm)	254	254
Cap-to-pin spacing (mm)	146	146
Leakage distance (mm)	305	390
Top surface area (cm^2)	691	620
Bottom surface area (cm^2)	908	1380
Total surface area (cm^2)	1599	2000
V_{50}, mean critical flashover (kV/ disk) and σ, standard deviation for ESDD = 0.07 mg/cm^2	13.63σ = 2.3% (8 laboratories)	17.9 σ = 5.2% (6 laboratories)
E_{50}, mean critical flashover stress (kV/m$_{leakage}$)	44.7	45.9

Source: Lambeth et al. [1987].

The clean-fog test method was accepted as an international standard after a second series of tests, carried out jointly by participants in CIGRE and IEEE [Lambeth et al., 1987]. They carried out clean-fog tests with controlled steam input rates that gave wetting in a 10–35-minute period, and also specified that the short-circuit current of the test source be greater than 6 A, with a narrow specification of up to 12 A for testing fog-type disks.

Two disk profiles were tested, as shown in Table 4-3. Once again, an IEEE standard disk was tested and the results were compared with a deep-skirt anti-fog profile.

The mean V_{50} of 13.63 kV per IEEE standard disk with 2.3% standard deviation at an ESDD of 0.07 mg/cm^2 compares favorably to the IEEE round-robin results in Table 4-2. The former results [Cherney et al., 1983] obtained 12.6 kV per standard disk and a relative standard deviation of 8% for the seven laboratories with the same ESDD, similar wetting rate, and source impedance.

The effect of profile on the mean critical flashover gradient was about 3%, which is considered to be within the experimental uncertainty.

4.7.5. Rapid Flashover Voltage Technique

Normally, the voltage in the clean-fog test is held constant for the entire test. If there is a withstand, the insulator is cleaned, recontaminated, dried, and

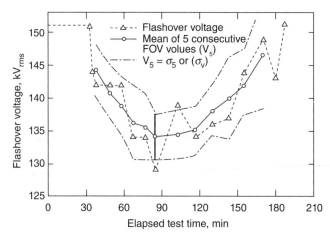

Figure 4-27: Observed flashover level on string of 10 IEEE standard disk insulators with ESDD of 0.07 mg/cm^2 in rapid flashover clean-fog test [from Lambeth, 1988].

retested at a higher voltage. Typically, at least five tests are needed to establish a withstand level and ten or more "valid" tests in an up-and-down sequence will give a critical flashover level, V_{50}.

The possibility of using different voltage levels during contamination tests has been explored by many researchers [e.g., Lambeth, 1988]. The advantage, similar to the quick flashover voltage technique in salt-fog testing, is that each test will deliver a value of V_{50}, reducing the cost of insulator testing by a factor of 10.

In a typical test result, Figure 4-27 shows the evolution of the flashover voltage of a string of 10 disks under ac clean-fog conditions. Once wetting is achieved after 30 minutes of fog, the flashover voltage drops from 145 kV to a constant level of about 135 kV, and then rises as the contamination continues to wash off the insulator surfaces.

The minimum flashover levels obtained with the rapid flashover process tend to be lower than those found in the standard tests on ceramic insulators. Also, multiple flashovers on polymer insulators tend to produce a flashover voltage level that declines continuously with test time, as shown in Figure 4-28.

A continuous decline of flashover voltage reflects a change in the polymer insulator surface condition from flashover arcing, which is not representative. More practically, this also means that the process never reaches the stable level needed to evaluate V_{WS} or V_{50}. For these reasons, the rapid flashover method has not been accepted as a standard test method.

4.8. OTHER TEST PROCEDURES

Both salt-fog and clean-fog tests have been accepted in insulator test standards used around the world, including IEC 60507 [1991] and IEEE Standard 4

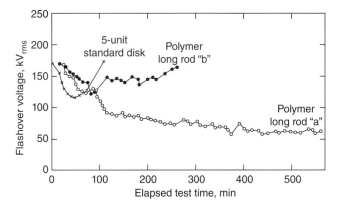

Figure 4-28: Comparison of results of rapid flashover clean-fog test method for 5-unit porcelain disk insulator and polymer long-rod insulators [from Lambeth, 1988].

[1995]. This does not mean that either test is fully appropriate for a particular local condition. A number of modifications of test procedures have been evaluated in order to either reduce cost or improve representativeness. This has been particularly important as polymer insulators have been developed for high-voltage lines, since for some their response to short-duration clean-fog tests gives an overoptimistic estimate of their long-term performance in polluted conditions.

4.8.1. Naturally Polluted Insulators

Artificial wetting of naturally polluted insulators is a hybrid test approach that controls the most important aspects of the wetting rate while allowing the insulators to respond to outdoor exposure [CIGRE 33.04.01, 2000]. This approach is especially appropriate for desert conditions, where there is a progressive accumulation of deposit with time and little influence of washing. In temperate conditions, it may be difficult to judge the most appropriate times to remove insulators from exposure, and a day of inclement weather can completely change the pollution levels.

4.8.2. Liquid Pollution Method

A "methylcellulose" method [Le Roy et al., 1984] strives to improve the repeatability of the clean-fog test method by using a wet rather than a dry pollution layer before the test. The contamination slurry consists of water, chalk, methylcellulose or kaolin, and a suitable amount of NaCl. The test voltage is applied within a few minutes of slurry application, just after the insulators stop dripping. With no drying period, there is no longer any need for a wetting phase. Leakage currents will build up within a few minutes and will either dry out the layer, giving a withstand, or proceed quickly to flashover.

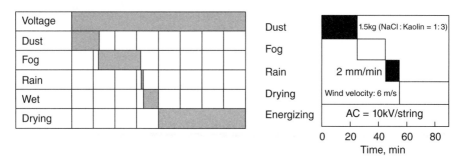

Eklund et al, [1995] (total cycle 120 min)   Ye et al, [2003] (total cycle 90 min)

Figure 4-29: Typical Time sequences in dust cycle methods.

4.8.3. Dust Cycle Method

Eklund et al. [1995] and Ye et al. [2003] have described experience with an insulator test method that attempts to address the various biases in slurry application to produce a solid pollution layer on ceramic versus nonceramic surfaces. Typical cycle of the *dust cycle method* (DCM) are shown in Figure 4-29.

Application of continuous service voltage to an insulator in its service orientation produces a pollution deposit that is characteristic of a local climate. In contrast with solid-layer methods, the cycle in Figure 4-29 would typically lead to a cycle-by-cycle increase in bottom-layer ESDD while the top insulator surfaces would remain relatively clean from the rain cycle.

Normally, the DCM is carried out on a reference insulator to establish the number of cycles needed to cause flashover for a particular simulated environment. Then, the same test cycles are repeated to rank the performance of alternatives.

Leakage current measurements play an important role in evaluating the results of DCM tests. Some materials, such as RTV silicone coatings on porcelain, show a progressive increase in leakage current with every cycle. Others (such as silicone insulators) have a highly repeatable and consistent increase and decrease in leakage current every time [Eklund et al., 1995].

4.8.4. Dry Salt Layer Method

As described by Engelbrecht et al. [2003], the dry salt layer (DSL) method is designed to reproduce the accumulation of marine pollution on nonceramic insulator surfaces near the seacoast. An injection system generates humid salt particles and injects them into the air flow of a specialized chamber. The fine salt particles are then blown by a uniform wind of 4–7 m/s past the energized

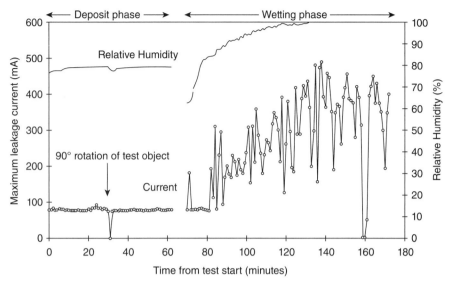

Figure 4-30: Typical time sequence in dry salt layer method (from Engelbrecht et al. [2003]).

insulator under test. A typical DSL test including a timed deposit phase at 80% relative humidity and a wetting phase, also at normal service voltage, is shown in Figure 4-30.

The DSL method leads to leakage currents of up to 500 mA on a hydrophobic insulator bushing, without flashover, matching observations of in-service performance.

In contrast, the high conductivity of spray in severe salt-fog tests [IEC 62217, 2005] inhibits the occurrence of large leakage currents. Peak levels of 0.13–0.2 mA with 40–80 kg/m^3 of salt were observed on the same test object [Engelbrecht et al., 2003].

4.8.5. Cold-Fog Test Method

Artificial tests in winter conditions could adapt either salt-fog or clean-fog test methods to evaluate the role of leakage distance and insulator shed shape on pollution performance. There has not been much interest in adapting the salt-fog method because the extremely low freezing temperature of seawater (0 °F or –18 °C) near the coast only occurs in areas of limited development.

Chilling of insulators to 0 °C was recognized by Cherney et al. [1983] as one method that could be used to improve consistency of wetting. This approach gave essentially the same flashover results as other wetting methods at

20–30 °C, but the electrical weakness of insulators in cold conditions just below the freezing point was not explored.

The standard clean-fog contamination test method was developed into a "cold-fog" test method to reproduce line voltage flashovers across insulator leakage distance in winter conditions. This method reproduces flashovers at normal service voltage at the melting point in the absence of heavy ice accretion or full bridging. The cold-fog test method [Chisholm et al., 1996; Chisholm, 2007] consists of:

- Precontamination using flow coating of kaolin/salt mixtures or dry-spray methods.
- Chilling of insulator to –2 °C.
- Application of line-voltage stress.
- Circulation of fog with 10-µm volume mean diameter and 3-m/s wind speed, similar to natural conditions, giving dew point depression of less than 2 C°.
- Controlled increase in temperature and dew point.
- Increase of voltage in 5% steps to flashover at measured dew point temperature intervals.
- Interpolation to establish 50% flashover voltage (CFO) for 30 min at dew point of 0 °C.

The cold-fog flashover process is essentially a freeze-frame snapshot of the more complex contamination flashover process. The fog layer, even while frozen and low in conductivity, still functions in the same way as a heavy nonsoluble deposit density in excess of $0.4 \, mg/cm^2$ [Chisholm, 2007]. Additional positive feedback in the arc propagation is introduced by the temperature–conductivity relationship of the ice-coated contaminated surface. Dry banding does not have time to develop and the entire cold-fog flashover process develops or dies within a few ac cycles.

4.8.6. Tests of Polymeric Insulator Material Endurance

There are a wide variety of test methods that have been established to simulate the short- or long-term environmental exposure of polymer insulators. With varying degrees of success, these tests cause surface degradation after exposure periods of 1000–5000 hours that can match many years of outdoor exposure. Examples of relevant electrical material endurance tests [CIGRE Task Force 33.04.07, 1999] include the following:

- A 1000- or 5000-hour exposure to continuous salt fog at a USCD of $34.6 \, mm/kVac_{l-g}$ as specified in IEC 62217 [2005].

- A 5000-hour exposure with continuous ac voltage and cycles of salt fog, rain, humidification, heating, and solar radiation, adapted by Electricité du France (EdF), the Electric Power Research Institute (EPRI) in the United States, and Ente Nazionale per L'Energia Elettrica (ENEL) in Italy. A similar process is used by FGH-GmbH in Germany for a 20 mm/kV dc stress.
- Inclined-plane tracking and erosion tests, such as ASTM D2303.
- Cyclic tracking and erosion tests, such as the CEATI LWIWG 1000-h tracking wheel test at 28.6 mm/kVac$_{l-g}$.
- Boiling tests of 42- or 100-h duration for end-seal integrity, evaluated with steep-front lightning impulse tests in CEATI LWIWG. The boiling test is also suitable for evaluating adhesion of silicone coatings to substrates.

The advice in the 1999 CIGRE summary of aging tests was that it was not possible to select a single standard method that gave predictions of satisfactory long-term performance of polymer insulators. Instead, a process of continuous comparison of laboratory test results with service experience was recommended as the best way to ensure that a selected method was representative for its service area.

4.8.7. Summary of Pollution Test Methods

Type	Test Method	Relevant Standard	Total Tests (duration)	Comments
Standard methods	Salt fog	IEC 60507 [1991], Sections 7–12 (ac) IEEE Standard 4 [1995], Clause 15.6 (ac)	5–10 (8 h)	Salinity 2.5–224 g/L
		IEC 61109 [1992]/62217 [2005], (ac)	1 (1000 h)	Endurance test, not pollution test
		IEC 61245 [1993] (dc)	5–10 (8 h)	Not recommended by IEEE
	Clean fog	IEC 60507 [1991], Section 13	5–10 (8 h)	Binder: kieselghur, kaolin, or tonoko
		IEEE Standard 4 [1995], Clause 15.3 (ac), 15.4 (dc) and 15.5 (both)		Binder: kaolin

Type	Test Method	Total Tests (duration)	Comments
Other methods	Quick salt-fog test	1 (4h)	Not representative; flashover levels low
	Rapid clean-fog test	1 (8h)	Not representative; flashover levels low
	Naturally polluted insulators in clean fog	1 (2h)	Representative pollution on top and bottom surfaces; not reproducible.
	Liquid pollution	5–10 (2h)	Representative only for exposure to polluted spray
	Dust cycle method	5–10 (12h)	Representative; relation to standard tests unclear
	Dry salt layer method	5–10 (3h)	Representative; leakage currents on polymer compared to natural exposure
	Cold-fog test	1 (15h)	Each test has 5 w/s, 5 f/o; results extrapolated to 0°
Polymeric insulator endurance tests	Polymer aging tests	1 (5000h)	Several variations to simulate local UV, temperature, and pollution exposure
	Inclined plane tracking and erosion test	4 (1h)	ASTM D2303 standard; used in electrical industry
	Tracking wheel test	1 (1000h)	CEATI LWIWG purchase specification
	Boiling test	1 (42 or 100h)	CEATI LWIWG

4.9. SALT-FOG TEST RESULTS

4.9.1. ac Salt-Fog Test Results

Results of ac salt-fog tests presented in Lambeth et al. [1970] and Figure 4-31 show a linear relation between flashover strength and insulator string length in salt-fog tests on strings up to 6.2 m in length. A flashover gradient of 77 kV/m of dry arc distance (36.6 kV/m$_{leakage}$) was found for the standard profile with 305-mm leakage distance per disk at a salinity of 7 g/L. At a salinity of 56 g/L, the flashover gradient on fog-type disks with 410 mm of leakage distance was 87 kV/m$_{dry\ arc}$ and 30.9 kV/m$_{leakage}$.

Figure 4-32, derived from Le Roy et al. [1984] and CIGRE Task Force 33.04.01 [2000] and the two test values from Lambeth et al. [1970], shows the average value of ac flashover gradient per meter of leakage distance to be expected on standard ceramic disk insulators in salt-fog tests.

The critical flashover stress E_c (kV/m) has a general relation with the salinity S (g/L or kg/m^3) of

$$E_c = K \cdot S^\alpha \qquad (4\text{-}4)$$

The value of $\alpha = 0.2$ was recommended by Lambeth et al. [1973] after evaluating 23 different insulator profiles in the same salt-fog test.

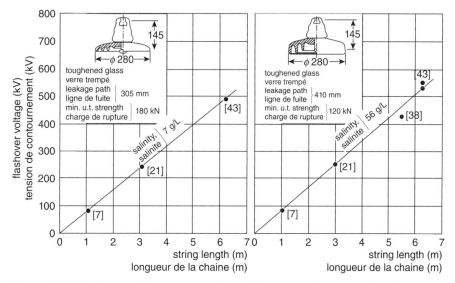

Figure 4-31: Flashover voltage versus strings of 7, 21, 38, or 43 cap-and-pin disk insulators in salt-fog tests (from Lambeth et al. [1970]; courtesy of CIGRE).

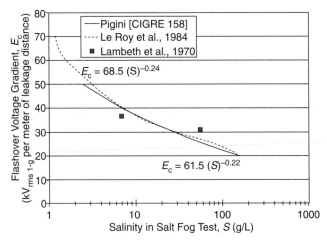

Figure 4-32: ac Flashover strength of suspension disk insulators in salt-fog conditions.

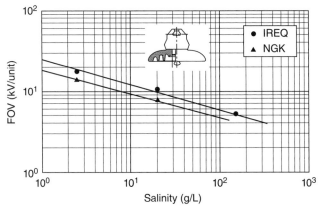

Figure 4-33: Results of 50% flashover voltage (FOV) versus salinity in salt-fog tests for two different laboratories with IEEE standard disks ($146 \times 254\,\text{mm}^2$, 305-mm leakage distance) (from Naito and Schneider [1995]).

4.9.2. dc Salt Fog Test Results

Naito and Schneider [1995] sponsored and reported round-robin tests of salt-fog tests on three different types of insulator using negative-polarity dc rather than ac. They found that the flashover voltage on IEEE standard disk insulators declined with increasing salinity with values of $\alpha = -0.30$ and $\alpha = -0.31$ for the NGK and IREQ results shown in Figure 4-33.

For the NGK results, the negative-polarity dc flashover stress can be approximated by the expression $E_{dc} = 62\,S^{-0.3}$ for E_{dc} (kV) and S (g/L). With 20 rather than 10 disks in the chain, the results from the IREQ tests were

$E_{dc} = 86\,S^{-0.31}$. This is a large discrepancy between the two laboratories. The lack of reproducibility for the dc salt-fog test has limited the acceptance of results such as Figure 4-33 as a tool for engineering of HVDC insulators.

4.10. CLEAN-FOG TEST RESULTS

4.10.1. ac Clean-Fog Tests

Tests of IEEE standard ceramic disk insulators with a relatively conservative bottom-surface profile have been carried out at many laboratories in the development of the clean-fog test method [IEEE, 1979; Rizk and Bourdages, 1985; Lambeth et al., 1987; Matsuoka et al., 1996b]. Baker et al. [1989] consolidated the IEEE results and these are shown in Figure 4-34, overlaid with the CIGRE Task Force 33.04.01 [2000] values.

Baker et al. [2008] assumed a standard deviation of 10% in the clean-fog test result and thus used a factor of 0.7 to convert the average critical flashover stresses in Figure 4-34 to withstand values at a −3 standard deviation (−3σ) level. This gave the following approximating curve for the withstand level:

$$V_{WS} = 12.6 \left(\frac{SDD}{1\,\mathrm{mg/cm^2}} \right)^{-0.36} \qquad (4\text{-}5)$$

where

V_{WS} is the maximum withstand voltage gradient (kV line-to-ground per meter of leakage distance), and

SDD is the salt deposit density (mg/cm²).

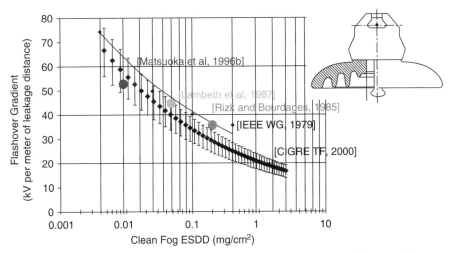

Figure 4-34: Power-frequency critical flashover voltage gradient (kV rms line-to-ground per meter of leakage distance) as a function of contamination level (ESDD) for porcelain disks subjected to clean-fog tests.

The critical flashover curve is plotted in Figure 4-34. The exponent of $\alpha = -0.36$ in Equation 4-5 reflects the lower range of contamination levels, compared to the CIGRE curve. This tendency toward higher values of α at lower contamination levels in clean-fog tests matches the observations of Le Roy et al. [1984] for salt-fog tests. The IEEE standard disk insulators fall somewhat above range of expected performance from CIGRE results for ESDD levels greater than $100\,\text{mg/cm}^2$. This suggests that the shallow rib profile of the IEEE standard disk, illustrated in Figure 4-34, works with relatively high efficiency in clean-fog tests.

Provided the ratio of leakage distance to dry arc distance is not too large, generally, clean-fog test results tend to be linear with regard to the number of insulators in a string and hence to the leakage and dry arc distances. For this reason, it is common to express test results as a critical flashover gradient, V_{50}. When the flashover stress is normalized to leakage distance in units of (kV rms of line-to-ground voltage per meter of leakage distance), this tends to correct for much of the performance difference found among disks of different profiles.

In ac clean-fog conditions, insulator flashover was observed [CIGRE Task Force 33.04.01, 2000] to have an exponent of $\alpha = -0.22$ in Equation 4-5, based on a synthesis of results from many different laboratories over a wider range of ESDD values than those considered by Baker et al. [2008]. The CIGRE value of α is conveniently the same exponent found in the CIGRE relation between salt-fog flashover levels and salinity. This coincidence allows an alignment of the test results to derive the equivalence between salt-fog and clean-fog severity levels as illustrated in Figure 4-35.

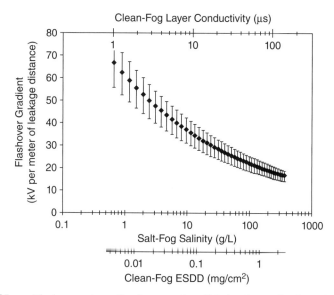

Figure 4-35: ac Flashover strength of suspension disk insulators in clean-fog and salt-fog conditions (adapted from Kawai and Sforzini [1974]).

The approximate relation between the salinity S in salt-fog tests and the ESDD in clean-fog tests is [Kawai and Sforzini, 1974]

$$\frac{S}{1\,\text{g/L}} = 140\frac{ESDD}{1\,\text{mg/cm}^2} \qquad (4\text{-}6)$$

In Figure 4-35, there is a modest difference in the results for clean-fog tests with wetting method A, where the layer conductivity is measured directly on the upper axis, and wetting method B, which relies on the salt deposit density on the lower axis. The exponents are $\alpha = -0.28$ and $\alpha = -0.22$ for methods A and B, respectively. The difference between these values suggests that it is the artificial clean-fog test limitations in fully wetting the salt layer, rather than the inefficient insulator profiles, that cause most deviation away from the expected $\alpha = -0.33$ for a perfectly wetted and uniform salt electrolyte surface.

4.10.2. dc Clean-Fog Tests

Baker et al. [1989] evaluated the dc flashover strength for 13 different porcelain post insulators, and also tested the 60-Hz ac strength in the same clean-fog conditions. One of these posts, No. 13, had semiconducting glaze and did not flash over under clean-fog conditions. The other posts had one of three general shed profiles—(I) uniform, (II) bottom rib, and (III) alternating diameter.

Baker and co-workers expressed the post insulator shapes in Figure 4-36 using the ratios of leakage distance to height, called "specific leakage" in Table 4-4. They carried out tests at ESDD of 0.005 and 0.05 mg/cm^2 (5 and 50 μg/cm^2), representing maximum levels in lightly and heavily polluted areas, respectively. Results here were expressed as the critical flashover gradient across the post height, which will in general be slightly larger than the dry arc distance. Figure 4-37 shows that there was a large improvement in ac performance from 50 to 80 kVrms/m for posts with a specific leakage of 4 compared to one with a specific leakage of 2. However, for negative-polarity dc, the improvement was only on the order of 15%, from 50 to 57 kVdc/m, for the same station post insulators.

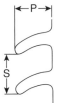
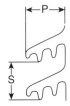

Type I: Uniform sheds   Type II: Ribbed bottom surface   Type III: Alternating diameter sheds

Figure 4-36: General shed profiles of station post insulators in ac and dc fog tests (from Baker et al. [1989]).

TABLE 4-4: Shed Dimensions for Station Post Insulators Tested by Baker et al. [1989]

Post Number	Post Type	Outer Diameter (cm)	Leakage Distance (cm)	Number of Sheds	Leakage per Shed (cm)	Spacing Between Sheds (cm)	Shed Projection (cm)	Specific Leakage
1	I	36	226	15	14.9	3.74	6.35	2.97
1A	I	36	191	12	15.9	4.76	6.35	2.51
2	I	36	157	9	17.4	6.55	6.35	2.06
3	I	40	204	9	22.4	6.27	8.26	2.68
4	I	42	241	10	23.7	5.57	9.52	3.16
5	I	42	178	7	25.3	8.36	8.52	2.33
6	II	37	254	11	22.7	4.98	6.67	3.33
7	II	37	221	9	24.2	6.22	6.67	2.90
8	II	40	249	9	27.3	6.22	8.26	3.27
9	II	43	295	9	32.3	6.22	9.52	3.87
10	II	43	216	6	35.5	9.52	9.52	2.85
11	III	42	274	9 + 9	21.5/8.7	1.9/6.4	9.21/4.29	3.60
12	III	42	251	9 + 9	21.4/6.2	1.9/6.4	9.21/3.02	3.29
13	IA	36	152	9	16.9	5.38	6.19	2.31

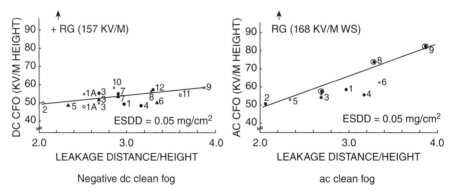

Figure 4-37: dc and ac critical flashover voltage per meter of station post height, versus ratio of leakage distance to height, for ESDD = 0.05 mg/cm^2 (from Baker et al. [1989]).

Figure 4-38 shows that the ac rms critical flashover stress is nearly constant at 23–25 kV rms per meter of leakage distance, while the dc stress falls to as low as 15 kV/pm.

These results illustrate that the processes of extinction and reignition in ac arcing take significantly better advantage of leakage distance than the negative dc flashover process. In this series of experiments, the exponent α relating critical flashover level to ESDD was found to be $\alpha = -0.36$ for both dc and ac

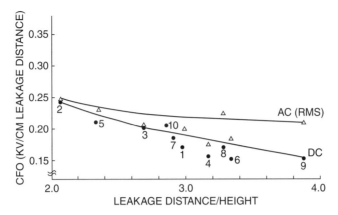

Figure 4-38: ac(rms) and dc Clean-fog critical flashover voltage per centimeter of leakage distance, versus ratio of leakage distance to insulator height, for ESDD of $0.05\,mg/cm^2$ (from Baker et al. [1989]).

tests, and it was also noted that the effect of post diameter in the range of 360–430 mm could be neglected.

Pargamin et al. [1984] compared the performance of disk insulators in dc tests using both salt and clean fog. Their results in Figure 4-39 are expressed in terms of critical flashover stress, kilovolts per meter of connection length between insulator fittings. Normally, the connection length is nearly equal to the insulator dry arc distance. Tests were carried out at a salinity of 28 g/L, which corresponds to an ESDD of about $0.2\,mg/cm^2$ using Equation 4-6. These flashover levels are considerably lower than those observed in clean-fog tests with an ESDD of only $0.07\,mg/cm^2$ $(70\,\mu g/cm^2)$.

Under dc conditions, the relative performance of the insulators does not match in the two test methods. Also, the salt-fog results indicate extremely low flashover stress levels, requiring more than 60 standard 146-mm disks (8.9-m dry arc distance) for a 400-kV dc system at this severe salinity level.

4.10.3. Impulse Voltage Clean Fog Tests

For switching surges, Macchiaroli and Turner [1971] carried out tests on ceramic strings and long-rod insulators at various levels of surface conductivity. They applied ac prior to the impulse test to develop dry bands, simulating the normal in-service conditions. The positive-polarity results in Figure 4-40 were consistently lower than the negative-polarity results except for the 165×318 glass disk with high 495-mm creepage distance.

They noted that the switching-surge flashover voltage of moderately contaminated insulator strings was 35–65% of the wet critical flashover level, making the switching-surge performance an important consideration in overvoltage coordination. CIGRE Working Group 33.07 [1992] provides additional discussion.

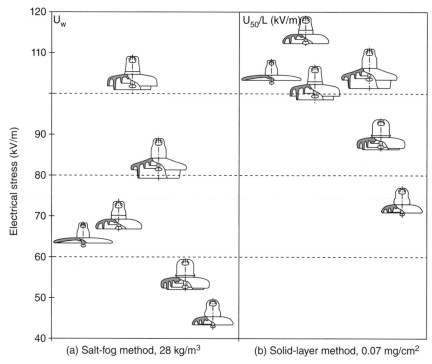

Figure 4-39: Relative ranking of dc pollution performance (50% flashover stress per meter of connection length) of suspension insulators in salt-fog and clean-fog tests (from Pargamin et al. [1984]; courtesy of CIGRE).

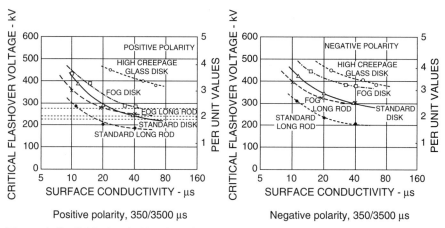

Figure 4-40: Critical switching impulse flashover voltage versus surface conductivity for disk and long-rod insulators (from Macchiaroli and Turner [1971]).

4.11. EFFECTS OF INSULATOR PARAMETERS

4.11.1. Leakage Distance and Profile

One of the most effective ways to improve the performance of an insulator in contamination tests is to introduce additional leakage distance. Table 4-5 shows some typical profiles on standard and fog-type disk insulators. The ratio of leakage distance to 146-mm cap-to-pin spacing ranges from 2:1 to 3:1.

Ratios of leakage to dry arc distance of up to 4:1 can be achieved on station post insulators, as shown previously in Table 4-4.

Flashover across the leakage distance is predicted by the Obenaus–Rizk model to have an approximate relation of the form

$$E\left(kV/m_{\text{leakage}}\right) = K\left(\frac{ESDD}{1\,\text{mg/cm}^2}\right)^{\alpha} \tag{4-7}$$

where

E is the flashover gradient along the leakage distance of the insulator, in kV rms line to ground per meter of leakage distance,

K is a constant in kV/m for each insulator,

$ESDD$ is the equivalent salt deposit density, expressed in mg/cm², and

α is an exponent, in the range of –0.2 to –0.4.

The exponent α relating ESDD and flashover gradient varies, depending on insulator type and voltage. Generally, insulators with relatively smooth shapes make full use of their leakage distance, while there is a tendency for

TABLE 4-5: Typical Profiles and Dimensions of Ceramic Suspension Disk Insulators

Insulator Designation	IEEE Disk	ANSI C29 2 Class 52-3	Typical Fog Type
Illustration			
Dimensions			
Led Diameter × Unit Spacing × Leakage Distance	254 mm × 146 mm × 305 mm (10 in. × 5¾ in. × 12 in.)	254 mm × 146 mm × 292.1 mm (10 in. × 5¾ in. × 11½ in.)	254 mm × 146 mm × 432 mm (10 in. × 5¾ in. × 17 in.)

Source: IEEE Working Group [1979].

flashovers across the air gaps if bottom-surface ribs are too deep. The variation in the value of α expresses this tendency. From a practical perspective, the value of the exponent relating ESDD to flashover voltage could be seen as a weighted average of the expected value of $\alpha = -0.33$ for the salt electrolyte surface and $\alpha = 0$ for the breakdown of air spaces between ribs or sheds. This average is a function of several factors, such as:

- The efficiency of the insulator profile. Deep sheds are more prone to breakdown than shallow skirts. With the same test procedure, Matsuoka et al. [1991] report values of $\alpha = -0.20$ for ac and $\alpha = -0.33$ for dc tests on 11 different profiles of bushing shells up to 13.5 m long.
- The efficiency of the wetting process in the artificial salt-fog or clean-fog tests. An efficient wetting process, or a good measure of the wetting such as the layer conductivity, will have a value of α that is closer to that expected for the saline layer.
- The pollution severity. As the layer conductivity decreases, the value of α may change, for example, from $\alpha = -0.33$ in the salinity range of $S = 0.6$–5 g/L down to $\alpha = -0.25$ for $S = 40$–160 g/L as reported by Le Roy et al. [1984] for salt-fog tests.

The question about whether there can ever be too much leakage distance has been addressed by many researchers. Perhaps the most convincing answer, given by Looms [1988], is the Lucas insulator, constructed by cementing multiple cones of thin polystyrene, to provide more than 10 m of leakage distance per meter of insulator dry arc distance. In spite of the poor electrical properties of the polystyrene material, samples of this insulator performed well in field trials.

Studies carried out by Schneider and Nicholls [1978] explored the linearity of contamination flashover voltage up to 25 m of leakage distance and 840 kV ac rms, line to ground. There is a modest reduction in the value of flashover gradient on both standard and anti-fog disk insulators in their results, plotted in Figure 4-41.

It is remarkable that the flashover stress for the two different insulators at the 0.04-mg/cm^2 ESDD is nearly the same. This convergence is a result of improving shed profile efficiency for higher pollution layer conductivity.

4.11.2. Effect of Small Diameter : Monofilaments and ADSS

A spider's web assembly was suggested by Looms [1988] as a future possibility that would take advantage of the ability of monofilaments to resist pollution flashover. For a constant insulating strand radius r, the form factor F is

$$F = \int_0^L \frac{dl}{2\pi r} = \frac{L}{2\pi r}$$

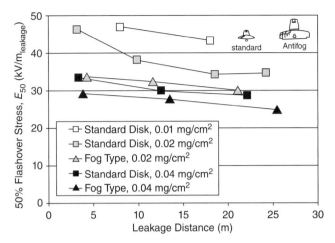

Figure 4-41: The 50% flashover stress (E_{50}) across leakage distance, versus leakage distance, for HV and UHV tests up to 1400 kV.

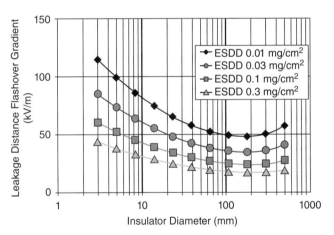

Figure 4-42: Predicted flashover gradient along small-diameter insulator thread from model of Topalis and Gonos [2001].

As the radius decreases, the form factor increases. The model of Topalis and Gonos [2001] suggests as well that the radius of the arc root on the insulator surface will decrease as the insulator diameter decreases.

As an insulator thread diameter shrinks below 100 mm, the flashover gradient at each pollution level is seen to increase considerably in Figure 4-42. At the level of about 3-mm strand diameter, however, direct observations of the arc root radius are needed to validate the extrapolated behavior from the Topalis and Gonos [2001] model, or to develop a new equation using a more appropriate geometry. With present materials having a tensile strength of not more than 3 GPa, it would take four, 3-mm insulating strands to support a

TABLE 4-6: Recommended Test Parameters for Dry-Band Arcing Qualification of All-Dielectric Self-Supporting Cables

Pollution District	Resistance of Pollution Layer on ADSS Cable (MΩ/m)	Series Impedance of Source (MΩ∠degree)
Heavy	0.1	5.9∠44.2°
Heavy	0.2	8.2∠45.0°
Heavy	0.5	13.0∠44.8°
Medium	1	18.6∠45.4°
Medium	2	26.1∠44.5°
Medium	5	41.6∠44.5°
Light	10	58.6∠44.2°
Light	20	83.9∠43.4°
Light	50	135.5∠42.4°
Very light	100	192.3∠41.1°

Source: Karady et al. [2006].

tension load of 7000 kg. These would still be too thick to give adequate flash-over strength along the leakage distance. However, there are other interesting reasons for studying the pollution behavior of insulating strands.

Insulating ropes of about 10-mm diameter are used in live-line work procedures. The cleanliness of the 10-mm rope in Figure 4-42 makes a factor-of-2 difference in the leakage distance flashover gradient. There have also been cases of live-line rope burn-down overnight as a result of leakage currents that may flow prior to flashover.

Another pollution problem became apparent when all-dielectric, self-supporting (ADSS) optical fiber cables of less than 10-mm diameter started to be used on overhead transmission lines in the 1980s. These cables often have a high standing ac potential, resulting from the mutual capacitance from ADSS cables to the nearby phase conductors. Damage to the cable can occur from dry-band arcing if the cable potential exceeds 10 kV and the available current exceeds 0.5 mA [Karady et al., 2006]. Standard application practice is to limit the space potential to 5 kV in areas of high contamination conditions, when buildup of a resistance layer can increase the available current.

Karady et al. [2006] recommend that ADSS cable samples be tested with series resistance and capacitance to weaken the source impedance as a function of pollution level, as shown in Table 4-6. This is the reason why the typical voltage is nonsinusoidal in Figure 4-43.

Accretion of electrically conductive ice on ADSS cable fibers is a severe form of pollution and the effects on dry-band arcing damage may also be important. This is an interesting area for future study, should problems arise in practical ADSS cable applications.

4.11.3. Influence of Average Insulator Diameter

In classical models for an arc in series with a pollution layer, any decrease in the resistance of the pollution layer per unit length will tend to reduce the

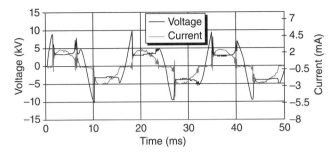

Figure 4-43: Typical voltage and current during dry-band arcing along insulating strand (all-dielectric self-supporting optical fiber cable) (from Karady et al., [2006]).

Type	Shed shape	Shed pitch, P, mm	Shed Projection, mm	Leakage distance per shed, L, mm	L/P
A	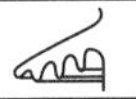	95	95	366	3.9
B		70	70	238	3.4
C	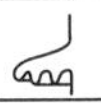	65	65	250	3.8
D		27.5 42.5	36 65	240	3.4
E	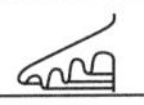	95	105	383	4.0

F		105	105	393	3.7
G	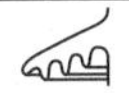	80	95	351	4.4
H		70	70	190	2.7
I		70	70	203	2.9
J	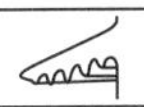	92	120	407	4.4
K		100	95	364	3.6

Figure 4-44: Profiles of station post insulators tested in ac and dc contamination conditions (from Matsuoka et al. [1991]).

flashover voltage. This means that any increase in the diameter of a cylindrical insulator should tend to reduce the flashover voltage.

Matsuoka et al. [1991] evaluated the influence of the diameter on both the ac and dc contamination flashover and withstand voltages of large station post insulators. They used a wide range of post profiles, shown in Figure 4-44, with diameters ranging from 200 to 1500 mm.

For station post insulators of more than 200-mm diameter, any additional increase in diameter will need increased leakage distance for the same electrical performance as a post with smaller diameter. The withstand level in Figure 4-45, using mm per kV of line-to-ground voltage, increases as $D_{avg}^{0.43}$, in line with values noted in Equation 4-10 after negating the exponent.

The dc results in Figure 4-46 do not show as much dependence on average insulator diameter as the ac results. For example, an exponent of $D^{0.30}$ rather than $D^{0.43}$ fits the central tendency of the dc results.

The increase in average diameter of the post insulators in both Figures 4-45 and 4-46 significantly lowers the contamination flashover voltage, and thus

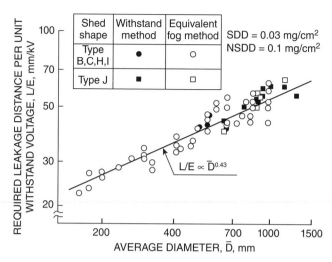

Figure 4-45: Relation between average diameter and required leakage distance per unit ac withstand voltage (from Matsuoka et al. [1991]).

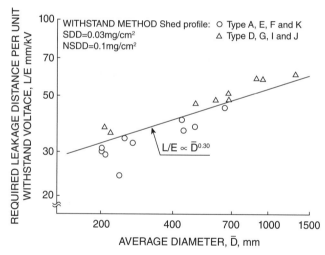

Figure 4-46: Relation between average diameter and required leakage distance per unit dc withstand voltage (from Matsuoka et al. [1991]).

calls for a significant increase in the unified specific creepage distance for the same contamination level. Matsuoka et al. [1991] also noted that the increase in station post diameter also lowered the ESDD on the insulator surface, with exposure levels being held constant. This would tend to increase the flashover strength for the larger-diameter posts. However, the combined effect of both factors for a given site is that the increase in the insulator diameter substantially reduced the contamination flashover and withstand voltages.

For long-rod, post and hollow core insulators with diameter larger than 300 mm, IEC 60815 [2008] offers a correction factor K_{ad} that is multiplied by the unified specific creepage distance of a post in the reference diameter range from 100 to 300 mm. This is given by

$$K_{ad} = 0.001D_a + 0.7 \qquad (4-8)$$

where D_a is a weighted average of the outer shed diameter. For alternating shed diameters, $D_a = (2D_t + D_{s1} + D_{s2})/4$ with outer shed diameter D_{s1}, inner shed diameter D_{s2}, and core diameter D_t. For uniform sheds, $D_a = (D_t + D_s)/2$, where D_s is the shed diameter and D_t is the core diameter.

For an 800-mm diameter insulator, the IEC method suggests that the USCD should be 50% higher than the value for a 300-mm average diameter. This is in reasonably close agreement with the variation observed experimentally by Matsuoka et al. [1991], who recommended a power-law relation of the form

$$USCD \propto D^q \qquad (4-9)$$

where

USCD is the unified specific creepage distance, mm of leakage distance per kV rms (or dc) of line-to-ground voltage,

D is the average insulator diameter (mm), and

q is an exponent, found to be $q = 0.43$ for ac and $q = 0.30$ for dc.

In general, the value of q will vary depending on insulator shape, material, test method including type of voltage, and pollution severity.

A review of the observed exponent that relates the USCD to the insulator diameter [CIGRE Task Force 33.04.01, 2000] suggests a mean value of $q = 0.43$ with a standard deviation of 0.08 for dc and ac clean-fog tests when Matsuoka's values are also included from Figures 4-45 and 4-46. For ac salt-fog tests, the value of q had a wider variation of 0.14–0.74 and the overall mean value was given as $q = 0.35$.

The insulator diameter also has a strong effect on the flashover performance of single insulator disks or distribution pin and post units. For example, Beauséjour [1981] carried out clean-fog tests on 11 different profiles of pin-type and suspension disks for distribution systems, with the types, profiles, and dimensions shown in Table 4-7.

Insulators were tested at three levels of ESDD in clean-fog conditions at the Research Institute of Hydro-Québec (IREQ). Insulator profiles were ranked in order of flashover strength, kVrms$_{l-g}$ per meter of leakage distance at each ESDD level as shown in Table 4-8 and Figure 4-47.

The effect of distribution insulator diameter on the critical flashover stress E_{50} is shown for both average and maximum dimensions in Figure 4-47.

As in the case of tests on station posts, insulators with larger diameter have lower critical flashover stress. The critical flashover stress, E_{50}, for distribution

TABLE 4-7: Dimensions[a] of Distribution Insulators Tested in Clean-Fog Conditions

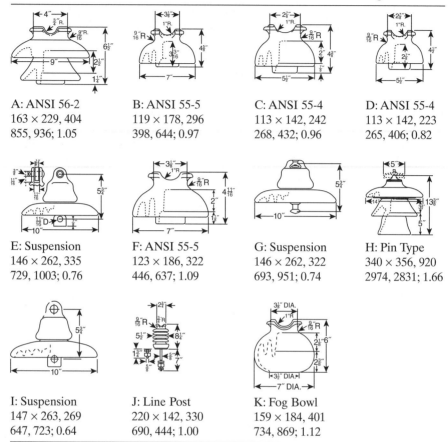

A: ANSI 56-2
163 × 229, 404
855, 936; 1.05

B: ANSI 55-5
119 × 178, 296
398, 644; 0.97

C: ANSI 55-4
113 × 142, 242
268, 432; 0.96

D: ANSI 55-4
113 × 142, 223
265, 406; 0.82

E: Suspension
146 × 262, 335
729, 1003; 0.76

F: ANSI 55-5
123 × 186, 322
446, 637; 1.09

G: Suspension
146 × 262, 322
693, 951; 0.74

H: Pin Type
340 × 356, 920
2974, 2831; 1.66

I: Suspension
147 × 263, 269
647, 723; 0.64

J: Line Post
220 × 142, 330
690, 444; 1.00

K: Fog Bowl
159 × 184, 401
734, 869; 1.12

[a]Legend:
  Code: Type
  Height H × Diameter D_{max}, Leakage Distance L (mm)
  Areas A_{top}, A_{bottom} (cm^2); form Factor F
Source: Beauséjour [1981].

TABLE 4-8: Ranking of Clean-Fog Flashover Performance for Insulator Profiles in Table 4-7

kV$_{rms}$/m$_{leakage}$ for ESDD (mg/cm^2)	D	J$_{Hrz}$	C	A	B	F	J$_{Vrt}$	K	H	I	G	E
0.05	47.1	44.2	41.7	40.3	35.1	36.0	31.8	32.7	31.0	30.1	28.9	26.3
0.15	35.9	34.2	33.9	30.2	29.7	28.9	27.6	26.4	24.6	24.2	23.6	21.5
0.40	27.8	27.3	28.1	23.3	25.3	23.6	24.5	21.9	20.0	19.7	19.9	18.2

Source: Adapted from Beauséjour [1981].

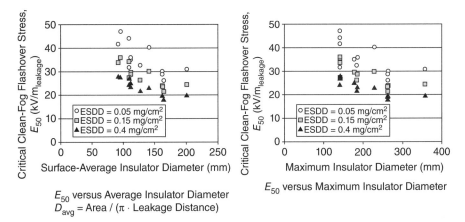

Figure 4-47: Critical clean-fog flashover stress versus distribution insulator diameter (data from Beauséjour [1981]).

insulators can be described empirically in terms of average or maximum insulator diameter, form factor, and ESDD using

$$E_{50} \approx 188 \cdot \left[\frac{ESDD}{1\,\mathrm{mg/cm^2}} \right]^{-0.2} \cdot \left[\frac{D_{max}}{1\,\mathrm{mm}} \right]^{-0.43} \cdot F^{0.12}$$

$$\approx 324 \cdot \left[\frac{ESDD}{1\,\mathrm{mg/cm^2}} \right]^{-0.2} \cdot \left[\frac{D_{avg}}{1\,\mathrm{mm}} \right]^{-0.58} \cdot F^{0.11} \qquad (4\text{-}10)$$

where

E_{50} is the clean-fog flashover stress (kV/$m_{leakage}$),

$ESDD$ is the equivalent salt deposit density (mg/cm^2),

D_{max} is the maximum insulator diameter (mm),

D_{avg} is the surface-average insulator diameter (mm) given by the insulator surface area (mm^2) divided by $\pi \cdot L$, where L is the leakage distance (mm), and F is the insulator form factor.

The role of form factor in prediction of flashover strength is discussed further in the next section.

4.11.4. Influence of Insulator Form Factor

The form factor of an insulator is sometimes thought to play a role in the contamination performance. For example, in Chapter 5, a model derived by Topalis and Gonos [2001] uses insulator form factor to calculate ac flashover voltage on single insulator disks. A consolidation of clean-fog test results on disks of four different profiles [Kontargyri et al., 2007] shows the effect of insulator form factor on the critical flashover stress relation to ESDD in Figure 4-48.

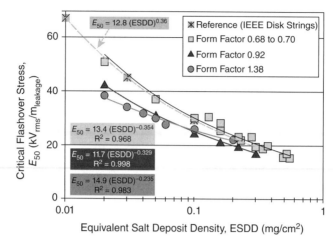

Figure 4-48: Influence of insulator form factor on critical flashover stress E_{50} (data from Kontargyri et al. [2007]).

The test results were converted from peak voltage to critical rms flashover stress. The E_{50} values for single insulators with a form factor of 0.68 or 0.70 show a good agreement with Equation 4-5 for strings of standard suspension disk insulators with similar form factor.

Insulators with a high form factor have deeper skirts, and this is suggested to have an effect on the exponent α relating flashover stress and ESDD. For insulators with low form factor of about 0.7 and relatively shallow ribs, shed efficiency is higher and the value of α approaches its theoretical value for a perfectly wetted surface of about –0.36. For insulators with deep ribs and a high form factor of about 1.4, the shed efficiency is lower and the exponent α drops to about –0.24.

It is interesting to expand on this work by applying the proposed flashover model incorporating insulator form factor [Topalis and Gonos, 2001] with the test results of Beauséjour [1981] on single pin-type distribution insulators. The predicted peak-of-ac-wave critical flashover voltages are plotted against the observations in Figure 4-49.

The Topalis and Gonos model does not describe the flashover voltages of the distribution insulators tested by Beauséjour particularly well. To get a sense for what is missing, it is reasonable to introduce the insulator height H (mm) into the empirical fit, recognizing that the form factor, maximum diameter, and height are interdependent and that several combinations of exponents can give adequate results. One reasonably accurate expression that includes insulator height H as well as maximum diameter D_{max} (mm) and form factor F is

$$E_{50} = 52 \cdot \left[\frac{H}{1\,\text{mm}} \right]^{0.2} \cdot \left[\frac{D_{max}}{1\,\text{mm}} \right]^{-0.4} \cdot F^{-0.1} \cdot \left[\frac{ESDD}{1\,\text{mg/cm}^2} \right]^{-0.24} \qquad (4\text{-}11)$$

where E_{50} is in kilovolts rms per meter of leakage distance. This empirical fit is seen to predict reasonable peak-of-ac-wave flashover voltages for both data

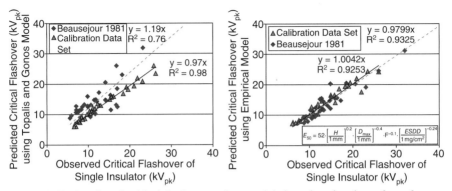

Figure 4-49: Predicted critical flashover using models based on insulator form factor.

sets in Figure 4-49, retaining the same units as the Topalis and Gonos [2001] model. The empirical fit suggests that the insulator diameter plays a more important role in flashover performance than either insulator height or form factor.

4.11.5. Influence of Surface Material

When the salt-fog and clean-fog test methods were developed, porcelain and glass were the only practical materials that demonstrated excellent long-term endurance in high-voltage applications. With the development of polymer materials such as ethylene-propylene-diene monomer (EPDM) and silicone rubber (SiR) materials to cover fibreglass rods, the nonceramic insulator was developed to a high state of reliability. The reduced number of mechanical connections—metal to insulator interfaces—is one reason why, on average, long-rod polymer insulators are now more reliable than strings of ceramic disks.

One advantage of some polymer materials, notably SiR, is that the surface resistance under contamination and fog conditions is considerably higher than porcelain. This can be a combination of the hydrophobicity of the material and also of the higher surface impedance. An example of the surface resistance measured in clean-fog tests of precontaminated ceramic and polymer insulators from Matsuoka et al. [1996a] is given in Figure 4-50.

The fog density of 3–5 g/m³ is seen to give the lowest surface impedance and this in turn would result in the lowest flashover stress. There is not much difference between the EPDM and porcelain materials in this case, and this is because the EPDM is typically hydrophilic, like the porcelain or glass. Figure 4-51 shows how the withstand voltage for all three insulators decreased with increasing salt deposit density (SDD).

The salt-fog and clean-fog test methods for nonceramic insulators generally act as combinations of aging and contamination test. For example, in the rapid flashover salt-fog test, Figure 4-28 showed that repeated flashovers caused a progressive reduction in polymer insulator strength and is thus biased heavily

Specimen Insulators	Number of sheds	Shed diameter mm	Trunk diameter mm	Leakage diameter mm	Unit diameter mm
Porcelain insulator	10	160	80	1020	585
Polymer insulator (SiR & EPDM)	7	126	26	980	623

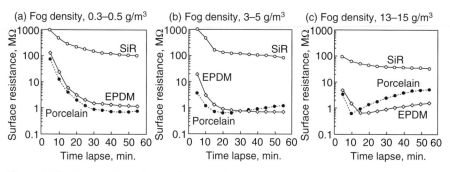

Figure 4-50: Change in surface resistance of long-rod porcelain and polymer insulators at three different fog densities (from Matsuoka et al. [1996a]).

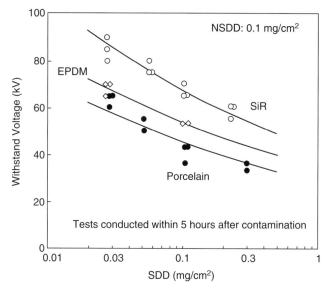

Figure 4-51: Effect of SDD on withstand voltage of contaminated insulators shown in Figure 4-50.

toward an aging test. Also, in order to achieve a uniform pollution deposit in the solid layer clean-fog test, some method to remove any surface oil is needed. The recovery time and conditions after this process also act more as a test of the ability of the material to recover after a stress than as a test of the response to the stress.

Treatment of every variable needed to define the contamination performance of a modern (or aged) nonceramic insulator is a worthy topic for a series of books, as the materials themselves have improved with time. However, in freezing conditions, which are the main focus of this work, the differences among polymer materials tend to be minimized below the freezing point. Ice adheres equally well to the polymer and ceramic materials, leaving the other insulator parameters—such as diameter and shed spacing—to delineate the real-world performance.

4.12. EFFECTS OF NONSOLUBLE DEPOSIT DENSITY

Recently, Jiang et al. [2007] described results of flashover tests on suspension disk insulators in Figures 4-52 and 4-53 that introduced two factors—the effect of insulator profile and the effect of nonsoluble deposit.

The value of $\alpha = -0.22$ for the change in ac flashover of porcelain disks with ESDD in Figure 4-52 matches the values reported [CIGRE Task Force 33.04.01, 2000]. However, the exponent for the outer-rib style of disk with smooth undersides was fitted to be $\alpha = -0.33$ in Figure 4-53.

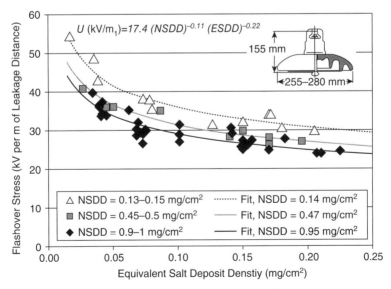

Figure 4-52: Observed flashover stress across leakage distance for three types of cap-and-pin insulator as function of ESDD with NSDD as a parameter.

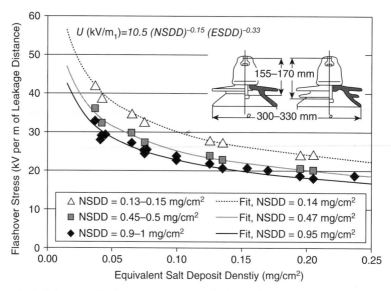

Figure 4-53: Observed flashover stress across leakage distance for two types of outer-rib cap-and-pin insulator as function of ESDD with NSDD as a parameter.

The relationships of flashover voltage with NSDD are also derived in Figures 4-52 and 4-53.

The observed value of $\alpha = -0.22$ for the cap-and-pin insulators in Figure 4-52 suggests that about a third of the overall leakage distance is not effective. A confirmation of this tendency is seen in the exponents that fit the relation with NSDD: the exponent for the outer-rib designs, -0.15, is higher than the value for the standard cap-and-pin insulator (-0.11) by a similar ratio.

The role of NSDD on the relative withstand voltage has also been evaluated by Matsuoka et al. [1996a] and CIGRE Task Force 33.04.01 [2000]. Normalized to an NSDD of $0.1\,mg/cm^2$, Figure 4-54 shows that the flashover strength with $1\,mg/cm^2$ NSDD is 78–82% of the original value, and the strength with NSDD of $10\,mg/cm^2$ is 65–70% of the reference.

4.13. PRESSURE EFFECTS ON CONTAMINATION TESTS

4.13.1. Standard Correction for Air Density and Humidity

The value of air flashover gradient itself changes with temperature, pressure, and humidity. The process to correct for this change is spelled out in IEC 60507 [1991] and IEEE Standard 4 [1995]. Details of the standard process for correcting high-voltage test results for ambient temperature, pressure, and humidity (expressed as dew point temperature) are described in Appendix B, but an overview is repeated here. The overall correction factor for air density and humidity, K, is

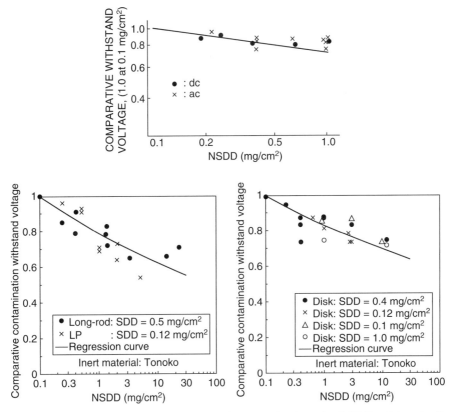

Figure 4-54: Withstand voltage relative to (withstand voltage for NSDD = 0.1 mg/cm²) for ac and dc Voltage (top figure from Matsuoka et al., [1996b]; bottom figures from CIGRE Task Force 33.04.01, [2000]; courtesy of CIGRE).

$$K = k_1 k_2 = \delta^m k^w \tag{4-12}$$

The factor K in Equation 4-12 is used to correct for the flashover strength of the air.

This process is used up to the point when condensation occurs, at which point the humidity correction is meaningless. Below this point, Figure 4-55 shows that the flashover strength of insulators increases with absolute humidity.

The relative air density δ at pressure b and temperature t is computed from the reference conditions of pressure ($b_0 = 101.3\,\text{kPa}$) and ambient temperature ($t_0 = 20\,°\text{C}$) using

$$\delta = \left(\frac{b}{b_0} \right) \left(\frac{273 + t_0}{273 + t} \right) \tag{4-13}$$

The exponents m and w for flashovers both depend on an overvoltage factor g, given by

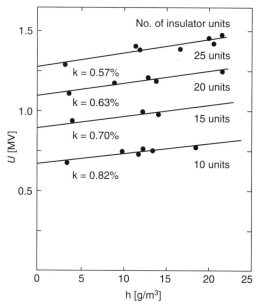

Figure 4-55: Switching impulse (220/2100) flashover level versus absolute humidity for strings of 146-mm × 254-mm suspension insulators (from Harada et al., [1971]).

$$g = \frac{V_{50\%}}{500 \text{ kV} \cdot L \cdot \delta \cdot k} \qquad (4\text{-}14)$$

where $V_{50\%}$ is the 50% flashover voltage in the contamination test.

Mercure [1989] fitted values of m for the pressure correction of flashover values to ac and dc, wet and dry test results as shown in Figure 4-56. For dry tests, the value of m is close to unity, while for wet tests, m varies from 0.29 to 0.5.

Contamination test results are usually obtained at nearly 100% relative humidity with values of $V_{50\%}$ less than about half of the 540-kV/m impulse flashover level. The nominal value of g is low and this suggests the exponents m and w should also be small. Taking a specific example, in clean-fog conditions, at normal pressure and saturated conditions (1013 mbar, 19 °C dew point, 20 °C ambient) with flashover at line voltage stress, the value of $g = 0.21$ along with an absolute humidity of 17.2 g/m³ give an atmospheric correction factor of $K = 1.0010$ that could be neglected. For $g < 0.2$, $m = 0$, meaning that the flashover would not vary at all with pressure. This is well outside the intended range for the standard correction factors in IEEE Standard 4 [1995] and IEC 60507 [1991]. Another process should be used instead.

Pressure effects on contamination flashover are outlined next in Section 4.13.2. Ambient temperature and humidity will affect the temperature,

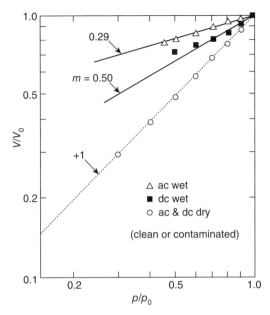

Figure 4-56: Ratio of V ($V_{50\%}$ at pressure p) to V_0 ($V_{50\%}$ at nominal atmospheric pressure p_0) [Mercure, 1989], showing typical fitted exponents m in Equation 4-12.

evaporation rate, and electrical conductivity of the pollution layer, as outlined in Section 4.14. The ambient temperature also affects the thermal balance of the arc. This means that there may be differences in the V–I characteristics of the arc itself, which would affect dc flashover. Mathematical modeling presented in Chapter 5 suggests that the reignition process of the arc, which is the important factor for ac contamination flashover, is practically independent of ambient temperature.

4.13.2. Pressure Corrections for Contamination Flashovers

Overwhelming experimental evidence demonstrates that ambient pressure plays an important role in contamination flashover, even though the values of g and m are low and standards suggest that correction is negligible. Kawamura et al. [1982] show in Figure 4-57 that, depending on contamination level, a 35% reduction in pressure leads to a 4–10% reduction in withstand voltage, taken as the $V_{5\%}$ level for dc tests. Mercure [1989] reviewed these results and derived exponents of $m = 0.35$ for negative dc and $m = 0.40$ for positive dc in Equation 4-12.

In his survey of major trends, Mercure also suggested a pressure dependence exponent of $m = 0.5$ for ac, considering the $V_{50\%}$ flashover strength of standard insulators at altitudes less than 4 km (600 kPa $< p <$ 1000 kPa). For

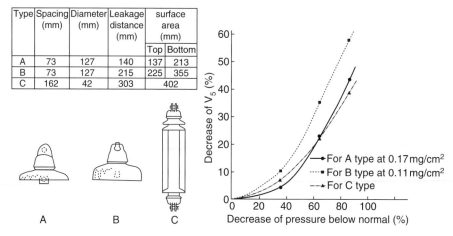

Type	Spacing (mm)	Diameter (mm)	Leakage distance (mm)	surface area (mm)	
				Top	Bottom
A	73	127	140	137	213
B	73	127	215	225	355
C	162	42	303	402	

Figure 4-57: Change in 5% flashover voltage ($V_{5\%}$) with decrease in pressure from 100 kPa for three types of insulator (from Kawamura et al. [1982]).

deep-rib anti-fog profile insulators, the value of m increased to 0.55. He combined the effects of pressure and pollution level in a single expression:

$$V_{50\%}(A_2, S_2) \approx \left[\left(\frac{\delta_2}{\delta_1} \right)^{0.5} \cdot \left(\frac{S_1}{S_2} \right)^{0.33} \right] \cdot V_{50\%}(A_1, S_1) \tag{4-15}$$

where

δ_1 is the air density at the altitude A_1 and severity S_1, and

δ_2 is the air density at the altitude A_2 and severity S_2.

Figure 4-58 shows that the fitted line in Equation 4-15 captures the central trend of experimental data from low and high altitudes, with about 10% scatter.

Rizk and Rezazada [1997] suggest, for example, that the median exponent of $m = 0.5$ for pressure summarized many clean-fog contamination flashover tests in the pressure range from 50 to 100 kPa, as well as their own mathematical modeling results. The experimental range was $0.28 < m < 0.8$, depending on insulator shape and pollution level. Higher levels of pollution, leading to lower flashover voltage, tend to have higher values of m, which contradicts the predicted effect that m should decrease with the overvoltage factor g.

The process for environmental correction of contamination test results at constant temperature was generalized by CIGRE Task Force 33.04.01 [2000] to use the altitude h (in km) using the form

$$V = V_0 (1 - k_i h) \tag{4-16}$$

where

V_0 is the flashover strength (kV) at sea level,

V is the flashover strength at the altitude h (km),

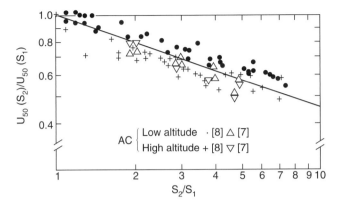

Figure 4-58: Ratio of $V_{50\%}$ at two severities S_1 and S_2 as a function of severity ratio (S_2/S_1) for low- and high-altitude test locations [Mercure, 1989], (adapted from Bergman and Kolobova [1983] and Rudakova and Tikohdeev [1989]).

k_i is a constant that depends on applied voltage (dc, ac, switching (SI), or lightning impulse (LI)) as follows: k_{dc+} and k_{dc-} = 0.035; k_{ac} = 0.05; k_{SI} = 0.05; k_{LI} = 0.10.

4.14. TEMPERATURE EFFECTS ON POLLUTION FLASHOVER

4.14.1. Temperatures Above Freezing

It is normal in artificial tests using steam fog that the temperature in the chamber rises over the duration of the test. In general, the electrical conductivity of pollution layers increases at the rate of about 2.2% per C° around 20 °C. As detailed in Chapter 3, the specific rate of change will be a function of the ionic content, but most of the common pollutants such as sulfate and nitrate have roughly the same rate of change as sodium chloride. This means that the pollution layer resistance will reduce by about 22% as the test chamber warms from 20 °C to 30 °C, a typical temperature difference [Cherney et al., 1983].

A 22% increase in the electrical conductivity of a pollution layer from a 10 °C increase in temperature should lead to a 6% decrease in flashover voltage, using the relation $1.22^{-0.33}$ = 0.94 from Equation 4-15.

Tests to evaluate the dc flashover of disk insulators as a function of ambient temperature were carried out by Ishii et al. [1984]. These support the idea that flashover strength decreases with increasing temperature (Figure 4-59).

The effect of ambient temperature on the dc flashover strength of insulators in salt-fog conditions was also demonstrated by Ramos et al. [1993] in Figure 4-60. Their change in flashover voltage observed over temperature was about ±10% in the range of 5–30 °C.

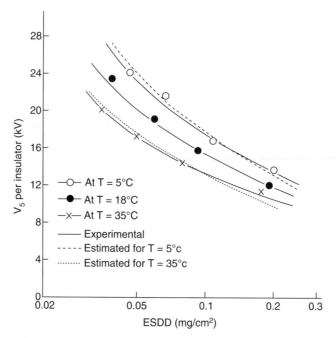

Figure 4-59: Change in withstand flashover voltage (taken as $V_{5\%}$) with ESDD at three different ambient temperatures (from Ishii et al., [1984]).

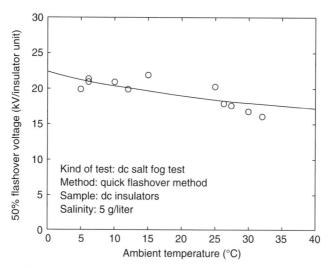

Figure 4-60: Effect of ambient temperature on dc flashover in salt-fog test (adapted from Ramos et al., [1993]; Mizuno et al., [1997]).

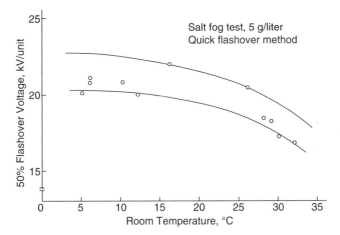

Figure 4-61: Flashover voltage ($V_{50\%}$) per disk in salt-fog conditions as function of test chamber temperature (from CIGRE Task Force 33.04.04 [1992]; courtesy of CIGRE).

The effect of decreasing flashover voltage with increasing temperature has also been noted in the results of salt-fog flashover tests on dc insulators [CIGRE Task Force 33.04.04, 1992]; see Figure 4-61.

In all three cases, the change in flashover strength can be fully explained by the change in layer conductivity with temperature. Mizuno et al. [1997] established that, theoretically, contamination flashover voltage should decrease by 0.4% and 0.66% per degree C for ac and dc, respectively. This agreed well with experimental data such as that shown in Figure 4-60. Near 20°C, the change in layer conductivity with temperature is about 2.2% per °C as shown in Figure 4-62.

In artificial fog tests, the temperature of the pollution layer is more important than, and not necessarily the same as, the ambient temperature. The factors affecting pollution layer temperature in artificial tests include:

- The rate of rise of temperature and dew point inside the chamber.
- The surface area, mass, and specific heat capacity of the insulator, combining to give a thermal time constant that causes a lag of insulator temperature relative to ambient temperature.
- The ambient temperature inside the test chamber.
- The altitude and dew point temperature of atmospheric air outside the test chamber.
- The temperature of the fog, if the drops are large and have not reached equilibrium with the chamber temperature.

4.14.2. Temperatures Below Freezing

Flashover levels obtained from the cold-fog test procedure are generally similar to those that have been measured with clean-fog tests at the same

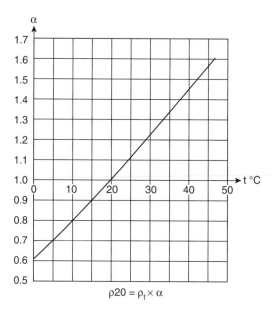

$$\rho20 = \rho_t \times \alpha$$

Figure 4-62: Correction factor for resistivity change with temperature (from Ishii et al., [1984]).

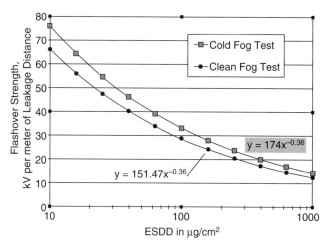

Figure 4-63: Comparison of clean-fog and cold-fog ac flashover strength of leakage distance as a function of contamination level, expressed as equivalent salt deposit density (ESDD) (adapted from Chisholm [2007]).

contamination level. With the restricted level of leakage current activity, the standard deviation of cold-fog test results is 4% rather than 7–12% for clean-fog tests. Figure 4-63 shows that the observed specific cold-fog flashover strength, expressed in kilovolts per meter of leakage distance, for cap-and-pin porcelain insulators [Chisholm, 2007], is about 15% higher than the clean-fog

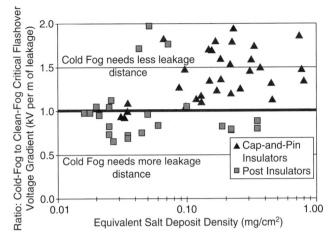

Figure 4-64: Ratio of cold-fog to clean-fog critical flashover gradient across leakage distance for ceramic cap-and-pin and station post insulators (from Chisholm, [2007]).

results for IEEE standard suspension disk insulators [Baker et al., 2008]. In both cases, the flashover voltage decreases nonlinearly with increasing ESDD, and in both cases, a power curve exponent of $\alpha = -0.36$ fits the data well. This suggests that the cold-fog flashover problem is indeed a specific type of contamination flashover problem. The high value of α also suggests that the cold-fog process fully wets the surfaces and that the flashover involves the entire insulator leakage distance.

Under ac conditions, cold-fog test results on station post insulators of large diameter were found to fall below the results for clean-fog conditions, while strings of cap-and-pin insulators, including apparatus types with 400-mm spacing, performed relatively well. Figure 4-64 shows that cap-and-pin insulators dimensioned properly for a contaminated environment in clean-fog conditions will perform adequately in the same environment in the winter.

However, the only ceramic post insulators that performed better in cold-fog than clean-fog conditions were small-diameter 44-kV units with 125 kV BIL and 240 mm of leakage distance. The 500-kV and 69-kV post insulators required as much as 40% more leakage distance in cold-fog conditions than in clean-fog conditions.

4.15. PRÉCIS

The electrical breakdown of air is a function of the local electric field. Separation distances among electric power system conductors make use of mathematical models that fit laboratory test results for a wide range of steady-state and transient overvoltage conditions.

The electrical breakdown of air in the immediate vicinity of polluted insulators was also studied in outdoor field tests, but proved to be more difficult to reproduce in the laboratory. Studies led to controlled salt-fog and clean-fog test methods using expensive power supplies that provide the necessary voltage and current. These test methods can reproduce specific pollution problems on ceramic insulators with hydrophilic surfaces that could be fully wetted. However, the tests have not been fully adapted for new hydrophobic polymer materials that retain an ability to bead water.

REFERENCES

Aleksandrov, G. N., V. Y. Kizvetter, V. M. Rudakova, and A. N. Tushnov. 1962. "The AC Flashover Voltages of Long Air Gaps and Strings of Insulators," *Elektrichestvo*, No. 5, pp. 27–32.

Aleksandrov, G. N., V. I. Ivanov, and V. P. Radkov. 1965. "Electrical Strength of Air Gaps Between EHV Line Conductors and Earth During Surges," *Elektrichestvo*, No. 4, pp. 20–24.

Arai, N. 1982. "AC Fog Withstand Test on Contaminated Insulators by Steam Fog," *IEEE Transactions on Power Apparatus and Systems*, Vol. PAS-101, No. 11, pp. 4316–4323.

Baker, A. C., L. E. Zaffanella, L. D. Anzivino, H. M. Schneider, and J. H. Moran. 1989. "A Comparison of HVAC and HVDC Contamination Performance of Station Post Insulators," *IEEE Transactions on Power Delivery*, Vol. 4, No. 2 (Apr.), pp. 1486–1491.

Baker, A. C., M. Farzaneh, R. S. Gorur, S. M. Gubanski, R. J. Hill, G. G. Karady and H. M. Schneider. 2008. "Selection of Insulators for AC Overhead Lines in North America with respect to Contamination," *IEEE Transactions on Power Delivery*, in press.

Beauséjour, Y. 1981. "Insulator Contamination Study for Distribution Systems." CEA R&D Report for Project 75-05. October; available from CEATI International Inc.

Bergman, V. I. and O. I. Kolobova. 1983. "Some Results of Investigation of the Dielectric Strength of Polluted Line Insulators in Conditions of Reduced Atmospheric Pressure," *Electrotechnika*, Vol. 54, No. 2, pp. 54–56.

Chang, J. W. and R. S. Gorur. 1994. "Surface Recovery of Silicone Rubber Used for HV Outdoor Insulation," *IEEE Transactions on Dielectrics and Electrical Insulation*, Vol. 1, No. 6 (Dec.), pp. 1039–1046.

Cherney, E. A., Y. Beausejour, T. C. Cheng, K. J. Lloyd, G. Marrone, J. H. Moran, K. Naito, L. Pargamin, E. Reynaert, J. D. Sakich, and C. F. Sarkinen. 1983. "The AC Clean-Fog Test for Contaminated Insulators," *IEEE Transactions on Power Apparatus and Systems*, Vol. PAS-102, No. 3 (Mar.), pp. 604–613.

Chisholm, W. A. 2005. "Ten Years of Application Experience with RTV Silicone Coatings in Canada," in *Proceedings of 2005 INMR*, Hong Kong, November.

Chisholm, W. A. 2007. "Insulator Leakage Distance Dimensioning in Areas of Winter Contamination using Cold-Fog Test Results," *IEEE Transactions on Dielectrics and Electrical Insulation*, Vol. 14, No. 6 (Dec.), pp. 1455–1461.

Chisholm, W. A., K. G. Ringler, C. C. Erven, M. A. Green, O. Melo, Y. Tam, O. Nigol, J. Kuffel, A. Boyer, I. K. Pavasars, F. X. Macedo, J. K. Sabiston, and R. B. Caputo.

1996. "The Cold Fog Test," *IEEE Transactions on Power Delivery*, Vol. 11, pp. 1874–1880.

CIGRE Task Force 33.04.04. 1992. "Artificial Pollution Testing of HVDC Insulators: Analysis of Factors Influencing Performance," *Electra*, No. 140 (Feb.), pp. 99–113.

CIGRE Working Group 33.07. 1992. "Guidelines for the Evaluation of the Dielectric Strength of External Insulation." Technical Brochure 72, Paris: CIGRE.

CIGRE Task Force 33.04.07. 1999. "Natural and Artifical Ageing and Pollution Testing of Polymeric Insulators." Technical Brochure 142, Paris: CIGRE.

CIGRE Task Force 33.04.01. 2000. "Polluted Insulators: A Review of Current Knowledge." Technical Brochure 158. Paris: CIGRE.

Cortina, R., E. Dabusti, and G. Marrone. 1976. "Experimental Research at Enel on Surface Insulation in Naturally Polluted Conditions," in *Conference on Partial Discharge in Electrical Insulation*, Bangalore, India. April.

Dakin, T.W., G. Luxa, G. Oppermann, J. Vigreux, G. Wind and H. Winkelnkemper. 1974. "Breakdown of Gases in Uniform Fields: Paschen Curves for Nitrogen, Air and Sulfur Hexafluoride," *Electra*, Vol. 32, pp. 61–82.

De La O, A., R. S. Gorur and J. Chang. 1994. "AC Clean Fog Test on Non-ceramic Insulating Materials and a Comparison with Porcelain," *IEEE Transactions on Power Delivery*, Vol. 9, No. 4 (Oct.), pp. 2000–2008.

Eklund, A., I. Gutman, and R. Hartings. 1995. "Dust Cycle Method: Fast Pollution Accumulation on Silicone Rubber/EPDM Insulators," in *Proceedings of IEEE, International Symposium on Electric Power Engineering*, Stockholm Power Tech, Vol. "High-Voltage Technology," pp. 293–297. June 18–22.

Ely, C. H. A., R. G. Kingston, and P. J. Lambeth. 1971. "Artificial- and Natural-Pollution Tests on Outdoor 400-kV Substation Insulators," *Proceedings of the IEE*, Vol. 118, No. 1 (Jan.), pp. 99–109.

Engelbrecht, C. S., R. Hartings, H. Tunell, B. Engström, H. Janssen, and R. Hennings. 2003. "Pollution Tests for Coastal Conditions on an 800-kV Composite Bushing," *IEEE Transactions on Power Delivery*, Vol. 18, No. 3 (July), pp. 953–959.

EPRI (Electric Power Research Institute). 2005. *Transmission Line Reference Book, 200 kV and Above*, 3nd edition. Palo Alto, CA: EPRI.

Forrest, J. S. 1936. "The Electrical Characteristics of 132-kV Line Insulators Under Various Weather Conditions," *IEE Journal*, Vol. 79, pp. 401–423.

Forrest, J. S., P. J. Lambeth, and D. F. Oakeshott. 1959. "Research on the Performance of High-Voltage Insulators in Polluted Atmospheres," in *IEE Proceedings*, Paper 3014S, pp. 172–196. November.

Fujitaka, S., T. Kawamura, S. Tsurumi, H. Kondo, T. Seta, and M. Yamamoto. 1968. "Japanese Method of Artificial Pollution Tests on Insulators," *IEEE Transactions on Power Apparatus and Systems*, Vol. PAS-87, No. 3 (Mar.), pp. 729–735.

Garcia, R. W. S, N. H. C. Santiago, and C. M. M. Portela. 1991. "A Mathematical Model to Study the Influence of Source Parameters in Polluted Insulator Tests," in *Proceedings of 3rd International Conference on Properties and Applications of Dielectric Materials*, Tokyo, Japan, July 8–12.

Gorur, R. S., E. A. Cherney, R. Hackam. 1987. "Performance of Polymeric Insulating Materials in Salt-Fog," *IEEE Transactions on Power Delivery* Vol. PWRD-2, No. 2 (Apr.), pp. 486–492.

Gutman, I., R. Hartings, U. Anström and D. Gustavsson. 2001. "Procedure to Obtain a Realistic Degree of Hydrophobicity on Artificially Polluted Silicone Rubber Insulators," *Proceedings of 12th International Symposium on High Voltage (ISH)*, Bangalore, India (Aug. 20–24), paper 5-2.

Harada, T., Y. Aihara, and Y. Aoshima. 1971. "Influence of Humidity on Lightning and Switching Flashover Voltages," *IEEE Transactions on Power Apparatus and Systems*, Vol. 90, pp. 1433–1441.

Hileman, A. R. 1999. *Insulation Coordination for Power Systems*. Boca Raton, FL: CRC Press, Taylor & Francis Group.

Holtzhausen, J.P. 1993. "Leakage current monitoring on synthetic insulators at a severe coastal site," *Proceedings of International Workshop on Non-Ceramic Outdoor Insulation*, Société des Electricients et des Electroniciens (SEE), 15–16 Apr., Paris.

Houlgate, R. G. and D. A. Swift. 1990. "Composite Rod Insulators for AC Power Lines: Electrical Performance of Various Designs in a Coastal Testing Station," *IEEE Transactions on Power Delivery*, Vol. 5, No. 4 (Oct.), pp. 1944–1955.

IEC 60815. 1986. *Guide for the Selection of Insulators in Respect of Polluted Conditions*. Geneva, Switzerland: Bureau Central de la Commission Electrotechnique Internationale. January 1.

IEC 60507. 1991. *Artificial Pollution Tests on High Voltage Insulators to Be Used in AC Systems*. Lausanne, Switzerland: IEC.

IEC 61109. 1992. *Composite Insulators for a.c. Overhead Lines with a Nominal Voltage Greater than 1000V—Definitions, Test Methods and Acceptance Criteria*. Geneva, Switzerland: Bureau Central de la Commission Electrotechnique Internationale. March.

IEC 61245. 1993. *Artificial Pollution Tests on High-Voltage Insulators to be Used in d.c. Systems*. Geneva, Switzerland: Bureau Central de la Commission Electrotechnique Internationale, October.

IEC 60071-2. 1996. *Insulation Co-ordination Part 2: Application Guide*. Geneva, Switzerland: Bureau Central de la Commission Electrotechnique Internationale.

IEC 62217. 2005. *Polymeric Insulators for Indoor and Outdoor Use with a Nominal Voltage >1000V—General Definitions, Test Methods and Acceptance Criteria*. Geneva, Switzerland: Bureau Central de la Commission Electrotechnique Internationale. October.

IEC 60815. 2008. *Selection and Dimensioning of High-Voltage Insulators Intended for Use in Polluted Conditions*. Geneva, Switzerland: Bureau Central de la Commission Electrotechniuqe Internationale, October.

IEEE Standard 4. 1995. *IEEE Standard Techniques for High Voltage Testing*. Piscataway, NJ: IEEE Press.

IEEE Task Force on Icing Performance of Line Insulators. 2007. "Selection of Line Insulators With Respect to Ice and Snow—Part II: Selection Methods and Mitigation Options," *IEEE Transactions on Power Delivery*, Vol. 22, No. 4 (Oct.), pp. 2297–2304.

IEEE Working Group on Insulator Contamination. 1979. "Application of Insulators in a Contaminated Environment," *IEEE Transactions on Power Apparatus and Systems*, Vol. PAS-98, No. 5 (Sept. 1Oct.), pp. 1676–1695.

Ishii, M., M. Akbar, and T. Kawamura. 1984. "Effect of Ambient Temperature on the Performance of Contaminated DC Insulators," *IEEE Transactions on Electrical Insulation*, Vol. EI-19, No. 2 (Apr.), pp. 129–134.

Jiang, X., J. Yuan, Z. Zhang, J. Hu and C. Sun. 2007. "Study on AC Artificial-Contaminated Flashover Performance of Various Types of Insulators," *IEEE Transactions on Power Delivery*, Vol. 22, No. 4 (Oct.), pp. 2567–2574.

Karady, G. G., M. Shaw, and R. L. Brown. 1995. "Flashover Mechanism of Silicone Rubber Insulators Used for Outdoor Insulation—I," *IEEE Transactions on Power Delivery*, Vol. 10, No. 4 (Oct.), pp. 1965–1971.

Karady, G. G., E. Al-Ammar, S. Baozhuang, M. W. Tuominen. 2006. "Experimental Verification of the Proposed IEEE Performance and Testing Standard for ADSS Fiber Optic Cable for Use on Electric Utility Power Lines," *IEEE Transactions on Power Delivery*, Vol. 21, No. 1 (Jan.), pp. 450–455.

Kawai, M. and D. M. Milone. 1969. "Tests on Salt-Contaminated Insulators in Artificial and Natural Wet Conditions," *IEEE Transactions on Power Apparatus and Systems*, Vol. PAS-88, No. 9 (Sept.), pp. 1394–1399.

Kawai, M. and M. Sforzini. 1974. "Problems Related to the Performance of UHV Insulators in Contaminated Conditions," in *CIGRE International Conference on High Voltage Electric Systems*, Paper 33–19. August 21–29.

Kawamura, T., M. Ishii, M. Akbar, and K. Nagai. 1982. "Pressure Dependence of DC Breakdown of Contaminated Insulators," *IEEE Transactions on Electrical Insulation*, Vol. EI-17, No. 1 (Feb.), pp. 39–45.

Knudsen, N. and F. Iliceto. 1970. "Flashover Tests on Large Air Gaps with DC Voltage and with Switching Surges Superimposed on DC Voltage," *IEEE Transactions on Power Apparatus and Systems*, Vol. PAS-89, No. 5–6 (May/June), pp. 781–788.

Kontargyri, V. T., A. A. Gialketsi, G. J. Tsekouras, I. F. Gonos, and I. A. Stathopulos. 2007. "Design of an Artificial Neural Network for the Estimation of the Flashover Voltage on Insulators," *Electric Power Systems Research*, Vol. 77, pp. 1532–1540.

Kuffel, E., W.S. Zaengl, and J. Kuffel. 2000. *High Voltage Engineering: Fundamentals*, Second Edition. Oxford: Newnes.

Lambeth, P. J. 1973. "Insulators for 1000 to 1500 kV Systems," *Philosophical Transactions of the Royal Society of London Series A, Mathematical and Physical Sciences*, Vol. 275, No. 1248 (Aug.), pp. 153–163.

Lambeth, P. J. 1988. "Variable-Voltage Application for Insulator Pollution Tests," *IEEE Transactions on Power Delivery*, Vol. 3, No. 4 (Oct.), pp. 2103–2111.

Lambeth, P. J. 1990. "Laboratory Tests to Evaluate HVDC Wall Bushing Performance in Wet Weather," *IEEE Transactions on Power Delivery*, Vol. PWRD-5, No. 4 (Oct.), pp. 1782–1793.

Lambeth, P. J, J. S. T. Looms, M. Sforzini, C. Malaguti, Y. Porcheron, and P. Claverie. 1970. "International Research on Polluted Insulators," in *Proceedings of CIGRE 1970 Session*, Paper 33.02. August 24–September 2.

Lambeth, P. J., J. S. T. Looms, M. Sforzini, R. Cortina, Y. Porcheron, and P. Claverie. 1973. "The Salt Fog Test and Its Use in Insulator Selection for Polluted Localities," *IEEE Transactions on Power Apparatus and Systems*, Vol. PAS-92, No. 6 (Nov.), pp. 1876–1887.

Lambeth, P. J., H. M. Schneider, Y. Beausejour, E. A. Cherney, D. Dumora, T. Kawamura, G. Marrone, J. H. Moran, K. Naito, R. J. Nigbor, J. D. Sakich, R. Stearns, H. Tempelaar, M. P. Verma, J. Huc, G. Perin, G. B. Johnson, and C. De Ligt. 1987. "Final Report on the Clean Fog Test for HVAC Insulators," *IEEE Transactions on Power Delivery*, Vol. PWRD-2, No. 4 (Oct.), pp. 1317–1326.

Le Roy, G., C. Gary, B. Hutzler, J. Lalot, and C. Dubanton. 1984. *Les propriétés diélectriques de l'air et les très hautes tensions.* Paris: Editions Eyrolles.

Looms, J. S. T. 1988. *Insulators for High Voltages.* London: Peter Peregrinus Ltd.

Macchiaroli, B. and F. J. Turner. 1971. "Switching Surge Performance of Contaminated Insulators," *IEEE Transactions on Power Apparatus and Systems,* Vol. PAS-90, No. 4 (July), pp. 1612–1619.

Matsuoka, R., S. Ito, K. Sakanishi, and K. Naito. 1991. "Flashover on Contaminated Insulators with Different Diameters," *IEEE Transactions on Electrical Insulation,* Vol. 26, No. 6 (Dec.), pp. 1140–1146.

Matsuoka, R., H. Shinokubo, K. Kondo, Y. Mizuno, K. Naito, T. Fujimura, and T. Terada. 1996a. "Assessment of Basic Contamination Withstand Voltage Characteristics of Polymer Insulators," *IEEE Transactions on Power Delivery,* Vol. 11, No. 4 (Oct.), pp. 1895–1900.

Matsuoka, R., K. Kondo, K. Naito, and M. Ishii. 1996b. "Influence of Nonsoluble Contaminants on the Flashover of Artificially Contaminated Insulators," *IEEE Transactions on Power Delivery,* Vol. 11, No. 1 (Jan.), pp. 420–430.

Meek, J.M. and J.D. Craggs. 1978. *Electrical Breakdown of Gasses.* Norwich, United Kingdom: Wiley.

Mercure, H. P. 1989. "Insulator Pollution Performance at High Altitude: Major Trends," *IEEE Transactions on Power Delivery,* Vol. 4, No. 2 (Feb.), pp. 1461–1468.

Mizuno, Y., H. Kusada, and K. Naito. 1997. "Effect of Climatic Conditions on Contamination Flashover Voltage of Insulators," *IEEE Transactions on Dielectrics and Electrical Insulation,* Vol. 4, No. 3 (June), pp. 286–289.

Naito, K. and H. M. Schneider 1995. "Round-Robin Artificial Contamination Test on High Voltage DC Insulators," *IEEE Transactions on Power Delivery,* Vol. 10, No. 3, pp. 1438–1442.

Naito, K., R. Matsuoka, and K. Sakanishi. 1990. "Investigation of the Insulation Performance of the Insulator Covered with Lichen," *IEEE Transactions on Power Delivery,* Vol. 5, No. 3 (July), pp. 1634–1640.

Naito, K., K. Izumi, K. Takasu and R. Matsuoka. 1996. "Performance of Composite Insulators under Polluted Conditions," *Proceedings of 1996 CIGRE Session,* Paper No. 33-301.

Naito, K., R. Matsuoka, T. Irie, and K. Kondo. 1999. "Test Methods and Results for Recent Outdoor Insulation in Japan," *IEEE Transactions on Dielectrics and Electrical Insulation,* Vol. 6, No. 5 (Oct.), pp. 732–743.

Nasser, E. 1972. "Contamination Flashover of Outdoor Insulation," *Electroteknik + Automation, (ETZ-A), VDE-Verlag,* Vol. 93, No. 6, pp. 321–325.

Olivier, G., G. Roy, R.-P. Bouchard and Y. Gervais. 1984. "Analytical Model of HV Cascade Connected Test Transformers," in *Electrical Machines and Converters— Modelling and Simulation,* H. Buyse and H. Robert (ed.), North-Holland: Elsevier.

Pargamin, L., J. Huc, and S. Tartier. 1984. "Considerations on the Choice of the Insulators for HVDC Overhead Lines," in *Proceedings of 30th CIGRE Session,* Paris, Paper No. 33-11. August 29–September 6.

Phillips, A. J., D. J. Childs, and H. M. Schneider. 1999. "Aging of Non-Ceramic Insulators Due to Corona from Water Drops," *IEEE Transactions on Power Delivery,* Vol. 14, No. 3 (July), pp. 1081–1089.

Pigini, A., G. Rizzi, E. Garbagnati, A. Porrino, G. Baldo, and G. Pesavento. 1989. "Performance of Large Air Gaps Under Lightning Overvoltages: Experimental Study and Analysis of Accuracy of Predetermination Methods," *IEEE Transactions on Power Delivery*, Vol. 4, No. 2, pp. 1379–1392.

Ramos, G., M. Campillo, and K. Naito. 1993. "A Study on the Characteristics of Various Conductive Contaminants Accumulated on High Voltage Insulators," *IEEE Transactions on Power Delivery*, Vol. 8, No. 4 (Oct.), pp. 1842–1850.

Rizk, F.A.M. and D.H. Nguyen. 1984. "AC Source-Insulator Interaction in HV Pollution Tests," *IEEE Transactions on Power Apparatus and Systems*, Vol. PAS-103, No. 4 (Apr.), pp. 723–732.

Rizk, F. A. M. and M. Bourdages. 1985. "Influence of AC Parameters on Flashover Characteristics of Polluted Insulators," *IEEE Transactions on Power Apparatus and Systems*, Vol. PAS-104, No. 4 (Apr.), pp. 948–958.

Rizk, F. A. M and A. Q. Rezazada. 1997. "Modeling of Altitude Effects on AC Flashover of High-Voltage Insulators," *IEEE Transactions on Power Delivery*, Vol. 12, No. 2 (Apr.), pp. 810–822.

Rizk, F. A. M. 1989. "A Model for Switching Impulse Leader Inception and Breakdown of Long Air-Gaps," *IEEE Transactions on Power Delivery*, Vol. 4, No. 1 (Jan.), pp. 596–606.

Rudakova, V. M. and N. N. Tikhodeev. 1989. "Influence of Low Air Pressure on Flashover Voltages of Polluted Insulators: Test Data, Generalization Attempts and Some Recommendations," *IEEE Transactions on Power Delivery*, Vol. 4 No. 1 (Jan.), pp. 607–613.

Ryle, P. J. 1931. "Two Transmission Line Problems—Suspension Insulators for Industrial Areas in Great Britain; Conductor Vibration," *IEE Journal*, Vol. 69, pp. 805–849.

Schneider, H. M. and C. W. Nicholls. 1978. "Contamination Flashover Performance of Insulators for UHV," *IEEE Transactions on Power Apparatus and Systems*, Vol. PAS-97, No. 4 (July/Aug.), pp. 1411–1420.

Spangenberg, E. and G. Riquel. 1997. "In Service Diagnostic of Composite Insulators: EDF's Test Results," in *Proceedings of 10th International Symposium on High Voltage Engineering (ISH)*, Montreal. August.

Topalis, F. V. and I. F. Gonos. 2001. "Dielectric Behaviour of Polluted Porcelain Insulators," *IEE Proceedings on Generation, Transmission and Distribution*, Vol. 148, No. 4, pp. 269–274.

Train, D. and P.E. Vohl. 1976. "Determination of Ratio Characteristics of Cascade Connected Transformers," *IEEE Transactions on Power Apparatus and Systems*, Vol. PAS-95, No. 6 (Nov./Dec.), pp. 1911–1918.

Ye, H., J. Zhang, Y. M. Ji, W. Y. Sun, K. Kondo, and T. Imakoma. 2003. "Contamination Accumulation and Withstand Voltage Characteristics of Various Types of Insulators," in *Proceedings of 7th International Conference on Properties and Applications of Dielectric Materials*, Nagoya, Japan, Paper S15-2. June 1–5.

CHAPTER 5

CONTAMINATION FLASHOVER MODELS

The likely chemical composition and the resulting electrical conductivity of a polluted and wetted insulator surface were discussed in Chapter 3. Then, test methods and results for contamination flashover in salt-fog, clean-fog, and other environments were described in Chapter 4. This chapter reviews the parameters of suitable mathematical models that have been suggested and developed to predict the voltage stresses that cause contamination flashovers at room temperature.

The models of contamination flashover strength as a function of precontamination and wetting take into account insulator dimensions, configuration, and material and can be derived from test results or theoretical models. The mathematical model is an essential component in the calculation of the outage rate of the overhead line or its sections due to pollution level. Pollution flashover models are used to evaluate the insulator performance under constant line voltage while exposed to the highly variable ESDD, which is estimated from:

- Number and duration of days without rain that end in wetting events.
- Distance from pollution sources.
- Height above ground.
- Wind speed, direction, and variability.

The pollution layer resistance is a function not only of the ESDD, but of the degree of wetting provided by the environment. To take advantage of a

Insulators for Icing and Polluted Environments. By Masoud Farzaneh and William A. Chisholm
Copyright © 2009 the Institute of Electrical and Electronics Engineers, Inc.

theoretical model for pollution flashover, a utility would generally need to establish or estimate the following input parameters:

- ESDD variations based on single, average, or statistically reliable measurements.
- Number of flashover hazards per year, based on wetting events.
- Number of insulators exposed to the same flashover hazards.
- Pollution flashover stress for insulators derived from laboratory tests or field tests to confirm predictions.

The specific details of insulation coordination for selecting a suitable leakage distance in the icing and polluted environments form the basis of Chapter 10.

5.1. GENERAL CLASSIFICATION OF PARTIAL DISCHARGES

The partial discharge processes associated with pollution flashovers can be described using a general system of partial discharge classification set out by Niemeyer [1995] as shown in Table 5-1 and Figure 5-1.

In general, the partial discharges associated with contamination and icing flashover have a medium surface contribution to ionization. The classification "g" in Figure 5-1, with partial discharges initiating in a region of higher electric stress and propagating toward the low-voltage electrode, is most common in the icing situation and is also the geometry considered in the mathematical models for contamination flashover. However, the classifications "f" and "h" in Figure 5-1 are more representative of the real situation on contaminated insulators, with multiple arcs across dry bands between pools of electrically conductive wet areas.

TABLE 5-1: Survey of Partial Discharge Types and Characteristics

Partial Discharge Type	Nature	Order of Magnitude	Relevance for Contamination Failure
Surface emission	Continuous	10^{-10} to 10^{-5} A/m^2	Low
Surface conductivity	Continuous	10^{-15} to 10^{-8} S	High
Glow discharge	Continuous		Low
Townsend discharge	Pulse	100 ns	Low
Streamer	Pulse	1–100 ns, >10 pC	Initiation level only
Leader	Pulse	10–1000 μs, 1–10 A	Initiation level only
Spark	1–1000 μs	0.1–10 A	Low
Partial arc	0.1–1 s	1–10 A	High

Source: Adapted from Niemeyer [1995].

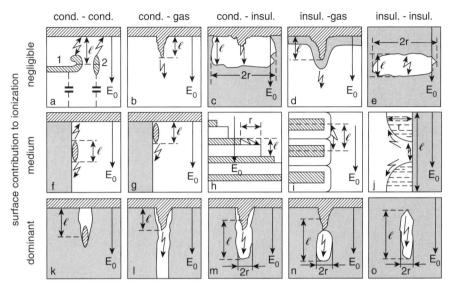

Figure 5-1: General classification matrix for partial discharge activity (from Niemeyer [1995]).

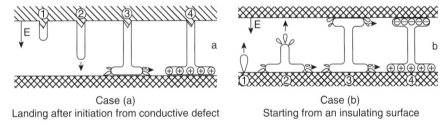

Figure 5-2: Discharge attachment modes to insulator surface (from Niemeyer [1995]).

Figure 5-2 shows schematically how the discharges can attach to an insulating surface in two ways. In case (a), there is a single "arc root" on the insulating surface. This can be treated as a disk-like contact to the underlying insulator, which in the general case will have a resistance given by the material resistivity ρ in the landing area divided by four times the arc root radius. For the partial discharge developing between two areas in (b), there will be two arc roots and any voltage drop across arc root resistance from the flow of partial discharge current will also be roughly doubled.

5.1.1. Discharges in the Presence of an Insulating Surface

The influence of a clean dielectric surface on the electric field near a metal electrode was studied in detail by Gallimberti et al. [1991]. Figure 5-3d shows that there was a 40% enhancement of the electric stress when a typical dielectric material—PVC—was placed near a rod electrode with

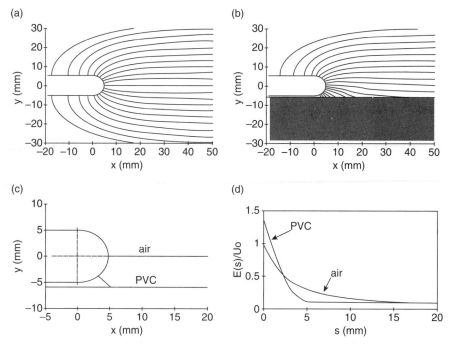

Figure 5-3: Comparison of electric field near high-voltage electrode in presence of air and dielectric surface: (a) electric field lines in the air reference gap, (b) field lines in the gap with PVC dielectric ($\varepsilon = 4$), (c) maximum stress lines with and without PVC dielectric, and (d) field distributions along the maximum stress lines (adapted from Gallimberti et al. [1991]).

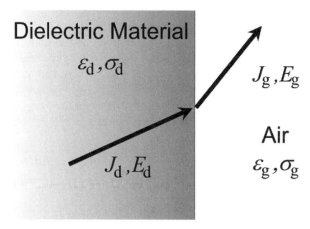

Figure 5-4: Electric fields at dielectric–gas interface.

hemispherical tip. The effects were also evaluated on glass, with its dielectric constant of 8 rather than 4.

Figure 5-4 shows the electric fields E_d and E_g (V/m) along with the current densities J_d and J_g (A/m²) at a dielectric–gas interface. Ohm's law dictates that $J_d = \sigma_d E_d$ and $J_g = \sigma_g E_g$, where σ_d and σ_g are the electrical conductivity of the dielectric material and the air (S/m). Ndiaye [2007] applied Gauss' law and conservation of charge to the dielectric–air interface shown in Figure 5-4 in quasi-static conditions, ignoring displacement current, to obtain:

$$\varepsilon_g E_{n,g} - \varepsilon_d E_{n,d} = q_s$$

$$\sigma_g E_{n,g} - \sigma_d E_{n,d} = -\frac{\partial q_s}{\partial t}$$

(5-1)

where

ε_d and ε_g are the electrical permittivity (F/m) of dielectric and gas,

$E_{n,d}$ and $E_{n,g}$ are the electric fields (V/m) normal to the interface,

σ_d and σ_g are the electrical conductivity (S/m) of dielectric and gas,

t is time (s), and

q_s is the surface charge density (coulomb/m²).

The equations can be combined to establish the rate of accumulation of charge σ_s on the air–insulator surface:

$$\frac{\partial q_s}{\partial t} + \frac{q_s \sigma_d}{\varepsilon_d} = \frac{\varepsilon_g \sigma_d}{\varepsilon_d} E_{n,g} - \sigma_g E_{n,g}$$

(5-2)

The normal electric field components force the charge accumulation. At equilibrium, for constant normal electric field components, the surface charge will reach

$$q_s = E_{n,g} \cdot \left[\varepsilon_g - \varepsilon_d \frac{\sigma_g}{\sigma_d} \right]$$

(5-3)

There are thus two conditions for surface charge accumulation. The normal electric field $E_{n,g}$, a resultant of applied voltage, and induced fields from nearby charges must be nonzero. Also, the term inside brackets in Equation 5-3 must be nonzero, as shown in Equation 5-4 [Jing, 1995; Ndiaye, 2007]:

$$E_{n,g} \neq 0 \quad \text{and} \quad \frac{\varepsilon_g}{\sigma_g} \neq \frac{\varepsilon_d}{\sigma_d}$$

(5-4)

A positive normal electric field will result in accumulation of negative surface charge.

Investigations [Verhaart et al., 1987; Gallimberti et al., 1991] have shown that, at the air–insulator interface, induced polarization charges cause field distortions that can affect avalanche growth close to the surface. Gallimberti and co-workers established that an insulating surface affects the coefficients of ionization α and attachment η in air. Comparing air and PVC, they noted a shift in the balance point from $\alpha_{air} = \eta_{air} = 8/\text{cm}$ at 29 kV/cm in air to

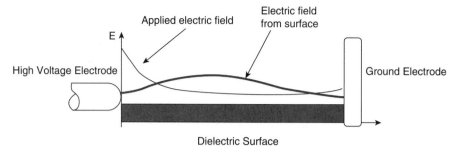

Figure 5-5: Schematic of tangential electric field along point-plane electrode near dielectric surface (from Ndiaye [2007]; courtesy of I. Ndiaye).

$\alpha_{PVC} = \eta_{PVC} = 14$/cm at 27 kV/cm on PVC. They also noted that the progressive accumulation of charge on the dielectric surfaces affected the form of the streamer discharge and the field at which it was initiated.

The velocity and peak current of positive streamers at insulator surfaces were noted to be higher than those in air gaps due to the additional contribution of photoelectron emissions from the surface. Propagation characteristics were thought by Allen and Faircloth [2003] to be controlled by increased photoionization and charge attachment coefficients at the surface, to an extent that depends on the nature of the gas and insulator. The insulator surface can also accumulate space charge, which may change the prebreakdown conditions. The resulting field distortions midway in the gap, shown in Figure 5-5, are also thought to reduce the dielectric strength [Jing, 1995; Jun and Chalmers, 1997].

5.2. DRY-BAND ARCING ON CONTAMINATED SURFACES

5.2.1. Wetted Layer Thickness and Electrical Properties

Once a contaminated insulator surface is fully wetted, its electrical surface resistivity tends to remain relatively constant as more water accumulates or evaporates. For example, taking a moderate equivalent salt deposit density of 0.06 mg/cm^2 (60 μg/cm^2) and a distilled water layer thickness of 100 μm, the salt concentration in the pollution layer is 0.06 mg/(1 cm × 1 cm × 0.01 cm), giving a salinity S_a of 6 mg/cm^3 or 6 g/L. The electrical conductivity of this salt solution at 20°C is given by Hewitt [1960]; and IEEE Standard 4 [1995]

$$\sigma_{20} = (1754\,\mu S/cm) \cdot \left(\frac{S_a}{1 g/L} \right)^{0.97} \tag{5-5}$$

The conductivity of the wetted salt layer with 100-μm thickness works out to 9970 μS/cm. If the water layer thickness is halved by evaporation to 50 μm but remains as a continuous film, the salinity will double and the conductivity will increase to $19,540\,\mu S/cm$. The surface conductance γ, defined as the product of layer thickness t and conductivity σ_{20} corrected to $20\,^{\circ}C$, will decrease by 2% as half the water is lost.

Wetting by fog, with typical median values of $300\,\mu S/cm$, contributes some additional ions to the water layer, compared to self-humidification by scavenging pure water vapor from the surrounding air. The contribution, however, is relatively minor. Wetting by fog would give a layer conductivity about 3% higher than self-wetting for a moderate ESDD level of $60\,\mu g/cm^2$ in the example. The influence of fog conductivity on wetted layer conductivity is stronger for very light pollution levels below $6\,\mu g/cm^2$ and negligible for heavy pollution.

5.2.2. Surface Impedance Effects

The surface impedance, measured between two electrodes with equal length and separation, is independent of electrode size but depends instead on the conductivity and thickness of the surface layer. To carry out this measurement, for example, a square mask is placed on the insulator surface and electrodes are painted on with silver paint or a suitable Electrodag™ product. A $1 \times 1\,cm^2$ geometry is suitable for measuring surface impedance on insulators and pollution layers. The layer cross-sectional area A will be the electrode separation L multiplied by the layer thickness. The resistance is given by $R = L/(\sigma \cdot A)$, which reduces to $R = 1/(\sigma \cdot t)$ in a square geometry. The values are reported as surface impedance, with a unit of Ω. Surface resistance is sometimes reported with units of ohms per square ($\Omega/\square$) to distinguish from ordinary resistance, but this is not a preferred unit.

Considering the case in the previous section of a water layer thickness of 5 or $10\,\mu m$, fully wetting a salt deposit density of $60\,\mu g/cm^2$, the surface resistances are computed to be $R = 10.24\,k\Omega$ and $10.03\,k\Omega$, respectively. The 2% difference is negligible, since the estimate or measurement of water layer thickness typically has higher uncertainty.

Often, when carrying out high-voltage measurements, the arcing and dry-banding processes lead to periods where there is no current flow. Rather than report infinite values of surface impedance (say, in $M\Omega$), the standard practice is to report a zero value for surface conductivity γ in μS.

5.2.3. Temperature Effects Leading to Dry-Band Formation

The temperature of the electrolyte on the insulator surface has an effect on its electrical conductivity and surface impedance. Near $20\,^{\circ}C$, for most salts, this effect works out to about 2% per $^{\circ}C$. As the temperature increases, conductivity decreases. This increases current, which increases power dissipation. The positive feedback in this process can lead to thermal runaway—until a part of the water film evaporates away to form a dry band.

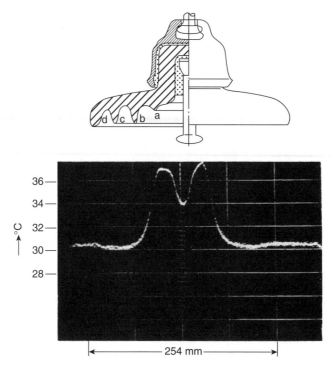

Figure 5-6: Temperature distribution on bottom surface of disk insulator (from Ishii et al. [1984]).

The temperature distribution on the bottom surface of the insulator will not be constant. Typical variations of 8 °C were reported by Ishii et al. [1984] with the maximum temperature occurring, as shown in Figure 5-6, in area "a" near the pin of the insulator.

The temperature rise near the pin of the insulator reflects the fact that this region has the highest current density and resistive power dissipation. This also means that the layer conductivity in this critical region will be elevated, by about 2.2% per °C or 37% at 37 °C using Ishii's observations, compared to the reference value of γ at 20 °C. Other researchers have introduced this factor for correcting the conductivity of the pollution layer. Wilkins [1969] introduced a factor of 1.35 for disk insulators like those in Figure 5-6 and recommended larger values for long-rod insulators. Guan and Zhang [1990] recommended a factor of $\gamma_e = 1.25\gamma_{20}$.

5.2.4. Dry-Band Formation

Textier and Kouadri [1986] studied the formation of dry bands using infrared photographs of the temperature profile of a uniform wetted salt layer, 230 mm long by 35 mm wide, stabilized by a high nonsoluble deposit density. Surface resistance was studied in the range of 650–32,000 Ω. Time evolution of leakage current was studied with voltages up to 2.5 kV rms. They found abrupt declines in current, corresponding to the formation of dry zones, when their applied voltage gradient exceeded 1.1 kV/m. On most electrical insulators, this gradient will always be exceeded, meaning that dry-band formation will always occur.

Without voltage, an evaporation rate was measured to be 28 mg/m^2 per second at room temperature and 60% relative humidity. Under voltage, the weight loss increased to about 530 mg/m^2 per second, fitted to 31 mg/m^2 per second per °C of difference between the pollution layer and ambient temperatures. When the voltage gradient was low, the entire surface warmed up to about 50–55 °C with a peak of 60 °C. When the voltage gradient was high enough to cause a dry band, there was a fast rise of temperature to above 70 °C in a narrow 20-mm wide zone, while the rest of the 230-mm long electrode surface remained at about 40 °C. Power and energy balances were performed, leading to the conclusion that most (60–87%) of the power supply energy went into evaporation of the water layer.

The formation of at least one dry band on the polluted and wetted insulator surface is an important precursor to the appearance of electrical arcs that can develop into flashover.

5.2.5. Arcing and Enlargement of Dry Bands

Once the current is interrupted by the formation of a dry band, the entire insulator potential will appear across this space. Initial dry bands are a few centimeters in size. The electrical strength of the gap will be on the order of 5 kV/cm, and with a typical voltage stress of 10–15 kV per insulator disk, it is likely that a partial arc will form. The current in this arc will be restricted by the high resistance of the pollution layer, but it will still have a strong acoustic noise signature and significant ultraviolet emission. At the two roots of the arc on the water surfaces, current density will be high and this will lead to local growth of the dry band.

Jolly and Poole [1979] evaluated the growth of dry bands and arc propagation speeds with time-lapse and high-speed photography at 1000 frames per second. Initially, their dry bands formed around a small-radius electrode simulating the pin of an insulator. Figure 5-7 shows how a typical dry band separated from the pin and moved to a region between the two electrodes after 45 minutes of fog wetting.

The same process of dry-band formation, starting near the area of highest electrical stress, occurs near the small-radius pin area on the bottom surface of typical suspension disk insulators. In their study of the dynamics of this process, Baker and Kawai [1973] concluded that the "effective leakage distance," corresponding to potentials of 10% and 90% of total cap-to-pin voltage, was 41–60% of the geometric leakage distance for the seven types of insulators tested. They found that the top surface of the insulator did not contribute to effective leakage distance, nor did the area close to the pin. Figure 5-8 shows that the area near the pin was easily bridged by streamers in typical contamination tests.

 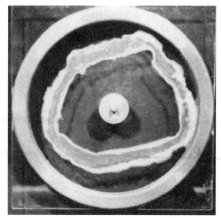

| Dry band (white region) appears near 44-mm anode immediately after fog application | Dry band (irregular annulus) after 45 minutes of fog application; overall diameter, 305 mm |

Figure 5-7: Time evolution of dry banding of salt solution on flat plate between circular electrodes (from Jolly and Poole [1979]).

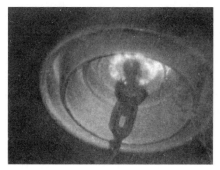 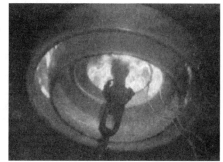

| Light activity | Heavy activity |

Figure 5-8: Photographs of quasi-stable arcing activity on bottom surface of 146-mm × 254-mm diameter cap-and-pin insulator with 430-mm leakage distance (from Baker and Kawai [1973]).

The multiple series and parallel zones of partial discharge activity complicate the mathematical modeling of the pollution flashover process and will be treated later in this chapter.

5.2.6. Nuisance Factors from Discharges on Wetted Pollution Layers

The leakage currents flowing on the polluted surface from partial discharge activity can cause some serious side effects on the power system and the local environment. These effects take four main forms:

- Annoyance from the sound associated with partial discharges.
- Broadband electromagnetic interference (EMI) from the interruption and resumption of arc current.
- Risk of ignition of semiconducting components such as wood crossarms from the flow of leakage currents.
- Buildup of ions from the arcing process, accelerating metal corrosion and erosion of insulator surfaces.

Acoustic Noise Ceramic disk insulators are particularly susceptible to audible noise complaints because each disk has a mechanical resonant frequency (1–2 kHz) with a narrow bandwidth. This is used when inspecting insulators in service. Ceramic disks in good condition respond to a moderate mechanical impact with a "ping" while those with mechanical defects, such as cracks in the body or head, respond with a "thud." Partial discharge activity excites these resonances repetitively at the peaks of ac voltage, leading to a distinctive and annoying sound that is centered in the most sensitive audible range. Generally, in fog and rain conditions for ac, noise power from insulators dominates the audible noise from conductor corona within about 50 m of towers. The same is true for dc: Yasui et al. [1988] report a sound pressure level of 55 dBA from partial discharge activity on polluted HVDC insulators at a distance of 50 m.

Electromagnetic Interference Power lines emit electromagnetic radiation that acts as an unwanted disturbance to the reception of a desired signal. Regulations for radio interference (540–1610 kHz) from insulators are governed internationally by CISPR Publications 18-2 [1986] and its amendments [CISPR, 1993, 1996]. These insulators are tested for radio noise influence voltage (RIV), expressed in dB above 1 μV/m, according to CISPR and IEC 60437, in clean and dry conditions.

- Insulators to be applied in *Type A* areas are clean, corresponding to very light or light site pollution severity (SPS) in IEC 60815 [1986]. These insulators meet a radio noise limit of 23 dB above the fair weather noise voltage produced by the conductor at 20 m from the outer phase of the line.
- Insulators in *Type B* areas experience periodic buildup and natural washing of pollution, corresponding to moderate SPS. They seldom experience dry-band arcing. These insulators meet a radio noise limit of 15 dB above the fair weather noise voltage produced by the conductor at 20 m from the outer phase of the line.
- Insulators in *Type C* areas support frequent formation of dry-band discharges. These areas correspond to heavy or very heavy SPS. At present, there are no test procedures and limits described in the relevant standards for Type C applications. Instead, general advice is to ensure adequate

radio noise levels by reducing voltage stress across leakage paths, using polymer insulators, greasing, or periodic washing of the insulators or by relocating antennas in problem areas.

Generally, insulator arcing activity is not a strong source of EMI disturbance to high-frequency signals for television or cellular telephone, compared to the response of the energized conductor under the same environmental conditions [CIGRE Working Group 36.01, 2000].

Pole Fires and Other Leakage Current Effects   When leakage current flows across insulators, it may enter wood components through metal end fittings. Wood insulation is often retained on medium-voltage lines to add lightning impulse strength. However, this choice for improved reliability in lightning conditions means that leakage currents will flow into the electrical resistance of the wood pole under contamination conditions.

At a metal-to-wood interface, the following conditions can lead to wood pole ignition [Darveniza, 1980]:

- Long-term natural aging and shrinkage of the contact area between metal and wood, leading to high local resistance and current density.
- Contamination of the ceramic insulator surfaces by industrial pollution, sea salt, or agricultural fertilizers and dust.
- Moisture on insulator surfaces from fine rain or mist.
- Leakage current of 8–10 mA at the metal–wood interface.
- Air movement providing fresh oxygen to a possible ignition area.

Vulnerable locations for pole fires include the suspension insulator eye-bolts and crossarm through the bolts.

Darveniza considered electrical characteristics of wood poles in the temperature range from 20 °C to 70 °C, but did not evaluate effects below freezing temperatures. Winter conditions involving long periods without rain, followed by winter fog, have produced widespread distribution system pole fires in Canadian urban environments including Toronto, Calgary, and Edmonton [e.g., Star, 1994].

For the winter condition, a more detailed model of the electrical resistance of a wood pole is described by a ladder network, for example, as outlined by Filter and Mintz [1990] in Figure 5-9. In this figure, the resistance of the rainwater layers, R_w, will instead be the resistance of the accumulated ice accretion per unit length. Another very important factor is related to the moisture content of the pole. The resistance terms for treated and untreated sapwood R_t' and R_u drop by two orders of magnitude as relative humidity changes from 12% to 16%.

The model in Figure 5-9 also shows the hand, foot, and body resistance of a person climbing the pole. For electrical safety evaluations, the resistance of

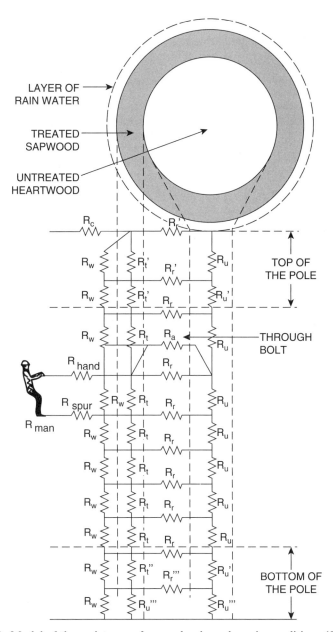

Figure 5-9: Model of the resistance of a wood pole under rain conditions (from Filter and Mintz [1990]).

the ice layer will be significantly higher than R_w and this can be an important mitigation factor. Conversely, greater care will be needed when traversing any air gaps in the ice layer or when climbing in melting conditions.

Generally, mechanical solutions to the pole fire problems such as gang nailing have proved to work loose over a 10-year period. Wrapping poles and crossarms with metal bands, also called "banding," has proved to be more effective for long-term prevention of pole fires from insulator leakage currents on ceramic insulators. Replacement with polymer coatings or insulators is another effective mitigation option.

Ozone Production   The partial discharges across dry bands will be burning in a mixture of steam and air. This will form ozone (O_3), which reacts with any nitrogen or hydrocarbons in air. The half-life of O_3 in dry air is about an hour, while in the presence of moisture this falls to about 20 minutes [CIGRE Working Group 36.01, 2000].

Near electrical discharge activity on polluted surfaces, some additional NO_X ions will form, and a fraction of these will dissolve in the water layer. Suginuma et al. [1993] measured concentrations of nitric acid in pure or salt water layers as a result of electrical arcing activity using typical arc currents of 14–22 mA. They report a change from an initial, neutral pH of 7.0 to values between 3 and 3.5 after accumulation of 5 coulombs of charge. The slow rate of deposit, reaching equilibrium in a minimum of 400 seconds, is still important enough to be considered in some pollution flashover calculations.

A nitric acid solution has an electrical conductivity given by

$$\sigma_{0°C} = \frac{240 \text{ S·cm}^2}{\text{mEq}}[H^+] + \frac{40 \text{ S·cm}^2}{\text{mEq}}[NO_3^-]$$

$$\sigma_{20} = \frac{324 \text{ S·cm}^2}{\text{mEq}}[H^+] + \frac{64 \text{ S·cm}^2}{\text{mEq}}[NO_3^-] \qquad (5\text{-}6)$$

$$\sigma_{50} = \frac{465 \text{ S·cm}^2}{\text{mEq}}[H^+] + \frac{104 \text{ S·cm}^2}{\text{mEq}}[NO_3^-]$$

where $[H^+]$ and the corresponding $[NO_3^-]$ are concentrations of 10^{-3} mole equivalent per liter ($\text{mEq}/1000\,\text{cm}^3$) corresponding to a pH of 3.0. The value of $\sigma_{20} = 388\,\mu\text{S/cm}$ at the reference temperature of 20°C increases to $\sigma_{50} = 705\,\mu\text{S/cm}$ at the pollution layer temperature of 50°C during formation of dry bands, and falls to 280 μS/cm at 0°C just before freezing. Like the contribution of ions in fog, the electrical conductivity from the nitric acid by-product is considerably less than the electrical conductivity of a layer of water on a moderate 60-μg/cm² deposit of salt, noted previously to be 10-20,000 μS/cm. However, nitric acid flux will have more effect on the electrical conductivity of natural rain or fog accretion on clean insulators than the effect of heat input from sustained arcing.

5.2.7. Stabilization or Evolution to Flashover

In each weather condition, the contaminated insulator will reach a balance between wetting and the drying process. In the desirable case for power system reliability, heat input from the partial discharge current flowing in the wetted pollution layer will dry the layer out completely, restoring the insulator strength to its dry value. If the wetting rate matches the evaporation rate, a dynamic equilibrium may persist for minutes or hours, consisting of quiescent periods followed by bursts of leakage current.

When the wetting rate overwhelms the evaporation rate, the arc will continue to grow and reignite reliably with each ac cycle. This condition is the most dangerous. Once a partial arc has bridged about two-thirds of the pollution layer, positive feedback causes rapid extension of the arc across the remaining surface. This will provide a low-impedance constant-voltage source (the power system) without the series resistance of the pollution layer. The arc will flash over and increase to hundreds or thousands of amperes of fault current. Once a flashover has occurred, protective equipment such as automatic circuit breakers or fuses will operate to clear the fault safely.

5.3. ELECTRICAL ARCING ON WET, CONTAMINATED SURFACES

Several theories have been proposed to explain the mechanisms underlying the propagation of local arc on polluted surfaces:

- An external force "pulling" the arc was hypothesized by Rahal and Huraux [1979]. External forces, such as the electrostatic force, are thought to move the arc along the polluted surface, resulting in flashover.
- An electrical breakdown theory was proposed by Jolly [1972a,b]. The high local electric field at the arc root results in the ionization of the air in front of the arc root. This ionization is thought to extend the arc to flashover.
- The thermal breakdown theory has been put forward by Wilkins and Al-Baghdadi [1971], Li et al. [1989], and Zhang et al. [1991]. The high temperature at the arc root results in the ionization of the air, which then elongates the arc and results in flashover.

Jolly analyzed the factors mentioned above and stated in a discussion of Rahal and Huraux [1979] that, while the electrostatic force was able to deflect an arc, it was too weak to elongate the arc over an electrolyte. Hence electrostatic force was not likely to be a main factor contributing to flashover on a polluted surface. Li et al. [1989] and Zhang et al. [1991] dismissed Jolly's hypothesis when they found that the electric field strength at the arc root was less than 15 kV/cm during most of the flashover period. This is below the level thought to describe the corona envelope of a discharge. This leaves the thermal ionization of air at the arc root as a likely reason for arc propagation.

5.3.1. Discharge Initiation and Development

The flashover on an insulator with a polluted surface starts with the initiation of an arc at a region of high local electrical stress and ends its propagation with increasing speed along a layer of semiconducting pollution. The Obenaus [1958] model was developed to describe flashover under dc conditions for a layer of uniform resistance per unit length. Researchers such as Claverie [1971] and Rizk [1971a,b, 1981] added the essential criteria to adapt the Obenaus model for ac flashover.

While it has only moderate success when applied to complicated insulators, such as chains of many disks under dry-band arcing conditions, the Obenaus model with Rizk [1981] ac reignition conditions has proved to be accurate and adaptable to a reasonable range of station post insulators, where the assumption of uniform pollution resistance per unit length is more valid.

The ac flashover on a polluted surface is considered as an arc in series with a residual resistance consisting of an ice layer that is not bridged by the arc as shown in Figure 5-10. The circuit equation for this model is as follows:

$$V_m = A \cdot x \cdot I_m^{-n} + I_m \cdot R_P(x) \tag{5-7}$$

where

V_m is the peak value of applied voltage (V),

A and n are arc constants,

x is the local arc length (cm),

I_m is the peak value of leakage current (A), and

$R_P(x)$ is the residual resistance (Ω) of pollution layer from the arc root at
  x to the ground electrode.

For dc conditions, an electrode voltage drop V_e is added and Equation 5-7 is expressed in terms of the constant dc-positive or dc-negative voltage and current.

5.3.2. Arc V–I Characteristics in Free Air

The arc constants A and n in Equation 5-7 relate the arc voltage to its current in a simple way. In free air, Peelo [2004] summarized the development

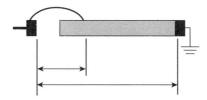

Figure 5-10: Obenaus model for flashover of uniformly polluted surface.

of this model from its origins in the 1880s with work by Steinmetz and Ayrton.

Ayrton's classical expression for the arc voltage of length up to 15 mm is

$$V = a + bx + \frac{c + dx}{I} \tag{5-8}$$

where

a is a constant representing the total electrode voltage drop at the arc interfaces to metal or surface,

bx is the constant voltage drop in the arc column of length x, and

$(c + dx)/I$ is the inverse characteristic of the arc of current I, where c and d are constants.

Nottingham revised Ayrton's expression for longer arc lengths by introducing a nonlinear exponent n as follows:

$$V = a + bx + \frac{c + dx}{I^n} \tag{5-9}$$

where n is a constant and the other variables are defined in Equation 5-8. For sufficiently long arc length x, the electrode voltage drops at the arc roots, a, can be neglected.

For low currents I, the second term in Equation 5-9 can be neglected, giving a simplified expression of the form $V = A \cdot x \cdot I^{-n}$, where A and n are constants and x is the arc length.

For high currents I, the third term in Equation 5-9 can be neglected and there is a constant voltage drop per unit of arc length x.

Peelo [2004] illustrated in Figure 5-11 that there is a 2:1 range in the predicted arc voltage gradient as a function of current, depending on the choice of equation in Table 5-2. This reflects the wide range of application, from short arcs of a few centimeters at high current to long arcs with low currents.

Extinction of arcs between the opening jaws of air-break disconnect switches is an important factor in the use of these devices to break "loop flow" currents in power systems. Typical duty requires interruption of 10–100 A with a resulting potential difference of only a few kilovolts when the power system reconfiguration is completed. Peelo noted that the nature of the V–I characteristic of arcs in air changed as a function of loop impedance in the range of 20–150 Ω, as shown in a series of 20 analyses in Figure 5-12.

The role of series loop impedance Z_{loop} in the V–I characteristics of an operating air-break disconnect switch was evaluated by Chisholm and Peelo [2007]. They fitted a single quadratic expression to arc voltage as a function of current by regression to all of the data in Figure 5-12:

$$V = AI^2 + BI + C \tag{5-10}$$

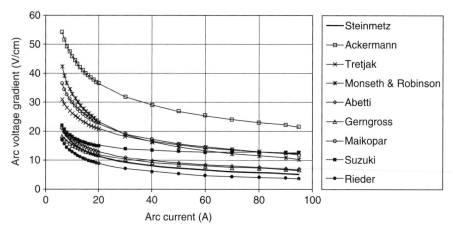

Figure 5-11: Variation of arc voltage gradient with arc current for some expressions in Table 5-2 (from Peelo [2004]; courtesy of D. Peelo).

TABLE 5-2: Summary of Arc Equations for Arc in Air

Researcher	Equation[a]	Comments
Steinmetz	$V = 51 \cdot x \cdot I^{-0.5}$	
Nottingham	$V = K \cdot x \cdot I^{-0.67}$	K depends on contact material
Ackermann	$V = 98 \cdot x \cdot I^{-0.33}$	
Eaton/Tretjak et al.	$V = 56 \cdot x \cdot I^{-0.33}$	
Warrington	$V = 286 \cdot x \cdot I^{-0.4}$	
Monseth and Robinson	$V = 104 \cdot x \cdot I^{-0.5}$	Used by Andrews for arc reach
Abetti	$V = 43 \cdot x \cdot I^{-0.4}$	
Gerngross	$V = 35 \cdot x \cdot I^{-0.36}$	
Browne	$V = K \cdot x \cdot I^{-1}$	K is a constant
Maikopar	$V = 75 \cdot x \cdot I_{\mathrm{P}}^{-0.4}$	I_p is the peak current
Suzuki	$V = 12 \cdot x + 60 \cdot x \times I^{-1}$	
Rieder	$V = 45.5 \cdot x \cdot I^{-0.55}$	
Terzija and Koglin	$V = 113 \cdot x \cdot I^{-0.4}$	

[a]V in volts, I in amperes, and x in centimeters.
Source: Adapted from Peelo [2004].

where

$$A = 97.3 - 15.7 \cdot Z_{\mathrm{loop}} \ (\mathrm{V/A}^2)$$
$$B = 1.449 - 0.138 \cdot Z_{\mathrm{loop}} \ (\mathrm{V/A})$$
$$C = 1281 - 2.87 \cdot Z_{\mathrm{loop}} \ (\mathrm{V})$$

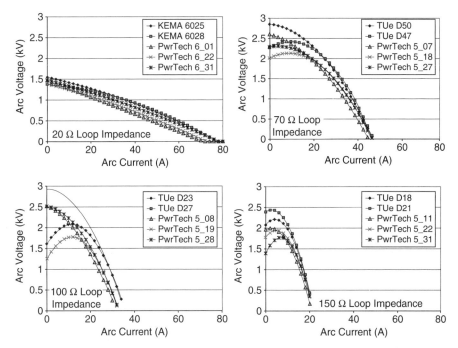

Figure 5-12: Effect of series impedance on *V–I* characteristics of arcs in air (data from Peelo [2004]).

The opening speed of the air-break disconnect jaws, typically making an air gap of 1 meter in 2 seconds, is similar to the overall speed of propagation of an arc on a polluted or iced surface. However, the series impedance of the polluted surface is two to four orders of magnitude higher than the highest 150-Ω loops tested in high-current laboratories.

When considered over a wider current range from 0.1 to 300 A, there is an overall agreement in the trend to lower arc resistance at higher peak arc current. Figure 5-13 shows the arc resistance per unit length calculated from the arc current and voltage for four different domains—pollution flashover [Obenaus, 1958], ice layer flashover [Farzaneh et al., 1997], typical air-break disconnect switch operation [Peelo, 2004], and radiated energy from faults in electrical panels [ArcPro v.2, 2000].

Given the ±2:1 range of variation among models for arc voltage as a function of current in each domain, the overall trend of the resistance per unit length of an evolving arc over the time scale of seconds is remarkably consistent from one application to the next.

5.3.3. Arc V–I Characteristics on Water or Ice Surfaces

In the original work, Obenaus [1958] selected values of *A* and *n* in the relation $V = A \cdot x \cdot I^{-n}$ for an arc over a polluted surface of $A = 102$ and $n = 0.7$ for arc

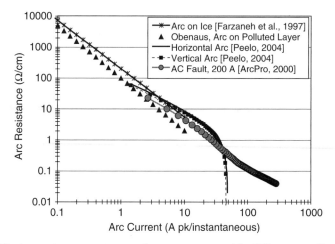

Figure 5-13: Arc resistance versus peak arc current used in different applications: pollution flashover [Obenaus, 1958], ice flashover [Farzaneh et al. 1997], air-break disconnect switch operation [Peelo, 2004], and radiated energy from electrical faults [ArcPro v.2, 2000].

TABLE 5-3: Typical Values[a] of A and n in the Relationship $V = AxI^{-n}$ for ac Arcs over Wet or Polluted Surfaces

Reference	A	n	Comments
Obenaus [1958]	102	0.70	Arc over pollution layer
Alston and Zoledziowski [1963]	63	0.76	Arc over pollution layer
Claverie and Pocheron [1973]	101	0.50	Arc over pollution layer
Jolly and Chu [1975]	80	0.62	Arc over tin-oxide layer
Rahal and Huraux [1979]	534	0.24	Arc over solution
Guan and Zhang [1990]	138	0.69	Arc over polluted layer
Ghosh and Chatterjee [1995]	270	0.66	Arc over $FeCl_3$ solution
Ghosh and Chatterjee [1995]	360	0.59	Arc over NaCl solution
Ghosh and Chatterjee [1995]	451	0.49	Arc over $CuSO_4$ solution
Ghosh and Chatterjee [1995]	462	0.42	Arc over $CaCl_2$ solution

[a]V in peak volts, I in peak amperes, and x in centimeters.

length x (cm), V (volts), and I (amperes). This leads to the arc resistance per unit length shown in Figure 5-13.

While the studies of the arc coefficients on the polluted surface do not have as long a history as those carried out for arcs in air, Table 5-3 and the corresponding plot in Figure 5-14 show there are considerable variations in these coefficients, depending on experimental conditions and methods.

Ghosh and Chatterjee [1995] carried out similar experiments to measure the V–I relation of arcs on polluted surfaces with different chemical com-

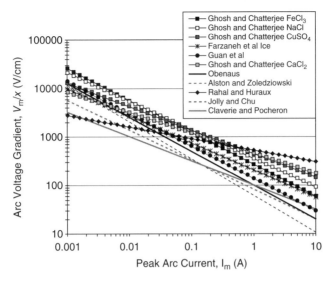

Figure 5-14: Arc voltage gradients measured on contaminated, wet, or iced surfaces from Table 5-3.

position. They found significant differences, depending on the pollution type. Farzaneh et al. [2005] also advanced a hypothesis on the role of sodium ions in activating the flashover process at low pressure across ice surfaces.

Rumeli [1976] studied the electrical flashover of water columns as a means to explore the details of the related physics. He established that the accuracy of the voltage–current model for the arc was important, and also reported that typical values of $A = 63$ and $n = 0.76$ in the relation $V = A \cdot x \cdot I^{-n}$ for water vapor were not suitable. Instead, the values $A = 518$ and $n = 0.273$ for an arc burning in steam gave satisfactory results. These coefficients were later updated to $A = 530$ and $n = 0.24$ in Peyregne et al. [1982a]. Corresponding values for arcs in water vapor and on ice surfaces are given in Table 5-4.

5.3.4. Dynamics of Arc Propagation

Jolly and Poole [1979] observed a discharge elongation speed of 1–2 m/s between two coaxial electrodes on a 305-mm glass plate. This slow speed suggests a plasma drift mechanism rather than a true air breakdown process, which is much faster.

Peyregne et al. [1982b] also studied the time to flashover on conducting films and found a 10:1 change in time to flashover, from 50 ms to 5 ms, on sample lengths of 8–12 cm. For applied voltage greater than 30% of the critical flashover level, time to flashover of 3 ms and sample length of 10 cm gave an average velocity of about 33 m/s on samples with series resistance $R(x)$ of 5 kΩ/cm.

TABLE 5-4: Typical Values[a] of A and n in the Relationship $V = AxI^{-n}$ for ac Arcs in Steam, Water, and Ice

Reference	A	n	Comments
Rumeli [1976]	518	0.273	Arc in steam
Peyregne et al. [1982a]	530	0.24	Arc in steam
Rumeli [1976]	63	0.76	Arc in water vapor
Guan and Zhang [1990]	138	0.69	Arc over polluted layer
Farzaneh et al. [1997]	205	0.561	Arc over wet NaCl ice
Farzaneh et al. [2005]	127	0.70	Arc over wet NaCl ice layer

[a]V in peak volts, I in peak amperes, and x in centimeters.

5.4. RESIDUAL RESISTANCE OF POLLUTED LAYER

The residual resistance consists of two terms, one being related to the local resistance of the ionized arc foot to the pollution layer and the other being the resistance of the pollution layer to infinity. Depending on the size of the arc roots, their separation from electrodes, and the thickness of the pollution layer, either term may be important.

5.4.1. Observations of Series Resistance of Pollution Layer

In the case where a pollution layer is spread uniformly along a surface and wetted consistently, it is generally possible to calculate the resistance from any point to the ground electrode. This is the residual pollution layer resistance $R_P(x)$ in Equation 5-7.

Before introducing mathematical models for the resistance of the wetted pollution layer, it is useful to review the results of actual measurements of series pollution layer resistance on insulators exposed in pollution conditions. Claverie [1971] carried out measurements of $R_P(x)$ on disk and long-rod insulators that were artificially polluted with salt. Figure 5-15 plots the values, normalized by the resistivity ρ ($= 1/\sigma_{20}$) of the polluting layer liquid, measured from the arc root position to the pin or lower (high-voltage) electrode.

In the hollows between bottom-surface ribs on disk insulators, marked by ④ and ⑥ in Figure 5-15, it was difficult to obtain consistent results. This is not a practical problem as the arc does not stabilize at these locations. The relations of $R_P(x)/\rho$ to x decay exponentially, $R_P(x) = R_0 e^{-x/x_0}$ with distance constants of about $x_0 = 8$ cm for the disk and $x_0 = 21$ cm for the long-rod insulator. The variation of $R_P(x)$ along the long-rod post insulator with small diameter is nearly linear, with a simple description of $R_P(x) = B(L - x)$, where L is the total leakage path and B is a constant of about $1100 \, \Omega/(\Omega\text{-cm})$ for this particular insulator.

While the vertical axes of Figure 5-15 are labeled as R/ρ, this ordinate is also defined as the insulator form factor, defined in Chapter 2.

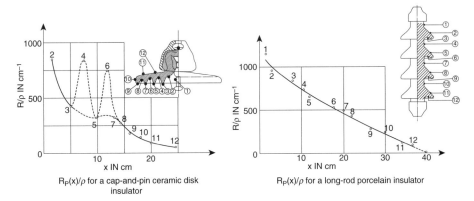

Figure 5-15: Variation in normalized $R_P(x)/\rho$ with ρ in Ω-cm for arc location x (from Claverie [1971]).

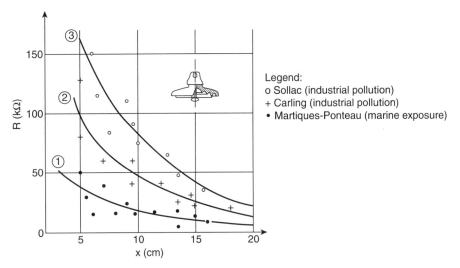

Figure 5-16: Measurements of $R_P(x)$ on naturally exposed disk insulators, with x measured from the insulator pin (from Claverie and Porcheron [1973]).

The measurements of $R_P(x)$ on naturally polluted insulators will show the effects of variation in surface convolutions, and also in variation in pollution accumulation on top and bottom surfaces. Claverie and Porcheron [1973] reported measurements of $R_P(x)$ for ceramic disk insulators at three separate locations, and combined the values in Figure 5-16.

The distribution of the pollution layer resistance was found to be a single function for each type of insulator, once the site-to-site differences in the pollution type and severity were factored out. This site normalization was done by dividing the measured pollution resistance by the cube of the critical

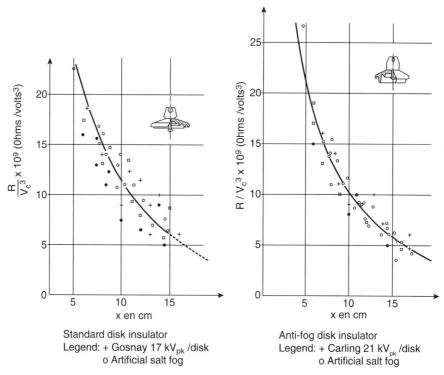

Standard disk insulator
Legend: + Gosnay 17 kV_{pk} /disk
o Artificial salt fog

Anti-fog disk insulator
Legend: + Carling 21 kV_{pk} /disk
o Artificial salt fog

Figure 5-17: Normalized resistance $R_P(x)$ of natural pollution layers at four exposure sites (from Claverie and Porcheron [1973]).

flashover voltage V_c. The choice of the cube power of V_c agrees with the theoretical model that the flashover strength under heavy pollution declines as the 1/3 power of layer salinity.

The normalized resistance functions using the flashover voltage values in Figure 5-17 follow the same trend in spite of differences in the type and severity of pollution, ranging from industrial chemicals and a central power plant (Gosnay), cement plant (Sollac), coke (Carling), and marine pollution at Martigues. The disks had 30–40 cm of overall leakage distance. There was a steep initial decline in resistance with distance from the pin x for these insulators. A high fraction of 75–80% of the pollution layer resistance was found in the first 15 cm of the leakage path, which represented only one-half to one-third of the total path length. The normalized resistance values $R_P(x)/V_c^3$ of all three disks also suggest an exponential decay with distance x:

$$R_P(x) = V_c^3 \cdot R_0 e^{-x/x_0} \qquad (5-11)$$

where R_0 is about $45 \times 10^9 \, \Omega/V^3$ and the distance constant $x_0 = 8$ cm. This same distance constant also fits all three curves ① to ③ in Figure 5-16.

5.4.2. Mathematical Functions for Series Resistance of Pollution Layer

Wilkins [1969] expressed the resistance of a rectangular, wetted polluted layer of length L (cm), width b (cm) and dry-band bridged by an arc of length x (cm) and radius r_0 (cm) as

$$R_P(x) = \frac{1}{4\pi\gamma} \sum_{n=-\infty}^{n=\infty} \ln \left[\frac{\left(1+\cosh\frac{\pi(nb+r_0)}{2L}\right)^2 + \tan^2\frac{\pi x}{2L}\sinh^2\frac{\pi(nb+r_0)}{2L}}{\left(1-\sinh\frac{\pi(nb+r_0)}{2L}\right)^2 + \tan^2\frac{\pi x}{2L}\sinh^2\frac{\pi(nb+r_0)}{2L}} \right] \tag{5-12}$$

For the case where the pollution layer is wide ($b > 3L$) relative to the leakage path L, the simplified expression is independent of b and becomes

$$R_P(x) = \frac{1}{\pi\gamma_e}\left[\ln\left(\frac{2L}{\pi r_0}\right) - \ln\left(\tan\frac{\pi x}{2L}\right) \right] \tag{5-13}$$

This also simplifies for case with $b < L$ to

$$R_P(x) = \frac{1}{\pi\gamma_e}\left[\frac{\pi(L-x)}{b} + N\ln\left(\frac{b}{2\pi r_0}\right) \right] \tag{5-14}$$

where N is the number of arc roots on the pollution layer. If one root of the arc is on a metal electrode, then $N = 1$. In the typical case shown in Figure 5-19, for an arc across a dry band in a wet, polluted layer, $N = 2$.

A worked example in Figure 5-18 with a high surface conductivity shows nearly linear variation in $R_p(x)$ as x varies along the length of the wetted pollution layer.

The expression for the resistance of the thin strip in Equation 5-14 provides a good match to Equation 5-12 through the area of greatest interest in arc development, between 50% and 80% of the total leakage path.

The role of arc roots (arc foot points) has been considered in other models as well. Rizk [1981] compared the Näcke and Wilkins models using the presentation in Figure 5-19.

According to Rizk [1981], Näcke divided the resistance into two terms. The geometry considered in Figure 5-19 was slightly different, leading to an additional constant term of $R_{Pi} = 1/\pi\gamma_e$ for the internal resistance of the arc root, added to an external resistance R_{Pe} that depended on the number of sheds in series and the width of the pollution layer b relative to its length L.

The resistance in series with the arc root in contact with a polluting layer can also be expressed as the sum of two terms [e.g., Claverie, 1971; Le Roy et al., 1984]:

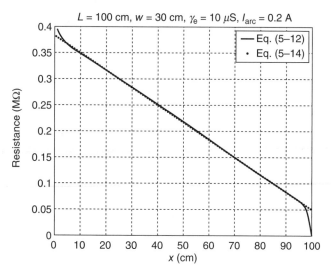

Figure 5-18: Resistance of 1-m pollution layer with $\gamma_e = 10\,\mu S/cm$, $L = 100\,cm$, $b = 30\,cm$, and $I = 0.2\,A$ (adapted from Tavakoli [2004]).

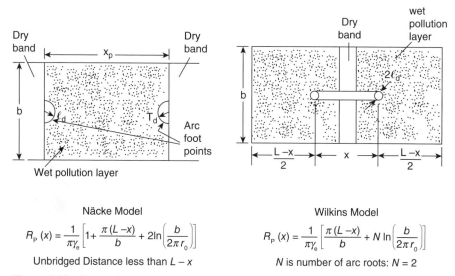

Näcke Model

$$R_p(x) = \frac{1}{\pi\gamma_e}\left[1 + \frac{\pi(L-x)}{b} + 2\ln\left(\frac{b}{2\pi r_0}\right)\right]$$

Unbridged Distance less than $L - x$

Wilkins Model

$$R_p(x) = \frac{1}{\pi\gamma_e}\left[\frac{\pi(L-x)}{b} + N\ln\left(\frac{b}{2\pi r_0}\right)\right]$$

N is number of arc roots: $N = 2$

Figure 5-19: Comparison of models for $R(x)$ of rectangular pollution layer (from Rizk [1981]; courtesy of CIGRE).

- An offset that takes into account concentration of current density around the arc root, independent of position x.
- A linear resistance that depends only on x.

Claverie [1971] proposed that the constant-resistance term be approximated by $A = \rho \cdot (50\,cm^{-1})$. This approach has been refined in later work, for example,

Guan and Zhang [1990]. A plane model of the insulator is defined, along with consideration of a pair of arc roots on the polluted surface.

The presence of two arc roots in Figure 5-20 is typical of the evolution of dc flashover and is more general than the case where a single arc root from the high-voltage electrode is considered. For a pair of arc roots, of radius a_1 and a_2 located at a separation d, the resistance R is given formally by

$$h_1 = \frac{d^2 + a_1^2 - a_2^2}{2d}, \quad h_2 = \frac{d^2 + a_2^2 - a_1^2}{2d}, \quad b_i = \sqrt{h^2 - a_i^2}$$

$$R = \frac{1}{2\pi\gamma} \ln\left(\frac{(b_1 + h_1 - a_1)(b_2 + h_2 - a_2)}{(b_1 - h_1 + a_1)(b_2 - h_2 + a_2)}\right)$$

(5-15)

For the case for two arcs of equal radius $a_1 = a_2 = r_0$ and $r \ll h$, $h_1 \cong h_2 \cong b$, and $d = (L - x)$, Guan and Zhang [1990] simplified Equation 5-15 to the following expression for $R(x)$:

$$R(x) = \frac{1}{\pi\gamma_e} \ln\left(\frac{L - x}{r_0}\right)$$

(5-16)

where

γ_e is the effective layer conductivity (μS) at the critical moment of flashover, which is recommended to be $1.25 \cdot \gamma_{20}$ based on an average electrolyte temperature of about $31\,^\circ$C,

L is the leakage path (cm),

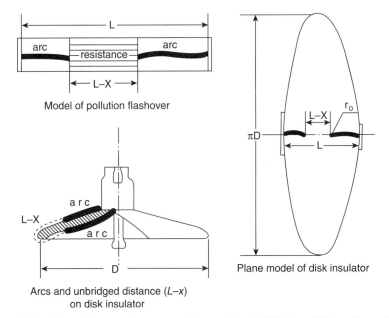

Model of pollution flashover

Arcs and unbridged distance ($L-x$) on disk insulator

Plane model of disk insulator

Figure 5-20: Dimensions and geometry for model of $R(x)$ for disk insulator (from Guan and Zhang [1990]).

x is the total length of the arc(s) (cm), and
r_0 is the arc root radius.

The Guan–Zhang expression is independent of diameter, which is considered to be large compared to the unbridged arc length $L - x$. Figure 5-21 shows that there are significant 30% differences between models for small and large arc length, with better agreement in the critical region at about two-thirds of the leakage path of 30.5 cm.

It seems that none of the mathematical models for the variation in pollution layer resistance with arc length have the exponential decay characteristics, with a distance constant x_0 of about 8 cm, that were reported by Claverie [1971] and Le Roy et al. [1984] on naturally polluted insulators.

The relationship between arc root radius r_0 (cm) and arc current I (A) used in Figure 5-21 for both models was established by Wilkins [1969] as

$$r_0 = \sqrt{\frac{I}{1.45\pi}} \qquad (5\text{-}17)$$

The next section compares the resistance of the arc root from the Wilkins model with values predicted from the general Korsuncev ionization model, derived from dimensional analysis, which applies over many physical scales.

5.4.3. Resistance of Arc Root on Conducting Layer

Korsuncev [1958] defined two dimensionless variables that relate the radius r of a hemispherical ionized zone under high current density to the conductivity

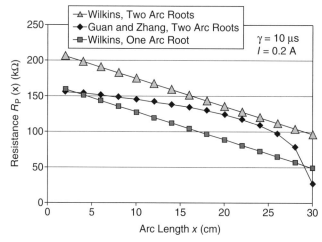

Figure 5-21: Comparison of rectangular [Wilkins, 1969] and ellipse [Guan and Zhang, 1990] models for pollution layer resistance $R(x)$ with constant arc root radius $r_0 = 0.21$ cm and $L = 30.5$ cm.

of a volume of material σ, a limiting ionization gradient E_0, the arc current I, and the resulting resistance R:

$$\Pi_1 = \sigma r R \approx \min\left(\Pi_1^0, 0.263\Pi_2^{-0.308}\right)$$

$$\Pi_1^0 = \frac{1}{2\pi}\ln\left(\frac{2\pi e r^2}{A}\right) \quad \Pi_2 = \frac{I}{\sigma E_0 r^2} \tag{5-18}$$

The ionization gradient E_0 at the edge of the ionized zone is normally taken as about 4000 V/cm for large electrodes. While this model was originally conceived for ionization of ground electrodes on the 10-m scale, it has also been used with success for laboratory evaluation of ionization in small samples of soil with 100 Ω-m resistivity (100 µS/cm conductivity) on a 10-mm scale. In the case of an arc root of about 2-mm radius on an infinite half-space of conductivity 100 µS/cm, also typical of natural precipitation, the general Korsuncev model predicts that ionization will start at a peak current of about 30 mA.

The Wilkins [1969] model for the arc root radius is independent of conductivity and is only a function of peak current. This model was developed specifically for flashover calculations along wet surfaces. The arc radius r_0 can be expressed in terms of current density leading to a value of 1.45 A/cm² [Rizk, 1981]. It is also possible to compute the resistance of the ionized arc root of radius r_0 in Equation 5-17 using

$$R = \frac{10^6}{4\sigma_{20}r_0} \tag{5-19}$$

where R (Ω) is the resistance of the disk under the foot of the arc root to infinity in the half-space with conductivity σ_{20} (µS/cm). Figure 5-22 shows that,

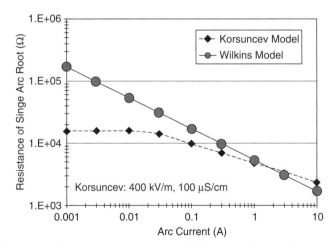

Figure 5-22: Comparison of Korsuncev and Wilkins models for resistance of arc root for σ_{20} of 100 µS/cm.

in the region of interest for flashover currents on polluted insulators, the two models agree quite well.

The close agreement between the Korsuncev and Wilkins models remains the same for all practical values of σ_{20}. This adds support to the physical basis of the Wilkins model, which has proved to be a fundamental component of the calculation of the resistance of polluted or iced layers.

Mercure and Drouet [1982] measured the current density of the arc root as it passed over miniature probes beneath electrolyte surfaces. Figure 5-23 shows the critical current I_c obtained from $(4\pi/J_0)\cdot(\chi E_0)$, where J_0 is the average current density of $2\,A/cm^2$. The gradient of $E_0 = 30\,kV/cm$ represents a fast breakdown regime. The stress of $E_0 = 3\,kV/cm$ corresponds to wet surface flashover. At $0.3\,kV/cm$, the arc propagates slowly with low probability of breakdown.

For thin layers of moderate pollution, to the right of the dashed line in Figure 5-23, the breakdown gradient is consistent with the Korsuncev and Wilkins models.

Jolly and Poole [1979] noted some differences in the positive and negative arc roots under dc conditions, using a relatively weak 600-mA source with the 230-mm annulus between two electrodes as shown Figure 5-7. They hypothesized that the diffuse arc root for positive polarity had higher resistance than the numerous cathode spots for negative polarity. Their research and mathematical modeling on a flat glass plate with concentric electrodes wetted with salt solutions led to an inverse relation between flashover voltage and surface conductivity γ as shown in Figure 5-24.

The positive dc flashover voltage in this controlled geometry was, on average, 26% higher than the negative dc level. This means that the mathematical models for dc flashover do need to use separate expressions for each polarity when evaluating the arc root radius as a function of leakage current.

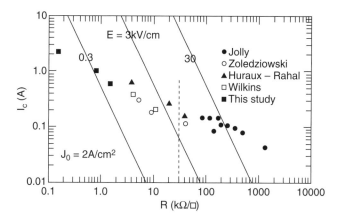

Figure 5-23: Comparison of flashover data with a breakdown-induced arc propagation model for arc root current density of $2\,A/cm^2$ versus surface resistivity (from Mercure and Drouet [1982]).

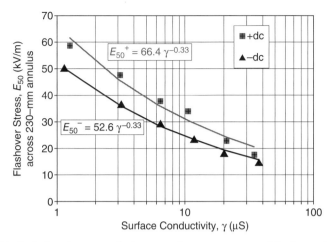

Figure 5-24: Relation between flashover voltage across coaxial electrodes on glass plate and surface conductivity (adapted from Jolly and Poole [1979]).

5.5. DC POLLUTION FLASHOVER MODELING

For calculation of dc contamination flashover levels, an electrode voltage drop V_e (V) is added to Equation 5-7, corresponding to the electrode voltage drops on the terminal electrodes. Also, the source voltage and current are constant for each position x (cm) along the leakage path L. This leads to the equation of applied voltage as follows:

$$V = V_e + A \cdot x \cdot I^{-n} + I \cdot R_P(x) \tag{5-20}$$

5.5.1. Analytical Solution: Uniform Pollution Layer

With all of the terms in Equation 5-20 now developed, it is possible to calculate the critical dc flashover voltage V_{50}. This can be derived mathematically for cases where the pollution layer resistance $R_P(x)$ is a differentiable function of arc length x. The derivative of Equation 5-20 with respect to x is equated to zero and checked to make sure it is a maximum. For the case of a uniform pollution layer $R_P(x) = B(L - x)$, the critical flashover stress $E_{50} = V_{50}/x$ becomes [Rizk, 1981]

$$E_{50} = A^{1/(n+1)} B^{n/(n+1)} \tag{5-21}$$

where A and n are the arc constants from Table 5-5 and B is the average resistance of the pollution layer (in Ω/cm). For typical values of $B = 400$ to $10,000\,\Omega/\text{cm}$, corresponding to ESDD values of about 1 and $0.03\,\text{mg/cm}^2$, respectively, the value of critical flashover gradient varies from 10 to 80 kV/ m_{leakage}.

TABLE 5-5: Typical Values[a] of A and n in the Relationship $V = AxI^{-n}$ for dc Arcs on Polluted Layers

Researcher	A	n	E_{50} (kV/m$_{leakage}$)
Nasser	63	0.76	$1.05 \cdot B^{0.43}$
Los	52	0.43	$1.58 \cdot B^{0.30}$
	44	0.67	$0.56 \cdot B^{0.50}$
Nottingham	31	0.985	$0.90 \cdot B^{0.40}$
	39.2	0.67	$0.35 \cdot B^{0.58}$
	20.3	1.38	$0.74 \cdot B^{0.48}$
Gers	46.05	0.91	$0.74 \cdot B^{0.48}$
	43.8	0.822	$0.80 \cdot B^{0.45}$
	59.64	0.773	$1.00 \cdot B^{0.44}$
Mayr	40.6	0.724	$0.86 \cdot B^{0.42}$
Swift	80	0.5	$1.86 \cdot B^{0.33}$
	60	0.5	$1.53 \cdot B^{0.33}$
Guan and Zhang	138	0.69	$1.85 \cdot B^{0.41}$
Sundararajan and Gorur	60	0.8	$0.97 \cdot B^{0.44}$

[a]V in volts, I in amperes, x in centimeters, E_{50} in kV/m, and B in kΩ/cm.

Source: Adapted from Chaurasia [1999].

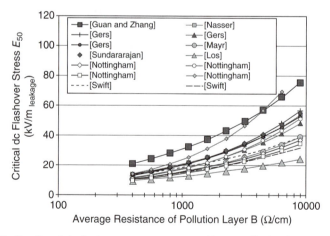

Figure 5-25: Predicted dc flashover stress versus uniform pollution layer resistance for various values of dc arc constants in Table 5-5.

For each pair of arc constants in Table 5-5, the empirical relation between dc flashover stress and uniform pollution layer resistance was established (Figure 5-25). The exponents in the power-law relation $E_{50} \propto B^\alpha$ are given by $\alpha = n/(n + 1)$ and for the most part fall in the range of $0.40 \leq \alpha \leq 0.45$ with a median of 0.43.

The discrepancies among the values of A and n reflect the environment in which the arc is propagating. Corrections for variation with temperature and pressure can be introduced to improve predictions. Also, the flashover gradients in Table 5-5 do not include the effect of electrode voltage drop V_e of nearly 1 kV, and this can be important for experiments that were carried out on a small physical scale.

5.5.2. Analytical Solution Using Insulator Form Factor

There have been many studies of the contamination flashover performance of the standard 146-mm × 254-mm disk insulators used in long strings for transmission lines. Each individual disk is relatively easy to handle and test, with a dry flashover level under ac conditions of about 80 kV. This means that a practical voltage source for flashover testing of single disks will fit within a one-story building with 3-m height. This height limit is an important consideration for many university researchers where a dedicated high-voltage laboratory with sufficient vertical height (10–20 m) is not available for studies on long insulator strings.

For individual disk insulators, Topalis and Gonos [2001] proposed a detailed ac flashover model following the Obenaus approach in Figure 5-10. Their choices for the model include a relation between ESDD and surface conductivity γ_s:

$$\gamma_s = (369.05 \cdot ESDD + 0.42) \times 10^{-6} \qquad (5\text{-}22)$$

where

γ_s is the surface conductivity in siemens (S or Ω^{-1}), equal to $\sigma\varepsilon$ where σ is the volume conductivity (μS/cm) and ε is the layer thickness (cm), and

$ESDD$ is the equivalent salt deposit density (mg/cm^2) on the disk insulator.

Normally, the relation between the product $\sigma\varepsilon$ and ESDD is governed by the relation for salt solutions, with $\gamma_s = 1820\,\mu$S per mg/cm^2 of ESDD on a flat surface for typical concentrations using Equation 5-5.

Using an insulator form factor F, Topalis and Gonos [2001] also derived the critical flashover voltage with Equations 5-23 to 5-27:

$$F = \int_0^L \frac{dl}{p(l)} \qquad (5\text{-}23)$$

where $p(l)$ is the perimeter of the insulator along the leakage distance.

The peak critical current I_c (A) that leads to flashover is

$$I_c = (\pi \cdot A \cdot D_m \cdot \gamma_s)^{1/(n+1)} \qquad (5\text{-}24)$$

where D_m is the maximum insulator diameter (cm).

Their corresponding peak ac critical flashover voltage V_c is

$$V_c = \frac{A}{n+1}(L + \pi \cdot n \cdot D_m \cdot F \cdot K) \cdot I_C^{-n} \qquad (5\text{-}25)$$

where K is defined as

$$K = 1 + \frac{n+1}{2\pi Fn} \ln\left(\frac{L}{2\pi r_0 F}\right) \qquad (5\text{-}26)$$

and r_0, the arc root radius (cm), is given by

$$r_0 = 0.469 \cdot \sqrt{I_C} \qquad (5\text{-}27)$$

Topalis and Gonos [2001] proposed arc constants of $A = 124.8$ and $n = 0.409$ to complete this model, and validated it with test results on four different shapes of disk insulators at nine levels of contamination.

5.5.3. Numerical Solution: Nonuniform Pollution Layer

Most of the expressions for $R_P(x)$ use an estimate of arc root radius, which in turn varies with leakage current I. This dependency means that an analytical solution for the flashover voltage is no longer convenient.

In these cases, the following approach is more suitable using a standard spreadsheet and its built-in Tools/Solver function.

- For each value of x in the range of 0 to L:
    Select a trial current I.
    Calculate the corresponding arc root radius r_0.
    Calculate the pollution layer resistance $R_P(x)$ from the arc root location to the insulator terminals.
    Calculate the total voltage V in Equation 5-20.
    Find the trial currents I that minimize V, the sum of arc voltage and voltage drop along the pollution layer, for each position x.
- Read out the maximum value of V for all values of x.

It is possible to form a single sum of all values of V for many values of x and to minimize this sum by changing all associated trial currents at once. At least 100 trial values of I can be converged simultaneously with this approach.

5.5.4. Comparison of Different Models for Pollution Layer

The results of the pollution flashover calculation are relatively sensitive to the values of A and n, as shown in Figure 5-25 for a uniform pollution layer. In contrast, the large differences between the pollution layer resistance values in Figure 5-26 lead to the same flashover gradient of about +31.5 kV/m$_{\text{leakage}}$.

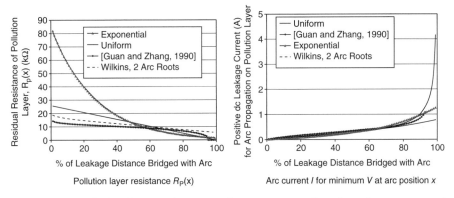

Figure 5-26: Pollution layer resistance and arc current as function of percentage of bridged leakage distance on standard disk insulator.

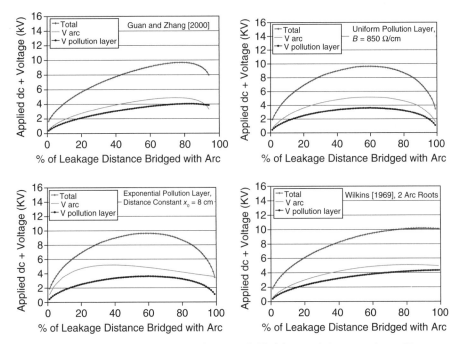

Figure 5-27: Contribution of arc voltage and $IR_P(x)$ to minimum voltage V versus percentage of bridged leakage distance on standard disk insulator. $V_e = 0.8\,\text{kV}$.

All four models give about the same variation of leakage current along the leakage distance, since they were all calculated with the same values of $A = 138$ and $n = 0.69$. When about 60% of the leakage distance is bridged by an arc, the models predict similar values of residual resistance. This leads to the close agreement in the maximum values of applied voltage in Figure 5-27.

The Guan–Zhang curve in Figure 5-27 terminates at about 93% of the leakage distance because the ratio of $(L - x)/r_0$ becomes less than unity, making the pollution layer resistance estimate from Equation 5-16 a negative number.

For a uniform pollution layer, the critical point with highest applied voltage is given by $L/(n + 1)$, where $n = 0.69$ is the arc constant exponent. This gives the peak stress at 59% of the leakage distance L in Figure 5-27. The Wilkins model for the resistance of the pollution layer with two arc roots suggests that this critical point x will be more than 80% along the total leakage path.

5.5.5. Introduction of Multiple Arcs in Series

During flashover testing in contaminated conditions, a dynamic equilibrium of arcing and dry-band growth is established. In general, there can be as many dry bands as there are insulators. More usually, the arcing activity is confined to a modest fraction of the insulators, focused near the areas of highest electrical stress.

The presence of two or more pollution layers in series, connected by partial arcs, is not treated by the mathematical models for pollution flashover as this is thought to be a precursor stage to the important phase, when several arcs merge to span a large fraction of the insulator leakage path. The general calculation of the resistance of the pollution layer under arcing conditions is affected by several factors that are somewhat random in nature. The most important factors include:

- Arc bridging between sheds and ribs of the leakage path.
- Formation of a number of consecutive arcs in series.
- Arc drift off the surface of the insulators.

The parameter N was already introduced as the number of arc roots in Figure 5-19, and electrode voltage drop V_e was introduced in Equation 5-20. A third parameter can also be introduced to improve the calculation of insulator voltage in Equation 5-7. This is a coefficient K_p, giving the ratio of the arc length to the leakage path length. Under some conditions, the arc may float off the insulator surface, leading to $K_p > 1$. For partial bridging of the leakage path across insulator sheds, $K_p < 1$.

$$V = V_e + A \cdot K_p x \cdot I^{-n} + \left[\frac{NI}{\pi \gamma_e} \right] \ln \left[\frac{L - x}{r_0} \right] \tag{5-28}$$

where

V_e is the electrode voltage drop (V),

A and n are arc constants,

K_p is the ratio of the partial arc length in air to the arc root location,

x is the position of the arc root along the leakage path (cm),

I is the leakage current (A),

N is the number of arc roots, which will be two times the number of pollution layers in series with the arcs,

γ_e is the equivalent layer conductivity γ (μS) at the critical moment,

L is the length of the leakage path (cm), and

r_0 is the arc root radius (cm) from Equation 5-17.

The number of pollution layers and arcs that form under dc conditions will
be a function of the wetting rate, surface conductivity, and uniformity of the
polluted layer.

5.5.6. dc Arc Parameter Changes with Pressure and Temperature

Novak and Ellena [1987] measured the parameters of dc arcs from 20 mA to
4 A at pressures of 20–150 kPa. They noted a change from glow discharge to
arcing at about 300 mA. For currents greater than $I = 1$ A, they found a gradient of $E = 57I^{-0.54}$, with E in (V/cm) and I in (A), independent of pressure in
the practical range of interest from 60 to 150 kPa. For lower currents, their E–I
exponent changed from $\alpha = -0.54$ to $\alpha = -0.81$ at 100 kPa.

Ishii et al. [1984] evaluated the V–I characteristic of arcs occurring on polluted insulators in dc conditions. Their typical results, shown in Figure 5-28,

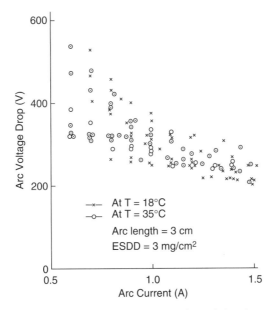

Figure 5-28: V–I characteristic of arc on contaminated insulator surface for –dc
(adapted from Ishii et al. [1984]).

were interpreted to separate the effect of temperature on the series resistance and on the arc parameters. Ishii et al. [1984] concluded that the dc arc parameters were independent of temperature in the range of interest.

5.6. AC POLLUTION FLASHOVER MODELING

In order to maintain an arc burning on a polluted surface, Equation 5-20 must be satisfied at the peak of the applied voltage wave. Also, in general, the electrode voltage drop V_e can be neglected and set to zero. This leads to a revised expression for the applied voltage:

$$V_m = A \cdot x \cdot I_m^{-n} + I_m \cdot R_P(x) \tag{5-29}$$

where

V_m is the peak of the applied voltage (V),

I_m is the peak of the leakage current (A),

x is the arc length (cm),

$R_P(x)$ is the residual resistance of the polluted surface that is not bridged by arcs, and

A and n are arc constants A and n from Tables 5-3 and 5-4.

Under ac conditions, the arc will also tend to extinguish at every half-cycle as the current passes through zero. However, a hot column of gas will remain, and this gas can easily be reignited by the applied voltage in the next half-cycle. For flashover to occur, the peak of the applied voltage V_m must also satisfy a reignition condition that is normally based on the amplitude of the peak current I_m in the previous ac half-cycle and the arc length x. The process of ac arc reignition is a fundamental difference between ac and dc pollution flashover, and the introduction of an appropriate ac reignition model [Rizk, 1981] dominates the mathematical description of the ac pollution flashover process.

5.6.1. ac Arc Reignition

For dc flashover, the criteria for arc motion and extension are sufficient to establish the conditions for flashover. Under ac voltage, however, the arc current reduces to zero at every half-cycle. A different set of conditions apply. It is necessary to solve the energy balance equation [Rizk, 1971b] of the residual hot gas, which starts with an arc temperature of typically 3000 K as the current goes to zero. The remaining hot gas in the arc cools by thermal convection, and one convenient way to model this process uses the arc boundary radius r_b. The arc radius is a function of the peak arc current, and this current is in turn a function of the Obenaus circuit model, including the quasi-static $V-I$ characteristics of the arc observed at the power system frequency.

The use of an arc boundary radius has the advantage that the mathematical predictions of r_b can be compared with visual observations.

Experimental support for a thermal balance model of arc reignition is found, for example, in Claverie [1971] and Claverie and Pocheron [1973]. They measured the peak reignition voltage as a function of the peak current in the previous half-cycle and described the relation with an empirical expression:

$$V_{cx} = \frac{800x}{\sqrt{i_m}} \tag{5-30}$$

where

V_{cx} is the peak reignition voltage (V),

x is the arc length (cm), and

i_m is the peak arc current (A) in the previous half-cycle of ac voltage.

The experimental data supporting this empirical fit are shown in Figure 5-29.

The empirical relation for reignition conditions is replotted on linear scales, along with the most recent mathematical model [Rizk and Rezazada, 1997] in Figure 5-30.

The mathematical model for reignition of an ac arc in series with a uniform pollution layer, based on the energy balance criterion [Rizk, 1971b] is

$$V_d = V_{da}\left\{1 + \frac{S_0/S_b - 1}{1 + \dfrac{4_a^*(S_0 - S_b)t}{r_b^2}}\right\}^{-1/\beta} \tag{5-31}$$

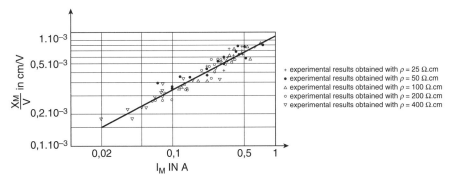

Figure 5-29: Experimental verification of Equation 5-30 (from Claverie [1971]).

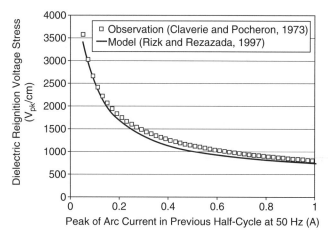

Figure 5-30: Comparison of empirical and mathematical models for the reignition of a 50-Hz arc.

where

V_d is the minimum breakdown voltage (V/cm) of the residual arc column,

V_{da} is the breakdown voltage stress of air, 5233 V/cm at the reference temperature (20 °C) and pressure (101.3 kPa),

S_0 is the thermal flux function at the initial arc temperature, 350.8 J/(m-s) at 3000 K,

S_b is the thermal flux function at ambient temperature, 5.34 J/(m-s) at 300 K,

a^* is a fitted constant, $3.78 \times 10^{-6}\,\mathrm{m^3/J}$,

β is a fitted constant, 1.778,

r_b is the boundary radius (cm) of the arc, calculated from the peak arc current (I_m) using $r_b = 0.497 \cdot I_m^{0.663}$, and

t is the time (s) to peak of ac wave, $t = 1/(4f)$, where f is the frequency in Hz.

Normally, an air gap would be expected to flash over at an applied voltage stress of $E_{50} = 520$–560 kV/m (5233 V/cm in Equation 5-31) for a fast lightning impulse or for the peak of the ac voltage wave. This flashover stress E_{50} (or V_{da} in Equation 5-31) will increase by 7% at 0 °C and decrease by 6% at 40 °C [IEEE Standard 4, 1995].

When the mathematical model for arc reignition [Rizk and Rezazada, 1997] is evaluated against the reference flashover stress of 523.3 kV/m at 20 °C, Figure 5-31 shows that leftover hot gas from arc currents of 1 A can reduce the air-gap strength by more than 80%.

The reduction in air-gap flashover strength from the leftover hot gas in Figure 5-31 is further quantified by doing the calculation for three different

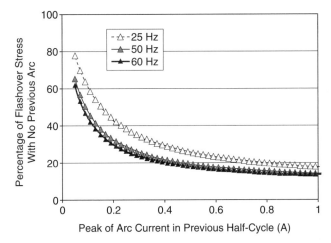

Figure 5-31: Percentage reduction in flashover gradient at 101.3 kPa versus peak arc current in previous ac half-cycle.

power system frequencies. For the longer quarter-cycle duration t of the legacy 25-Hz systems, the arc cools to a lower temperature and arc reignition is more difficult. The implication in Figure 5-31 is that, for critical arc currents of 0.2 A, pollution performance will be better for 25-Hz systems than for 50- or 60-Hz systems.

The original Rizk thermal model for the ac arc reignition was refined [Rizk, 1981; Rizk and Rezazada, 1997] and applied to establish the minimum voltage necessary for ac flashover of a polluted insulator. This work pointed out that the reignition criterion of the residual arc following a current zero of the critical leakage current, rather than the static arc extension, was the limiting condition for ac flashover at 50 or 60 Hz. This is reiterated to be a fundamental difference between mathematical models for ac and dc flashover [Obenaus, 1958].

5.6.2. ac Reignition Conditions Versus Ambient Temperature

In the mathematical model for arc reignition [Rizk and Rezazada, 1997], ambient temperature affects the thermal flux function S_b, given approximately by

$$S_b = 0.00022 \cdot T^\beta \qquad (5\text{-}32)$$

where T is the temperature in kelvin units and S_b is in units of J/(m-s). The appropriate values of S_0 and S_b are substituted into Equation 5-31, along with the corresponding values of V_{da}, at 0 °C, 20 °C, and 40 °C, of 562, 523, and 490 kV/m to establish the reignition gradient as a function of ambient temperature.

Figure 5-32 shows how the reignition gradient varies with peak arc current in the previous half-cycle for these three temperatures.

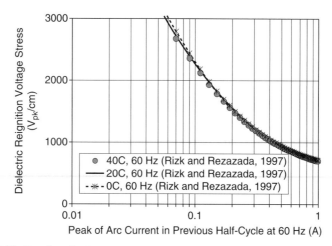

Figure 5-32: Predicted effect of temperature on 60-Hz reignition voltage gradient.

For the limited pressure range of 600–1013 kPa (0.6–1 atm) and for the arc temperature range of 300–3000 K, Rizk and Rezazada [1997] suggest that the effect of pressure on the thermal conductivity of air can be neglected. Also, the initial temperature of the leftover arc gas at the start of the dielectric recovery period would be constant at 3000 K. Therefore the values of S_0, S_b, and β defined in Equation 5-31 do not change with pressure. This leaves correction of the breakdown voltage stress V_{da} using, for example, IEEE Standard 4 [1995] as the only suggested step for environmental correction of reignition conditions.

5.6.3. Mathematical Model for Reignition Condition

Within the range of 0.05–1 A, Rizk and Rezazada [1997] performed a regression analysis of their updated model for the reignition process and then recommended that the minimum reignition gradient be expressed as

$$V_m + \frac{kx}{I^b} = \frac{716x}{I_m^{0.526}} \tag{5-33}$$

where V_m and I_m are peak voltage (V) and current (A) and x is the arc length (cm). This expression is used as a second and usually limiting condition that must be satisfied when calculating the ac flashover voltage of a polluted insulator.

The method for solving for the balance point at each location x along a leakage path of length L is as follows:

• For each value of x in the range of 0 to L:

　　Select a trial current I_m.

　　Calculate the corresponding arc root radius r_0.

Calculate the pollution layer resistance $R_P(x)$ from the arc root location to the insulator terminals.

Calculate the total voltage V_m in Equation 5-29.

Calculate the reignition voltage V_m corresponding to I_m in Equation 5-33.

Find the trial currents I_m that minimize the squares of the differences between the two values of V_m for each position x.

- Read out the maximum value of V_m for all values of x.

It is possible to form a single sum of all squared differences between values of V_m (Equation 5-29–Equation 5-33)^2 for many values of x. This error sum can be minimized in a spreadsheet (Tools/Solver function) by changing all associated trial currents at once. At least 100 trial values of I_m can be converged simultaneously with this approach.

5.6.4. Comparison of dc and ac Flashover Models

The difference between ac and dc flashover models can be illustrated by comparing the numerical results for linear and exponential functions of $R_P(x)$, which were already shown in Figure 5-26.

To facilitate the comparison, the same values of arc constants were used in ac and dc calculations. Normally, the arc constants for ac will differ somewhat from the dc constants. For example, Guan and Zhang [1990] recommend $A = 140$ and $n = 0.67$ for ac compared to their values of $A = 138$ and $n = 0.69$ that were used in the computations for Figure 5-33. There is also some evidence that the dc– constants differ from the dc+ values, depending on the geometry of the insulator.

The current required for ac reignition is about two times larger than the current needed for dc arc propagation on the polluted layer. In the dc case, the arc voltage and the voltage drop along the pollution layer are nearly equal at each position x. In the ac case, the voltage drop along the pollution layer is a much larger fraction of the total applied voltage. The flashover gradients E_{50} are 41–43 kV$_{pk}$/m$_{leakage}$ for the ac case, with values of 31.5 kV/m$_{leakage}$ computed for dc arcs on the same pollution layers with the same arc constants.

The Wilkins model for the pollution layer was used to compute critical flashover stress for a 30.5-cm leakage path typical of a standard disk insulator. In this situation, two arc roots represent the normal state of the dc arc across a dry band that extends over the insulator edge. Figure 5-34 shows the computed values, along with empirical fits to the results using power-law relations.

Figure 5-34 also includes the empirical description of the ac flashover strength of insulator strings in clean-fog conditions [Baker et al., 2008] from Figure 4-63, converted from rms to peak. The Wilkins model for pollution layer resistance with two arc roots, along with the reignition conditions from Rizk and Rezazada [1997], gives a single-disk flashover stress that tracks the string flashover stress closely over a suitably wide range of ESDD values.

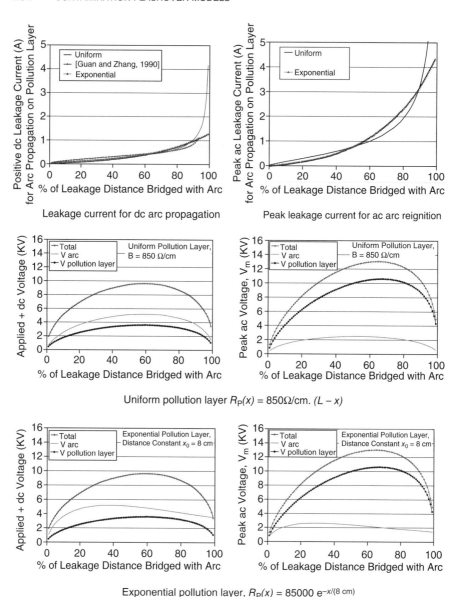

Leakage current for dc arc propagation

Peak leakage current for ac arc reignition

Uniform pollution layer $R_P(x) = 850\Omega/\text{cm}. (L - x)$

Exponential pollution layer, $R_P(x) = 85000\ e^{-x/(8\ \text{cm})}$

Figure 5-33: Comparison of dc and ac flashover voltages and currents for exponential and linear pollution layers using the same arc constants $A = 138$ and $n = 0.69$.

5.7. THEORETICAL MODELING FOR COLD-FOG FLASHOVER

One of the successes in the application of the Obenaus model for dc flashover, adapted by Rizk [1981] for ac reignition, was for predicting flashover along leakage distance in freezing conditions [Chisholm, discussion to

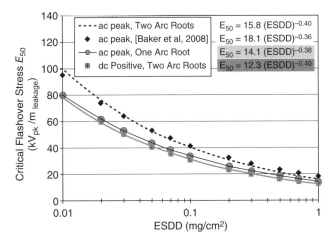

Figure 5-34: Critical flashover stress versus ESDD from mathematical model of pollution flashover using Wilkins estimate of $R_P(x)$.

Farzaneh et al., 1997]. This was achieved with the use of arc coefficients ($A = 205, n = 0.561$) for a suitably large arc length, along with the corresponding ac reignition coefficients, measured to be $k = 1118$ and $b = 0.528$ compared to $k = 716$ and $b = 0.526$ in Equation 5-33 at room temperature. These coefficients were measured on 30-cm triangular ice surfaces, but their extrapolation to use on the 1000-cm leakage distance along EHV insulators led to remarkably good prediction of cold-fog flashover levels on polluted insulators. Adjustments to the surface conductivity are also needed for accurate results in freezing temperatures.

The details of the modeling of icing flashovers along the insulator leakage and dry arc distances, depending on level of ice accretion, are covered in more detail in Chapter 7.

5.8. FUTURE DIRECTIONS FOR POLLUTION FLASHOVER MODELING

The assumption of a static arc model with constant pollution layer resistance ignores the input of electrical energy into the pollution layer. However, a large fraction (60–87%) of the power supply energy [Textier and Kouadri, 1986] goes into heating and evaporation processes. The resistance of the pollution electrolyte also has strong temperature dependence. Thus a dynamic model for the increase in wetted layer temperature and the corresponding decrease in surface conductivity with the progress of the arc root during each half-cycle of ac reignition would seem to be an important area of future development.

The temperature of the insulator will not be uniform [Abonneau, 2000]. Instead, the tips of insulator sheds will tend to have higher temperatures, especially in non-steady-state conditions.

One of the unresolved issues in this adaptation was the large increase in the reignition coefficient k in freezing conditions compared to the value at 20 °C. As already explored, this magnitude of change is not predicted by the heat balance model of the arc. This may be resolved with direct and dynamic measurements of the arc temperature on the polluted surface at various ambient temperatures to support analytical models.

The arc constants depend to a large extent on the environmental conditions, including the nature and conductivity of the pollution layer, relative humidity, and air density including both temperature and pressure effects. While there have been convincing demonstrations that the arc coefficients change with type of pollution, there has not been much supporting work to explain how different ionization potentials or other factors cause these changes. Dynamic measurements of the light spectra from arcs may provide additional experimental support to develop suitable models for these effects.

5.9. PRÉCIS

A pollution layer on an insulating surface adds distributed resistance per unit length to its inherent capacitance. This changes the electric field distribution along the insulator near metal electrodes. Dry banding may follow from the flow of leakage current, resulting in the initiation of arcing. The flow of current in the arc may be continuous under dc or may be sufficiently high to reignite in a quasi-stable way after ac zero crossings. Dry bands may grow to span most or all of the insulator leakage distance, bringing the insulator to flashover.

Mathematical models for the pollution flashover process rely on two major components: the nonlinear voltage–current characteristic of the arc and the resistance of the polluted layer. The arc characteristic is a linear function of the arc length x. The residual resistance of the pollution layer is a monotonic and often logarithmic function of $(L - x)$, the unbridged part of the overall leakage distance L. A search for the maximum voltage supported by this series of nonlinear elements can provide a good estimate of the critical flashover voltage of insulators as a function of their surface pollution level, expressed as an equivalent salt deposit density.

The modeling suggests that the ac flashover strength is a bit higher than dc flashover, comparing peak of the ac wave to positive polarity dc predictions. In both cases, flashover strength decreases at about the 0.4 power of surface contamination level. When two or more arc roots form on the pollution surface, flashover strength will increase compared to a situation with a single arc root.

REFERENCES

Abonneau, L. 2000. "Modélisation du comportement dynamique d'isolateurs sous pollution." Doctoral thesis, l'Ecole Centrale de Lyon, No. ECL-2000-10, April 12.

Allen, N. L. and D. C. Faircloth. 2003. "Corona Propagation and Charge Deposition on a PTFE Surface," *IEEE Transactions on Dielectrics and Electrical Insulation*, Vol. 10, pp. 295–304.

Alston, L. L. and S. Zoledziowski. 1963. "Growth of Discharges on Polluted Insulation," *Proceedings of IEE*, Vol. 110, No. 7 (July), pp. 1260–1266.

ArcPro v.2. 2000. "Electrical Arc Hazard Assessment Software." H.D. Electric Company Bulletin No. ARC-V2000.

Baker, A. C. and M. Kawai. 1973. "A Study on Dynamic Voltage Distribution on Contaminated Insulator Surface," *IEEE Transactions on Power Apparatus and Systems*, Vol. PAS-92, No. 5 (Sept.), pp. 1517–1524.

Baker, A. C., M. Farzaneh, R. S. Gorur, S. M. Gubanski, R. J. Hill, G. G. Karady and H. M. Schneider. 2008. "Selection of Insulators for AC Overhead Lines in North America with respect to Contamination," *IEEE Transactions on Power Delivery*, in press.

Chaurasia, D. C. 1999. "Scintillation Modelling for Insulator Strings Under Polluted Conditions," in *International Symposium on High Voltage Engineering (ISH)*, IEE Conference Publication No. 467, Vol. P2, pp. 4.224–4.227. August 22–27.

Chisholm, W. A. and D. F. Peelo. 2007. "MV and HV Air Break Disconnect Switch Arc Reach Study." CEATI Report No. T033700-3027.

CIGRE Working Group 36.01. 2000. "Electric Power Transmission and the Environment: Fields, Noise and Interference." CIGRE Technical Brochure 74.

CISPR. 1986. *IEC CISPR 18-2 Radio Interference Characteristics of Overhead Power Lines and High-Voltage Equipment—Part 2: Methods of Measurement and Procedure for Determining Limits*, Edition 1.0. Geneva, Switzerland: International Electrotechnical Commission.

CISPR. 1993. *IEC CISPR 18-2 Amendment 1—Radio Interference Characteristics of Overhead Power Lines and High-Voltage Equipment. Part 2: Methods of Measurement and Procedure for Determining Limits*. Geneva, Switzerland: International Electrotechnical Commission.

CISPR. 1996. *IEC CISPR 18-2 Amendment 2—Radio Interference Characteristics of Overhead Power Lines and High-Voltage Equipment. Part 2: Methods of Measurement and Procedure for Determining Limits*, Edition 1.0. Geneva, Switzerland: International Electrotechnical Commission.

Claverie, P. 1971. "Predetermination of the Behaviour of Polluted Insulators," *IEEE Transactions on Power Apparatus and Systems*, Vol. PAS-90, No. 4 (July), pp. 1902–1908.

Claverie, P. and Y. Pocheron. 1973. "How to Choose Insulators for Polluted Areas," *IEEE Transactions on Power Apparatus and Systems*, Vol. PAS-92, No. 3 (May/June), pp. 1121–1131.

Darveniza, M. 1980. *Electrical Properties of Wood and Line Design*. St. Lucia, Queensland: University of Queensland Press, pp. 124–128.

Farzaneh, M., J. Zhang, and X. Chen. 1997. "Modeling of the ac Arc Discharge on Ice Surfaces," *IEEE Transactions on Power Delivery*, Vol. 12, No. 1 (Jan.), pp. 325-338.

Farzaneh, M., J. Zhang, and Y. Li. 2005. "Effects of Low Air Pressure on ac and dc Arc Propagation on Ice Surface," *IEEE Transactions on Dielectrics and Electrical Insulation*, Vol. 12, No. 1 (Feb.), pp. 60–71.

Filter, R. and J. D. Mintz. 1990. "An Improved 60 Hz Wood Pole Model," *IEEE Transactions on Power Apparatus and Systems*, Vol. 5, No. 1 (Jan.), pp. 442–448.

Gallimberti, I., I. Marchesi, and L. Niemeyer. 1991. "Streamer Corona at an Insulating Surface," in *Proceedings of 7th International Symposium on High Voltage Engineering* (ISH), Dresden, Germany, Paper 41.10. August.

Ghosh, P. and N. Chatterjee. 1995. "Polluted Insulator Flashover Model for ac Voltage," *IEEE Transactions on Dielectrics and Electrical Insulation*, Vol. 2, No. 1, pp. 128–136.

Guan, Z. and R. Zhang. 1990. "Calculation of dc and ac Flashover Voltage of Polluted Insulators," *IEEE Transactions on Electrical Insulation*, Vol. 25, No. 4 (Aug.), pp. 723–729.

Hewitt, G. F. 1960. "Tables of the Resistivity of Aqueous Sodium Chloride Solutions." UK Atomic Energy Research Establishment Report AERE-R3497.

IEC 60815. 1986. *Guide for the Selection of Insulators in Respect of Polluted Conditions*. Geneva, Switzerland: IEC.

IEEE Standard 4. 1995. *IEEE Standard Techniques for High Voltage Testing*. Piscataway, NJ: IEEE Press.

Ishii, M., M. Akbar, and T. Kawamura. 1984. "Effect of Ambient Temperature on the Performance of Contaminated DC Insulators," *IEEE Transactions on Electrical Insulation*, Vol. EI-19, No. 2, pp. 129–134.

Jing, T. 1995. "Surface Charge Accumulation: An Inevitable Phenomenon in DC GIS," *IEEE Transactions on Dielectrics and Electrical Insulation*, Vol. 2, pp. 771–778.

Jolly, D. C. 1972a. "Contamination Flashover, Part 1: Theoretical Aspects," *IEEE Transactions on Power Apparatus and Systems*, Vol. PAS-91, No. 6 (Nov.), pp. 2437–2442.

Jolly, D. C. 1972b. "Contamination Flashover, Part II: Flat Plate Model Tests," *IEEE Transactions on Power Apparatus and Systems*, Vol. PAS-91, No. 6 (Nov.), pp. 2243–2451.

Jolly, D. C. and S. T. Chu. 1975. "Surface Electric Breakdown of Tin Oxide Coated Glass," *Journal of Applied Physics*, Vol. 50, pp. 6196–6199.

Jolly, D. C. and C. D. Poole. 1979. "Flashover of Contaminated Insulators with Cylindrical Symmetry Under DC Conditions," *IEEE Transactions on Electrical Insulation*, Vol. EI-14, No. 2 (Apr.), pp. 77–84.

Jun, X. and I. D. Chalmers. 1997. "The Influence of Surface Charge upon Flashover of Particle-Contaminated Insulators in SF6 Under Impulse-Voltage Conditions," *Journal of Applied Physics D*, Vol. 30, pp. 1055–1063.

Korsuncev, A. V. 1958. "Application on the Theory of Similarity to Calculation of Impulse Characteristics of Concentrated Electrodes," *Elecktrichestvo*, No. 5, pp. 31–35.

Le Roy, G., C. Gary, B. Hutzler, G. Lalot, and C. Dubanton. 1984. *Les propriétés diélectriques de l'air et les très hautes tensions*. Paris: Editions Eyrolles.

Li, S., R. Zhang, and K. Tan. 1989. "Measurement of the Temperature of a Local Arc Propagating," in *Proceedings of 6th International Symposium on High Voltage Engineering*, New Orleans, Louisiana, Paper No. 12.05.

Mercure, H. P and M. G. Drouet. 1982. "Dynamic Measurements of the Current Distribution in the Foot of an Arc Propagating Along the Surface of an Electrolyte," *IEEE Transactions on Power Apparatus and Systems*, Vol. PAS-101, No. 3 (Mar.), pp. 725–736.

Ndiaye, I. 2007. "Approche Physique du Développement de Streamers Positifs sur une Surface de Glace." Doctoral Thesis, Université du Québec à Chicoutimi. May.

Niemeyer, L. 1995. "A Generalized Approach to Partial Discharge Monitoring," *IEEE Transactions on Dielectrics and Electrical Insulation*, Vol. 2, No. 4 (Aug.), pp. 510–528.

Novak, J. P. and G. Ellena. 1987. "Arc Field Measurement with a Simple Experimental Apparatus," *Journal of Physics D: Applied Physics*, Vol. 20, No. 4 (Apr.), pp. 462–467.

Obenaus, F. 1958. "Fremdschichtüberschlag und Kreichweglänge," *Deutsche Elektrotechnkik*, Vol. 4, pp. 135–136.

Peelo, D. F. 2004. *Current Interruption Using High Voltage Air-Break Disconnectors.* Eindhoven: Technische Universiteit Eindhoven, Proefschrift.

Peyregne, G., A. Rahal, and C. Huraux. 1982a. "Flashover of a Liquid Conducting Film, Part 1: Flashover Voltage," *IEEE Transactions on Electrical Insulation*, Vol. EI-17, No. 1 (Feb.), pp. 10–14.

Peyregne, G., A. Rahal, and C. Huraux. 1982b. "Flashover of a Liquid Conducting Film, Part 2, Time to Flashover—Mechanisms," *IEEE Transactions on Electrical Insulation*, Vol. EI-17, No. 1 (Feb.), pp. 15–19.

Rahal, A. M. and C. Huraux. 1979. "Flashover Mechanism of High Voltage Insulators," *IEEE Transactions on Power Apparatus and Systems*, Vol. PAS-98, No. 6, pp. 2223–2231.

Rizk, F. A. M. 1971a. "Analysis of Dielectric Recovery with Reference to Dry-Zone Arcs on Polluted Insulators," in *IEEE PES Winter Power Meeting*, New York, Paper No. 71CP134-PWR.

Rizk, F. A. M. 1971b. "A Criterion for AC Flashover of Polluted Insulators," in *IEEE PES Winter Power Meeting*, New York, Paper No. 71CP135-PWR.

Rizk, F. A. M. 1981. "Mathematical Models for Pollution Flashover," *Electra*, No. 78, pp. 71–103.

Rizk, F. A. M and A. Q. Rezazada. 1997. "Modeling of Altitude Effects on AC Flashover of High-Voltage Insulators," *IEEE Transactions on Power Delivery*, Vol. 12, No. 2 (Apr.), pp. 810–822.

Rumeli, A. 1976. "Flashover Along a Water Column," *IEEE Transactions on Electrical Insulation*, Vol. EI-11, No. 4 (Dec.), pp. 115–120.

Star. 1994. "Thousands in Metro Suffer Power Outages: 'Frozen Fog' Creates Havoc for Both Drivers and Football Fans," Headline article, *The Toronto Star*. January 24.

Suginuma, Y., T. Takahashi, M. Nogaki, and J. Matsushima. 1993. "pH, Conductivity and Nitric Acid Concentration of Aqueous Electrodes Sustaining Weak Current Atmospheric Discharges," in *Proceedings of 8th ISH*, Yokohama, Japan.

Tavakoli, C. 2004. "Dynamic Modeling of ac Arc Development on Ice Surfaces." Doctoral Thesis, Université du Québec à Chicoutimi. November.

Textier, C. and B. Kouadri. 1986. "Model of the Formation of a Dry Band on an NaCl-Polluted Insulation," *IEE Proceedings Part A*, Vol. 133, No. 5, pp. 285–290.

Topalis, F. V. and I. F. Gonos. 2001. "Dielectric Behaviour of Polluted Porcelain Insulators," *IEE Proceedings on Generation, Transmission and Distribution*, Vol. 148, No. 4, pp. 269–274.

Verhaart, H., A. Tom, L. Verhage, and C. Vos. 1987. "Avalanches Near Solid Insulators," in *Proceedings of 5th International Symposium on High Voltage Engineering*, Braunschweig, Germany, Paper 13.01. August 24–28.

Wilkins, R. 1969. "Flashover Voltage of HV Insulators with Uniform Surface Pollution Films," *Proceedings of the IEE*, Vol. 116, pp. 457–465.

Wilkins, R. and A. A. J. Al-Baghdadi. 1971. "Arc Propagation Along an Electrolyte Surface," *Proceedings of IEE*, Vol. 118, No. 12, pp. 1886–1892.

Yasui, M., Y. Takahashi, A. Takenaka, K. Naito, Y. Hasegawa, and K. Kato. 1988. "RI, TVI and AN Characteristics of HVDC Insulator Assemblies Under Contaminated Conditions," *IEEE Transactions on Power Apparatus and Systems*, Vol. 3, No. 4 (Oct.), pp. 1913–1921.

Zhang, J., L. Gu, and C. Sun. 1991. "Calculation and Measurement of Surface Field of Polluted Post Insulator During Flashover," in *Proceedings of the 7th International Symposium on High Voltage Engineering*, Braunschweig, Germany. August.

CHAPTER 6

MITIGATION OPTIONS FOR IMPROVED PERFORMANCE IN POLLUTION CONDITIONS

Insulator monitoring has three main purposes:

- Long-term assessment of site pollution severity.
- Normalization of laboratory test results based, for example, on salt-fog salinity, ESDD, NSDD, and other factors.
- Short-term input to initiate maintenance actions.

The long-term exposure characteristics have been discussed in Chapter 3, with Chapter 4 providing examples of how insulator electrical strength depends on the pollution levels. This chapter discusses how a program of insulator monitoring can be used to initiate maintenance actions, such as washing or cleaning. Once cleaned, the insulators have improved electrical performance compared to their contaminated state. The combined process of monitoring and periodic cleaning has proved to be one of the most important and effective mitigation options when applying insulators in polluted conditions.

In cases where maintenance intervals prove to be too frequent, it may be necessary to apply mitigation options that actually improve contamination performance. These may only need to be applied in restricted areas. This is the focus of the second part of the chapter.

Insulators for Icing and Polluted Environments. By Masoud Farzaneh and William A. Chisholm
Copyright © 2009 the Institute of Electrical and Electronics Engineers, Inc.

6.1. MONITORING FOR MAINTENANCE

A proactive program of insulator condition monitoring and maintenance actions, usually high-pressure washing, can provide good long-term reliability. This choice of mitigation is often applied in stations where there are many insulators in the same, severe contamination environment and the consequence of faults is more severe.

CIGRE [Task Force 33.04.03, 1994] summarized utility practices in monitoring of insulators. Their resumé of the main characteristics of monitors was as follows:

- One country used corrosion rate to infer insulator performance.
- Two countries reported prototype optical measurements of surface pollution.
- Three countries used ESDD measurements on exposed insulators.
- Six countries used air pollution monitoring methods rather than measurements on insulators.
- Ten countries used leakage current measurements at normal service voltage stress.
- Eleven countries used surface conductivity measurements at voltage low enough to prevent arcing.

For nine of these countries, maintenance rather than site characterization or insulator performance evaluation was the primary driver for the insulator monitoring program.

6.1.1. Insulator Pollution Monitoring

Direct measurements of contamination levels on energized insulators are difficult to obtain. The line or station element must be removed from service. A quantity of distilled water must be handled with care so that contamination from other sources—hands, dust, or even the wash rag itself—does not bias the results. The detailed processes used to measure the ESDD and NSDD on insulator surfaces are described in Appendix B.

It is tempting to deploy passive insulator test stands to obtain ESDD measurements. This means that results can be obtained at any time, and by using a large number of portable insulators, it is more feasible to set up a program of routine sampling. However, the relation between the ESDD on a portable insulator and the value on an energized station post insulator is loose at best and this approach is not endorsed without a calibration phase.

In situ measurements on operating insulators that are deenergized for the measurement can be valuable. For example, Schneider [1988] carried out ESDD measurements on HVDC post insulators, to establish the roles of polarity and field stress on the pollution accumulation rates. As a second illustration, field trials were carried out over two winters to compare the ESDD values on three different profiles of removable insulators, in comparison with energized station post insulators and unenergized posts of the same shape, located within 2 m. Table 6.1 shows the station post insulator profile, along with the shapes of distribution post, bell, and aerodynamic insulators used as surrogates. The small insulators were transported indoors for ESDD measurements, while the values from sheds of the ac-energized station posts were measured in situ every week. Typical results are shown in Figure 6-1.

It is clear from Figure 6-1 that there is a good relation between the ESDD measurements on the energized and unenergized station post insulators. The relation between the average energized-post $ESDD_{avg}$ ($\mu g/cm^2$) and the values U_t and U_b ($\mu g/cm^2$) from top and bottom surfaces of the unenergized post was

$$ESDD_{avg} = 0.513 \cdot U_t + 0.650 \cdot U_b + 1.2, \quad R^2 = 0.96 \qquad (6\text{-}1)$$

The regression coefficient R^2 of 0.96 indicates close agreement. The sum of the coefficients for the top and bottom surface (0.513 + 0.650) and the low offset value suggest that the average ESDD on the energized post is simply 16% higher than the value on the unenergized post.

The measured values of ESDD on the top surfaces of the insulators rose and fell more quickly than the values on the bottom surfaces. Also, levels measured on the aerodynamic disks were consistently a factor of 2 (0.487 + 1.548 in Table 6.1) lower than those found on energized station posts. The other insulator profiles in Table 6.1 were also tested to see whether their top and bottom surface ESDD values could also predict the energized-post values.

The conclusion from the cross-calibration values in Table 6.1 was that, while there was a significant linear relation in each case, none of the removable samples gave satisfactory accuracy.

In another season's results plotted in Figure 6-2, the energized-post ESDD levels were again about 10% higher than those measured on the unenergized posts of the same shape. The dielectrophoresis forces acting to increase the pollution levels on the ac-energized insulators are thus relatively small in winter field conditions.

The conclusion from this evaluation was that it is possible and practical to predict the ESDD on ac-energized station post insulators of 2-m length from measurements on small passive insulators. The best estimates were obtained using ESDD values from top and bottom surfaces of unenergized post sheds of the same shape, placed close to the energized posts. The inference is that, provided there is a calibration phase, a similar passive sampling process, with unenergized disks of the same shape and location as the energized insulators, will also give good results for transmission lines. The electric field effects are stronger for EHV insulators, and the calibration factors may deviate well away from the 10% level shown in Figure 6-2.

ESDD values from insulator monitoring can be used in two ways. If the values exceed design limits, insulator washing or other remedial actions can

TABLE 6-1: Prediction of Energized Station Post Insulator ESDD Levels Using Measurements from Passive Distribution Post, Bell, and Aerodynamic Disk Insulators

Insulator Profiles	Predicted Versus Observed ESDD (μg/cm^2) of Station Post Insulator

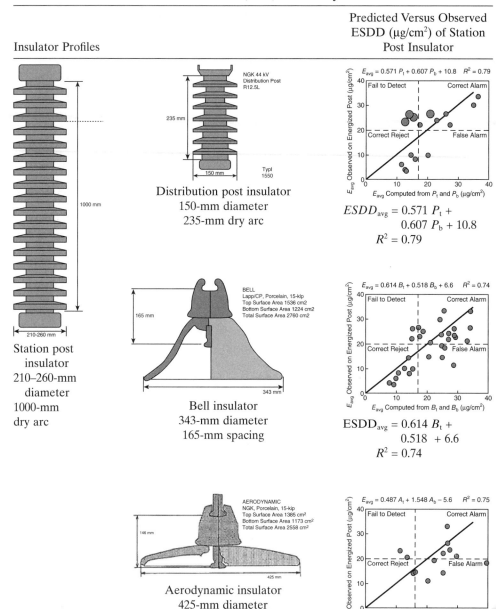

Station post insulator 210–260-mm diameter 1000-mm dry arc

Distribution post insulator 150-mm diameter 235-mm dry arc

$$ESDD_{avg} = 0.571\,P_t + 0.607\,P_b + 10.8$$
$$R^2 = 0.79$$

Bell insulator 343-mm diameter 165-mm spacing

$$ESDD_{avg} = 0.614\,B_t + 0.518 + 6.6$$
$$R^2 = 0.74$$

Aerodynamic insulator 425-mm diameter 146-mm spacing

$$ESDD_{avg} = 0.487\,A_t + 1.548\,A_b + 5.6$$
$$R^2 = 0.75$$

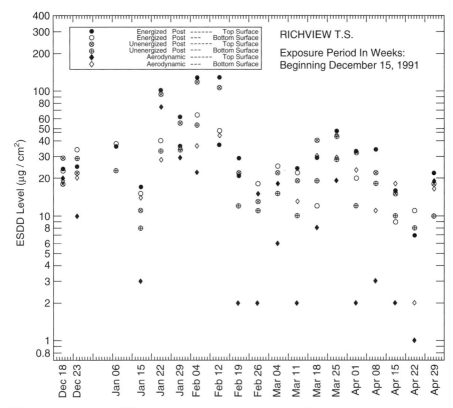

Figure 6-1: Average ESDD on energized and unenergized station posts and aerodynamic disks over winter season, expressed in μg/cm² and corrected for temperature using IEEE Standard 4 [1995]. (Courtesy of Kinectrics.)

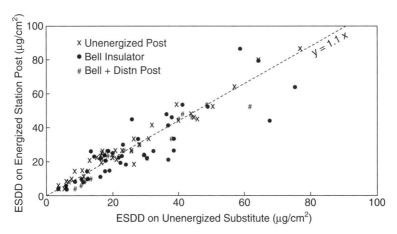

Figure 6-2: Relations between ESDD on ac-energized station post and ESDD on unenergized substitutes. (Courtesy of Kinectrics.)

be initiated immediately in a reactive program. This process can be extended to be proactive by evaluating the rate of change of ESDD with time, measured from the most recent washing period or heavy rain.

6.1.2. Condition Monitoring Using Leakage Current

Insulator leakage current is a convenient way to assess the combined effects of pollution severity and state of wetting on insulators in laboratory tests. It is also desirable to manage insulator leakage currents in the outdoor environment to a low level for a number of reasons, including power loss, audible noise, EMI, and (to a limited extent) ability to anticipate flashovers. Several commercial systems monitor pollution buildup through the changes in the leakage current and alert power system operators when predefined thresholds are exceeded. These systems have traditionally been based on concepts of:

- Surge counting—for example, like the original work of Forrest [1936, 1942], who modified electromechanical telephone relays to advance one count every time a current greater than 20 mA passed into the ground end of the insulator string under test
- Maximum leakage current—measured in a variety of ways but now mainly with a shunt resistor and digitizer.

After receiving a suitable indication from automatic systems, operators can then dispatch maintenance personnel to wash the insulators. This will reduce the frequency of unplanned outages from contamination flashovers and improve reliability. Additionally, it is argued that leakage current monitoring can optimize the time interval between washing cycles, compared to a fixed schedule.

Leakage currents can be inferred from the intensity of partial discharge activity and its associated ultraviolet, acoustic, and thermal signatures. However, it is more reliable to measure leakage current directly on insulators that have been set up specifically for this purpose. In recent work, such as Richards et al. [2003], the roles of relative humidity and of high-frequency components in the leakage current have been highlighted. They carried out long-term exposure tests on pin-type distribution insulators precontaminated with 40 g/L kaolin and NaCl to give an initial ESDD of 0.37 mg/cm^2. Figure 6-3 shows how their typical low-frequency and high-frequency components of leakage current varied from one ac cycle to the next.

The distortion in leakage current from a pure sine wave in Figures 6-3 and 6-4 can be modeled with third and fifth harmonic content. This approach was exploited by both Richards et al. [2003] and Amarh [2001]. The time evolution of the ratio of third harmonic to fundamental for several tests is shown in Figure 6-5. There is a downward trend at 30–60 minutes for insulators that withstood medium contamination (0.05–0.150 mg/cm^2 ESDD) at high-voltage

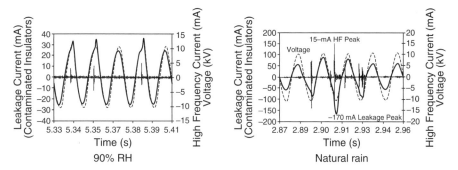

Figure 6-3: Typical leakage currents across surface of contaminated distribution insulators under natural wetting conditions (from Richards et al. [2003]).

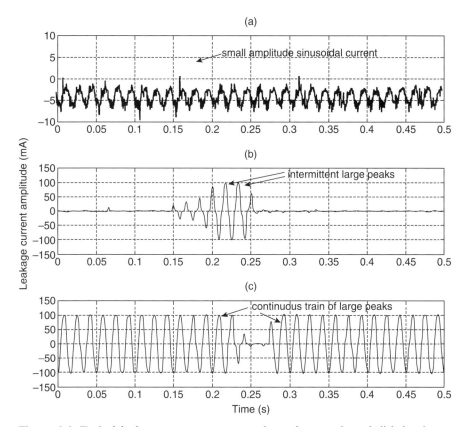

Figure 6-4: Typical leakage currents across surface of contaminated disk insulators under artificial clean-fog wetting conditions (a) initial stages, (b) mid-stages, and (c) final stages (from Amarh [2001]; courtesy of PSERC).

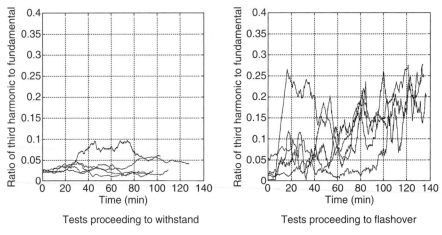

Tests proceeding to withstand | Tests proceeding to flashover

Figure 6-5: Time evolution of the ratio of third harmonic to fundamental 60-Hz leakage current for polluted insulators in clean-fog test (from Amarh [2001]; courtesy of PSERC).

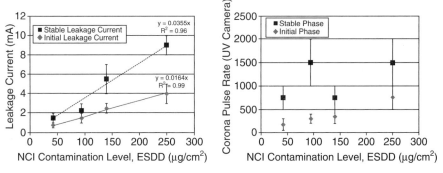

Figure 6-6: Leakage current (left) and corona-camera pulse counts (right) for clean-fog tests on precontaminated nonceramic insulators (data from Yizheng et al. [2007]).

stress of 41 kV per meter of leakage distance, or unified specific creepage distance (USCD) of 24 mm/kV.

A comparison of leakage current and corona camera pulse rate as indicators of the ESDD and arcing activity on nonceramic insulators is shown in Figure 6-6. The corona camera technique gives an independent measure of arcing activity and is discussed further in the next section. In this case, a distinction was drawn between the initial corona near the line end and the stable activity later in the artificial clean-fog test with controlled wetting.

In the controlled wetting conditions of the clean-fog test, the relation between leakage current and contamination level in this experiment was highly linear. This is a typical result and forms a fundamental basis of the commercial systems now deployed to monitor insulator leakage current in real time on power systems.

The strongest criticism of monitoring using leakage current activity to estimate insulator contamination level and to infer the possibility of flashover is the sensitivity of wetting [Richards and Renowden, 1997] and leakage current [Richards et al., 2003] to local climate conditions. Leakage current monitored in any way, including pulse amplitude, number of pulses, rms, or charge, is not a reliable indicator of flashover. Local climate also needs to be monitored to ensure consistent results. The example of a partially obscured mirror in a bathroom after a short shower gives a clear picture of how sensitive the condensation process is to temperature gradient. The same process occurs in the field, with the additional complication that wind speed will be increasing at about the one-seventh power of line height over grassy terrain. This means that conditions at the bottom phase of a line may be somewhat different from those at the top phase, and that activity from tower to tower may vary as well.

Kawai and Milone [1969] reported on the sensitivity of flashovers to relative humidity in their outdoor tests on insulator strings, which were artificially contaminated by spraying them with a slurry of sodium chloride and kaolin, and then energized at a line-to-ground voltage of 350 kV. Figure 6-7 shows that their flashovers occurred at relative humidity greater than 80%, rather than in fog wetting conditions.

Figure 6-8 shows that there can be daily cycles in leakage current activity for naturally wetted and contaminated insulators, keeping step with the daily variation in relative humidity. The daily variation in leakage current can be practically binary, with a consistent level of 15 mA rms when the humidity was

Outdoor test at 350 kv line to ground

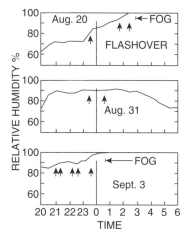
Occurrence of flashovers at night

Figure 6-7: Variation of relative humidity in natural humidity tests at 350 kV (from Kawai and Milone [1969]). Arrows indicate occurrence of flashovers.

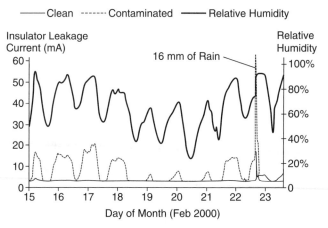

Figure 6-8: Variation in leakage current on clean and contaminated insulators in outdoor environment (from Richards et al. [2003]).

over 75% and minimal leakage current activity when the RH was less than 60%. This behavior is expected from the water activity level of NaCl, given in Table 3-15 as $a_w = 0.76$ at 5 °C, for insulators precontaminated with salt and kaolin.

Each insulator in Figure 6-8 is dissipating about 120 W (15 mA × 8 kV) when the RH exceeds 75%. With about 50 insulators per three-phase kilometer of distribution line, power losses reach 6 kWh/km for every hour with wetting and heavy pollution. This loss from leakage current exceeds the typical I^2R heat loss from the series resistance of the distribution system conductors.

Some utilities, notably on isolated islands such as Crete in saltwater oceans and seas, have been able to monitor the distribution network power losses from leakage current. This can provide a wide-area contamination monitoring function but usually leads to another mitigation path. With sufficient resolution in these power-loss measurements, it is also possible to quantify the economic value of a leakage current mitigation program. Leakage currents on polluted ceramic insulators are normally eliminated by greasing, applying an electrical-grade RTV silicone coating, or by replacing ceramic insulators with hydrophobic polymer units. When an insulator leakage current mitigation program meets a business case for reasonable payback interval, this investment has the advantage of improving the flashover performance at the same time. Other benefits of RTV coating are evaluated in Section 6.3.

Leakage current monitoring needs to be calibrated for each specific insulator. As an example, tests on 3.5-m ceramic insulator strings with three different insulator profiles were carried out at voltages up to 400 kV rms line to ground [Schneider and Howes, 1988]. Peak leakage currents varied with each insulator shape and voltage stress as shown in Figure 6-9.

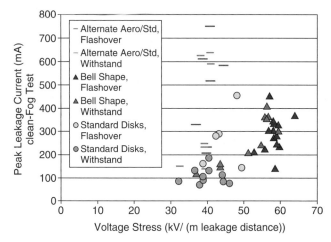

Figure 6-9: Relation between peak leakage current and voltage stress along leakage distance for three different 3.5-m strings of disk insulators in clean-fog test, 20 °C, ESDD of 0.017–0.023 mg/cm² (17–23 µg/cm²).

The insulator string in Figure 6-9 with alternating standard and aerodynamic insulators had the worst performance, expressed as a flashover level in kilovolt per meter of leakage distance, and also the highest leakage currents at voltage stress above $40 \, \text{kV/m}_{\text{leakage}}$. Generally, leakage current is seen to be a relatively uncertain predictor of flashover or withstand, although by eye, the 200-mA peak current level could be identified as a threshold of concern in Figure 6-9. In contrast, Amarh [2001] found the number of times the waveform crossed 60 mA to be a better predictor of flashover on four-unit strings of standard insulator disks, each with 305-mm leakage distance.

Insulators with semiconducting glaze can be used to measure pollution levels using their rms current rather than the peak leakage current. Typically, the glaze current of 1 mA will increase to levels as high as 20 mA under heavy pollution conditions and full surface wetting.

6.1.3. Condition Monitoring Using Corona Detection Equipment

Leakage current activity associated with wetting of surface pollution can occur in fog conditions and even when the relative humidity is about 80%, depending on the pollution type. This activity is easy to detect visually at night and also has strong acoustic, electromagnetic, and ultraviolet signatures.

Normally, partial discharge activity from corona would be expected in the air space surrounding the metal line-end fittings of insulators nearest the conductor, or those at the ground end near the tower, since these regions have the highest operating voltage stress. However, for polluted insulators, the potential distribution along the units in a string or the surface of the post is more linear. This means that partial discharge activity, which is distinct from

Figure 6-10: Ultraviolet (left) and normal photographs of leakage current activity on 115-kV transformer bushing during inclement weather (from Driscoll and Ballotte [2007]; courtesy of Northeast Utilities Service Company).

corona activity, can occur on any insulator surface in a contaminated string or anywhere along a post, not just at the terminals.

Ultrasonic microphones have been used for many years to find corona and partial discharge sources. They reduce the reliance on operator experience with the various sounds and increase sensitivity of detection. The microphones typically have parabolic reflectors to make the reception stronger and more directional. The converted audio signals are played through headphones and some instruments can count pulse rates and store results. Improved localization of corona can be achieved with a stereo pair of ultrasonic microphones, and signal-to-noise level can be improved with noise-canceling headphones. Correct operation of any ultrasonic system is tested by rubbing two fingers together as they move through the sound field of the microphone(s).

Cameras that detect ultraviolet radiation (Corocam, Ofil) also provide localization of corona and partial discharge sources on insulators, with 10-cm resolution as shown in Figure 6-10. This technology provides adequate signal-to-noise level even in daylight with response to individual photons now being a well-established technology. These cameras also have added built-in UV pulse-rate counting features. With experience and accurate weather monitoring, it seems on first view that the UV pulse rate could serve as a noncontact, quantitative indicator of leakage current pulse rate and intensity.

The first difficulty with applying ultraviolet or acoustic inspection to pollution monitoring is that corona activity, from positive and negative streamers, negative Trichel pulses, and glow discharge, also produces ultraviolet and acoustic energy.

- Most equipment of high quality can detect the ultraviolet or acoustic signatures of corona. UV equipment can detect Townsend discharges at current levels of $10\,\mu A$. Generally, corona has a relatively *constant voltage* for a wide range of partial discharge current from $10\,\mu A$ to $10\,mA$, similar to the clipping voltage characteristic of a surge arrester.

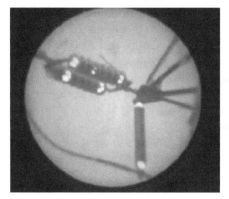 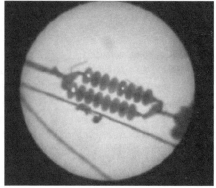

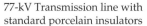

77-kV Transmission line with
standard porcelain insulators

77-kV Transmission line with
semiconducting glaze insulators

Figure 6-11: Comparison of partial discharge activity on standard and high-resistance semiconducting glaze insulators using night-vision camera (from Shinoda et al. [2005]).

- Arcs are different from corona. They have a *decreasing V–I curve*, with negative resistance, in the range of *100 mA to 100 A*. This negative-resistance characteristic is what makes them more likely to cause flashovers when fed from a constant-voltage source.

When inspecting insulators, the line voltage and location of the ultraviolet or acoustic activity are used to discriminate whether UV signatures are corona (at EHV) or arcing (on distribution lines). It is fair to say that many examples of corona on metal hardware are benign but that arcing always indicates a pending problem that calls for immediate attention.

Standard night-vision equipment has also been used by Shinoda et al. [2005] to detect corona activity on a 77-kV transmission line with twin strings of eight 250-mm diameter high-resistance semiconducting glaze disks. During the 2-year observation period, the insulators were checked for audible noise from partial discharge at design ESDD levels up to 0.5 mg/cm^2. While no audible noise was apparent on the semiconducting glaze disks, Figure 6-11 shows that small regions of discharge activity can indeed be detected with this equipment.

A common problem for semiconducting glaze suspension disk insulators has been glaze deterioration around the pin area. It is thus possible that the partial discharge activity in Figure 6-11 is unrelated to contamination. This degradation can be checked by observing the glaze temperature or measuring the insulator resistance.

6.1.4. Condition Monitoring Using Remote Thermal Monitoring

Infrared cameras have some ability to measure the occurrence of dry-band arcing. For example, a peak temperature of about 25 °C above a 9 °C ambient was reported on a contaminated 115-kV bushing as shown in Figure 6-12. This bushing is located about 25 m from ocean water in the Piscatqua River in New Hampshire, USA. Generally, sustained leakage currents in the 10-mA range are needed to provide enough thermal power to produce heat signatures.

Multiple areas of stable temperature rise have been observed on 400-kV post insulators, 20 minutes after drizzle near the freezing point in Figure 6-13 [Chrzan and Moro, 2007]. These discharges were not visible in daylight.

Infrared inspection measures cumulative energy rather than peak levels. This makes it a good complement to detection of discharge activity with acoustic, ultrasonic, ultraviolet, or leakage current monitoring technologies. However,

Figure 6-12: Infrared (left) and normal photographs of temperature rise during leakage current activity on 115-kV transformer bushing during inclement weather (from Driscoll and Ballotte [2007]; courtesy of Northeast Utilities Service Company).

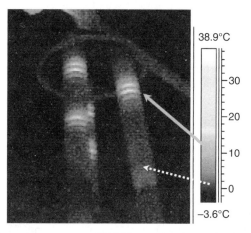

Figure 6-13: Stable warm areas on 400-kV post insulators in Czech Republic near freezing point (from Chrzan and Morro [2007]; courtesy of J. Dusbaba).

the temperature rise will be affected by relative humidity in the same way that was demonstrated for leakage current. This means that, while temperature rise indicates the presence of contamination, a lack of temperature rise does not mean that there is no contamination.

6.2. CLEANING OF INSULATORS

When one of the insulator monitoring methods is applied consistently and indicates a problem, there are a number of possible actions that could be taken. These actions need to be cost effective and they need to reduce the number of contamination outages without adversely affecting safety or long-term life of components.

6.2.1. Doing "Nothing"

Taken at face value, waiting for natural rain to wash the pollution from insulators may not seem to be a desirable option, while it is obviously inexpensive. However, there are a number of considerations that could be factored into this evaluation. The following examples illustrate practical implementations of this approach. While there are no direct actions to reduce the risk of contamination flashover, there are indirect actions that do or could mitigate the consequences.

Planning for "Safe Posture"   Power systems can be configured to be relatively insensitive to the loss of individual lines or stations. Using the NERC reliability classification in Table 1-2 of Chapter 1, these would be Class D extreme events resulting from two or more elements removed or cascading out of service. Configuration to ensure a stable bulk electrical system for Class D events can be done by increasing spinning reserve, redispatching generation resources to provide alternate paths around critical stations, or other modifications to the load flow. These "safe postures" often carry economic costs that can be reduced with a day or more of advanced planning to allow time to run up thermal resources to operating temperatures or to establish agreements for support from interconnected utility partners.

Climate studies suggest that fog will occur within ±2 h of sunrise. Thus if monitoring shows that the insulators are heavily contaminated, planning for a safe posture can be initiated for problems within a ±2-h window of sunrise for the next few days or weeks, until natural rain occurs.

Planning for Outages   Real-time lightning data are used at many utilities as an input in line maintenance and outage planning. If a lightning storm is approaching, it is prudent to delay the initiation of work and also to cancel ongoing work that has temporarily removed an element from service.

Utilities with contamination monitoring programs sometimes miss important implications when planning outages. As an example, winter salt levels built up quickly on one station providing an overhead link to a high-voltage cable feeding the downtown area of a major Canadian city [Hydro One Networks, 2004]. The other major cable feed was removed from service on this day for scheduled maintenance. When winter flashovers on 115-kV bushings removed the single feed from service, a 90-minute blackout occurred in the downtown area. Disseminating the internal knowledge that the power system was at risk of contamination flashovers from adverse local winter weather more widely to the outage planning groups might have averted this problem.

6.2.2. Insulator Washing: Selecting an Interval

One of the traditional measures used to mitigate the buildup of contamination on insulator surfaces is washing with clean water. This was traditionally more practical for station insulators than for lines, since major transmission stations often had permanent maintenance staff who could monitor and carry out the process. Nowadays, with both lines and stations having high maintenance costs related to remote access, insulator washing programs tend to be more expensive.

Washing programs adjust the interval to suit the local rate of pollution buildup, and the tolerance of the selected insulators to this pollution level. The rate of pollution buildup is measured in terms of an ESDD escalation rate ($\mu g/cm^2/day$), and a tally of the number of days without rain. Adjustments to washing intervals with this approach may lead to a requirement for washing in the middle of the peak contamination season, which in many areas is the middle of winter.

Other factors in IEEE 957 that are used to establish a washing interval that cleans the insulators before they reach a critical level include the following:

- *Results of ESDD Monitoring Programs:* Measurements of ESDD from deenergized test insulators, sensors, or energized insulators themselves can be compared with the allowable equivalent salt deposit density (ESDD) from the design process. Annual or quarterly measurements of ESDD have not shown any value in this process, but excellent results have been obtained in programs using weekly or biweekly measurement intervals when coordinated with the number of days without rain prior to each measurement.
- *Results of Other Monitoring Programs:* The onset and degree of partial discharge activity can be evaluated with leakage current monitoring— visual, thermal, or ultraviolet inspection during damp weather conditions from the ground or from the air by helicopter within 2 hours of sunrise.
- *Previous Experience:* Past experience on periods between flashovers, or pole fires, is often related to weather conditions. The dangers of flashover

and pole fires are correlated and both tend to increase after a long, dry period, either in winter or summer, followed by a light drizzle or fog condition.

- *Customer Complaints:* Insulator partial discharge activity will often occur within 2 hours of sunrise. The acoustic noise, electromagnetic interference, and in some cases light from the activity often generate customer complaints in summer conditions. These data should be logged and consolidated to guide the selection of a suitable washing interval.

6.2.3. Insulator Washing: Methods and Conditions

Natural washing of insulators does not give perfect results, especially on the bottom surfaces of disk insulators, and it cannot be called into play when needed. For this reason, utilities have opted to develop maintenance methods that use low- or high-pressure water streams (Table 6-2) to wash the pollution off insulators prior to wetting in fog conditions.

If the line or station insulators can be deenergized, low-pressure washing with manual scrubbing of ceramic insulators can be effective in areas of low labor cost, or if it is carried out at the same time that the insulators are being inspected for other defects such as cracks or surface damage. Typical tap water pressure is at least 600 kPa (90 psi) and conductivity is usually less than 700 μS/cm. According to the IEEE Standard 957 [2005] for cleaning insulators, they may be hand washed with rags or wiping cloths in mild detergent water. This should be followed by a low-pressure flood rinse with clean water to remove any residue. Solvents or harsh abrasives are normally not recommended. Wetting agents or additives can be used to improve the washing action of the cleaning water. Solvents may be used, provided all cleaning residue is removed by the final clean water rinse and only after manufacturer approval.

Most utilities that wash ceramic insulators make use of high-pressure water streams, using equipment operated from the ground. The high-pressure range

TABLE 6-2: Pressure Ranges and Conversion Table for Insulator Washing

Pressure	Units			
	atmospheres	kPa	kg/cm^2	psi
Low	1	101	1.0	15
	10	1013	10.3	147
Medium	20	2027	20.7	294
High	30	3040	31.0	441
	40	4053	41.3	588
	50	5066	51.7	735
	60	6080	62.0	882

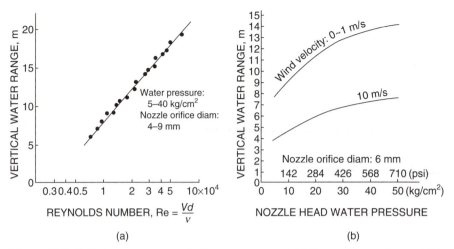

Figure 6-14: Relation between vertical water range and (a) Reynolds number and (b) nozzle water pressure (from Yasuda and Fujimura [1976]).

of 3000–7000 kPa (30-70 atm or 450–1000 psi) is used for insulator washing. The technical reason for this is explained in Figure 6-14 using the Reynolds number, which is a dimensionless parameter given by the water velocity V (m/s) times the nozzle orifice diameter d (m), divided by the dynamic viscosity coefficient of air v (m²/s).

For water pressures of 5–40 kg/cm² and practical water jet nozzle diameters of 4–9 mm, high pressure of more than 50 kg/cm² (700 psi) is needed to obtain a vertical water reach of 14 m in still air. With a wind speed of 10 m/s (36 km/h), the vertical range is cut in half for the same pressure.

The discharge quantity of the nozzle, Q (L/min), is related to the nozzle orifice diameter d (mm) and the water pressure p (kg/cm²) by a "flow coefficient" C that is fixed for each nozzle shape. Up to a water pressure of 80 kg/cm², Yasuda and Fujimura [1976] give the following relation:

$$Q = 0.66Cd^2\sqrt{p} \tag{6-2}$$

The duration of washing for each post establishes how much pollution is removed. Figure 6-15 shows that 5 seconds remove more than half of the salt at an ocean exposure in strong wind, with a 30-second period removing more than 85% of the SDD compared to the initial value. In a discussion to Yasuda and Fujimura [1976], Ely and Lambeth noted in Figure 6-15 that a 2-second wash did much more harm than good on a 400-kV vertical multicone post insulator.

Automatic pressure washing with two sets of spray heads at a 90° angle have been used at 500-kV stations near the Boso and Shinkoga substations in Japan (Figure 6-16).

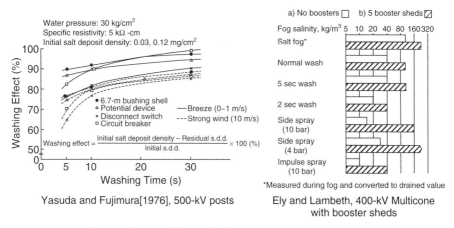

Figure 6-15: Washing effectiveness versus washing time.

Figure 6-16: Automatic live-line washing of 500-kV insulators in Japan (from Yasuda and Fujimura [1976]).

6.2.4. Case Study: Southern California Edison, 1965–1976

In 1976, Johnson reported that Southern California Edison (SCE) had been using live-line methods to wash insulators on power lines, substations, and generating stations with good success for 11 years. The service territory covers a wide geographic range, including contamination exposure to the Pacific Ocean on the west and long periods without rain toward the Nevada and Arizona desert borders to the east.

One successful component in the SCE live-line washing program was noted to be the provision of deionized water, through a combination of on-site filters

TABLE 6-3: Historic Values for Southern California Edison Safe Washing Distances Using Deionized Water of Conductivity Less than 50 µS/cm

Nominal Phase-to-Phase Voltage, kV	Safe Distance with Tap Water, m (feet)	Safe Distance with Deionized Water, m (feet)
4	2.1 (7)	1.2 (4)
12	2.7 (9)	1.5 (5)
16	3.0 (10)	1.5 (5)
33	3.4 (11)	1.8 (6)
66	3.7 (12)	2.1 (7)

Source: Johnson [1976].

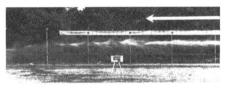 

Deionized water reach at 500 psi: 8.2 m (27 ft)   "Slick" water reach at 500 psi: 13.1 m (43 ft)

Figure 6-17: Comparison of deionized and "slick" water columns for pressure washing at 500 psi (from Johnson [1976]).

and wash trailers with capacities up to 5700 liters (1500 gallons). The desired resistivity was expressed as "greater than 20 kΩ, measured on a cube of water," meaning 20 kΩ-cm resistivity or 50 µS/cm conductivity. The deionized water gave a significant improvement in the safe distance for washing substations, as shown in Table 6-3.

Another innovation was the use of "slick water," composed of 340 g of a polyethylene glycol (Polyox FRA) and 3.8 liters of alcohol per 6800 liters (1800 gallons) of wash water, to provide greater reach and better cleaning action (Figure 6-17).

The present regulations on release of organic materials may no longer allow the use of "slick" water but it is important to note that there are additives that can improve performance. The commitment of SCE to high-pressure insulator washing was illustrated by the provision of on-side deionizing equipment, dedicated washing trucks, and trailers shown in Figure 6-18.

This case study illustrates that insulator pressure washing has been evolving for many years into a proven and safe method for improving contamination performance, as described in the most recent IEEE Standard 957 [2005] with some details in the next section.

6.2.5. Insulator Washing Using Industry Standard Practices

High-pressure washing is normally carried out with a hand-held nozzle. The line worker raises the hose and nozzle to the wash position by climbing the

On-site supply of deionized water

Wash trailer

Dedicated wash truck

High-pressure wash gun

Figure 6-18: Equipment used by Southern California Edison for insulator washing (from Johnson [1976]).

tower or by using an aerial lift. Substation insulators may also be washed using a hand-held nozzle from the ground. The water stream between the worker and the energized conductor or bus must be qualified as an electrical barrier for live-line work, following industry guidelines such as IEEE Standard 957 [2005] and IEEE/ANSI C2 [2007]. The IEEE guidelines call for control of the electrical conductivity of the water in the tank.

In order to control leakage currents along the water stream and improve operator comfort, some utilities specify wire braid conductive hose or drainer wires and bond these to the tower or station ground. Like any bonded connection, continuity is checked prior to the start of the job and connections are checked for corrosion and cleaned, as required.

Washing trucks or trailers may acquire a relatively high potential if non-conductive wash hoses are used. The nuisance shocks should be identified in work planning. A good practice is to deny access to, and restrict access near, the equipment during the actual washing operation. Alternately, the washing equipment can be grounded right next to the pumping system. Work planning should also consider whether the windows and doors are closed on any nearby buildings or parked cars.

Insulator washing starts when the line worker directs the pump operator to increase the water pressure. If the unit is equipped with a demand throttle, the pressure (revolutions per minute) will be increased automatically when the gun is opened. The water is directed away from the insulator string until full pressure has been achieved. The line worker on the tower then directs the wash stream at the insulator, respecting the minimum distance set out in IEEE Standard 957 [2005], which is an industry standard that is widely accepted. Greater nozzle-to-conductor distance for a given nozzle diameter results in lower leakage currents, but is less effective at removing pollution, calling for a longer wash time.

In winter conditions, demineralized water of less than $20\,\mu$S/cm conductivity ($>50{,}000\,\Omega$-cm resistivity) should always be used. This high water resistivity also can be qualified as an electrical safety barrier in cases where the wash distance is limited by the tower dimensions. Examples of the qualification process for live-line washing can be found in Johnson [1976], Perin et al. [1995], and Figure 6-19. In one case, use of water with $50{,}000$-Ω-cm resistivity increased the ac critical flashover strength of the 4-m water stream to $900\,$kV.

In the absence of an economical source of fresh or deionized water, insulators have been washed off-potential with salt water but this should never be done in freezing conditions.

Suspension insulator strings are washed by first directing the stream of water at the insulator nearest the energized conductor in such a manner as to take advantage of both the impact and the swirling action of the water to remove deposits. After the bottom insulators in the string are washed, the wash stream is moved up a few units. These units are washed and the stream then is directed on the clean units below to re-rinse them. This process is repeated, moving up a few insulators at a time until the entire string is clean. Failure to re-rinse lower insulators before moving further up the string can lead to flashover.

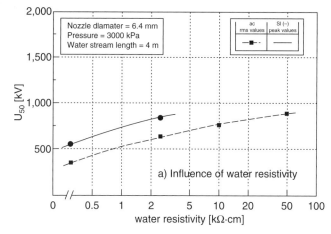

Figure 6-19: Increase in 50% flashover voltage of 6-4-mm diameter, 4-m water stream at 3000 kPa with water resistivity (from Perin et al. [1995]).

The stream must be moved away from any energized part of the insulators before the water pressure is reduced, or if the pressure falls on its own. Care should be taken to prevent the spray from wetting nearby dirty insulators, particularly in stations. Dead-end insulators must be washed carefully to keep overspray from causing flashover. Washing begins on the downwind end of the insulator string and then proceeds upwind.

While it is learned mainly from experience and training, partial discharge activity often starts on the insulators during washing. The operator does not have much time to make these observations and may not be able to see arcing on the back end of an insulator, so a spotter is sometimes used when washing highly contaminated insulators. As a practical guide:

- A harmless blue corona arc usually appears during washing. This typically indicates that the wet insulator is sufficiently clean. However, if the blue corona persists for 30–60 seconds after the water stream is removed, the insulator should be rewashed from a different direction as some surfaces have not been completely cleaned.
- An orange arc during washing indicates that the insulator is not clean, and power washing should continue.
- A yellow flame arc that moves off the insulator indicates that the arc current has increased to an unacceptable level. The water stream should be used to extinguish any yellow flame arc before it develops further into a complete flashover arc.

Part of the safety planning process for live washing in IEEE Standard 957 [2005] may include the use of hold-offs—blocking of automatic circuit breaker reclosing—to limit the magnitude of switching surges should an insulator flash over to ground during the process. Safe limit distances for washing normally provide protection against arc burns, but flash goggles and arc-resistant clothing of the appropriate rating should always be used in accordance with local regulations and practices.

Even if the insulators are being washed off-potential, there can be a considerable time saving in using live-line washing procedures. This is related to the extra time needed to apply and prove grounds on the bus work or phase conductors before starting the washing. Applying and removing the safety grounds may take several minutes, sometimes doubling the time needed to wash each insulator.

6.2.6. Insulator Washing: Semiconducting Glaze

Normally, ceramic insulators with resistive or semiconducting glaze should not be washed at all, as their superior contamination performance makes this costly maintenance unnecessary. Incorrect washing practice, including drift overspray from nearby washing work, may overwhelm resistive glaze properties, making them susceptible to wet flashover [Sharp et al., 2001]. They found that the leakage currents with these insulators were relatively constant, compared to those in Table 6-4 measured on posts or cap-and-pin insulators with standard glaze. This problem can be mitigated by washing the semiconducting glaze insulators off-potential.

As an additional disincentive to washing insulators, it was found in a 14-year application study that the only time semiconducting glaze station post insulators flashed over in winter conditions was shortly after they had been pressure-washed to improve appearance.

6.2.7. Insulator Washing: Polymer Types and RTV Coatings

The surfaces of polymer insulators and of RTV silicone coatings on ceramic insulators are more resistant to contamination flashover, but they are also much softer. Some contaminants also accumulate faster on the surfaces and natural washing is less effective at cleaning upward-facing surfaces. This means that it may be necessary to carry out periodic washing or cleaning to restore hydrophobicity.

Some manufacturers of polymer insulators only recommend low-pressure washing or flooding. This is mainly a function of manufacturing method and should be respected especially for insulators with unbonded construction. Polymer insulators with molded designs are generally more resistant to power-washing at pressures up to 4000 kPa (600 psi). Consequences of ignoring the manufacturer's recommendations for both pressure and volume include damage to the end fittings, shed puncture, tearing, and problems with wet flashover strength [Burnham et al., 1995].

RTV silicone that is well bonded to the substrate insulator, whether porcelain, glass, or polymer, can be washed at high pressure without damage. If the RTV is not properly bonded, high-pressure cleaning will blow the rubber off the surface. This is a sign that the original application method was not satisfactory.

The high-pressure washing process at 4000 kPa will start to thin a standard RTV coating on ceramic materials if the wash wand is held in place for 15 seconds or more. This means that it is not a practical way to strip the coating if it has expired. Holding the water stream in a single point also increased the surface damage of polymer insulators in tests carried out by Burnham et al. [1995].

High-pressure water washing can remove cement-like accumulations on RTV rubber coated insulators. Normally, these deposits, listed in Table 6-5, can only be removed from ceramic insulators with dry abrasive methods.

Pressure washing will normally reduce the hydrophobicity of silicone insulators or coatings. This will recover in a few days at 20°C, or in a few weeks at 0°C. If necessary, the hydrophobicity can be restored immediately by wiping or spraying the insulator with a light molecular weight silicone fluid.

TABLE 6-4: Effect of Washing Overspray on Insulator Leakage Current

Type	Profile and Dimensions (mm)	Leakage Current Envelope (mA)	Leakage Current (mA)	
			Fog	Clean
Semiconducting glaze post	Height 1143 Leakage 3300	Max. 24 mA	24	11
Normal glazed station post		Max. 250 mA	250	16
Normal glazed cap-and-pin apparatus	Height 1473 Leakage 3350	Max. 390 mA	390	35

Source: Sharp et al. [2001].

6.2.8. Insulator Washing: Procedures in Freezing Weather

An early mention of the need for washing insulators in winter conditions is found in Forrest [1936]: "Consequently, insulator cleaning should be concentrated during November, December and January, although these are the most inconvenient months from an operation point of view."

Insulator washing improves the performance of insulators in freezing conditions for both cold-fog and icing conditions. For this reason, details will be discussed in Chapter 9. Demineralized or deionized water with a conductivity of less than 20 μS/cm is recommended for washing in winter conditions. If water of higher conductivity is used, any water that freezes on the insulator surface will lower the insulator flashover voltage when it melts. Also, as discussed previously, the use of water with low conductivity can qualify as an electrical safety barrier when there is sufficiently high flashover strength of the water column compared to other flashover path lengths.

6.2.9. Insulator Cleaning: Dry Media

Some insulator pollution deposits are not soluble in water. The most common example is cement, but IEEE Standard 957 [2005] lists a number of other problems in Table 6-5.

For insulator surface deposits that cannot be removed with high-pressure washing, the alternatives of hand or dry media methods can be used. These have a much higher cost and may only be carried out a few times before the insulator surfaces are damaged.

Hand cleaning of insulators is tedious and labor intensive, but can also be thorough and effective. Hand cleaning of deenergized insulators can be the only choice to reach some station insulators that are difficult to access with heavy vehicles.

Depending on the type of contamination, a cloth rag, nylon, Scotch-Brite™ or scrubbing pad, steel wool, or wire brush may be used. Any foreign material

TABLE 6-5: Types of Insulator Contaminant and Suitable Cleaning Methods

High-Pressure Washing	Hand or Abrasive Methods
Sodium chloride (ocean or road salt)	Cement or lime
Urea and liquid manure	Fertilizer dust
Metallic dusts	Volcanic ash
Sulfates, nitrates, and other pollution (smog)	Baked-on defecation
Calcium, magnesium, potassium salts	Carbon
Cooling tower, irrigation water with high dissolved solids	Mold or algae
Earth dust	Paint

Source: IEEE Standard 957 [2005].

should be removed from the insulators. Solvents may also be used to loosen heavy deposits, but these should always be removed with a low-pressure water rinse afterwards.

Some automotive polishes and rubbing compounds are specified to leave no residual deposit prior to painting. Others leave a light coating of silicone oil that results in "fish-eyes" in paint and also interferes with adhesion of RTV silicone coatings. Both materials can be very effective for hand polishing of insulator surfaces, but in an area where RTV application may be feasible, only those products that are compatible with automotive paint refinishing should be used. For heavily contaminated insulators, it may be necessary to clean the insulator with a coarse grade of compound and to polish it with a finer grade, or with a glass polish.

Cleaning with dry media is an alternative to remove heavy deposits of pollution with less physical labor. Processes that use blast media that are harder than the porcelain glaze or glass surface—metal shot, sand, or glass beads—are not recommended. These materials are hard enough to damage the porcelain insulator glaze and shatter glass disks. However, softer abrasive materials such as walnut shells and ground corn cobs have been used with greater success on ceramic insulators. However, corn cob media can literally cut nonceramic insulators in half with 3 seconds of incorrect procedure.

Dry media can be used to clean energized insulators if suitable equipment has been fitted to keep the air stream free of both moisture and oil. The moisture content of the dry media must also be controlled to a level below 20%. As the fresh bags of typical corn-cob grit have a humidity of 8%, the bags should remain sealed until they are ready to be used.

The organic blast media are sometimes mixed with a small fraction of limestone to improve performance. Limestone should never be used alone as it is too conductive for energized cleaning.

The insulated wand for dry media cleaning, shown in Figure 6-20, is typically longer than minimum requirements for hot sticks used for live-line work. The tip of the wand stands away from the insulator surface by about 100 mm. If the flow of dry media stalls and backs up in the wand, the air/media flow should be stopped and the wand should be removed from the insulator surface.

Organic blast media can be left on transmission rights-of-way to compost, but they tend to contaminate the gravel in substations and promote the growth of grass and weeds and may be combustible. This means that ground cover must be used to capture the spent abrasive when corn-blasting or shell-blasting methods are used in substations.

Insulators can also be cleaned with ice pellets, either from frozen water or from dry ice. The use of dry ice is well established for cleaning printed circuit boards with 100-cm^2 surface areas, but scaling up the equipment to clean insulators with 1-m^2 surfaces has proved to be expensive. The advantages of no residue from the dry-ice process and limited damage to insulator surfaces mean that this method is a potential candidate for live-line work and in cases where other methods cannot be used.

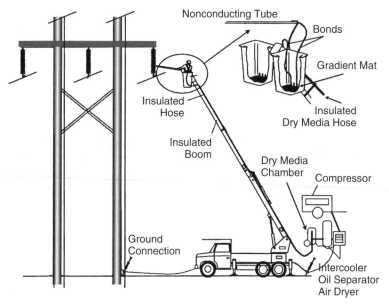

Figure 6-20: Typical arrangement for dry media cleaning of transmission line insulators.

Loss of hydrophobicity on silicone insulator after corn blasting

Immediate recovery after wiping with 1-centistoke polydimethylsiloxane(PDMS) fluid

Figure 6-21: Effect of abrasive cleaning on silicone insulator.

Nonceramic insulators can also be cleaned with abrasive methods. The cost of cleaning versus replacing the insulators can make this impractical. One problem, related to the loss of hydrophobicity after abrasive cleaning, can be addressed with the use of light molecular weight silicone fluid as shown in Figure 6-21.

The viscosity of PDMS fluids is related to the length of the DMS chains. A fluid with 1 centistoke (cSt) viscosity has a chain length of about six DMS units, making this a "light molecular weight" material. The immediate improvement in hydrophobicity in Figure 6-21 suggests that this treatment could be a useful maintenance option. The silicone fluid is environmentally friendly to the point of being edible: it is also used to control foaming in the deep fryers of fast-food restaurants.

6.3. COATING OF INSULATORS

Sometimes, there are locations where the frequency of washing or cleaning is too frequent to be a practical mitigation option. These are areas where it may be possible to improve the insulator performance by using a nonceramic coating of some sort on existing insulators. There has been a long-term evolution in these coatings, from the original oil bath types to greases to the modern RTV silicone coatings used today.

6.3.1. Oil-Filled Insulators

An insulator with insulating oil in the leakage path will suppress leakage current and reduce the incidence of pollution or icing flashovers, but it will have many other practical problems.

Johnson and Phillips [1876] patented a ceramic bulb telegraph insulator in the United Kingdom with a pin immersed in oil and sheltered from insects. The Lauffen–Frankfurt 10-kV transmission line, constructed in 1891, used an oil-filled insulator that was common on the communication circuits in Germany at that time.

With problems related to oil ignition from lightning surge flashovers and power system arcs, and a periodic need to change the oil blown away by wind action, this technology fell into disuse in most areas. However, Forrest [1936] established that insulators with a 25-mm annulus of oil around the cap of 146-mm × 254-mm disks had excellent leakage-current performance in comparison tests with porcelain disk and long-rod insulators. In the 1936 discussions, oil-filled insulator technology from the 1890s was noted to be "way before its time" for power system application, and the combination of oil-filled insulators with line surge arresters was proposed.

Most oil-bath insulators are now found in museums, as shown, for example, in Figure 6-22. However, these types of insulators are still found occasionally on power systems today in desert areas with low lightning ground flash density, such as Morocco [Zimmerman, 2000].

6.3.2. Greases

Many hydrocarbon and silicone greases have been evaluated for their ability to improve ceramic insulator performance in pollution conditions. These materials are hydrophobic in their pure state and also encapsulate pollution. After a long exposure, however, the greases tend to harden up, at first losing their

Figure 6-22: Oil-bath insulator manufactured in Japan in 1963. (Courtesy of TEPCO Electric Power Historical Museum.)

benefits and eventually degrading the insulator performance from its original state. This means that greases are a form of periodic maintenance, rather than a permanent solution.

An IEEE Task Force [Gorur et al., 1995] evaluated the advantages of hydrocarbon and silicone greases and summarized their results in Table 6-6.

Greases are applied by hand, brush, or spray on clean insulators. Hand cleaning is usually needed to prepare a greased insulator for regreasing.

The silicone greases can have anywhere from 1 to 10 years of life, depending on the local contamination environment and climate. Based on general field experience, greases have to be applied every few months in severely contaminated areas, or every 2 years in areas where contamination is not so severe [Gorur et al., 1995]. After the loss of hydrophobicity, leakage currents and dry-band discharges will decompose the grease quickly, releasing the filler as an additional source of contamination. At this point, the surface must be cleaned and regreased immediately, whether convenient or not, to prevent insulator damage.

A major problem with the use of grease is the problem of disposal when it reaches its end of life. At one time, grease was simply put into land fill, but it is now regulated as a fluid and must be incinerated. Since the grease has a silicone base, the incineration temperature is very high and thus disposal is expensive.

Greasing of polymer or nonceramic insulators is not advisable [IEEE Standard 957, 2005] as channeled arcing in the grease may lead to tracking damage of the polymer surfaces.

TABLE 6-6: Comparison of Petroleum Jelly and Silicone Grease

Comparison Parameter	Petroleum Jelly	Silicone Grease
Basic constituents	Hydrocarbon oils, petroleum, and synthetic waxes	Dimethylsiloxane or phenylmethylsiloxane fluid, coupling agents, fillers, and solvents
Useful temperature	0–60 °C	–50 °C to 200 °C
Melting point	60–90 °C	>200 °C
Recommended spraying temperature	90–115 °C	–30 °C to 30 °C
Encapsulation rate at 20 °C	Slow	Rapid
Ease of application	Difficult, especially in cold weather	Easy
Ease of removal	Labor intensive	Labor intensive
Arc resistance, ASTM D495	n/a	80–150 s, depending on formulation, fluid, and filler
Material cost	Low	Moderate

Source: Gorur et al. [1995].

6.3.3. Silicone Coatings

The oil-filled insulator from 1891 had an excellent pollution encapsulation process. Water and dirt accumulation sank to the bottom of the porcelain trough. Fresh oil floated to the top. If sufficient pollution accumulated or the protection against water accumulation was inadequate, eventually the oil reservoir would flood, giving an indication that it was time to clean and replace the oil.

Oil-filled insulators for electric power systems have not proved to be successful, because of the high maintenance, environmental consequences, and problems related to lightning and power system arcs. However, the operating principle serves as an important illustration to the function of the modern nonceramic insulator with a silicone surface. Rather than an external supply of insulating oil, the silicone material has short cyclic chains of dimethylsiloxane groups. These cyclic groups bleed out of the rubber and are normally considered to be a disadvantage in building sealants because they interfere with paint adhesion and collect dirt. Also, some creation of light molecular weight silicone fluid with linear chain lengths of six to ten PDMS groups will occur as a result of dry-band arcing on insulator surfaces. With the low surface energy and hydrophobic nature of the silicone-oiled surface, most water beads up and runs off the insulator. The PDMS fluid readily flows over and encapsulates pollution as it accumulates.

According to an IEEE Outdoor Service Environment Committee 32-3 Task Force [Gorur et al., 1995], commercially available room-temperature vulcanized (RTV) silicone coatings fulfill the need for a material with high resistance to ultraviolet and dry-band arcing. The RTV coatings are single-component liquids that are applied, mainly to ceramic insulators or bushings, by spraying, dipping, or painting. The humidity in the air cures the material into a thin, flexible rubber.

RTV insulator coating systems consist of polydimethylsiloxane (PDMS) polymer, alumina trihydrate or alternate filler for increased tracking and erosion resistance, catalysts, and a crosslinking agent. Manufacturers may also formulate their products with adhesion promoters, reinforcing fillers, and pigments.

For good adhesion of the RTV coating, the surfaces of the insulators must be thoroughly cleaned and dried. In most instances, high-pressure water washing will be sufficient. Insulators contaminated with cement materials listed in Table 6-5 may need to be cleaned with abrasive methods.

Greased insulators are very difficult to prepare for silicone coating and replacement is usually a better option. They must be thoroughly hand-cleaned, and wiped repeatedly with a solvent to ensure complete removal of the grease. The well-cleaned surface will exhibit poor hydrophobicity before it is dried.

Best results are normally obtained when insulators are coated using high-volume, low-air-pressure (HVLP) spraying systems that put a high pressure on the liquid product in a special pressure pot. This technology uses a high volume of air at a low (10 psi) pressure to reduce overspray and improve the surface finish. Two hoses supply air and liquid to a spray head, as shown in Figure 6-23.

High-volume low-pressure spray system

Typical surface finish of RTV coating

Application of RTV coating to bushing

Application of RTV coating to post insulator

Figure 6-23: High-volume low-pressure method for applying RTV silicone coating.

Silicone coating adhesion should be tested by boiling sample insulators for 100 h in salt water with a conductivity of 2000 μS/cm. This boiling test is used to evaluate the end-seal integrity of polymer post insulators. If the coating forms blisters or loses adhesion at the end of the test, as shown in Figure 6-24, the preparation method should be changed.

While RTV coatings are normally applied to ceramic insulators, they can also be used to improve the performance of nonceramic insulators. Electrically sound polymer materials such as EDPM or Protectolite™ that are not naturally hydrophobic can be treated in the same way as porcelain or glass (Figure 6-25).

Generally, the endurance of the RTV coating in a new application on a polymer material is tested using a tracking wheel test or an IEC 61109 salt-fog test, using the same evaluation standards as those applied to polymer insulators.

Pressure washing with deionized water: Wipe with 1,1,1-trichloroethane on paper cloth:
          successful result   failed result

Figure 6-24: Pass and fail results from 100-h boiling test on RTV silicone coating, comparing pressure washing (left) and 1,1,1-trichloroethane wipe (right).

Polymer "button" insulator   Polymer "button" insulator   Tracking wheel endurance
for 600-Vdc subway system   with silicone coating   test

Figure 6-25: Application of RTV silicone to nonceramic insulator. (Courtesy of Kinectrics.)

According to the IEEE Task Force [Gorur et al., 1995], RTV coatings from this period were thought to have a useful life of 7–10 years. Some utilities such as Pacific Gas and Electric and Bonneville Power have more than 30 years of experience with RTV silicone coatings at locations with salt fog from the ocean, cement, industrial, and agricultural pollution. Other utilities have increased the water washing interval to 1 year or more for lines close to the sea and thermal power plants.

In one case, a utility coated all of the bushings at their most heavily contaminated 230-kV station in 1994 and has obtained improving hydrophobicity with time, leading to 14 years of perfect service. To date, there has not been any remarkable reduction in coating thickness from annual pressure washing at 4000 kPa (600 psi). The local pollution contains a fraction of carbon-black soot with small particle sizes that may be contributing to the improving hydrophobicity.

An RTV application at the seacoast in Connecticut, USA [Carberry and Schneider, 1989] was initiated because tests showed a remarkable improvement in breaker support insulator flashover levels, from 209 kV unprotected to 300 kV with the coating. However, Eldridge et al. [1999] noted that the useful life of the selected coating was limited to about 4 years from results of acid water depolymerization to a putty-like material.

Many of the benefits of RTV coatings are reduced in freezing conditions. These include the ability to encapsulate pollution, the ability to repel water, and the significant increase in surface impedance that is a consequence of these factors. Chapter 9 will discuss the positive and negative roles of RTV coatings as a possible mitigation option under icing conditions.

6.4. ADDING ACCESSORIES

There are a number of options, such as booster sheds, creepage extenders, bird guards, or corona rings, that can be fitted to insulators or bushings in order to improve their contamination performance. Most of these accessories extend the leakage distance path. Corona rings can improve performance by improving the distribution of electric field. While they do not specifically improve the contamination performance, arcing horns (or rod gaps) are accessories that should also be considered when modifying the contamination performance of any insulator or bushing to maintain coordination with cables, breakers, or transformers.

6.4.1. Booster Sheds

The "booster shed" [Ely et al., 1978a] was developed to improve the flashover performance under heavy rain conditions, including high-pressure washing under energized conditions. Figure 6-26 shows a C-shaped, flexible booster shed with peg-and-hole fasteners. The rubber shed slips around the insulator without disassembling the connections and is snapped together. In some designs, RTV silicone is used as an adhesive and sealant.

Some designs of booster sheds include spacers that push the shed off of the insulator surface. These spacers are desirable because they allow high-pressure wash water to reach between the booster shed and the insulator shed.

Most designs of booster sheds tend to interfere with natural rain washing of the insulators by diverting rain off the surfaces underneath. In one set of tests, shed-by-shed sampling of ESDD on station post insulators showed a factor-of-7 increase in the five sheds directly sheltered, compared to levels measured lower down on the insulator. In the typical application, with a booster shed every five skirts in Figure 6-26, the bushing ESDD will be considerably higher than the untreated insulator.

While booster sheds may increase the ESDD levels, they also tend to give a significant improvement in the flashover performance of post insulators under all live-line washing conditions, including rain, impulse wash, overspray, and 2- and 20-second pressure washing [Ely et al., 1978b]. There can be high electric stress at the junction of the booster shed and the insulator, sometimes as a direct consequence of nonuniform pollution deposit and sometimes, as illustrated in Chapter 9, from nonuniform ice deposit. This can lead local arcing activity that, over the long term, can damage the porcelain glaze surfaces or erode the booster shed itself over time. Glaze damage on sheds is a minor concern. Erosion of semiconducting glaze or glaze on the central core of the porcelain shell should be avoided, and in-service inspections to identify audible noise, ultrasonic noise and /or corona activity may be necessary for successful long-term service.

6.4.2. Creepage Extenders

With ongoing concerns about the possibility of glaze damage from corona activity between the booster sheds and the insulator surfaces, and also the

Figure 6-26: Typical booster sheds applied to transformer bushing (left) and post insulator (right). (Courtesy of Tyco.)

Creepage extenders on
station surge arresters

Creepage extenders with bird excrement
on station post insulator

Figure 6-27: Typical creepage extender applications. (Courtesy of Tyco.)

increased frequency of washing, an alternative of a creepage extender has been developed. Creepage extenders are an interesting alternative to booster sheds, especially for smaller-diameter insulators. They increase the shed diameter and leakage distance, and can be effective for improving performance in conditions of both heavy and light wetting. The band is attached to the edge of the insulator sheds, as shown in Figure 6-27. The creepage extender is bonded with a mastic material to the outside edge of the shed. The shed material shrinks to form a tight fit when heat is applied with a torch or heat gun. The correct inner diameter of creepage extender must be selected to fit over the largest sheds if the insulator has an alternating-diameter profile.

With the relatively narrow 20-mm margin between the outer diameter of the insulator and the inner diameter of the creepage extender, it is necessary to disassemble the bus work at the top of the insulator. While they have some flexibility, sometimes the creepage extenders may not fit gracefully over the metal end fittings of insulators or surge arresters and this should be considered in the selection process.

One style of slip-on creepage extender with an open gap of about 30 mm that is small compared to the insulator shed circumference. This style also uses mastic to hold the accessory in place. These are normally arranged so that open gaps face opposite directions in an alternating pattern (Figure 6-28). In laboratory and outdoor exposure tests, dry-band arcing activity in light wetting and fog conditions tends to be focused at the gaps in Figure 6-28, rather than distributed randomly around the circumference of the insulator. Flashover arcs on slip-on creepage extenders also found their way through the gaps rather than over the edges. This gapped design may be less effective per mm of added creepage distance than the gapless heat-shrink style in Figure 6-27, but this is typically compensated by providing a larger rib thickness that adds more leakage distance.

Figure 6-28: Slip-on creepage extender with open gap. (Courtesy of Tyco.)

The mastic materials for accessories of either style are sometimes used incorrectly in the field, for example, by disregarding the manufacturers' instructions to apply them to clean and dry insulator surfaces to ensure adequate adhesion. Also, test results showed that the flashover arcs on slip-on creepage extenders found their way through the gaps rather than over the edges, meaning that the increase in leakage distance is not as large as with the design in Figure 6-27.

In a comparison test of the heat-shrinkable and slip-on creepage extenders, Metwally et al. [2006] concluded that the use of two slip-on extenders on 33-kV porcelain line posts increased the effective dry arc distance and lightning impulse flashover by 15%, but did not materially improve the contamination performance. Posts fitted with two heat-shrinkable creepage extenders gave 23% higher ac flashover levels for very heavy pollution. Their conclusion, however, was that replacement of the porcelain posts with silicone polymer posts was more effective than either creepage extender option.

With the installation of any creepage extender calling for considerable labor and the significant cost of each extender, the application of these devices is relatively rare. They are most effectively applied on bushings and surge arresters, rather than post insulators.

6.4.3. Animal, Bird or "Guano" Guards

Birds that tend to roost above insulators produce streamers of excrement that can flash over EHV insulation under some conditions. The birds can be discouraged from roosting by the use of sharp metal spikes. An alternative is to divert the stream over the surface of a polymer bird shield, such as the one illustrated in Figure 6-29. This diverts the contamination away from the insulator, rather than improving the electrical performance, but the net result is mitigation of the problem.

Figure 6-29: Assembly of bird shield for suspension Insulator (from Sundararajan et al. [2004]; courtesy of Tyco).

Figure 6-30: Bird streamer protection system. (Courtesy of Tyco.)

These shields have an assembled diameter of 450–600 mm (18–24 in.) and are attached with plastic bolts. Once they are fully loaded with bird excrement, the guards can be replaced, rather than requiring cleaning of the top few insulators using abrasive methods.

Full protection against animal contacts may include large bird shields and insulating conductor covers as shown in Figure 6-30. This arrangement may be used when protecting lines against flashover when large birds roost in the line-to-ground gap near the insulator.

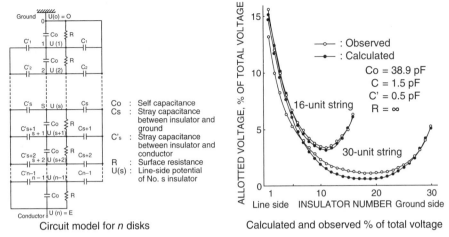

Circuit model for *n* disks   Calculated and observed % of total voltage

Figure 6-31: Calculated and observed potential distribution along strings of standard disk insulators (from Fukui et al. [1974]).

The animal guard accessories in this case have a relatively minor effect on the overall contamination performance of the insulators. At present, there are no standards that deal with the suitability of materials for wildlife protective device durability. This is the subject of a current IEEE Task Force and Standards Project (P1656).

6.4.4. Corona Rings

Corona rings are metal toroids that are typically used to improve the electric field distribution near the ends of insulators. They are commonly applied at the line end to ensure good long-term reliability of nonceramic insulators at voltages above 115 kV. The use of corona rings at the ground end of the insulator string also becomes necessary as length increases. The calculated and measured electric field per disk, without corona rings, is shown in Figure 6-31 for 16 and 30 insulator disks [Fukui et al., 1974].

The main purpose of the corona rings is to eliminate partial discharges under fair weather conditions by reducing the electric field in the areas of high electric field stress below the corona threshold. Corona is a nuisance source of noise, corrosion, and electromagnetic interference on ceramic insulators but has several bad effects on polymer insulators, related to:

- Physical action of the discharge, known as corona cutting of the rubber sheath over the central fiberglass rod, especially at mold lines and also at the seals on insulators that have the housing molded over end fittings.
- Production of nitric and other acids that can reach the core of the fiberglass insulator, leading to brittle fracture or accelerated corrosion of metal components.

Figure 6-32 illustrates the maximum electric field stress computed with and without a corona ring along a polymer insulator next to the sheath, and passing through the sheds.

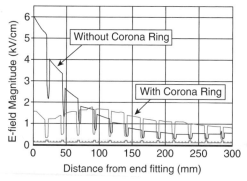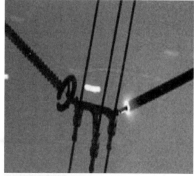

Figure 6-32: Electric field along a nonceramic insulator (left) and partial discharge activity (right) with and without corona ring (Phillips et al., [2008]).

The area of greatest stress in Figure 6-32 is moved from just next to the metal hardware, at a gradient of 600 kV/m, to a position 100 mm (four sheds) along, at a level of 180 kV/m that is more than three times lower, with the corona ring. The corresponding visual signatures of partial discharge also reflect this difference in local electric field.

Continuous corona serves as an ignition source for the contamination flash-over process, causing a reduction in electrical strength. Conversely, under contamination conditions, the suppression of the onset of arcing resulting from reduced electrical stress leads to improved electrical performance. This is one reason why corona rings that have been extensively damaged by power-follow current after flashover should be replaced.

Some line designs using either ceramic or nonceramic insulators, rough hardware, or small conductors may produce corona under fog conditions at normal service voltage. If these lines also have poor contamination perfor-mance, one effective option for improvement is to fit corona rings to eliminate the corona near the insulator surfaces.

Kawai [1970] evaluated the effect of electric field grading on the contamina-tion performance of EHV insulators, using a precursor of the clean-fog test at Project UHV. Kawai's tests used a slow wetting condition to simulate the transition of the atmosphere from dry daytime conditions to humid conditions in night or early morning, rather than duplicating natural fog. The surface impedance of contaminated insulators changed gradually as a function of the surface rates of wetting and drying. The initial nonlinear voltage distribution remained throughout the tests, and caused a nonlinear relation in flashover voltage with leakage distance. Kawai's results showed that the use of a large 1270-mm diameter corona ring improved the flashover performance of high-creepage fog disks in the slow wetting tests by a considerable margin, and that a shield screen at the ground end to increase nonlinearity had the opposite effect. The shield ring also reduced the nonlinearity that Kawai observed in flashover voltage with string length (Figure 6-33).

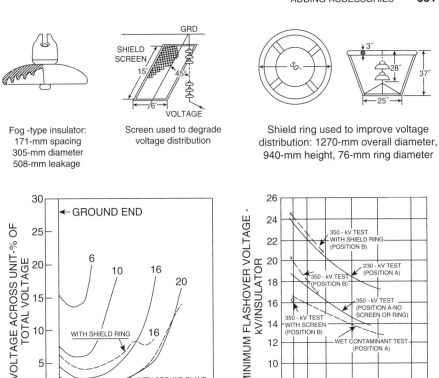

Fog -type insulator:
171-mm spacing
305-mm diameter
508-mm leakage

Screen used to degrade
voltage distribution

Shield ring used to improve voltage
distribution: 1270-mm overall diameter,
940-mm height, 76-mm ring diameter

Voltage distribution along insulator string
with and without shield ring

Minimum contamination flashover
test results with and without shield ring

Figure 6-33: Effect of shield ring on contamination performance of EHV insulator string (from Kawai [1970, 1973b]).

As additional evidence of the shield ring benefits, Kawai noted that scintillation predischarges normally present on the string almost disappeared, and that flashover took place suddenly.

6.4.5. Arcing Horns

Arcing horns are used in several countries to provide a path for the power-frequency flashover arc that is off the insulator surfaces. A recent CIGRE paper [de Tourreil and Schmuck 2009] provides a comprehensive summary of the application of devices for power arc protection of transmission line insulators. Most applications are presently made across ceramic insulators, but arcing horns are also recommended as a way to take power-follow current off the vulnerable end fitting interfaces of polymer insulators. In cases where there may be many contamination, icing, or lightning flashovers, it may be prudent to fit arcing horns in order to protect the mechanical integrity of the insulators as a mitigation option.

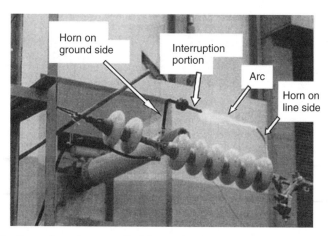

Figure 6-34: Operation of current-limiting arcing horn for 77-kV system (from Chino et al. [2005]).

Chino et al. [2005] reported the development of a current-limiting arcing horn for 77-kV systems (Figure 6-34). In addition to protecting the insulators from power-arc damage, an ablative material in the interruption portion provides gas that cools and quenches the power arc in a half-cycle of ac, preventing reignition. This means that the power quality disturbance from any contamination flashovers can possibly be mitigated without a circuit breaker operation.

The current-limiting arcing horn was designed mainly to function for lightning impulse flashovers, and its use for mitigating contamination flashovers after the fact should be explored in future research.

There are other important reasons to evaluate the use of arcing horns in contaminated areas. In areas of high contamination, it is common to retrofit distribution or subtransmission lines with insulators that have greatly increased leakage distance, or to add more disks to the strings of transmission lines. This increases the dry arc distance as well. If too many disks are added, there is a possibility of poor coordination with terminal equipment—for example, exceeding the rated BIL of circuit breakers. Since surge arresters carry their own risks of flashover, puncture, and thermal runaway in heavy contamination conditions, the use of rod gaps or arcing horns on station insulators may be a better choice in some cases to limit incoming surge voltages.

6.5. ADDING MORE INSULATORS

The historical, logical, and field-proven way to improve contamination performance is to use more disks in a chain of line insulators or taller posts in a station. This increases both the leakage distance and the dry arc distance.

Contamination flashover performance tends to scale linearly, so adding three new disks to a string of ten identical disks can improve the flashover strength by 30%. The benefits to using more disks for improved flashover performance translate into increased intervals between cleanings—two to three times longer for this specific upgrade.

There are limits to adding more insulators of the same type. On transmission lines, each additional standard suspension disk will lower the conductor height by 146 mm. At an average change of sag with temperature of 30 mm per C° this means that each disk derates the maximum operating temperature of the line by 5 C°. With some lines from the 1930s being designed for a 49 °C maximum temperature, there is not much margin before the line violates electrical clearance standards.

At many substations, fitting an additional apparatus-style insulator or an additional post insulator is much more difficult. The insulators and bus work must be reengineered for the new cantilever stresses under short-circuit conditions and checked for electrical clearances to adjacent components. Once these have been verified the bus work must be disassembled and reassembled at the new height.

For very slow wetting of precontaminated insulators, Kawai [1970] has shown that the linearity of pollution flashover does not hold perfectly for extra-high-voltage tests using single conductors at the line end. This is not a preferred arrangement and Kawai's photos of arcing at night show the conductors are in heavy corona. Kawai's results, replotted in Figure 6-35, are expressed as the ratio of EHV flashover to the level found for the same insulator and contamination level at a dry arc distance of 2 m.

Kawai [1973b] also noted that the linearity of the contamination performance in slow wetting tests improved to 80% efficiency for tests on long strings of A-11, G-2, and T-2 insulators of 5.8–7.3-m dry arc distance when the tests were carried out with a six-conductor bundle rather than a single conductor.

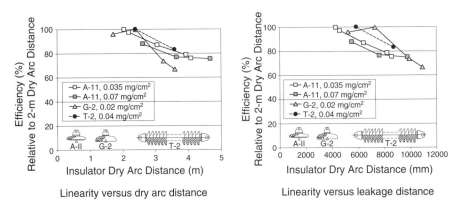

Linearity versus dry arc distance   Linearity versus leakage distance

Figure 6-35: Linearity of contamination flashover for UHV tests with single conductor and slow wetting rate (data from Kawai [1973a]).

6.6. CHANGING TO IMPROVED DESIGNS

The standard insulator profiles are meant to be applied in areas of very light to medium pollution. Wherever the local sources (industrial or seacoast) and the exposure duration (desert or winter) combine to give high or very high severity, it may be necessary to use disk insulation with improved designs, such as aerodynamic or anti-fog profiles, or posts with alternating shed shapes.

Most utilities have well-developed work methods for routine replacement of ceramic insulator strings that have been damaged by repeated flashovers or gunshot. It is relatively simple to use this process to replace standard disks with disks that have the same connection length, diameter, and cap-and-pin spacing.

Fog profiles have deeper ribs on their bottom surfaces, providing up to 50% more leakage distance. Extra leakage distance will reduce the electrical stress to accommodate the anticipated local pollution levels. A good anti-fog design will increase contamination flashover voltage by the same ratio. The shape of the insulator bell or post can be adjusted in order to increase the ratio of leakage distance to dry arc distance. Standard disk insulators have a ratio of 305 mm/146 mm or 2.1:1. Fog-type disk profiles may have a ratio of 3:1 and some profiles on station posts can be as high as 4:1.

As an alternative to increasing the leakage distance, designs that do not accumulate as much pollution may be substituted. These aerodynamic designs offer some additional merit in areas where washing of insulators is impractical.

6.6.1. Anti-Fog Disk Profile with Standard Spacing and Diameter

Insulators with a high ratio of leakage distance to dry arc distance were first developed for good performance in marine fog conditions. These tend to retain a smooth upper surface but have a number of deep ribs on the bottom. The ribs are relatively fragile and more easily damaged in transport, and the additional weight also adds to the insulator cost. These disks came to be known as "fog-type insulators" or "anti-fog disks." They are presently applied as an upgrade wherever increased leakage distance is required to prevent contamination-related outages, not just in coastal areas. Fog-type insulators are also used where increased string lengths of standard suspension insulators may reduce tower clearances, violate ground clearance codes, or simply cause installation problems.

The benefits of anti-fog insulators were demonstrated conclusively in a series of international tests set up to establish the reproducibility of the clean-fog test method [Lambeth et al., 1987]. Two disk profiles, shown in Table 6-7, were tested at a moderate contamination level of 0.07 mg/cm² (70 µg/cm²).

It is possible to increase the flashover voltage under contamination conditions by 31% (123 versus 93 kV/m) based on this result, with a modest

TABLE 6-7: Characteristics of Suspension Disks in Clean-Fog Tests by Joint CIGRE/IEEE Task Force

Characteristics	Standard Disk	Anti-Fog Disk
Shed diameter (mm)	254	254
Cap-to-pin spacing (mm)	146	146
Leakage distance (mm)	305	390
Top surface area (cm^2)	691	620
Bottom surface area (cm^2)	908	1380
V_{50}, mean critical flashover (kV/disk) and σ, standard deviation for ESDD = 0.07 mg/cm^2	13.63 $\sigma = 2.3\%$ (8 laboratories)	17.9 $\sigma = 5.2\%$ (6 laboratories)
E_{50} (kV per meter of leakage distance)	44.7	45.9
E_{50} (kV per meter of string length)	93	123

Source: Lambeth et al. [1987].

cost penalty for handling the fragile deep-rib design. This means that, in areas of moderate contamination, the tolerable level of ESDD will increase by a factor of $(1.31)^{(1/0.36)} = 2.1$, using the value of α from Equation 3-5 in Chapter 3.

6.6.2. Aerodynamic Disk Profile

Open or aerodynamic insulator profiles have no ribs on bottom surfaces. These insulators offer a different approach to pollution problems: they mitigate the rate of pollution accumulation, rather than increasing the leakage distance. When they are not directly exposed to salt spray, the aerodynamic insulators have often showed lower levels of ESDD that translate into better electrical performance for the same connection length.

Studies in areas with extended dry conditions, such as the desert [Akbar and Zedan, 1991] and the Ontario winter, as noted in Table 6.1, show that somewhat less pollution does tends to accumulate on the aerodynamic insulators. This, along with electrical performance, led Akbar and Zedan to give the aerodynamic profiles their highest rank in a comparative evaluation with long-rod, standard, and fog-profile designs. Four of these profiles are described in Table 6-8.

TABLE 6-8: Typical Dimensions of Standard and Aerodynamic Profile Disks

Characteristics	SP1 Disk	AG1 Aerodynamic	FP1 Anti-Fog	FG1 Anti-Fog
Shed diameter (mm)	280	390	320	320
Cap-to-pin spacing (mm)	170	160	160	160
Leakage distance (mm)	370	360	545	545
Ratio: leakage/ spacing	2.18	2.25	3.41	3.41
Top surface area (cm²)	800	1177	1355	1152
Bottom surface area (cm²)	1370	1224	2225	2349
Form factor	0.82	0.61	1.03	1.04

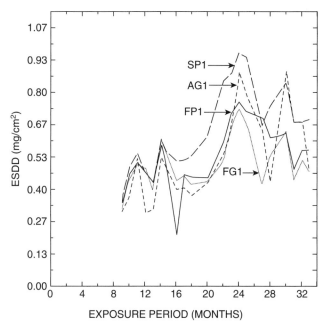

Figure 6-36: Observed bottom-surface contamination level (ESDD) versus months of exposure in desert environment (from Akbar and Zedan [1991]).

The observed values of ESDD on the bottom surfaces were all found to increase linearly with time after each rainstorm. Figure 6-36 shows that there is no strong trend toward higher or lower values for standard and aerodynamic profiles, with leakage/height ratios of 2.2:1, compared to fog-type disks with ratios of 3.4:1.

6.6.3. Alternating Diameter Profiles

A series of clean-fog tests were carried out [Schneider and Howes, 1988] on insulator strings using 146-mm × 267-mm disks and aerodynamic disks with dimensions given in Table 6-9. An alternating profile gives benefits in icing performance, which will be discussed in Chapter 9. The clean-fog tests reported here were carried out to establish the conventional contamination performance.

Figure 6-37 illustrates typical results in extra-high-voltage clean-fog testing of precontaminated insulators at 400 kV.

The relative standard deviation of the flashover and withstand values in the up-and-down test method is about 8% for light pollution on conventional porcelain disks. Note that this standard deviation matches their median value obtained from a large number of clean-fog tests. The relative standard deviations of clean-fog test results for bell-shape insulator strings and for strings of alternating standard and aerodynamic-disk profiles are lower. This leads to an interesting application characteristic in Table 6-10.

In this case, the low standard deviation of the flashover voltage for the alternating string leads to a higher withstand voltage gradient than the standard disk insulators, even though the 50% flashover gradient is higher.

The overall conclusion from this result is that, even for smooth surfaces of aerodynamic disks and for EHV transmission levels, the contamination performance scales linearly with leakage distance.

TABLE 6-9: Dimensions of Insulators Tested for 3.5-m Strings of Uniform and Alternating Profile

Characteristics	Reference Disk	Aerodynamic	Alternating Reference/ Aerodynamic
Shed diameter (mm)	267	430	267/430
Cap-to-pin spacing (mm)	146	146	146
Leakage distance (mm)	395	390	392.5
Top surface area (cm^2)	894	1598	
Bottom surface area (cm^2)	1463	1480	
E_{50} (kV per meter of string length) corrected to ESDD = 0.02 mg/cm^2	122	n/a	113

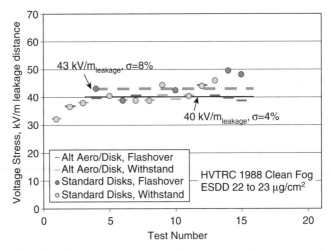

Figure 6-37: Results of clean-fog tests on 3.5-m strings of standard and alternating 430-mm aerodynamic/267-mm diameter standard disk insulators.

TABLE 6-10: Calculation of Withstand Voltage Gradient from 50% Flashover Gradient and Standard Deviation for Strings of 22–25 Standard or Alternating Aerodynamic/Standard Disk Insulators

Voltage Gradient ($kV/m_{leakage}$)	Alternating Aerodynamic and Standard Disk Insulators	Standard Disk Insulators
50% Flashover gradient, μ	39.8	42.9
Standard deviation, σ	1.8	3.6
Withstand gradient, $\mu - 2.2\sigma$	35.6	35.0

6.6.4. Bell Profile with Larger Diameter and Spacing

Insulators with deep skirts that enclose a conical volume have interior, downward-facing surfaces that are sheltered from the effects of precipitation and accumulation of pollution that settles on upward-facing surfaces. These insulators tend to perform better in contamination tests than standard disks, or aerodynamic disks with a smooth, horizontal downward-facing surface.

Bell-shaped insulators were tested against strings of standard disks by Schneider and Howes [1988] at low levels of contamination for UHV applications. The dimensions of these bells are shown in Table 6-11.

The authors noted that the bell-shaped insulators performed better in artificial clean-fog tests. An explanation was offered: the slurry used to precontaminate the insulators drips away, leaving the average ESDD at $18\mu g/cm^2$

TABLE 6-11: Disk Dimensions for 3.5-m Strings of Long-Leakage Disk and Bell Insulators

Characteristics	Reference Disk	Glass Bell	Porcelain Bell
Shed diameter (mm)	267	343	425
Cap-to-pin spacing (mm)	146	165	165
Leakage distance (mm)	395	430	445
Top surface area (cm^2)	894	1344	1828
Bottom surface area (cm^2)	1463	1235	1762
E_{50} (kV per meter of leakage distance) corrected to ESDD = 0.02 mg/cm^2	45	55	53
E_{50} (kV per meter of string length) corrected to ESDD = 0.02 mg/cm^2	122	143	144

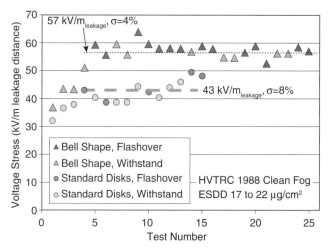

Figure 6-38: Results of clean-fog tests on 3.5-m strings of standard disk and bell insulators (data from Schneider and Howes [1988]).

compared to 22 µg/cm^2 for standard disks and 26 µg/cm^2 for the smooth surfaces of the 430-mm aerodynamic disks. This would typically change the test results by about 7%, using the empirical exponent $\alpha = -0.36$ in Equation 3-5 [Baker et al., 2008] to relate the ratio of flashover voltages. However, the observed margin of E_{50}, 143–144 kV per meter of string length for bell-shaped insulators in Figure 6-38 and Table 6-9, is corrected for this factor using an exponent of $\alpha = -0.30$. Hence the deep bell shape with its large sheltered leakage distance does offer a significant initial advantage in contamination performance.

Field experience, for example, Forrest [1936], suggests that the improved performance margin of bell-shaped insulators degrades over a period of years as pollution accumulates on the bottom surfaces. The interior surfaces are not washed naturally, and washing programs for transmission lines are much more expensive than those for stations. This has limited the application of deep-bell profiles to cap-and-pin apparatus insulators, which have performed very well in areas of high contamination.

An anti-contamination insulator with excellent performance in artificial contamination tests was developed by applying a semiconducting glaze to the deep bell shape [Nigol et al., 1974]. The combination of bell shape and semiconducting glaze to suppress leakage provided a double advantage. Unfortunately, the bell-shaped semiconducting glaze insulator proved to be difficult to manufacture compared to standard disks. As an alternative, the application of silicone coatings to deep-skirt apparatus insulators and deep-shed bushings has been remarkably effective. In one field study, long-term electrical performance of silicone-coated deep-skirt apparatus insulators and bushings proved to be better over a 10-year exposure period than semiconducting-glaze post insulators under identical, severe contamination exposure with annual washing [Chisholm, 2005].

6.6.5. Anti-Fog Disk Profiles with Larger Diameter and Spacing

An increase in diameter will reduce the per-unit-length surface resistance of the insulator. This may offset some of the advantage of a higher path length. Also, for high ratios of leakage distance to dry arc distance, the deep skirts or ribs are bypassed more easily by arcing. These considerations make larger-diameter disk profiles a bit less effective than fog profiles, and they are not as commonly used as a pin-compatible improved design.

An indication of the combined effect of insulator diameter and form factor was drawn from the experimental results consolidated by Kontargyri et al. [2007] and Beauséjour [1981], presented in Chapter 4. An empirical model, based on insulator height, diameter and form factor, was fitted to the combined results as Equation 6-3. This model suggested that increased insulator height or spacing H was a minor advantage, and that increased form factor was a modest disadvantage to electrical performance:

$$E_{50} = 50 \, \text{kV}/\text{m}_{\text{leakage}} \cdot \left(\frac{H}{1 \, \text{mm}}\right)^{0.2} \cdot \left(\frac{D_{\text{max}}}{1 \, \text{mm}}\right)^{-0.4} \cdot F^{-0.1} \cdot \left(\frac{ESDD}{1 \, \text{mg}/\text{cm}^2}\right)^{-0.24} \tag{6-3}$$

where

H is the insulator spacing or height (mm),
D_{max} is the insulator disk diameter (mm),
F is the insulator form factor, and
$ESDD$ is the contamination level (mg NaCl/cm^2).

If the insulator height variable is removed from Equation 6-3 and the coefficients are refitted, then the dependence of E_{50} on insulator form factor in the empirical model also reduces to zero. The simplified model fitting the combined results is

$$E_{50} = 75 \text{ kV/m}_{\text{leakage}} \cdot \left(\frac{D_{\text{max}}}{1 \text{ mm}} \right)^{-0.28} \cdot F^0 \cdot \left(\frac{ESDD}{1 \text{ mg/cm}^2} \right)^{-0.24} \qquad (6\text{-}4)$$

The pair of empirical expressions predicts that an insulator with 420-mm diameter will have a value of E_{50} that is 13–18% lower than an insulator with 254-mm diameter, if they both have the same leakage distance.

Normally, large-diameter bells have higher mechanical strength ratings as shown in Table 6-12. It is possible to change out standard insulator strings for strings of large-diameter bells with greater cap-to-pin spacing with profiles that provide more leakage distance. For example, Schneider and Nicholls [1978] found that a 3.5-m string of 220-mm × 420-mm disks had the same contamination performance as a 4.8-m string of 33 standard 146-mm × 254-mm disks. With the larger diameter, it was possible to have 726 mm of leakage distance on each anti-fog disk, compared to 305 mm on each standard disk. The larger insulator diameter also gives a large increase in weight and cost per disk.

The flashover voltage under light contamination conditions increased by 37% (107 versus 78 kV/m) based on this result, albeit at a considerable cost penalty for the large-diameter design. This means that, in areas of moderate

TABLE 6-12: Comparison of Standard and Large-Diameter Insulator Strings with Equal Pollution Performance

Characteristics	Standard Disk	Large-Diameter Anti-Fog
Mechanical rating	7,500 kg (15,000 lb)	40,000 kg (90,000 lb)
Dimensions (mm)	146 × 254	220 × 420
Units and string length (m) for CFO of 375 kV	33 units 4.82	16 units 3.52
String leakage distance (mm)	10,065	11,616
E_{50} (kV/m$_{\text{leakage}}$) from test results	37	32
E_{50} (kV/m$_{\text{leakage}}$) from Equation 6-3	38[a]	32[a]
E_{50} (kV per meter of string length)	78	107

[a]ESDD = 0.026 mg/cm^2.

Source: Schneider and Nicholls [1978].

contamination, the tolerable level of ESDD will increase by a factor of $(1.37)^{(1/0.24)} = 3.7$, using the value of $\alpha = 0.24$ from Equation 6-3 or Equation 6-4.

In this case, with five closely spaced ribs on the bottom surface, the E_{50} flashover stress per meter of leakage distance for the anti-fog design is 15% lower than the standard disk. The five-rib profile does not offer as much of an improvement as the two-rib profile of the 146-mm anti-fog disk shown above in Table 6-12, and this is a function mainly of the increased disk diameter.

6.6.6. Station Post and Bushing Profiles

It is far less common to change out station post insulators or bushings in stations as a routine maintenance function, and bushings are usually changed out at an indoor maintenance facility. Generally, station posts are available with a wide range of leakage profiles as shown in Figure 4-44. The electrical performance under contamination conditions tends to scale with the leakage distance in the same way that has been noted for suspension disks. However, the influence of post diameter is also important. The contamination flashover stress per meter of leakage distance in Figure 4-45 varied from 25 to 60 kV per meter of leakage under ac conditions, depending on the insulator diameter and profile.

If a post-type insulator has failed as a result of contamination flashovers, it is reasonable to replace it with a unit that has a unified specific creepage distance (USCD) of the next-highest classification in IEC 60815 [2008]. These levels are given in Table 6-13.

At many utilities, there are programs to replace older cap-and-pin apparatus insulators with modern station post designs. The mechanical problems associated with expansion of cement over time have been a weakness of the deep-skirt multicone and cap-and-pin station insulator designs. However, utilities should be aware that these older deep-skirt designs have been giving improved contamination flashover performance, compared to station posts. Some of this relates to the high leakage distance, and some relates to the improved performance of the deep-bell shape compared to the standard profile as shown in Figure 6-38. In a contaminated area, if the cap-and-pin

TABLE 6-13: Recommended Unified Specific Creepage Distance as a Function of Site Pollution Severity

Pollution Level	Very Light	Light	Medium	Heavy	Very Heavy
Recommended USCD (mm/kV$_{l-g}$)	22	28	35	44	55

Source: IEC Standard 60815 [2008].

apparatus insulators have given occasional flashover problems in years of service, then station posts with at least 20% more leakage distance should be selected as replacements. The additional margin for icing and cold-fog conditions is discussed in Chapter 9.

6.7. CHANGING TO SEMICONDUCTING GLAZE

Semiconducting glaze insulators can provide a bolt-in upgrade that requires no other mechanical changes. Many studies, such as Sangkasaad et al. [2000], suggest that the ac contamination withstand performance of semiconducting glaze disk insulators is two to three times higher than the performance of ordinary or fog-type glazed or glass insulators. This would be a strong motivation to replace existing disk insulators that have flashed over from contamination with physically identical semiconducting glaze units—if only the problem of thermal runaway could be fully solved for disk insulators as it has been for post insulators.

6.7.1. Semiconducting Glaze Technology

The production of conventional glaze on ceramic insulators involves the application of slurries made from raw materials with variable composition. Quality control at the slurry stage ensures low levels of electrically conductive impurities such as lead. Control of the slurry viscosity and the firing temperature profile, known as a "firing curve" in the ceramics industry, will then be selected to produce a glaze with the desired thickness. The mixing and firing process for conventional glaze is fairly straightforward, as there is a wide range of tolerance in the glaze thickness: coverage must be complete, but excessive thickness simply increases material cost slightly. This is not, however, the case for semiconducting glaze.

Forrest [1942] proposed the use of a semiconducting surface to mitigate dry-banding and arcing under pollution conditions. Research in the 1970s, such as that by Moran and Powell [1972], Fukui et al. [1974], and Nigol et al. [1974], explored temperature stability of glaze dopants such as tin, antimony, and iron, and also evaluated the performance of semiconducting glaze on disk, post, and bell-type insulators.

The conductivity of semiconducting glaze is sensitive to glaze composition, which can be well controlled, but is also influenced by the firing curve, including both the rates of temperature changes and the soak temperature(s). Manufacturers of semiconducting glaze insulators guard their formulations and process steps as trade secrets. The thickness of the semiconducting glaze must be carefully controlled. This is because resistance per unit length is a product of the glaze conductivity and cross-sectional area. If the glaze is too thin on areas with a high radius of curvature, the desired arc suppression will not occur.

All is not lost if a batch of semiconducting glaze insulators emerges from firing with glaze conductivity that is out of tolerance. Insulators can and often are re-fired to bring the conductivity into specification. The extra testing, handling, and firing are costly and this is the main reason why semiconducting glaze insulators cost more than conventional insulators of the same mechanical rating. Tolerance on glaze resistivity of station post insulators can be as close as ±10%. Typical lot-to-lot variation in production of semiconducting glaze insulators [Mizuno et al., 1999] was stated to be much wider, $13 \pm 5\,M\Omega$ per 146-mm disk, which may be another reason why the technology has not achieved commercial success for cap-and-pin insulators.

According to Ullrich and Gubanski [2005], the dielectric response of antimony-doped tin oxide semiconducting glaze is dominated by three processes: low-frequency dispersion, dc conduction, and interfacial relaxation polarization. The glaze structure consists of two layers that have different electrical properties. The bulk glaze is covered by a thin 1–5-µm glassy surface layer, which contains less tin oxide and is therefore less conductive. Aging of the glaze affects both the glassy surface layer as well as the bulk of the glaze and causes three types of changes in the dielectric response characteristics compared to new samples:

- Increase in low-frequency dispersion due to moisture trapped within the damaged surface.
- Decrease in the relaxation polarization due to the appearance of an alternative current path between electrodes and bulk glaze.
- Decrease in dc conductance related to aging changes in the bulk, meaning that glaze current will tend to decrease with time.

Most semiconducting glaze or resistance-graded technologies are applied on station post insulators and bushings. The semiconducting glaze approach has also been studied for use on suspension disk insulators. This application tends to be more difficult than post insulators because there is a significant variation in the typical radius of the insulator. Two insulator profiles were studied by Akizuki et al. [2002]. Both had 146-mm cap-to-pin spacing and 250-mm diameter. A standard profile had 280 mm of leakage distance and a fog profile, shown in Figure 6-39, had 430 mm of leakage. (see also Figure 6-40.)

Several generations of semiconducting glaze disks have been developed and field tested. A commercial design that successfully resists thermal runaway in all conditions, including heavy contamination, has not yet been developed. The possibility of bistable insulator coatings that only conduct when needed, under high local electric fields, may also be a future direction of research.

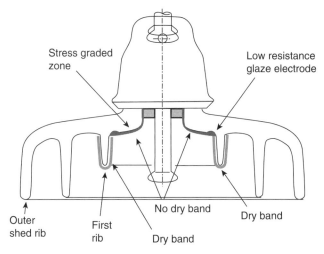

Figure 6-39: Schematic of insulator with graded electrical stress near pin (from Akizuki et al. [2002]).

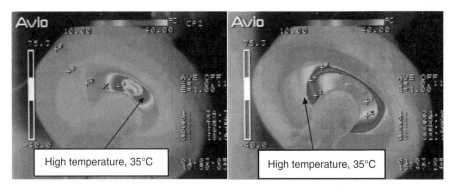

Figure 6-40: Comparison of dry-band location and temperature on conventional 250-mm diameter insulator (left) and stress-graded disk insulator of same shape (right).

6.7.2. Heat Balance: Clean Semiconducting Glaze Insulators

The improved performance of insulators with semiconducting glaze is the result of two benefits—uniform heat dissipation and uniform electric field distribution along the insulator surface.

The minimum resistance of semiconducting glaze insulators is normally set by the leakage current within the glaze. The glaze conductivity tends to increase with increasing temperature and may be constant or may decrease with time, especially in the first year of operation. Typically, manufacturers have settled on an asymptotic glaze current of about 1 mA as an optimal trade-off between continuous power dissipation and superior contamination performance. This means that a clean 69-kV post-type insulator with 40-kV rms line-to-ground voltage will have a resistance of about $40\,M\Omega$ and will normally dissipate 40W when clean and/or dry.

A horizontal cylinder with diameter $D = 200\,mm$ with 680-mm dry arc distance can be used to approximate the shape of the tested semiconducting glaze ceramic posts, ignoring end effects and the effect of insulator sheds as cooling fins. If 40W is dissipated from the surface of this cylinder, it is dissipating $(40\,W/0.68\,m)$ or 59W/m. In winter conditions with no sun, the insulator would reach a static heat balance temperature that can be computed as follows [IEEE Standard 738, 2006]:

$$q_{cn} = 0.0205\rho_f^{0.5} D^{0.75} (T_{ins} - T_a)^{1.25}$$

$$q_{c1} = \left[1.01 + 0.0372\left(\frac{D\rho_f V_{wind}}{\mu_f}\right)^{0.52}\right] k_f K_{angle} (T_{ins} - T_a)$$

$$q_r = 0.0178 D\varepsilon\left[\left(\frac{T_{ins} + 273}{100}\right)^4 - \left(\frac{T_a + 273}{100}\right)^4\right]$$

$$\rho_f = \frac{1.293 - 1.525 \times 10^{-4} H + 6.379 \times 10^{-9} H^2}{1 + 0.00367 \cdot T_{film}}$$

$$k_f = 2.424 \times 10^{-2} + 7.477 \times 10^{-5} \cdot T_{film} - 4.407 \times 10^{-9} \cdot T_{film}^2$$

$$\mu_f = \frac{1.458 \times 10^{-6} (T_{film} + 273)^{1.5}}{T_{film} + 383.4}$$

$$P = \max(q_{cn}, q_{c1}) + q_r$$

where

D is the insulator diameter (mm),

ε is the emissivity of glazed porcelain (0.92),

q_r, q_{cn}, and q_{c1} are the radiated, natural, and low-speed forced convection heat loss (W/m),

k_f is the thermal conductivity of the air at T_{film} (W/(m-°C)),

K_{Angle} is $1.194 - \sin(\beta) - 0.194\cos(2\beta) + 0.368\sin(2\beta)$, where β is the perpendicular angle of the wind to the insulator; for $\beta = 90°$, $K_{Angle} = 1$,

H is the altitude (m),

μ_f is the dynamic viscosity (Pa-s) of the air at T_{film},

V_{wind} is the wind speed (m/s),

ρ_f is the air density, $1.147\,kg/m^3$ at 1000m and 0°C, and

T_{film} is the average film temperature near the insulator, $T_{film} = 0.5(T_{ins} + T_a)$.

This approximate approach can be improved with use of a more detailed set of equations, including cylinder orientation, found in Morgan [1990]. Normally, the simple heat balance $(P - \max(q_{cn}, q_{fl}) - q_r)$ is set up in a spreadsheet and the "Solver" tool is then used to adjust T_{ins} to a value that makes this sum equal to zero.

Figure 6-41 shows the static temperature rise for a dry semiconducting glaze post insulator, modeled as a cylinder. The temperature rise decreases for larger insulator diameter. Also, the correct operating condition of the insulators, indicated by their temperature rise above ambient, is easier to verify using infrared temperature monitoring for wind speeds less than 2 km/h.

The insulator temperature rise problem is more complicated when a term is introduced to model the evaporative cooling from a wetted pollution layer. An additional heat loss term, q_e, will start to affect the heat balance of the insulator. The heat loss will be a function of the evaporation rate, the heat of vaporization of water (2260 J/g), the heat of fusion of ice (334 J/g), and the heat capacity of water (4.18 J/g).

To calculate the change in temperature from ambient to steady state, an additional term is introduced based on the specific heat capacity of the insulator, with excess heat stored in the insulator as $mC_s \, dT/dt$. For a 146-mm by 254-mm semiconducting glaze porcelain disk, Mizuno et al. [1999] report a mass $m = 5.8$ kg, a heat capacity of $C_s = 865$ J/(kg-K), and a thermal time constant of about 1 hour to reach 63% of the steady-state temperature in still air. Also, they incorporated a model for the temperature dependence of the glaze, using a thermal constant B as follows:

$$R = R_0 \exp\left[B\left(\frac{1}{T} - \frac{1}{T_0}\right)\right] \tag{6-5}$$

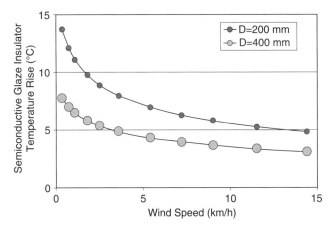

Figure 6-41: Calculated temperature rise of semiconducting glaze post insulator dissipating 58 W/m at altitude of 1000 m, $T_a = 0\,°C$.

where

R_0 is the insulator resistance at a reference temperature T_0 in kelvin units,

B is a thermal constant for each type of glaze: $B = 1400\,\text{K}$ for SnO_2, $B = 2800\,\text{K}$ for Fe_2O_3, and $B = 950\,\text{K}$ for semiconducting glaze, and

R is the insulator resistance at an operating temperature T in kelvin units.

Baker et al. [1990] noted that there was a long-term exponential decay in one formulation of semiconducting glaze resistivity, reaching a level that was about 60% of the initial resistivity after continuous energization at line voltage. This dry-aging characteristic can be controlled in the manufacturing process by setting the glaze resistivity about 40% above the desired long-term level.

6.7.3 Heat Balance: Contaminated Semiconducting-Glaze Insulators

When a semiconducting glaze insulator is covered with a wetted contamination layer, the rms current and power dissipation will increase. As an example, Figure 6-42 shows that the resistance of semiconducting glaze post insulators with 680-mm dry arc distance and 1940-mm leakage distance, covered with layers of dilute sulfuric acid (H_2SO_4). As described in Chapter 4, the sulfuric acid layer tends to be highly hygroscopic, leading to full wetting even at temperatures as low as $-17\,°C$. In Figure 6-42, the effect of saturated (fog) versus unsaturated conditions on the resulting pollution layer resistance was not found to be significant.

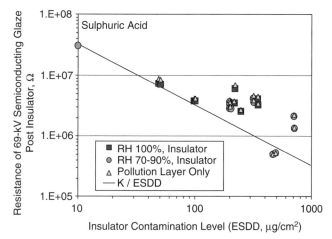

Figure 6-42: Resistance of 69-kV (680-mm dry arc) semiconducting glaze post with sulfuric acid contamination at $-12\,°C$, 40 kV. Points at 480 and 710 µg/cm² were measured at reduced voltage.

A minimum ESDD of $10\,\mu g/cm^2$ leads to a modest reduction in resistance from 40 to $30\,M\Omega$ on the insulator. This means that the resistance of the pollution layer is about $120\,M\Omega$. As ESDD increases, the resistance of the stabilized pollution layer decreases. This in turn causes greater rms leakage current and power dissipation.

At a certain point (about $200\,\mu g/cm^2$ in Figure 6-42) the power dissipation of the stabilized pollution layer starts to warm up the semiconducting glaze insulator temperature. It can reach several tens of degrees above ambient. This will overwhelm the fog accretion rate and the ability of the pollution layer to scavenge moisture from the environment, and some parts of the surface will dry out. Based on the measured resistances of the clean and contaminated 680-mm semiconducting glaze post insulator, and Equation 6-6, the fraction k of the insulator that remained wet was between 83% and 98%.

$$R = \frac{60\ M\Omega}{m_{\text{dry arc}}}\left[(1-k)+k\cdot\frac{8\ \mu g/cm^2}{ESDD}\right] \tag{6-6}$$

The ability of the semiconducting glaze to provide a continuous rms current even when partially wetted and contaminated is a good demonstration of its ability to suppress arcing activity. If there were partial discharges across the insulator dry bands, these would show up visibly, audibly, and also as highly distorted leakage currents with large ratios of peak to rms value.

For some of the higher contamination levels in Figure 6-42, measured power dissipation of 1200–1700W at 10-km/h wind speed led to calculated insulator temperatures of 135–175 °C. Thermal shock from these extreme temperatures can break the porcelain sheds or cause other damage. For this reason, laboratory measurements of semiconducting glaze post resistance at high contamination levels need to be taken at reduced voltage to avoid insulator damage.

Cold switch-on performance issue may also be important when restoring electrical service after an extended period of contamination buildup in the outdoor environment. Lambeth [1983] noted that, when initially energized in wet conditions, the flashover performance of a semiconducting glaze insulator is similar to that of a conventional insulator. Figure 6-43 shows that, in the seconds after switch-on, the margin of the semiconductive glaze insulators increases from 50% to 300%, while the flashover performance of the conventional-glaze unit degrades to its steady-state value within a minute.

In the case of semiconducting glaze long-rod or post insulators, the margin of cold switch-on to steady-state flashover strength increases to a factor of 100% [Lambeth, 1983].

6.7.4. On-Line Monitoring with Semiconducting Glaze Insulators

The semiconducting glaze post insulator resistance can easily be measured with reduced voltage, for example, by putting several posts or disks in series

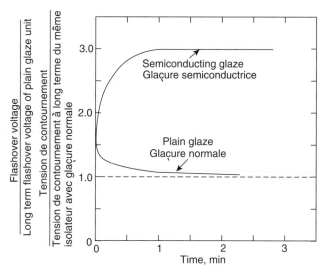

Figure 6-43: Comparison of cold switch-on flashover voltage of conventional and semiconducting glaze (from Lambeth [1983]).

and measuring the rms current through them at line voltage. Operation at reduced voltage will minimize the effects of power dissipation on the results. This suggests a simple way to monitor the pollution level, using a simple measurement of rms current, rather than by capturing complicated statistics about leakage current pulses on conventional porcelain insulators as described later in this chapter. Normally, a calibration phase to establish the relative levels of ESDD on the sample and target insulators would be needed along with measurements of temperature and relative humidity in a practical system.

6.7.5. Role of Power Dissipation in Fog and Cold-Fog Accretion

Typical winter fogs have median and maximum densities of 0.1 and $0.3\,g/m^3$. Considering the median value, along with a typical wind speed of 3 m/s (10.8 km/h) and an insulator diameter of $D = 0.2$ m, fog will deposit at the rate of $(0.1\,g/m^3 \times 3\,m/s \times 0.2\,m)$ or 0.06 g/s per meter of insulator dry arc distance. To evaporate this deposit, a power dissipation of $(0.06\,g/s \times 2260\,J/g)$ or 136 W would be required. The heat input from a typical semiconducting glaze insulator dissipating about $60\,W/m_{dry\ arc}$ in clean conditions will have a modest influence on the fog accretion rate. This means that the equilibrium temperature of the fog to the ambient (which is typically the dew point temperature) will be the dominant heat flux that controls the insulator temperature.

In freezing conditions, cold-fog accumulation as hoarfrost on the insulator surfaces will have an additional heat of fusion of ice, of about 334 J/g. This is much smaller than the heat of vaporization (2260 J/g) and will have only a minor impact on the insulator temperature rise.

6.7.6. Considerations for Semiconducting Insulators in Close Proximity

The electrical conductivity of semiconducting glaze insulators should have a well-controlled tolerance for three reasons.

1. *Thermal Runaway:* The semiconducting glazes normally have a positive temperature coefficient, meaning that the conductivity increases with temperature. If the glaze conductivity (or the operating voltage) is too high, it will conduct more current than planned. This will raise the glaze temperature, lower the conductivity further, and lead to a degenerative positive feedback process. Semiconducting glaze insulators are designed with a margin between the operating voltage and the thermal runaway voltage. A large positive tolerance on the glaze conductivity can erode this safety margin. Semiconducting glaze insulators must thus be used at the voltage for which they were designed.

2. *Matching of Stacking Station Posts:* The second reason for tight control of semiconducting glaze tolerance concerns its use on stacking station post insulators (typical at or above 230 kV) or strings of semiconducting disks. If a conductivity tolerance of ±50% is adopted, then the conductivity of individual sections in the stack can vary by a factor of 3. Considering a two-section post, if the bottom section has 50% of nominal resistance and the top section has 150% of nominal, then the potential at the central metal flange will be $V/4$ rather than $V/2$ as planned, where V is the line voltage. The top section will be prone to thermal runaway. Also, the voltage distribution across the stack will be less linear than if semiconducting glaze was not used at all.

3. *Matching Between Stacking Station Posts:* The third reason for tight control of semiconducting glaze tolerance relates to the use of pairs of stacking station post insulators in close proximity (typically on 0.5 m center-to-center spacing) to support air-break disconnect switches and other specialized apparatus. Consider the example from the previous paragraph for one of the stacking posts. With the ±50% glaze tolerance, the adjacent post could have the positions reversed (high conductivity above, low conductivity below), giving a central flange voltage of $3V/4$. The potential difference *between the flanges*, separated by an air space of less than 0.5 m, would be $V/2$. This would be a source of arcing, audible noise, and electromagnetic interference.

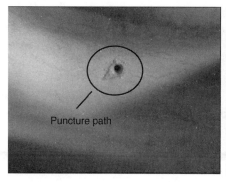 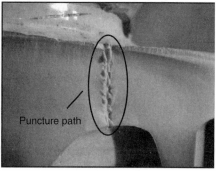

Insulator surface view   Cross section

Figure 6-44: Puncture path due to thermal runaway of semiconducting glaze disk insualtor in highly polluted conditions (from Shinoda et al. [2005]).

In a set of experiments [Hydro Quebec 735 kV], two practical mitigation methods for potential difference between flanges of semiconducting glaze insulators were:

- Bond the flanges electrically.
- Specify sufficient tolerance on glaze conductivity to ensure equal potentials on each flange.

6.7.7. Application Experience

Generally, in the test laboratory, the semiconducting glaze insulators fail from excessive power dissipation under extremely high pollution conditions (>1 mg/cm^2) before they will flash over. This problem may have occurred in the field as well. Problems with semiconducting glaze insulator failure under cold switch-on proved to be disappointing to one utility in the US Midwest.

An example of the kind of damage caused by thermal runaway on semiconducting glaze disks [Shinoda et al., 2005] is found in Figure 6-44.

On the bell-shaped semiconducting glaze insulators described by Nigol et al. [1974], the life of the electrical contact between cap and glaze proved to be about 10 years. The contacts were restored using silver paint and RTV silicone cover, giving a repair lifetime of about 5 years. More recent use of semiconducting glaze on post insulators, and improved glaze-to-metal contact systems, has mitigated this problem.

6.8. CHANGING TO POLYMER INSULATORS

Polymer or nonceramic insulators, as described in Chapter 2, are constructed with a fiberglass core and metallic end fittings to provide the mechanical support function, while rod sheathing and weather sheds provide the required electrical leakage distance and also protect the rod. Polymer insulators are designed to withstand specified service stresses for the lifetime of the unit.

Laboratory tests and field installations have both demonstrated that polymer insulators have better short-term resistance to contamination flashover than conventional porcelain or glass insulators of the same connection length and leakage distance. This means that it is sometimes possible to substitute polymer suspension or tension insulators directly for porcelain strings, in the same way that semiconducting glaze insulators can provide a bolt-in upgrade that requires few mechanical changes.

6.8.1. Short-Term Experience in Contaminated Conditions

The superior initial performance of polymer insulators is attributed [IEEE Standard 987, 2001] to the differences in geometry—polymer insulators having a much smaller core diameter—and differences in the ability of the surfaces to bead water. Like the fresh wax coating on an automobile, however, the hydrophobicity does not last forever. The performance of the polymer insulator deteriorates as it loses this desirable property through surface aging, as discussed in the next section.

The IEEE recommendations suggest that polymer insulators provide better performance than ceramic insulators in areas with heavy phosphate, cement, pulp, lime, or ash pollution. They have also demonstrated exceptional performance in areas where carbon dust and aerosol is present, such as near coal piles. However, failures with distribution polymer insulators have also occurred within 500 m downwind of a few phosphate, cement, pulp, and lime processing plants [Cherney et al., 1984]. Since damage may occur in any severe environment, the IEEE recommends periodic inspection to ensure that the polymer insulators are remaining in serviceable condition. The inspections should focus on damage rather than ESDD and NSDD measurements, since the polymer insulators retain the pollution they have encapsulated.

There are a number of chemicals that may attack polymer materials, causing swelling, embitterment, or even depolymerization [Cherney and Stonkus, 1981]. In particular, frequent inspections may be needed when using polymer insulators near hydrocarbon vapors, diesel fumes, or wood pole preservative. Damaged portions of an insulator may have lower (or higher) surface impedance than undamaged areas, leading to changes in the surface resistance. Areas with high surface resistance will be more prone to attack from leakage currents and dry-band arcing.

Surface aging can be accelerated by exposure to some sorts of contamination conditions. The contaminants themselves cover the surface, and also any arcing activity creates intense ultraviolet and thermal damage. However, unlike the wax coating on a painted steel automobile, the bulk polymer material itself may be hydrophobic. Polymer molecules can diffuse toward the surface, where they interact with pollutants and reduce the conductance of the pollution layer. The relaxation time needed for polymer molecules to reorient to stable positions on the surface varies dramatically between different polymers, within generic polymer systems, and with temperature. Some insulator surfaces such as silicone can even recover their original hydrophobicity with enough rest, even in damaged areas. This is one of the most important reasons why silicone materials, in particular, have better long-term contamination behavior than ceramic insulators.

6.8.2. Long-Term Performance in Contaminated Conditions

All polymers will age under long-term exposure to weathering, contamination, sunlight, and temperature. Aging under electrical stress in these outdoor conditions is aggravated by corona dry-band discharge activity. The aging may cause changes in the external housing of a polymer insulator [IEEE Standard 987, 2001], such as:

- Surface oxidation and chemical reactions, which change the polymer surface and volume structure.
- Surface crazing and cracking due to the effects of radiation, temperature, ozone, and water.
- Erosion or tracking caused by surface discharge activity.

Aging of the polymers may affect the insulator's functional properties including its contamination retention and flashover levels. However, polymer aging may simply cause material changes such as discoloration that do not affect the performance.

It is main focus of application standards such as IEEE Standard 987 [2001] to evaluate the effects of material aging on the performance of the polymer insulator system. This standard determines dominant insulator failure mechanisms through a consideration of materials, processing, environment, and stress on the insulators, and also defines which changes in material properties are likely to result in a reduction of the insulator's in-service performance.

Generally, long-term service experience with polymer insulators has been satisfactory. Some have been in service for more than 30 years on both transmission and distribution systems. Most applications have been on distribution systems. Polymer insulators have proved to be an excellent substitute for ceramic pin or post insulators to mitigate pole fire problems. At the transmission level, polymer insulators manufactured in the 1980s were found to have good performance in 78% of power system applications, acceptable in 18%, and poor in 4%. Of the failures reported, approximately 17% were identified as electrical. The electrical failures included flashover due to surface contamination, bird droppings, salt spray, and icing [IEEE Standard 957, 2005].

Most of the polymer insulators from the 1980s were installed in North America. With their higher levels of NSDD in Europe, studies on long-term contamination performance may have influenced the slower rate of adoption. It is difficult to draw many conclusions about the long-term performance of these insulators because there has been a process of continuous improvement in materials and assembly technologies.

6.8.3. Interchangeability with Ceramic Insulators

Unlike semiconducting glaze technologies, which are applied to the same post and disk insulators that receive conventional glazing, there are small differences between the dimensions of polymer insulators and ceramic strings. This leads to problems of mechanical, electrical, and dimensional interchangeability.

The first problem to be faced with a bolt-in replacement is that the polymer insulator typically has longer end fittings that reduce the dry arc distance. Table 6-14 shows that this has been a problem in applying polymer designs to replace porcelain disks, beginning more than 25 years ago.

The contamination performance in Table 6-14 is expressed as the critical flashover voltage V_{50} (kV) and also as a flashover stress E_{50} (kV per meter of leakage distance). The porcelain insulators in the heavy contamination condition of $0.2\,mg/cm^2$ have a gradient of 34–$35\,kV/m_{leakage}$. The epoxy insulator, which is also hydrophilic, has about the same flashover performance too. The ethylene-propylene rubber (EPR) material performed a bit better, at about 42–$44\,kV/m_{leakage}$, but the advice from 1984 was that equal contamination performance could be expected on the basis of equal leakage distance.

The lightning impulse flashover strength of an insulator will be very closely related to the dry arc distance of the insulator. For the same connection length, a polymer insulator may reduce this dry arc distance by 10%. Typically, a 10% decrease in BIL will lead to a 30% increase in lightning backflashover rate, which could be a concern in areas of high lightning ground flash density.

For higher transmission voltages, the necessary use of corona rings will further reduce the dry arc distance. Also, the electric fields around polymer insulators are more nonlinear than those around cap-and-pin insulators, since there are no internal grading capacitances from caps to pins. This means that an exact swap of polymer for porcelain will reduce the lightning impulse flashover level even further, with a possible adverse overall effect on line reliability in spite of improved contamination performance.

IEEE Standard 957 [2005] recommends that polymer insulators should be applied with leakage distances equal to or greater than those recommended for ceramic units. Reduced leakage polymer insulators in heavily contaminated areas may have a short service life, caused by increased leakage currents and dry-band arcing attacks. This advice to use the same or more leakage distance on polymer insulators should certainly be respected when replacing ceramic strings to improve contamination performance. When selecting a replacement polymer insulator, it is often convenient to use a larger value of

TABLE 6-14: Characteristics and Comparative Electrical Performance of Standard Porcelain and Polymer Insulators for Distribution Applications

| | Type of Insulator | | | | | | | |
| | Porcelain | | Bisphenol A Epoxy, 69% ATH Filler A | | | EPR, 77% ATH Filler B | | |
Material	Three Disks 6¾ in. × 4½ in. ANSI 52-9A	Three Disks 5¾ in. × 10 in. ANSI 52-3	15 kV	25 kV	34.5 kV	15 kV	25 kV	34.5 kV
Dry arc distance (mm)	430	489	200	335	440	200	320	395
Connection length (mm)	477	438	318	450	572	320	440	520
Dry flashover (kV)	≈180	215	104	152	192	98	160	214
Wet flashover (kV)	≈90	130	82	129	161	82	119	149
Positive impulse CFO (kV)	≈300	355	144	222	282	160	244	278
Leakage distance (mm)	514	876	410	680	970	480	900	1030
V_{50} (kV)(E_{50}(kV/m$_{leakage}$))	30	52.5	25	>45	>45	>45	>45	>45
ESDD = 0.02 mg/cm²	(58)	(60)	(61)	(>66)	(>46)	(>94)	(>50)	(>44)
V_{50} (kV)	20.1	35.1	20	30	>45	25	>45	>45
(E_{50}(kV/m$_{leakage}$))	(39)	(40)	(49)	(44)	(>46)	(52)	(>50)	(>44)
ESDD = 0.07 mg/cm²								
V_{50} (kV)	18	30	15	26	35	20	40	>45
(E_{50}(kV/m$_{leakage}$))	(35)	(34)	(37)	(38)	(36)	(42)	(44)	(>44)
ESDD = 0.2 mg/cm²								
High-humidity surface resistivity (Ω)								
As received			2 × 10⁹			1 × 10⁸		
Boil test 100 h			4 × 10⁷			4 × 10⁷		
Heat, 90 days 120°C			4 × 10⁹			6 × 10⁷		
UV, 1000 h			1 × 10⁷			6 × 10⁷		
Abraded surface			1 × 10⁷			2 × 10⁶		

Source: Cherney et al. [1984].

unified specific creepage distance (USCD) that will give lower stress per meter of leakage distance, because of the higher ratios of leakage to dry arc distance available with the polymer designs.

Utility policies on live-line work also play a role in the interchangeability evaluation for transmission systems. Every porcelain disk is proof-tested at high voltage at the factory, while polymer insulators are not. If the line upgrade is to be carried out with live-line work methods, then this could be a factor. Also, once installed, it is no longer possible to check the electrical integrity of the polymer insulator in the way that "buzz-stick" testing can be carried out on individual disks of an energized string. Infrared camera inspection can indicate the presence of internal polymer insulator defects between the fiber-glass core and the sheathing that calls for deenergized work. However, this equipment is not widely available and the thermal inspections for temperature rise may not find every electrical defect in polymer insulators.

6.8.4. Case Study: Desert Environment

With the presence of high NSDD, along with calcium chloride and sodium chloride salts, peak levels of more than $0.6 \, mg/cm^2$ have been observed in desert areas [Akbar and Zedan, 1991]. As mentioned in CIGRE Task Force 33.04.01 [2000], exposure of nonceramic insulators to this environment has been unsatisfactory. Microcracks occur under the intense ultraviolet, and these collect more pollution than the smooth surface. The high level of nonsoluble deposit also tends to overwhelm the natural hydrophobicity transfer process of the silicone oil [Rizk et al., 1997].

6.9. PRÉCIS

There are many reasons why insulators in an area may have unacceptable pollution performance. Demands for increased power quality with reduced labor cost tend to shift the perception of what constitutes "acceptable" performance. Local conditions may have changed from industrialization or urbanization. The insulators themselves may have been selected for the wrong environment, and it has simply taken many years to realize that an improvement is needed.

There are as many, or more, options for improving insulator pollution performance. Maintenance options such as SMART washing, on demand when insulator pollution levels reach targets, have been effective. Monitoring of leakage currents has been a promising approach since the 1930s but further development is still needed. Insulator surfaces can be coated with silicone rubber, or accessories can be added to extend leakage distance. Additional cap-and-pin disks can sometimes be added in transmission lines. Semiconducting glaze posts can be substituted for conventional porcelain posts of the same dimensions in station insulators. Silicone insulators offer good long-term ability to bead water (hydrophobicity) that mitigates pollution problems. Insulators with improved designs, perhaps an aerodynamic profile to reduce the rate of pollution accumulation, or a convoluted profile with an extensive and efficient leakage path, can also be substituted.

REFERENCES

Akbar, M. and F. Zedan. 1991. "Performance of HV Transmission Line Insulators in Desert Conditions, Part III. Pollution Measurements at a Coastal Site in the Eastern Region of Saudi Arabia," *IEEE Transactions on Power Delivery*, Vol. 6, No. 1 (Jan.), pp. 429–438.

Akizuki, M., O. Fujii, S. Ito, T. Irie, and S. Nishimura. 2002. "A Study on Anticontamination Design of Suspension Insulators—A Stress Grading Insulator," IEEE Paper 0-7803-7519-X/02, pp. 300–306.

Amarh, F. 2001. "Electric Transmission Line Flashover Prediction System: Ph.D. Thesis and Final Report." Power Systems Engineering Research Center Publication 01-16, Arizona State University. May.

Baker, A. C., J. W. Maney, and Z. Szilagyi. 1990. "Long Term Experience with Semiconducting Glaze High Voltage Post Insulators," *IEEE Transactions on Power Delivery*, Vol. 5, No. 1 (Jan.), pp. 502–508.

Baker, A. C., M. Farzaneh, R. S. Gorur, S. M. Gubanski, R. J. Hill, G. G. Karady and H. M. Schneider. 2008. "Selection of Insulators for AC Overhead Lines in North America with respect to Contamination," *IEEE Transactions on Power Delivery*, in press.

Burnham, J. T., J. Franc, and M. Eby. 1995. "High-Pressure Washing Tests on Polymer Insulators," in *Proceedings of IEEE ESMO-95*, Paper CP-11.

Carberry, R. E. and H. M. Schneider. 1989. "Evaluation of RTV Coating for Station Insulators Subjected to Coastal Contamination," *IEEE Transactions on Power Delivery*, Vol. 4, No. 1, pp. 577–585.

Cherney, E. A. and D. J. Stonkus. 1981. "Non-ceramic Insulators for Contaminated Environments," *IEEE Transactions on Power Apparatus and Systems*, Vol. PAS-100, No. 1 (Jan.), pp. 131–142.

Cherney, E. A., J. Reichman, D. J. Stonkus, and B. E. Gill. 1984. "Evaluation and Application of Dead-End Type Polymeric Insulators to Distribution," *IEEE Transactions on Power Apparatus and Systems*, Vol. PAS-103, No. 1 (Jan.), pp. 121–132.

Chino, T., M. Iwata, S. Imoto, M. Nakayama, H. Sakamoto, and R. Matsushita. 2005. "Development of Arcing Horn Device for Interrupting Ground-Fault Current of 77 kV Overhead Lines," *IEEE Transactions on Power Delivery*, Vol. 20, No. 4 (Oct.), pp. 2570–2575.

Chisholm, W.A. 2005. "Ten Years of Application Experience with RTV Silicone Coatings in Canada," World Congress and Exposition on Insulators, Arresters and Bushings, Hong Kong, Nov.

Chrzan, K. L. and F. Moro. 2007. "Concentrated Discharges and Dry Bands on Polluted Outdoor Insulators," *IEEE Transactions on Power Delivery*, Vol. 22, No. 1 (Jan.), pp. 466–471.

CIGRE Task Force 33.04.03. 1994. "Insulator Pollution Monitoring," *Electra*, No. 152 (Feb.), pp. 79–89.

CIGRE Task Force 33.04.01. 2000. "Polluted Insulators: A Review of Current Knowldege," CIGRE Brochure No. 158 (June).

deTourreil, C. and F. Schmuck. 2009. "On the Use of Power Arc Protection Devices for Composite Insulators on Transmission Lines," CIGRE WG B2.21 Report, in press.

Driscoll, J. and C. Ballotte. 2007. "Infrared/Corona Camera Tools to Detect Contamination and Examples of NU's Application of the Corona Camera," in *Proceedings of 2007 Ultraviolet Inspection Users Group Meeting*, Orlando, FL, Vol. 5. April 30 to May 2.

Eldridge, K., J. Xu, Yin Weijun, A.-M. Jeffery, J. Ronzello, and S. A. Boggs. 1999. "Degradation of a Silicone-Based Coating in a Substation Application," *IEEE Transactions on Power Delivery*, Vol. 14, No. 1 (Jan.), pp. 188–193.

Ely, C. H. A., P. J. Lambeth, and J. S. T. Looms. 1978a. "The Booster Shed: Prevention of Flashover of Polluted Substation Insulators in Heavy Wetting," *IEEE Transactions on Power Apparatus and Systems*, Vol. PAS-97, No. 6 (Nov.), pp. 2187–2197.

Ely, C. H. A., P. J. Lambeth, J. S. T. Looms, and D. A. Swift. 1978b. "Discharges over Wet, Polluted Polymers: The 'Booster Shed'," in *Proceedings of CIGRE General Meeting*, Paris, France, Paper 15–02.

Forrest, J. S. 1936. "The Electrical Characteristics of 132-kV Line Insulators Under Various Weather Conditions," *Journal of the IEE*, Vol. 79, pp. 401–423.

Forrest, J. S. 1942. "The Characteristics and Performance in Service of HV Porcelain Insulators," *Journal of the IEE*, Vol. 89, Part 1, pp. 60–92.

Fukui, H., K. Naito, T. Irie, and I. Kimoto. 1974. "A Practical Study on Application of Semiconducting Glaze Insulators to Transmission Lines," *IEEE Transactions on Power Apparatus and Systems*, Vol. PAS-93, No. 5 (Sept.), pp. 1430–1443.

Gorur, R. S., E. Cherney, C. de Tourreil, D. Dumora, R. Harmon, H. Hervig, B. Kingsbury, J. Kise, T. Orbeck, K. Tanaka, R. Tay, G. Toskey, and D. Wiitanen. 1995. "Protective Coatings for Improving Contamination Performance of Outdoor High Voltage Ceramic Insulators," *IEEE Transactions on Power Delivery*, Vol. 10, No. 2 (Apr.), pp. 924–933.

Hydro One Networks. 2004. "S92 Downtown Cables Application—Leave to Construct Underground Transmission Line from John TS to Esplanade TS." Ontario Energy Board docket EB-2004-0436.

IEC 60815. 2008. *Selection and Dimensioning of High-Voltage Insulators intended for Use in Polluted Conditions—Part 2: Ceramic and Glass Insulators for ac Systems*. Geneva, Switzerland: Bureau Central de la Commission Electrotechnique Internationale, October.

IEEE Standard 738. 2006. *IEEE Standard for Calculating the Current Temperature Relationship of Bare Overhead Conductors*. Piscataway, NJ: IEEE Press.

IEEE Standard 4. 1995. *IEEE Standard Techniques for High Voltage Testing*. Piscataway, NJ: IEEE Press.

IEEE Standard 957. 2005. *Guide for Cleaning Insulators*. Piscataway, NJ: IEEE Press.

IEEE Standard 987. 2001. *Guide for Application of Composite Insulators*. Piscataway, NJ: IEEE Press. May 2002.

IEEE/ANSI Standard C2. 2007. *National Electrical Safety Code, 2007 Edition.* Piscataway, NJ: IEEE Press.

Johnson and Phillips. 1876. UK Patent #3534.

Johnson, J. C. 1976. "Insulator Hot Washing with Deionized Water," *IEEE Transactions on Power Apparatus and Systems*, Vol. 95, No. 3 (May), Part 1, pp. 864–869.

Kawai, M. 1970. "Flashover Tests at Project UHV on Salt-Contaminated Insulators, Part II," *IEEE Transactions on Power Apparatus and Systems*, Vol. PAS-89, No. 8 (Nov./Dec.), pp. 1791–1799.

Kawai, M. 1973a. "Research at Project UHV on the Performance of Contaminated Insulators, Part I: Basic Problems," *IEEE Transactions on Power Apparatus and Systems*, Vol. PAS-82, No. 3 (May), pp. 1102–1110.

Kawai, M. 1973b. "Research at Project UHV on the Performance of Contaminated Insulators, Part II: Application to Practical Design," *IEEE Transactions on Power Apparatus and Systems*, Vol. PAS-82, No. 3 (May), pp. 1111–1120.

Kawai, M. and D. M. Milone. 1969. "Tests on Salt-Contaminated Insulators in Artificial and Natural Wet Conditions," *IEEE Transactions on Power Apparatus and Systems*, Vol. PAS-88, No. 9 (Sept.), pp. 1394–1399.

Kontargyri, V. T., A. A. Gialketsi, G. J. Tsekouras, I. F. Gonos, and I. A. Stathopulos. 2007. "Design of an Artificial Neural Network for the Estimation of the Flashover Voltage on Insulators," *Electric Power Systems Research*, Vol. 77, pp. 1532–1540.

Lambeth, P. J. 1983. "The Use of Semiconducting Glaze Insulators," *Electra*, Vol. 86 Jan., pp. 89–106.

Lambeth, P. J, H. M. Schneider, Y. Beausejour, E. A. Cherney, D. Dumora, T. Kawamura, G. Marrone, J. H. Moran, K. Naito, R. J. Nigbor, J. D. Sakich, R. Stearns, H. Tempelaar, M. P. Verma, J. Huc, D. Perin, G. B. Johnson, and C. De Ligt. 1987. "Final Report on the Clean Fog Test for HVAC Insulators," *IEEE Transactions on Power Delivery*, Vol. PWRD-2, No. 4 (Oct.), pp. 1317–1326.

Metwally, I. A., A. Al-Maqrashi, S. Al-Sumry, and S. Al-Harthy. 2006. "Performance Improvement of 33 kV Line-Post Insulators in Harsh Environment," *Electric Power Systems Research*, Vol. 76, pp. 778–785.

Mizuno, Y., K. Naito, Y. Suzuki, S. Mori, Y. Nakashima, and M. Akizuki. 1999. "Voltage and Temperature Distribution Along Semiconducting Glaze Insulator Strings," *IEEE Transactions on Dielectrics and Electrical Insulation*, Vol. 6, No. 1 (Feb.), pp. 100–104.

Moran, J. H. and D. G. Powell. 1972. "Resistance Graded Insulators: The Ultimate Solution to the Contamination Problem?" *IEEE Transactions on Power Apparatus and Systems*, Vol. 91, No. 6, pp. 2452–2457.

Morgan, V. T. 1990. *Thermal Behaviour of Electrical Conductors.* Hoboken, NJ: Wiley.

Nigol, O., J. Reichman, and G. Rosenblatt. 1974. "Development of New Semiconductive Glaze Insulators," *IEEE Transactions on Power Apparatus and Systems*, Vol. 93, No. 2, pp. 614–620.

Perin, D., A. Pigini, I. Visintainer, Channakeshava, and M. Ramamoorty. 1995. "Live-Line Insulator Washing: Experimental Investigations to Assess Safety and Efficiency Requirements," *IEEE Transactions on Power Delivery*, Vol. 10, No. 1 (Jan.), pp. 518–525.

Phillips, A. J., J. Kuffel, A. Baker, J. Burnham, A. Carreira, E. Cherney, W. Chisholm, M. Farzaneh, R. Gemignani, A. Gillespie, T. Grisham, R. Hill, T. Saha, B. Vancia, and J. Yu. 2008. "Electric Fields on AC Composite Transmission Line Insulators:

IEEE Taskforce on Electric Fields and Composite Insulators," *IEEE Transactions on Power Delivery*, Vol. 23, No. 2, pp. 823–830.

Richards, C. N. and J. D. Renowden, 1997. "Development of a Remote Insulator Contamination Monitoring System," *IEEE Transactions on Power Delivery*, Vol. 12, No. 1 (Jan.), pp 389–397.

Richards, C. S., C. L. Benner, K. L. Butler-Purry, and B. D. Russel. 2003. "Electrical Behavior of Contaminated Distribution Insulators Exposed to Natural Wetting," *IEEE Transactions on Power Delivery*, Vol. 18, No. 2 (Apr.), pp 551–558.

Rizk, M. S., A. Nosseir, B. A. Afrafa, O. Elgendy, and M. Awad. 1997. "Effects of Desert Environment on the Electrical Performance of Silicone Rubber Insulators," in *Proceedings of 10th International Symposium on High Voltage Engineering (ISH)*, Montreal, Canada, Vol. 3, pp. 133–136. August.

Sangkasaad, S., B. Staub, B. Marangsree, and N. Pattanadech. 2000. "Investigation on Electrical Performance of Semiconducting Glazed Insulators Under Natural Pollution in Thailand," in *Proceedings of International Conference on Power System Technology, PowerCon 2000*, Vol. 3, pp. 1229–1232.

Schneider, H. M. 1988. "Measurements of Contamination on Post Insulators in HVDC Converter Station," *IEEE Transactions on Power Delivery*, Vol. 3, No. 1 (Jan.), pp. 398–404.

Schneider, H. M. and C. W. Nicholls. 1978. "Contamination Flashover Performance of Insulators' for UHV," *IEEE Transactions on Power Apparatus and Systems*, Vol. PAS-97. No. 4, (July/Aug.), pp. 1411–1420.

Schneider, H. M. and D. R. Howes. 1988. "Ice Flashover and Clean Fog Tests on Contaminated Insulators." HVTRC Report for Ontario Hydro. April.

Sharp, R., S. Yokoi, M. Akizuki, and J. Burnham. 2001. "Washing Withstand Voltage Tests on Station Insulators," in *Proceedings of IEEE T&D Expo*, Vol. 2, pp. 765–770. October 28 to November 2.

Shinoda, A., H. Okada, M. Nakagami, Y. Suzuki, S. Ito, and M. Akizuki. 2005. "Development of High-Resistance Semi-conducting Glaze Insulators," in *Proceedings of IEEE Power Engineering Society 2005 Annual Power Meeting*, Toronto, Ontario, Paper 0-7803-9156-X. August.

Sundararajan, R., Chair, IEEE Task Force on Reducing Bird Related Power Outages. 2004. "Preventive Measures to Reduce Bird-Related Power Outages—Part II: Streamers and Contamination," *IEEE Transactions on Power Delivery*, Vol. 19, No. 4 (Oct.), pp. 1848–1853.

Ullrich, H. and S. M. Gubanski. 2005. "Electrical Characterization of New and Aged Semiconducting Glazes," *IEEE Transactions on Dielectrics and Electrical Insulation*, Vol. 12, No. 1 (Feb.), pp. 24–33.

Yasuda, M. and T. Fujimura. 1976. "A Study and Development of High Water Pressure Hot-Line Insulator Washing Equipment for 500-kV Substation," *IEEE Transactions on Power Apparatus and Systems*, Vol. PAS-95, No. 6 (Nov./Dec.), pp. 1919–1928.

Yizheng, D., C. Yong, W. Xiong, and X. Meng. 2007. "Application of Ultraviolet Detection in China's Power System," in *Proceedings of 2007 Ultraviolet Inspection Users Group Meeting*, Orlando, FL, Vol. 5. April 30 to May 2.

Zimmerman, M. 2000. "Vandalism Problems Move Moroccan Utility Increasingly Toward Composite Insulators," *INMR Magazine*, Nov./Dec.

CHAPTER 7

ICING FLASHOVERS

Ice is an interesting material and many researchers have studied its characteristics in its purest forms. There are fewer evaluations of the electrical properties of polluted ice, but enough to show that these properties are exquisitely
sensitive to temperature near the freezing point.

This sensitivity to temperature is a common feature among most of the
descriptions of power system problems under various icing conditions, ranging
from a very light coating on all insulator surfaces through to heavy accretion
that leads to full bridging of insulator strings by icicles. The survey of utility
problems is presented in chronological order, starting in the mid-1930s, and is
sorted into four categories of very light, light, moderate, and heavy icing
severity.

Laboratory tests to establish the performance of ice-covered insulators in
controlled conditions have taken many years to perfect. The early literature
shows that the role of the "freezing water conductivity" used to form the ice
caps and icicles was recognized, as was the importance of a melting phase in
obtaining the lowest flashover voltage for a given accretion under constant
service voltage. The laboratory test results are also organized into four groups
of icing severity and results with common experimental procedures are consolidated using an icing stress product of applied water conductivity times
weight of accreted ice per unit of leakage path.

Modeling of the icing flashover process has been able to take advantage of
progress in modeling of pollution flashovers. A relatively straightforward
adaptation of the pollution model has met with good success, especially for

Insulators for Icing and Polluted Environments. By Masoud Farzaneh and William A. Chisholm
Copyright © 2009 the Institute of Electrical and Electronics Engineers, Inc.

very light and heavy icing cases. The success of the same mathematical model for both cases is a result of the improved ability to establish fundamental arc characteristics on the ice layers under the relevant range of temperature and test conditions.

7.1. TERMINOLOGY FOR ICE

Applied Water Conductivity: The electrical conductivity of water used to simulate ice, snow, or cold-fog accretion on insulators, corrected to 20 °C.

Atmospheric Icing: Atmospheric icing is a complex phenomenon involving many basic processes affected by large variations over time and space and influenced by topography. It results from precipitation icing such as freezing rain and wet snow accretion or from in-cloud icing where suspended, super-cooled droplets freeze immediately upon impact on an object exposed to the airflow.

Conductivity: A factor σ that relates the conduction-current density (J) to the electric-field intensity (E). In general, σ is a complex tensor quantity. For an isotropic medium such as ice or snow, the tensor conductivity reduces to a complex scalar conductivity σ, in which case $J = \sigma E$.

Freezing Drizzle: Drizzle that falls in liquid form but freezes on impact to form a coating of glaze upon the ground and on exposed objects.

Freezing Rain: Rain that falls in liquid form but freezes on impact to form a coating of glaze upon the ground and on exposed objects.

Glaze Ice (Clear Ice): Type of precipitation icing resulting in pure ice accretion of density 0.7–0.9 g/cm³, sometimes with the presence of icicles underneath the wires. It very strongly adheres to objects and is difficult to knock off.

Hard Rime: Type of in-cloud icing resulting in ice accretion of density 0.3–0.7 g/cm³ characterized by a homogeneous structure with inclusions of air bubbles. It takes a pennant shaped aspect against the wind on stiff objects, but a more or less circular shape on flexible cables. It strongly adheres to objects and is more or less difficult to knock off, even with a hammer.

Hoarfrost: A type of low-density (<0.1 g/cm³), low-adhesion precipitation, with interlocking ice crystals (hoar crystals) formed by direct sublimation of moisture from saturated air below the freezing point onto objects of small diameter.

Ice Thickness: The radial thickness of ice accumulation measure on a monitoring cylinder.

Icing Stress Product (ISP): The product of the ice, snow, or rime accretion per centimeter of insulator length, and the electrical conductivity of the melted accretion corrected to 20 °C, (g/cm) · (µS/cm).

Liquid Water Content (LWC): The mass of water per unit volume of air, typically expressed in g/m^3 for natural and artificial fog and mist.

Melted Water Conductivity: The electrical conductivity of water used to simulate ice, snow, or cold-fog accretion on insulators, melted after accretion and corrected to $20\,°C$. The melted water conductivity can be higher than the applied water conductivity when accretion occurs on contaminated insulator surfaces. Melted water conductivity can also be lower than applied water conductivity for glaze ice accretion on clean insulators, through the process of water purification by crystallization.

Monitoring Cylinder: A cylinder for measuring reference ice accretion, typically 25–30 mm in diameter, either rotating at 1 rpm or fixed.

Return Period (of Icing Events): The predicted number of years between icing exposures that give a specified level of accretion on a reference cylinder, based on a statistical analysis of historical climate data and observations for a given location.

Supercooled: Cooled below the normal freezing temperature of a liquid, such that any drops freeze on impact with a solid surface.

Violet Arc: A continuous, luminous, and violet arc discharge of less than a few milliamperes that may initiate in air gaps between ice layers or from electrodes to ice layers. The violet arc may be quasi-stable, reaching the same extent for every ac cycle, or may move slowly along the surface of the ice.

White Arc: A continuous, luminous, and white arc discharge that may initiate between the electrode and an ice layer. The white arc has a current that exceeds a few tens of milliamperes. It may be quasi-stable, reaching the same extent for every ac cycle, or it can move with increasing speed across the surface of the ice. If the white arc traverses a large fraction of the dry arc distance, it evolves into a flashover arc.

7.2. ICE MORPHOLOGY

Morphology in general is the study of the shape and form of things. Ice morphology defines ice crystal shapes and sizes, and is important on several scales. Stabilization of the water molecules into the particular lattice structure of ice leads to the hexagonal symmetry of snowflakes. Other aspects of ice morphology have direct interest for electrical flashovers. Electrons pass through the bulk and across the surface of the ice layers, in the presence of contaminated ions or air inclusions, in ways that establish an alternate, and weaker, flashover path than the normal insulator.

7.2.1. Crystal Structure

At temperatures above $-100\,°C$ and standard pressures, ice forms in a hexagonal structure classed as Ih. The molecules of water in ice arrange themselves according to the following rules [Bernal and Fowler, 1933]:

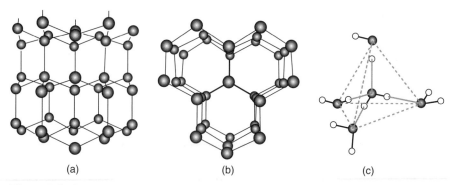

(a)   (b)   (c)

Figure 7-1: Crystal structure of Ih ice [Hobbs, 1974; Petrenko and Whitworth, 1999]. (a) Atoms of oxygen in matrix; (b) perpendicular view of the matrix; (c) 109° tetrahedral arrangement of oxygen (•) and hydrogen (∘). (Courtesy of Oxford University Press.)

- There are two hydrogen atoms adjacent to each oxygen atom.
- There is one hydrogen atom between every pair of oxygen atoms.

The rules force the arrangement of four neighboring oxygen atoms in a tetrahedral pattern of 109° with about 275 pm (picometers) between oxygen atoms. The tetrahedral angle is close to the natural 105° angle between hydrogen bonds in liquid water. Extension of the tetrahedral pattern leads to a hexagonal lattice as shown in Figure 7-1 [Hobbs, 1974; Petrenko and Whitworth, 1999].

7.2.2. Supercooling

Glaze ice is generally produced in saturated conditions with raindrops passing through a layer of air that is below 0°C. The heat exchange from the raindrops of relatively large diameter (100 μm) leads to supercooling to the ambient temperature, below freezing. As soon as the water drops come in contact with a surface, a certain mass dm will freeze on contact. Hobbs [1974] gives an expression for the fraction dm/m as

$$\frac{dm}{m} = \frac{T_s C_w}{L_f} \qquad (7\text{-}1)$$

where

dm is the mass of supercooled droplets that freeze on impact,

m is the mass of the drops,

T_s is the drop temperature, <0°C,

C_w is the specific heat of water, 1 cal/g, and

L_f is the latent heat of fusion of water, 79.7 cal/g.

Generally, the droplet size has a strong effect on the type of ice that forms. For example, laboratory studies show that drops of 15-μm mean diameter will freeze on contact to form rime ice that is white, opaque, of density 0.4–0.6 g/cm³, and with poor adhesion to surfaces. When the mean diameter is increased to 50–80 μm, the drops do not fully freeze on contact. The remaining liquid flows over the ice or insulator surfaces, forming ice caps. Depending on conditions, some liquid water may spill over the edges to form icicles. The ice that forms from large droplets in humid conditions is clear, has a density of 0.87–0.90 g/cm³, and has excellent adhesion to most surfaces, including Teflon™ and Kapton™.

The lattice structure of ice gives it a high degree of stability. This shows up as a high latent heat of melting (5987 joules per mole), a very high latent heat of sublimation (50,911 J/mol) and optical refractive index of 1.31.

7.2.3. Lattice Defects from Pollution

Many electrolytes dissociate fully or partially in water. Under freezing conditions, most of these ions are rejected to the liquid surface. This concentration process is known variously as freeze–thaw partitioning or purification by crystallization. Most of those ions that remain in solution form defects in the hexagonal ice crystal lattice structure, or inclusions of highly conductive liquid as found in seawater ice.

Hydrogen fluoride (HF) and ammonium (NH_3) can take the positions of water molecules in the crystal lattice, to form Bjerrum defects. While industrial release of HF acid is tightly controlled, NH_3 is widely used as fertilizer and is often noted as a contaminant on insulator surfaces. The role of the NH_3 Bjerrum defects on electrical conductivity of ice has not been studied extensively.

Weak solutions of strong electrolytes such as HCl [Young and Salomon, 1968] dissolve to some extent in ice. The small H^+ and Cl^- ions reside in the 0.275-nm spaces between oxygen molecules. In contrast, when NaCl dissolves in ice, Na^+ ions and an equivalent number of OH^- are pushed into the surface water layer, even though the sodium cations are smaller in diameter than chlorine anions.

7.3. ELECTRICAL CHARACTERISTICS OF ICE

The flashover process on an ice surface is a function of the distributed resistance per unit length of an ice layer. The ice layer builds up in parallel with the leakage resistance of an insulator. The electrical conductivity of the ice surface is much higher than the insulator and provides the main path for development of leakage current, electrical discharge, and local arcs that can evolve into flashovers. Ice conductivity is very sensitive to temperature around the freezing point, to the ion content of the applied water, and to the process of purification by crystallization that tends to force conductive ions to the ice surface as pure water is frozen into the bulk.

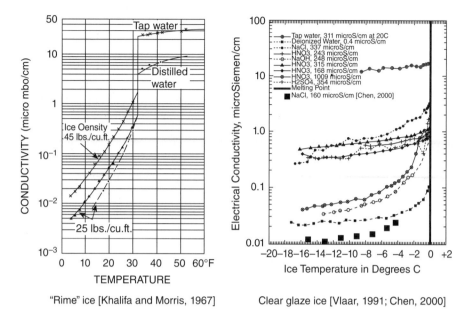

"Rime" ice [Khalifa and Morris, 1967]   Clear glaze ice [Vlaar, 1991; Chen, 2000]

Figure 7-2: Electrical conductivity of clean and contaminated bulk ice below 0 °C.

7.3.1. Conductivity of Bulk Ice

The power frequency electrical conductivity of ice varies considerably just below the freezing point. Vlaar [1991] measured a capacitive current in air-free ice at 60 Hz using a coaxial geometry. Figure 7-2 shows results using ice with varying conductivity and ion content in the freezing water for several researchers.

The ice conductivity increased by a factor of 7 as the temperature increased from –15 °C to 0 °C. Most of this change occurred in the narrow temperature range from –2 °C to 0 °C. However, peak conductivity of the ice samples at 0 °C remained a median of 187 times lower than the conductivity of the solution at 20 °C.

7.3.2. Conductivity of Ice Surface

Knowledge of the bulk electrical conductivity of the ice, σ_{ice}, is not sufficient to calculate the resistance per unit length $R(x)$ of an ice layer, to be used in pollution flashover modeling [Wilkins, 1969]. For many ice layer shapes, such as a cylinder or crescent accretion on an insulator, $R(x)$ has an important contribution from the surface water film conductivity σ_{water} at the melting point.

For a uniform water film on an ice layer with length L and width W, the overall resistance can be expressed as the parallel combination of the resistance of the bulk and surface layers as follows:

$$R = \frac{L}{W}\left[\frac{1}{\sigma_{ice}t_{ice} + \sigma_{water}t_{film}}\right] \qquad (7\text{-}2)$$

where

σ_{ice} is the conductivity (μS/cm) of the ice,

t_{ice} is the thickness (cm) of the ice layer,

σ_{water} is the conductivity (μS/cm) of the surface film of water, and

t_{film} is the thickness (cm) of the surface film of water on the ice.

The term $\sigma_{water}\,t_{film}$ is normally expressed as a product, defined as the equivalent surface conductivity γ_e. For typical ice layers, this term will dominate the equivalent bulk conductivity $\sigma_{ice}t_{ice}$.

Measurements of the resistance of melting ice using dc typically show a relaxation time constant on the order of 10 minutes as well as a resistance term that appears at power frequency. The power frequency value is important for both dc and ac flashover. Farzaneh et al. [1997] determined the equivalent surface conductivity γ_e from measured values of the resistance $R(x)$ of a thin triangular ice sample. These parameters are conveniently related by

$$R(x) = \int_{x+2}^{30} \frac{3 \times 10^6}{2x\gamma_e} dx = \frac{1.5}{\gamma_e} \ln\left(\frac{30}{x+2}\right) \times 10^6$$

$$(7\text{-}3)$$

$$\gamma_e = \frac{1.5}{R(x)} \ln\left(\frac{30}{x+2}\right) \times 10^6$$

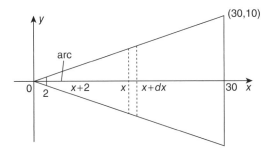

In Equation 7-3, the factor of 10^6 converts γ_e (μS) to siemens so that $R(x)$ is in ohms.

A series of tests were carried out to measure $R(x)$ on ice samples formed from melted water conductivity σ_{20} in the range of 1–200 μS/cm. Conversion of these values to the desired equivalent surface conductivity γ_e gave the relation shown in Figure 7-3:

$$\gamma_e = 0.0675\sigma_{20} + 2.45 \qquad (7\text{-}4)$$

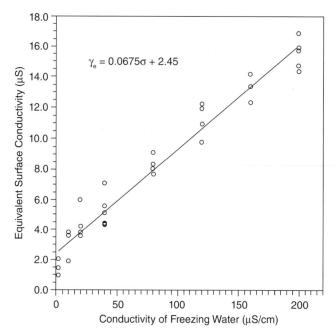

Figure 7-3: Relationship between equivalent surface conductivity γ_e and σ_{20}, conductivity of the water used to make the triangular ice surface (from Farzaneh et al. [1997]).

where

γ_e is the equivalent surface conductivity (μS), and

σ_{20} is the freezing water conductivity (μS/cm), corrected to 20 °C.

An icing stress product (ISP) model [Chisholm and Farzaneh, 1999] assumes that the relation in Figure 7-3 passes through the origin. For the practical range of freezing water conductivity causing flashovers, over 50 μS/cm, this is a reasonable simplification but there are some deviations for low values of σ_{20}.

The relation between γ_e and σ_{20} on a well-controlled geometry can be applied to calculation of the resistance per unit length of other geometries. For an approximation to a half-cylinder of ice that forms around a typical insulator, the resistance $R(x)$ is [Wilkins, 1969; Farzaneh and Zhang, 2007]

$$R(x) = \frac{1}{2\pi\gamma_e}\left[\frac{4(L-x)}{D+2t} + \ln\left(\frac{D+2t}{4r}\right)\right] \tag{7-5}$$

where

L is the length (dry arc distance) (cm) of the half-cylinder of ice,

D is the diameter (cm) of the insulator,

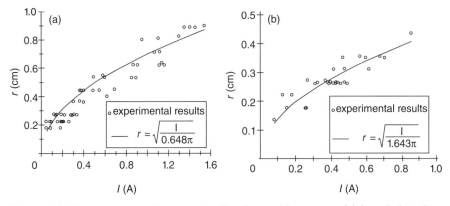

Figure 7-4: Measurements of arc root radius for positive arc on (a) ice–air interface and (b) arc inside ice layer (from Zhang and Farzaneh [2000]).

TABLE 7-1: Arc Root Radius as Function of Leakage Current[a]

Radius	Positive Arc	Negative Arc	Alternating Arc
Inner arc	$r = 0.440\sqrt{I}$	$r = 0.425\sqrt{I}$	$r = 0.361\sqrt{I_m}$
Outer Arc	$r = 0.701\sqrt{I}$	$r = 0.714\sqrt{I}$	$r = 0.603\sqrt{I_m}$

[a]I (A), I_m (peak A), r (cm).

Source: Adapted from Farzaneh [2000].

t is the thickness (cm) of the ice layer, and
r is the radius of the arc root on the ice surface.

The arc root radius r on an ice layer is typically about 0.3 cm [Farzaneh et al., 1997] with a 4:1 variation with arc current as shown in Figure 7-4.

The arc root radius also varies with polarity as shown in Table 7-1.

With the combination of the power frequency experimental results given here and the mathematical models [Wilkins, 1969], it is possible to compute $R(x)$ for a wide range of ice shapes on insulators or in experimental geometries.

7.3.3. High-Frequency Behavior of Ice

Bernal and Fowler [1933] showed that the dipole moments of individual water molecules have random orientation. This means that the hexagonal crystal structure does not restrain the ability of the water molecules to reorient in an electric field. According to Fletcher [1970], this explains why the low-frequency dielectric constant of polycrystalline ice at −10 °C is 97–98, compared to 80.3

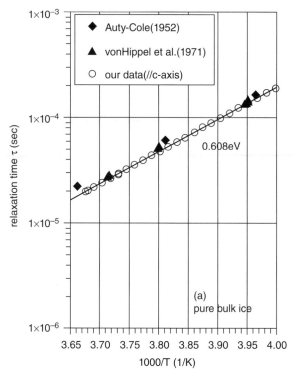

Figure 7-5: Relaxation time of ice as function of inverse temperature with 1000/(273 K) = 3.65 and 1000/(263 K) = 3.80. (from Takei [2007]; courtesy I. Takei).

for water at 20 °C and 87.7 for water at 0 °C. The dielectric relaxation time (τ) affects the high-frequency response of ice. Takei [2007] showed that τ changes from 20 μs at 0 °C to 60 μs at −10 °C as shown in Figure 7-5.

The relaxation time τ leads to maximum dielectric losses at 3 kHz at −10 °C and to a drastic reduction in dielectric constant at 100 kHz, of about 3.2. Usually, dielectric data for materials like contaminated water, ice, or soil are reported using a complex permittivity k^* defined as follows:

$$k^* = k' - jk'' = k_\infty + \frac{\Delta k_1}{1 + (j\omega\tau_1)^{\alpha_1}} + \frac{\Delta k_2}{1 + (j\omega\tau_2)^{\alpha_2}} - j\frac{\sigma}{\omega\varepsilon_0} \qquad (7\text{-}6)$$

where

k' and k'' are the real and imaginary parts of the permittivity,
j is the imaginary unit $\left(\sqrt{-1}\right)$,
ε_0 is the electrical permittivity of vacuum, 8.854×10^{-12} F/m,
ω is the angular frequency ($2\pi f$),

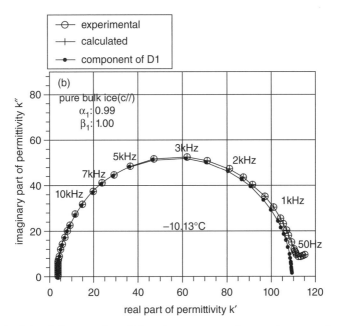

Figure 7-6: Cole–Cole complex dielectric response of pure ice crystal at –10 °C (from Takei [2007]); courtesy I. Takei).

Δk_i are the dispersion strengths of poles,

τ_i are the relaxation times, and

$\alpha_i = 1$ for Debye-type response and $0 < \alpha_i < 1$ for the Cole–Cole type.

The response plot of Cole and Cole [1941], plotting k' against k'', is commonly used in electrochemistry, biology, and geophysics. Takei [2007] shows in Figure 7-6 that ice at –10 °C has a simple Debye-type response.

The residual value of k'' at low frequency gets higher, and the plot is less circular, when the ice sample has significant conductivity σ in Equation 7-6. Snow and hoarfrost also have more complicated Cole–Cole plots with significantly lower values of k' and nonzero k'' at 100 Hz.

7.4. ICE FLASHOVER EXPERIENCE

In Chisholm [1997], a survey of power system operating problems in winter conditions noted that there were 16 separate incidents, severe enough to be reported to the National Electric Reliability Council (NERC) as significant events in annual reports such as NERC [1990]. The median significant event had line-to-ground flashovers at a 345-kV station with 650 MW of load loss, affecting 180,000 customers for a minimum of 50–150 minutes.

A CIGRE survey [Yoshida and Naito, 2005] of 35 utilities in 18 countries indicates that ice and snow electrical flashover problems occurred worldwide, with 83 events (69 line, 14 substation) reported by 17 of these utilities on their EHV transmission lines from 400 to 735 kV in relatively clean conditions.

Ice and snow flashover problems are not unique to EHV transmission voltage levels. In stations, 115 and 230-kV insulators have performed poorly in contaminated areas. Traditionally, freezing fog episodes also have led to widespread pole fires on distribution lines as the insulators carry small leakage currents (5–20 mA) during extended periods.

In addition to these utility surveys, there have been many individual reports of problems that are grouped here in terms of very light, light, moderate, or heavy icing as given in Tables 7-2 and 7-3.

A visual guide to the ice accretion levels in laboratory tests, based on ice accretion t in Table 7-2, is found in Chapter 9.

In Québec, the classification of icing flashovers includes those associated with insulator phase-to-ground faults as well as those associated with phase-to-phase clashes of conductors during galloping or ice release. Galloping is a large-amplitude skipping motion of the entire span and can reach amplitudes nearly equal to the conductor sag. Sleet jump is a sudden increase in conductor height when a large section of ice cover falls away, which happens sooner for the phase conductors than for the overhead groundwires. In the years 1981–1988, for example, 200 short circuits were attributed to ice or snow events. Among these, about 40 were attributed to insulator flashover [Drapeau, 1989].

7.4.1. Very Light Icing

The earliest reports about poor 132-kV insulator performance under very light icing were given by Forrest [1936]. Under fog conditions in polluted areas, a

TABLE 7-2: Classification of Ice Accretion Severity Based on Energized Porcelain Insulators

Ice Accretion Level t	Approximate Ground-Level Ice Accumulation	Reference Ice Level on Rotating Cylinder	Deposit on Uniform-Profile 300-mm Diameter Station Post with 50-mm Shed Spacing	Deposit on Standard 146-mm × 254-mm Cap-and-Pin Insulator String
Very light	<2 mm	<1 mm	No icicles; thin ice layer on all surfaces	No icicles; thin ice layer on all surfaces
Light	<12 mm	<6 mm	Partially bridged with icicles	Partially bridged with icicles
Moderate	12–20 mm	6–10 mm	Fully bridged with icicles	Partially bridged with icicles
Heavy	>20 mm	>10 mm	Fully bridged with icicles	Fully bridged with icicles

TABLE 7-3: Typical Utility Experience with Flashovers in Icing Conditions

Location, Date	IceType	Melting Phase	Accretion Level	Surface Precontamination	Altitude	System Voltage
UK, 1935–1936	Hoarfrost	Yes	Very light	Heavy	Low	132 kV
UK, 1962–1963	Hoarfrost	Yes	Very light	Heavy	Low	275 kV
Milan Italy January 27–28, 1964	Winter	—	—	Heavy	Low	145 kV
Ohio, USA 1980–1983	Freezing rain	Yes	Heavy	Unknown	Low	345 and 765 kV
Québec, Canada 1981–1988	Freezing rain	Some	Heavy	Light	Low	Mostly 735 kV
Ontario, Canada, 1986	Ice	Yes	Light	Moderate	Low	500 kV only
Chicago, IL USA 1986	Fog	Yes	Very light	Heavy	Low	138-kV breaker bushings
Norway, 1987	Thin rime ice and fog	Mostly Yes	Very light	Heavy	Medium	300 kV
Yugoslavia, 1989	Contaminated ice/fog	Yes	Heavy	Heavy	600 m	400 kV
Ontario, Canada, 1989	Ice	Yes	Light	Heavy	Low	230 and 500 kV
Norway, 1993	Thin rime ice, fog	Yes	Very light	Yes	Medium	300 and 420 kV
Idaho, USA 1994	None stated	Yes	n/a	Yes	1400 m	345 kV
Ontario, Canada, 1994	Freezing drizzle	Yes	Very light	0.04 mg/cm^2	Low	230 kV
Ontario, Canada, 1994	Ice	Yes	Light	Heavy	Low	115 and 230 kV
Ontario, Canada, 1997	Contaminated ice (road salt)	Yes	Moderate	None	Low	230 kV semiconducting glaze insulators
Czech Republic 2002	Contaminated ice (cooling tower)	n/a	Heavy	Heavy	Low	110, 220, and 400 kV
Alberta, Canada 2006	Freezing fog	−2°C	Very light	Long period without rain	Medium	500-kV line
China, 2000–2007	Heavy ice, moderate conductivity	Yes	Heavy	Not a factor	High	HVDC

condition he labeled "fog surging" was taking place. Steady values of leakage current less than 1 mA were observed on insulators, but at irregular intervals, the leakage current increased suddenly to levels of up to 100 mA for fractions of a second. Forrest noted that the worst fog surging took place for combinations of fog and frost, and for cases where melting snow was blown against polluted insulators by high winds. The characteristic of quiescent leakage current for many minutes, with rapid buildup of current leading to flashover or flash evaporation of the ice layer, is a recurring feature of insulator contamination response under cold-fog conditions.

Jolly [1972] gave a detailed description of cold-fog-type flashover problems, with thin ice layers stabilizing pollution on all surfaces of insulators without significant growth of icicles as follows:

> As transmission voltages have increased the contamination problem has become worse, despite intensive research on the problem. In the winter of 1962–63, for example, the British National Grid suffered the most serious dislocation in its history. An abnormally dry autumn had allowed airborne deposits to accumulate on insulator surfaces. For several nights in succession during January there were freezing fogs, and hoarfrost formed on the insulator surfaces. On the evening of January 25, 1963, a thaw set in and the frost melted, forming a conducting film on the insulator surface. There were 130 flashovers and the normally interconnected grid was split into four separate sections. Four thousand MW of load was disconnected of a total demand of 23,000 MW.

Jolly also reported that there were 281 line flashovers and 122 station flashovers in the two months of December 1962 and January 1963, attributed to these cold-fog and pollution problems. There had been an average of 9 line and 7 station problems in the previous 5 years. Forrest [1969] also noted that there had been many outages on 400-kV lines in England, caused by a combination of ice and contamination.

Commonwealth Edison [Schaedlich, 1987] reported that 200,000 customers were blacked out for 12 hours starting in the early morning of December 23, 1986 as a result of 14 flashovers on 138-kV breaker bushings. Ten of the flashovers punctured the bushings. The station was in a contaminated area and was subject at that time to a program of insulator washing. Wind speed was 13 km/h and visibility was 400 m at the time of the blackout. The flashovers were preceded by three consecutive days with fog. Problems actually started the previous evening when the temperature reached 0–1 °C with dew point temperature of –1 °C.

From 2300 on January 12, 1994 to 0300 the next day, there were four hours of freezing drizzle in the Toronto, Ontario, Canada area that led to numerous flashovers at an urban 230-kV transformer station located next to a major 16-lane expressway. Temperature during this period ranged between –0.7 °C and 0.1 °C, with dew point at a median –1.8 °C. There had been rain 12 days earlier, and the dry period led to insulator ESDD levels of about 0.04 mg/cm^2 (40 µg/cm^2).

7.4.2. Light Icing

Gorski [1986], Schaedlich [1987], and Boyer and Meale [1988] described the weather leading up to a series of 57 flashovers on the Ontario Hydro network. There was accumulation of about 7 mm of mixed freezing rain over a period of 10 hours. This was not sufficient to bridge transmission line insulators with icicles, but ice bridging was reported on post insulators at one 500-kV station. Typical ice and fog levels during this event are shown in Figure 7-7.

Eight hours after the accumulation, there had been six flashovers on the 500-kV system with none on lower-voltage networks. As the ambient and dew point temperatures passed through 0 °C, there were a total of 24 flashovers in the first hour and 27 more in the next two hours. Overall, line-to-ground insulation faults occurred in roughly equal numbers, 30 on 500-kV lines and 27 at stations.

Analysis by Farzaneh and Melo [1990] compared two ice storms of equal intensity with light ice accretion. One caused 500-kV flashovers and the other did not. Figure 7-8 shows the important features of the melting phases on March 10, 1986 and March 13, 1988.

After dawn on March 10, 1986, Figure 7-8 shows that most flashovers occurred as temperature increased above freezing, with the relative humidity remaining near saturation and visibility in fog of less than 3 km. In contrast,

  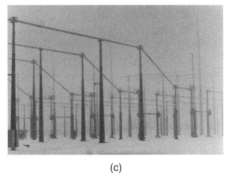

(c)

Fog conditions at 500-kV station,
Trafalgar, ON

Line: Bruce to Milton, ON   Station Post: Milton, ON

Ice accretion on 500-kV Insulators after Flashover

Figure 7-7: Weather conditions and ice accretion on 500-kV system, March 10, 1986. (Courtesy of Kinectrics.)

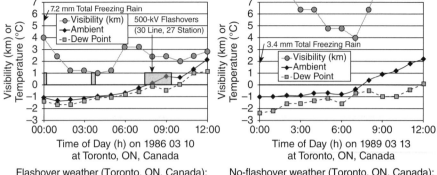

Flashover weather (Toronto, ON, Canada):   No-flashover weather (Toronto, ON, Canada):
500-kV system affected after 390 h without rain   No outages after 86 h without rain

Figure 7-8: Comparison of weather events that did, and did not, lead to widespread 500-kV power system outages after freezing rain.

sunrise on March 13, 1988 led to a divergence of temperature and dew point with visibility greater than 7 km.

There was another important weather factor that distinguished these two events. Previous to March 10, 1986, there was a period of 17 days (390 h) without rain, compared to a period of 86 h prior to March 13, 1988. This means that contamination levels on March 10, 1986 may have been four times higher than levels on March 13, 1988, based on the linear increase of ESDD with exposure duration observed for this area in later field studies.

While buildup of contamination over a long period of time may be a necessary condition for flashovers, it is not sufficient without a melting phase in fog. One important role of fog in March 10, 1986 was the accretion of a thin ice layer on all (top and bottom) surfaces of insulators, which were not fully bridged by icicles. This, combined with the moisture provided by fog to keep the air saturated during the melting phase, led to stable wetting of the surface pollution that reduced the withstand strength, to levels well below the ac service voltage stress. Figure 7-9 shows that a one-degree difference leading to rain before fog can make all the difference between 4.3 system-minutes of outages and none.

The flashover weather that was predicted to cause insulator flashovers, but did not, occurred on February 18, 1991. With 785 hours without rain, problems similar to December 31, 1989 were anticipated when a freezing rainstorm moved into the area. These did not materialize. With a minor shift in ambient temperature over the course of the day, there was no significant accumulation of freezing rain. Instead, rain started at an ambient temperature of +1 °C as shown in Figure 7-9. The period of dense fog starting 2 hours later had no effect. Overnight on December 30 and 31, 1989, in Figure 7-9, 9 hours of freezing rain occurred with fog and dew point within 0.6 °C of ambient. The rate of freezing rain accumulation was relatively low. Photographs of the resulting ice accretion in Figure 7-10 show that there were icicles but no ice

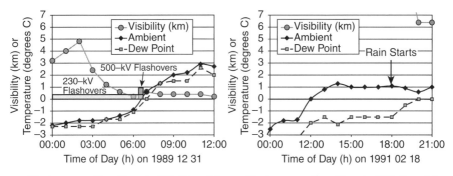

Figure 7-9: Comparison of weather events that did, and did not, cause flashovers at melting point.

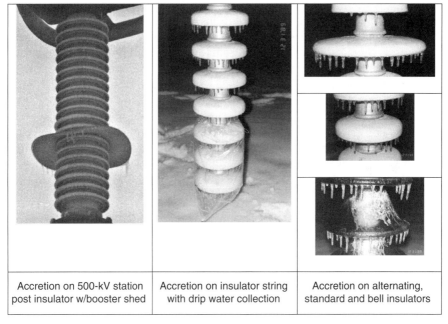

| Accretion on 500-kV station post insulator w/booster shed | Accretion on insulator string with drip water collection | Accretion on alternating, standard and bell insulators |

Figure 7-10: Observations of ice accretion on insulators, December 31, 1989 (from Chisholm and Tam [1990]; courtesy of Kinectrics).

bridging on either station post or suspension insulators. The wind speed was a median of 8 km/h during this period, and this resulted in formation of vertical icicles that did not bridge the dry arc distance of post or cap-and-pin insulators.

There had been a period of 33 days without rain, from November 28, 1989, prior to the event along with cold weather and several snow events. Outages

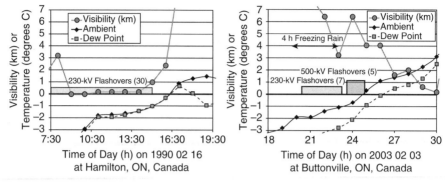

Flashover weather (Hamilton, ON, Canada):
3.4 system-minutes lost at high-contamination
230-kV station after 34 h without rain

Flashover weather (Buttonville, ON, Canada):
500-kV station had estimated ESDD = 15 μg/cm²
after 620 h without rain

Figure 7-11: Significant flashover events with temperature near melting point.

occurred mainly at transformer stations near urban expressways where extensive road salting was underway [PSOD, 1990]. This was confirmed by comparing chemical analysis of ice samples with ion content of drip water during the initial melting phase, collected as shown in Figure 7-10. More than 90% of the ion content in the melted ice was the result of surface precontamination with road salt.

The concept and term of "sunshine melting" is not appropriate for these flashover conditions as the solar input in dense fog is negligible. Two other significant flashover events, in Figure 7-11, show that the melting phase causing serious problems with repeated insulation flashover can occur at any time.

High relative humidity and the presence of visible fog near 0 °C are thus the weather factors that play a fundamental role in the susceptibility of both 230- and 500-kV insulators to flashovers under light icing conditions.

7.4.3. Moderate Icing

An example of moderate ice accumulation that illustrates the role of temperature in the flashover process occurred early in February 1992. A total of 10 mm of ice was measured at ground level at the Toronto Airport. Ice also accumulated on 500-kV insulators to the north at higher elevation on February 14, and caused a series of seven line-to-ground faults at ambient and dew point temperatures of ±1 °C. Then, the temperature dropped. The ice hardened and remained intact on the insulator strings for 4 days (Figure 7-12). This was enough time to allow for a complete photographic record of accretion at the faulted towers, which were marked in two cases by damaged insulators and a dropped bundle conductor.

As dew point and ambient temperatures warmed to above-freezing conditions 4 days later, there were two additional line-to-ground flashovers. This

Figure 7-12: Ice accretion on standard 500-kV insulator string during icing event, February 14–18, 1992 (from Farzaneh et al. [2007b]).

was a definitive indication of the sensitivity of EHV insulator flashover to melting and led to the adoption of testing methods that simulate a melting phase with high relative humidity.

7.4.4. Heavy Icing

Accretion of 10–30 mm of ice on electrical insulators can lead to flashover, both during the icing period and later during a melting period, at electrical stresses well below 100 kV/m.

What Khalifa and Morris [1967] called "rime" ice accretion with a density of 20–50 lb/ft^3 (0.32–0.8 g/cm^3) was stated to have caused some transmission line outages in winter seasons in the West Coast areas of Canada and the United States. It was also noted that a reduction in the operating voltage from 315 to 280 kV kept the lines in service under adverse winter conditions.

According to Kawai [1970], transmission outages were experienced in winter in northern, southern, and West Coast states in the United States. In a West Coast area, some outages were reported on a warm morning after an ice storm. In the Northeast, flashovers were recorded during snow or ice storms. A major disturbance at an EHV substation in the Southern area was reported as caused by an ice flashover.

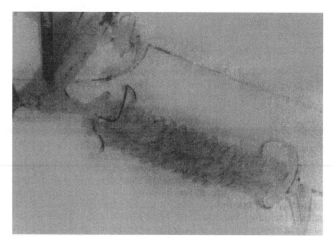

Figure 7-13: Photograph of ice accretion on pair of strain insulators (from Watanabe [1978]).

Watanabe [1978] illustrated heavy ice bridging on horizontal insulators. While the photo quality is poor, it is clear that there are no icicles, only heavy glaze ice caps on the top of each disk (Figure 7-13). This is a characteristic of glaze ice that forms at the relatively low temperature of −7 °C to −10 °C, similar to some episodes of heavy icing including the 1998 ice storm in Québec.

The Detroit Edison Company [Charneski et al., 1982] became concerned with icing flashovers in 1976 when an ice storm caused numerous 345-kV station post insulators to flash over in a station with low contamination levels. In the same station, adjoining 120-kV buses with lower voltage stress per meter of dry arc distance remained in service—flashover free. Similar problems were experienced on February 1, 1980 when their Monroe Power Plant 345-kV switchyard insulators were subjected to icing from freezing rain. It was suggested by Charneski et al. [1982] that the 345-kV station post insulators were prone to icing flashovers because their voltage stress was 43% greater than the stress on the 120-kV insulators.

Vuckovic and Zdravkovic [1990] noted that eastern Serbia was a region of frequent and heavy ice accretion, accompanied by high levels of air pollution. Their studies of polluted ice or wet snow, removed from 400-kV system insulators, showed a median conductivity of 200 μS/cm, compared to a median of 37 μS/cm for the natural precipitation, with sulfates being the dominant ions.

Shu et al. [1991] mentioned flashover incidents observed on EHV lines in China caused by the joint effects of ice and pollution, compounded by low atmospheric pressure at high altitude.

Matsuda et al. [1991] reported a survey of flashover problems caused by ice on 154- and 275-kV networks in Japan. The investigation of the electrical performance of transmission line insulators covered with snow or ice began in Japan in 1956 as a result of electric faults in the 154-kV Tadami transmission system of the Tohoku Electric Power Company (TEPCO) Inc. In January 1981, faults extending over nine TEPCO transmission lines occurred simultaneously in the 154-kV Niigata system, on the west coast of Japan. Identical faults recurred in this system and in five transmission lines of the 154-kV Akita system, just to the north, in January 1985. In addition, a ground fault occurred on a 275-kV transmission line in February 1985. Inspection of these fault lines revealed that Niigata's faults were due to horizontal tension insulator assemblies covered with a large amount of snow, while Akita's faults were due to suspension assemblies coated with ice.

Fikke et al. [1993] mentioned the destructive effect of cold precipitation and insulator contamination on cap-and-pin insulators in Statnett, the Norwegian grid company. The damage was caused by flashover and arcing. On February 8, 1987, a permanent phase-to-ground failure occurred on a 300-kV line in southwestern Norway, about 120 km from the seacoast. A total of 106 insulator disks were exchanged after inspections found arc burn damage on porcelain surfaces. This observation of arcing activity and flashovers with thin rime ice layers was noted to be different from the many cases with thicker accretions without similar failures, leading to questions about the role of the ion content of the ice in a mountain region, with no local pollution sources and too far from the coast to be affected by sea salt. There were other problems in this network in 1993.

Farzaneh and Drapeau [1995] reported that, out of four winter seasons in Québec from October 1989 to April 1992, seven series of valid ice accumulation observations were made on various parts of the Hydro-Quebec network. The observations led to the identification of two forms of ice accretion: the icicle form and the curtain form, the latter being an extension of the former.

Figure 7-14 shows the curtain form of accretion as it appeared on a 161-kV vertical post insulator and a 735-kV horizontal quadruple string. The curtain

| Curtain form of accretion on a 161-kV potential transformer | Icicle form of accretion on a dead-end insulator of four chains of disks for 735-kV line |

Figure 7-14: Ice accretion on Hydro-Québec insulators (from Farzaneh and Drapeau [1995]).

form of accretion was observed only once out of seven events and resulted from a freezing rain precipitation of 30–40 mm at ground level, at an ambient air temperature of about –1 °C.

7.5. ICE FLASHOVER PROCESSES

The widespread but sporadic history of flashovers under icing conditions gives rise to many questions about which factors are important [Kawai, 1970]:

- Is a precontamination condition required to cause flashover of the insulator?
- If this precontamination is necessary to explain the flashover, then what type of contamination is predicted in mountainous areas where ice flashovers have been reported?
- What is the mechanism of ice formation on long insulator strings under voltage application?
- What is the effect of leakage current over the insulator string on the type of ice formed?

Many of these questions have been addressed and answered by a combination of laboratory testing and ongoing field experience. For example, in a survey of the literature, Farzaneh and Kiernicki [1997] refined the factors that influence service-voltage flashover under heavy icing conditions as follows:

- What type of ice—wet or dry grown, with ice caps or icicles, with what density—is accreted on the insulators?
- How uniform is the ice accretion? For long insulators, how many air gaps appear when icing takes place under service-voltage conditions?
- How much ice builds up?
- What is the electrical conductivity of the ice layer and the water that drips from the surface?
- What is the effect of voltage polarity in dc conditions?

It is now realized [CIGRE TF 33.04.09, 1999] that flashover on an ice surface is an extremely complex phenomenon resulting from the interaction between the following factors:

- Electric field distribution.
- Wetness and contamination of the ice surface.
- Number and position of the air gaps.
- Environmental conditions.
- Shape and dimensions of the ice layer on the insulator geometry.

TABLE 7-4: Typical Input Parameters in Modeling of Flashover in Icing Conditions

Parameter	Unbridged Length $(L - x)$		Maximum Number of Air Gaps N_a per Meter of Dry Arc Distance	
	Station Posts and Polymers	Standard Cap-and-Pin Disks	Station Posts and Polymers	Standard Cap-and-Pin Disks
Shed spacing	50 mm	146 mm	50 mm	146 mm
Very Light	L = Leakage distance		n/a	n/a
Light			$N_a \approx 20$	$N_a = 7$
Moderate	L = Dry arc distance	L = Leakage distance	$N_a \approx 1$	$N_a = 7$
Heavy	L = Dry arc distance		$N_a \approx 1$	$N_a \approx 1$

Among these factors, ice shape, density, amount, and distribution, as well as applied water conductivity, have the greatest influence on the flashover voltage of ice-covered insulators. These factors are used in the calculation of the residual resistance $R(x)$ of the fractions $(L - x)$ of the ice layer that are not bridged by arcs, with x being the location of the arc root length along the layer and L being defined in Table 7-4.

The number of air gaps N_a for a particular ice layer is a factor that also affects the flashover voltage. The number of gaps depends on the insulator shape, shed-to-shed spacing, voltage, and other parameters. Table 7-4 also shows how the maximum value of N_a varies with icing level for two main types of insulators.

7.5.1. Icing Flashover Process for Very Light and Light Ice Accretion

In its frozen and poorly conducting state, accretion of pure cold fog, soft rime, hard rime, or dry snow onto contaminated insulator surfaces can be treated to first order with the addition of a nonsoluble deposit density (NSDD) to a preexisting soluble deposit density (ESDD). Arguments in favor of this view [Farzaneh et al., 2007a] are the following:

- These types of atmospheric ice do play a strong role in stabilizing the ESDD, just as found for NSDD.
- A thin, pure cold-fog layer of typically 0.05 mm will add an NSDD of 5 mg/cm^2, which is at the upper limit considered by IEC recommendations for characterizing site pollution severity [IEC Standard 60815, 2008].

The calculation process for very light and light icing flashover consists of computing the resistance per unit leakage distance.

Under light icing conditions, the leakage current passes across the ice cap and then tends to pass between icicles to the bottom surface of the insulators.

The pollution that was on the surface of the insulator is normally rejected to the surface of the ice layer through freeze–thaw purification. This leads to an electrical resistance per unit length that is very similar to the calculations that model the very light accretion condition. The main effect of a light ice layer is to stabilize the temperature of the insulator by adding substantial thermal mass.

7.5.2. Icing Flashover for Moderate Ice Accretion

The moderate ice deposit changes the shape of the insulator and redistributes the pollution from some insulator surfaces to the ice surface. Under moderate icing conditions, the ice cannot be modeled simply as a form of stabilizing deposit [Farzaneh et al., 2007a]:

- Wet ice growth may dissolve the contamination, and some soluble ions will be rejected to the ice surface. In this process, some of the soluble contamination may flow away and refreeze on icicle surfaces. This process leads to a highly nonuniform distribution of contamination, concentrating most ions in the icicles and leaving the iced surface of the insulators relatively clean.
- Even a thin ice accretion of 1 mm represents an additional NSDD of $100 \, mg/cm^2$, which is well above the upper limit of $4 \, mg/cm^2$ considered by the IEC Standard 60815 [2008].
- Ice or snow bridging under heavier 6–10-mm accretion will effectively change the insulator shape and dimensions.

The flashover model for the case of moderate icing must be selected based on the question of whether the particular insulator is bridged with icicles. The cases that cause the greatest risk of flashover for moderate ice accretion have included:

- A single, thin 3–10-mm diameter icicle with high conductivity that bridges most of the dry arc distance of the insulator or bushing. The icicle may hold up to 90% of the surface contamination that was originally on the top surfaces of the insulator.
- Accretion that has 10–30 icicles, each of which partially or fully bridges the shed-to-shed distance of station post insulators, as illustrated in Figure 7-15.

An icicle–ice plane gap at a temperature of −5 °C has a corona inception behavior that is quite different from that of a metal electrode with the same separation and 2.5-mm radius of curvature. However, near the melting point, the corona inception behavior changes. Under positive dc at 0 °C, inception

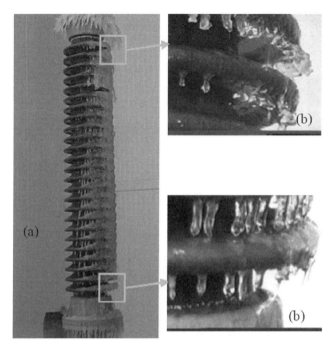

Figure 7-15: Partial discharge along station post insulator with moderate icing: (a) ice-covered insulator with air gaps and (b) inception of corona discharges in ice-free zones (from Yu et al. [2007]).

voltage increased from 19 to 24 kV dc for an icicle–ice plane gap of 3–9 cm. The repetition rate of impulses was also similar for ice near $0\,°C$ and for the metal electrode of the same shape and gap size.

The icicle–ice plane gap has an average pulse charge of 600 pC for a 5-cm space, compared to about 1400 pC at the same voltage and separation for the metal electrode. If the ice temperature is reduced to $-5\,°C$, the pulse repetition rate and 200-pC charge per pulse drop significantly compared to the metal electrode [Yu et al., 2007].

Flashover on insulators with moderate ice accretion will follow the process for heavy ice accretion, once corona inception activity evolves into formation of local arcing activity.

7.5.3. Icing Flashover Process for Heavy Ice Accretion

According to most observations and reports, flashover of heavily iced insulators is caused by the formation and development of local arcs along air gaps as shown in Figure 7-16.

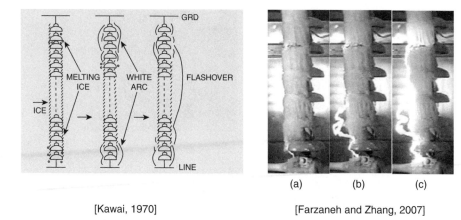

(a)　　　(b)　　　(c)

[Kawai, 1970]　　　　　　　　[Farzaneh and Zhang, 2007]

Figure 7-16: Steps in flashover process of suspension disk insulator.

Generally, heavy ice accretion along an insulator is not uniform if it is energized at service voltage. In addition, ice is formed only on the windward side of the insulator if the wind speed is high during either natural or artificial icing conditions. Some parts of the insulator tend to remain free of ice. These ice-free air gaps, occurring about once every meter of dry arc distance as noted in Table 7-4, are caused by the heating effect of discharge activities, ice shedding, and ionic wind effects.

While the ice surface is cold and thus dry, the electrical performance of the insulators is not significantly reduced. However, the presence of a highly conductive water film on the surface will shift the potential distribution along the ice layer so that most of the applied voltage appears across the air gaps. The result near the maximum withstand voltage level, V_{WS}, is a quasi-stable electrical discharge in these areas with minutes of local violet arc activity as shown in Figure 7-16a.

If the applied voltage is higher than V_{WS}, the local violet arc can change within seconds to a white arc [Hara and Phan, 1979] and extend along the ice surface as shown in Figure 7-16b. When the arc reaches a critical length of about two-thirds of the dry arc distance, this may result in a complete flashover, shown in Figure 7-16c.

7.6. ICING TEST METHODS

The in-service problems with icing conditions differ noticeably from one area to the next. Variables include the length of the icicles, the weight of the ice deposits on the insulators, or the thickness of the accumulated ice on a monitoring cylinder. Due to such difficulties and, from a practical point of view, the impossibility of controlling experimental parameters during field studies, most of the research work in this section was the subject of laboratory investigations.

Laboratory investigations made it possible to simulate most of the ice types built up on insulators as observed in the field observations. When high-voltage laboratories were modified to do this, studies on the flashover performance of ice-covered insulators proceeded with many advantages when compared with field studies or outdoor exposure tests:

- Research work is able to proceed without any interruption due to the changing seasons.
- Simulated atmospheric conditions can be controlled precisely.
- The effect of each parameter on the electrical behavior of insulators may be investigated separately.

The historic test methods to simulate each problem had different approaches as well. In some cases, only a single voltage such as service voltage stress had been applied, so that the maximum withstand voltage is not known. In other reports, the methods used to establish withstand and critical flashover levels were identical to those used for contaminated insulators.

7.6.1. Standard Electrical Tests of Insulators

The evolution of artificial icing test methods took advantage of the existence and concurrent development of standard test methods for pollution and heavy rain flashovers. IEEE Standard 4 [1995] and IEC 60060 [1989] now define standard electrical test methods for insulators under three types of contamination test, along with the technical requirements such as source impedance, pollution levels, and other factors that ensure repeatability. These are:

- Salt-fog tests, where salt aerosol with conductivity ranging from 4300 to 200,000 μS/cm is sprayed onto energized insulators.
- Clean-fog tests, where precontaminated insulators with salt deposit density of 25–400 μg/cm^2 are energized and then wetted with steam or water aerosol.
- Heavy rain tests, where water with controlled 100-μS/cm conductivity and rain rate is applied for a period of several minutes, then the insulator is energized.

The icing test methods are in the process of standardization at this time [IEEE PAR 1783, 2008].

The use of salt-fog and clean-fog test methods in normal and freezing conditions was discussed in Chapter 4. Chapter 10 will show how the polymer and ceramic insulator test results are used in selecting appropriate insulator leakage distance.

Of the three standard tests, the heavy rain test is closest to a method that could be adapted to simulate the deposit of glaze or rime ice in freezing conditions. With this approach, existing test levels, equipment, and setup experience can reduce setup time and costs. As an example, laboratories may already provide a suitable quantity of water with $\sigma_{20} = 100\,\mu S/cm$ to their simulated rain systems. This same conductivity is recommended as part of a standard icing test defined by the IEEE [Farzaneh et al., 2003; IEEE PAR 1783, 2008] as it is both practical and representative. Power supply requirements to ensure adequate voltage regulation under arcing conditions are also carried over.

7.6.2. Standard Mechanical Ice Tests for Disconnect Switches

In addition to electrical tests, some switchgear are also tested with icing to ensure proper function of outdoor disconnect switches under icing conditions. Outdoor disconnect switches are tested using IEEE C37.34 [1994] and IEC 62271-102 [2001].

The procedures for measuring ice accretion on reference cylinders are recommended for electrical tests. Measurements of radial ice thickness on a fixed reference cylinder are used to establish clear ice-breaking ratings in increments of 6.3 mm (¼ inch) starting at 12.6 mm (½ inch), 19 mm (¾ inch), and so on. For electrical tests, ice accretion should be measured on two separate 25–30-mm diameter cylinders, one fixed as in IEEE C37.34 [1994] and one rotating at 1 rpm at the same exposure location.

The inclusion of an ice-hardening phase is a second example of a phase in the mechanical tests that is also recommended [Farzaneh et al., 2003; IEEE PAR 1783, 2008] for the electrical performance tests.

7.6.3. Natural Icing Tests in Outdoor Test Stations

Most insulator test stations are located in temperate areas near the sea. The winter conditions that lead to flashovers on contaminated insulators also include accumulation of freezing rain or wet snow. It has thus proved to be impractical to establish insulator response to repeatable icing conditions from most insulator exposure sites. Specialized sites have been established in Québec [Beauséjour, 1999], Gaustatoppen, Norway at 1800 m [Fikke et al., 1993], Anneberg, Sweden and Kleinburg, Ontario (Canada), mainly to monitor leakage currents at a fixed supply voltage. These sites have all shown occasional activity near 0 °C similar to the "fog surging" of Forrest [1936].

In some cases, visual observations of discharges and arcs were recorded on video systems. There have been some results regarding withstand stress levels, obtained by overstressing the insulators in a string and isolating disks in series with fused links.

The IEEE Task Force on Icing Test Methods [Farzaneh et al., 2003] endorsed field tests, both to identify the worst conditions specific to each concerned utility and to develop experience with the performance of different insulator types under adverse winter conditions.

7.6.4. History of Laboratory Ice Testing

The methods used for determining the flashover voltage of iced insulators vary from one laboratory to another. The following methods are a few examples of the methods used by different researchers.

Khalifa and Morris [1967] sprayed tap or distilled water on porcelain insulators chilled in a freezer to as low as –18°C (0°F). The ice-coated insulators were then tested with an ac high-voltage supply having a series resistance of 100 kΩ, effectively limiting short-circuit current to about 400 mA at 40 kV.

Meier and Niggli [1968] simulated packed snow/ice accretion of 0.8 g/cm^3 by placing a mold around long-rod insulators that gave an ice mantle of about 3 cm on all sides. They used a water conductivity of σ_{20} = 56 μS/cm, corrected to 20°C. After the ice had solidified, the mold was removed, service voltage was applied, and the resistive leakage current was monitored. They focused their results on the time to the first exponential increase of leakage current from 0.1 to 3 mA, which then proceeded to flashover. They tested at –20°C (3 h to first flashover), –10°C (70 min), and –5°C (3 min).

Kawai [1970] formed the ice on nonenergized insulators, and then applied test voltage from a 400-kV source with a series resistance of 15 kΩ, giving an adequate short-circuit current capability of 26 A. Small-radius wire rather than multiconductor bundles was used to supply high voltage to the line end of the chains of suspension insulators. Outdoor test voltage on the ice-covered insulators was held constant until melting activity was initiated by the warming action of the sun. The leakage current and flashover behavior after the start of melting on long insulator strings outdoors were found to be similar to those occurring on short string tests indoors in a small refrigerated room.

Renner et al. [1971] tested iced insulators at Bonneville Power Authority (BPA) with a 500-kV dc source with 1.5-A continuous rating. They carried out ice accretion outdoors under energized conditions, using a system set up for rain testing. Some typical ice deposits from this approach are shown in Figure 7-17.

Flashover and withstand levels were tested both with and without a melting phase. Ice buildup under the test voltage was continued until a flashover occurred. The specimen was then reenergized to a lower voltage, and the icing process was continued until another flashover was obtained. This procedure was repeated until the lowest flashover level was obtained. When possible, a withstand level was defined for the maximum obtainable ice bridging. Leakage current activity usually limited this accretion level to slightly less than the full dry arc distance. The Renner et al. [1971] approach has much in common with the "ice progressive stress" method discussed in Farzaneh et al. [2003].

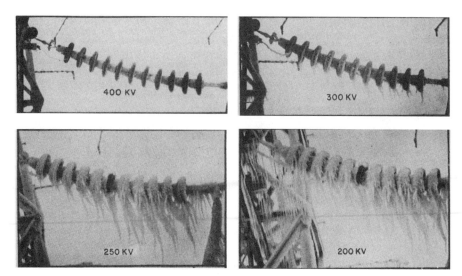

Figure 7-17: Effect of icing with water of 238 μS/cm on minimum dc flashover of 10-unit dead-end assembly (from Renner et al. [1971]).

Schneider [1975] tested UHV insulators with a 400-kV supply with a 12-kΩ series resistance, providing a minimum short-circuit current of 18 A. Icing with a fine mist of 29-μS/cm water was carried out at temperatures less than 3 °C under energized conditions. Icing continued until flashover occurred, or until 4 hours elapsed. Severity of the glaze accretion was established by measuring the ice layer on a 25-mm fixed reference cylinder, as mandated in test standards for mechanical performance of high-voltage switchgear. Ice thickness on insulators was also measured after test completion. V-strings had better performance than, and composite insulators did not perform as well as, standard vertical porcelain insulator strings.

The good performance of V-strings in icing conditions was also confirmed by Lee et al. [1975] at an applied water conductivity of $\sigma_{20} = 202$ μS/cm. They also tested horizontal and vertical porcelain disk strings and polymer insulators at 500 kV.

Fujimura et al. [1979] tested insulators in a cold room at −20 °C. Ice was formed to a density of 100–130 mg/cm^2 with water having σ_{20} of 50–500 μS/cm, and then a constant-voltage method was used to test the withstand or flashover level. Test voltages up to 600 kV were reported but the source current was not noted. The role of a melting phase in the test results was explored.

Cherney [1980] applied a constant voltage to insulators and submitted them to artificial freezing rain using an indoor test facility (Figure 7-18). The power supply was a 370-kV transformer with 1666 kVA. Icing was carried out at 10% above nominal 230- or 500-kV service voltage, using tap water with σ_{20} corrected to 320 μS/cm. Rain rate was adjusted to 8–10 mm/h. The test was terminated after a flashover or recorded as a withstand after 3 hours. This test

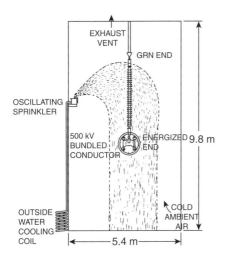

Dimensions of test chamber

Arrangement of 500-kV string
with four-conductor bundle

Figure 7-18: Icing test arrangement (from Cherney [1980]).

method did not have a melting phase. Ice accretion was measured on the top surfaces of the 25-mm conductors in the bundle. Insulators had heavy precontamination of $0.4\,mg/cm^2$, as Cherney did not observe flashovers on clean insulators.

Charneski et al. [1982] determined the critical length of insulators at which 10 flashovers occurred during a 1-hour period of ice accretion at $-12\,°C$ with tap water having $\sigma_{20} = 200\,\mu S/cm$. Insulators were evaluated on the basis of how many times flashover occurred in 10 trials of up 1-hour duration at a given voltage stress. As each withstand was obtained, the length of the insulators was reduced until 10 out of 10 tests gave flashovers. Their test arrangement in Figure 7-19 simulated the normal condition of asymmetrical icing on one side of post insulators, and used a 75-kV power supply with 6.7-A short-circuit current.

Phan and Matsuo [1983] used a $5 \times 3 \times 3\,m^3$ cold room along with a 120-kV ac supply with 4A of short-circuit current. They measured ice accretion on a rotating 38-mm monitoring cylinder rather than a fixed conductor. Icing was carried out at a constant service-voltage stress. The type of ice accretion was controlled using chamber temperature. Ice density was measured by weighing samples and obtaining the volume by displacement of mineral oil. Tests used soft rime, hard rime, and glaze ice. The value of σ_{20} was maintained between 54 and $63\,\mu S/cm$. At the end of the icing period at $-12\,°C$, voltage was increased in 5-kV steps with 5-minute periods between steps until flashover. Withstand levels were confirmed using the flashover level, reduced by 5kV.

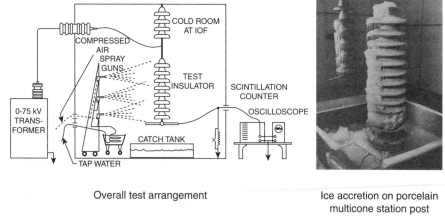

Overall test arrangement   Ice accretion on porcelain
                                            multicone station post

Figure 7-19: Icing test arrangement (from Charneski et al. [1982]).

Perhaps the first to test nonceramic insulators under icing conditions with a melting phase were Nourai and Peszlen [1984] and Nourai and Pokorny [1986]. These tests were carried out in response to problems with EHV and UHV system flashovers at 345 and 765 kV. Heavy ice accretion was applied by hand spraying to an NCI that was then warmed up to 0 °C in a process controlled by the number of inches that the environmental chamber door was opened. The measured difference between ambient and insulator temperature was about 5.5 C°. Most improvements in test methods since this work was completed have followed similar principles with attention to more precise control of the warming period, and to reducing the difference between insulator and ambient temperatures to better simulate natural conditions.

In response to a series of 500-kV insulator flashovers [Erven, 1988], a contamination-ice-fog-temperature (CIFT) insulator test was established to simulate flashover weather conditions in a specially designed fog chamber. Precontaminated insulators were dried and chilled to −5 °C. Intermittent water spray with $\sigma_{20} = 50\,\mu S/cm$ was supercooled by air flow and applied to vertical insulators at normal service voltage stress for 1.5 h at −5 °C to grow ice caps. Then, ambient temperature was increased at a controlled rate from −5 °C to −2 °C in 2 h, with continuing ice accretion at normal service voltage to form icicles to partial bridging in light fog. At the end of the ice accretion phase, a slow temperature rise from −2 °C to +1 °C in 4 hours, with dense fog, characterized the melting phase. An increase in test voltage to establish flashover or withstand was made at 0 °C.

With a single test result per day of testing, the CIFT test was simplified to eliminate the ice accretion phase, making a CFT or cold-fog test [Chisholm et al., 1996]. This process used a similar setup, with precontaminated insulators energized at normal service voltage, but used dense fog with 10-μm volume

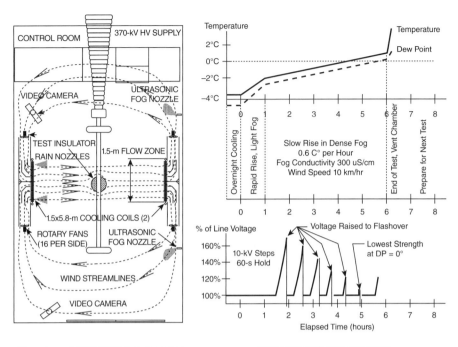

Plan view of CIFT/CFT test chamber   CFT or "cold fog" test method

Figure 7-20: Facility and test method for cold-fog test (from Chisholm et al. [1996]).

mean diameter, 3-m/s wind speed, and control of the spread between ambient and dew point temperature to a difference of less than $2\,C°$. The voltage application method shown in Figure 7-20 involved establishing 5-minute withstand levels at regular time intervals of about half an hour during the warming and melting phase.

Both the CFT and CIFT tests used a power supply with 370 kV and 1666 kVA, with a short-circuit current of more than 16 A at typical test voltages. This factor, along with control of the melting phase, made it possible to reproduce service-voltage flashover problems from 69 to 500 kV.

The icing test method used at the University of Quebec in Chicoutimi [Farzaneh et al., 1993; Farzaneh and Drapeau, 1995; Farzaneh, 2000] evolved from a method for establishing maximum withstand voltage of contaminated insulators [IEC, 1991]. This method is presented in Figure 7-21.

After an icing period, t_0, the water spray was stopped and the voltage applied to the insulators was turned off. The insulators were photographed and some preparations for flashover testing were made in a short period Δt_0 of 150 s. After this short period, water still continued to drip from the insulator. At this point, Figure 7-21 shows that two different possibilities were evaluated.

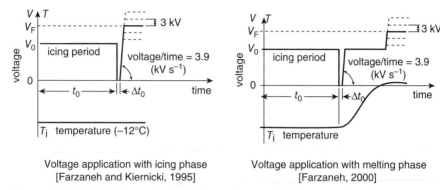

Voltage application with icing phase
[Farzaneh and Kiernicki, 1995]

Voltage application with melting phase
[Farzaneh, 2000]

Figure 7-21: Method of maximum withstand voltage measurement.

In the flashover test with an icing phase, voltage was reapplied once again to the insulator string and increased at a constant rate of 3.9 kV/s, until the estimated value of flashover voltage, V_F, was reached. The voltage was held constant until a flashover or withstand result occurred. The tests used a 120-kV source with 28-A short-circuit current, but did not include a melting phase.

In the flashover test with a melting phase, the service voltage V_0 was reapplied and the ambient temperature was increased at a controlled rate from T_i to about 1 °C. This initiated ice melting with a surface film of water. During the melting period, the voltage may be raised or lowered, depending on the level of leakage current activity.

In either case, with or without a melting phase, when the voltage V_F was reached, the maximum withstand voltage V_{WS} was established using IEC 60507 [1991]. Voltage steps of 5% of the initial voltage V_F were used and the voltage was held for a period of 15–30 minutes. Each time a flashover occurred, the ice layer was removed and the experiment reinitiated. Maximum withstand voltage was considered the maximum level of applied voltage at which flashover did not occur for a minimum of three of four tests under the same experimental conditions. The minimum flashover voltage V_{MF} corresponds to a voltage level that is one step (5%) higher than V_{WS}.

For fundamental research studies, the 50% withstand level V_{50} was also established according to IEC 60507 [1991]. After ice accretion, the procedure altered V_F in 5% steps using the up-and-down method. Insulators were subjected to a minimum of 10 "useful" tests, where the first useful test in a series yields a result that differs from the previous test at the same conditions. The value of standard deviation $\sigma_{50\%}$ was also established from the tests to obtain $V_{50\%}$.

Kannus et al. [1998] tested surge arresters with 78-kV MCOV under icing conditions. Their process used a descending chamber temperature. Artificial rain with σ_{20} of 200–2000 μS/cm was applied under service voltage. This turned to glaze ice below the freezing point. Tests continued for 5.5 hours or until

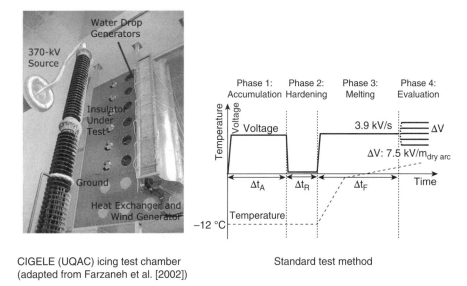

CIGELE (UQAC) icing test chamber   Standard test method
(adapted from Farzaneh et al. [2002])

Figure 7-22: Improved icing test chamber and test method for study of EHV system flashover.

external flashover occurred. Arrester leakage currents were monitored with a wideband system from a 350-kV, 350-kVA source.

A revised test method was published by Farzaneh et al. [2002], based on developments throughout the 1990s and improved design of a specialized facility, commissioned in 2000, and shown in Figure 7-22. In addition to an improved high-voltage supply with minimum 32-A short-circuit at 350 kV, the chamber featured precise 0.5° control of temperature, and an integrated wind and rain system to produce glaze ice and icicles at a mean 53° angle from droplets of 80-µm volume mean diameter.

The test procedure in Figure 7-22 has four phases. Ice accretion takes place under normal service voltage using water of controlled conductivity to achieve the desired thickness on a monitoring cylinder that rotates at 1 rpm. An ice-hardening or curing phase of 20 minutes with no voltage takes place after the ice accretion is completed, using constant wind and temperature to bring the ice layer to equilibrium with the environment. Then, the service voltage stress is reapplied and the chamber is warmed up at a controlled rate in two steps, −12 °C to −2 °C at 14 C°/h and then −2 °C to +2.5 °C at a rate of 3.5 C°/h.

The maximum withstand and minimum flashover voltage levels were established using the same approach as IEC 60507 [1991] and Figure 7-21. The test voltage in the evaluation phase is increased or decreased at a critical moment in the melting phase, when probability of flashover is highest. The factors that indicate this point include an increase in ice surface gloss (indicating presence of a water film), presence of water drops on the tips of icicles, and/or an

Figure 7-23: STRI ice progressive stress test facility with three, 420-kV class insulators. (Courtesy of I. Gutman.)

increase in the peak value of leakage current to more than 15 mA. Each withstand was established by holding the test voltage constant for at least 15 minutes. Experience with collection of drip water has established that each ice sample can be tested only once, in common with the CIFT test procedure with a similar melting phase.

Su and Jia [1993] addressed the problem of ice flashovers on 27-unit insulator strings of 170-mm × 280-mm XP-21 disks. They applied uniform ice on strings up to 5 m in length, building uniform ice by rotating the string at 1 rpm. Flashover and withstand were tested in an adjacent high-voltage test laboratory with good regulation at 2.7-A leakage current.

An ice progressive stress (IPS) test method was described by Gutman et al. [2002]. The tests were performed in a climate hall with a diameter of 18 m and a free height of 20 m as shown in Figure 7-23.

The large dimensions allowed testing of three, full-sized 420-kV insulators at the same time. Ice accretion took place at maximum service voltage, using a constant temperature of −7 °C and two or three nozzles, about 3.5 m away from each test object. Freezing water conductivity in the range of 27–270 μS/cm was tested. After accretion, voltage was ramped up rapidly to flashover, without a melting phase.

7.6.5. Recommended Icing Test Method

Generally, there are four phases to an icing test method: preparation, ice accretion, melting, and evaluation. The most representative test method simulates a melting phase, but takes more laboratory time. An "icing regime" test method with reduced test time is described as an alternative.

Preparation Phase Prior to any type of icing test, the insulator surfaces and joints should be cleaned carefully to remove all traces of dirt and grease, unless an insulator received from the field is to be tested in its as-received condition.

Ceramic surfaces should be cleaned thoroughly by washing with deionized water with a neutral detergent such as trisodium phosphate (Na_3PO_3). Adequately prepared surfaces will normally demonstrate poor hydrophobicity with large continuous areas of wetting after rinsing with deionized water.

Cleaning of nonceramic insulator surfaces should normally reduce hydrophobicity to give the desired wetting. It is important to select and use a cleaning material such as mild dishwasher soap that is effective but does not degrade the polymer surface.

If it is appropriate for the application, the object may be precontaminated using the methods described in Chapter 3 for the clean-fog test. After cleaning and precontamination with the solid-layer method, insulators should be dried at about 20 °C before they are installed in the desired test position.

Insulating parts of the test object, whether ceramic or polymer, should not be touched by hand after cleaning and/or precontamination. This means that the test object should be handled only by its metal parts, and using a lifting device such as an overhead crane. If it is not practical to move the insulator without touching the insulating surfaces, they should be rewashed when the test object has been installed in its desired test position and location. Any accessories such as corona rings should be installed, and the correct size of phase conductor, bus, or bundle should be fitted at the high-voltage end of the test object.

Once the test object is installed, it should be cooled immediately to the same temperature as the ambient temperature of the planned test. At the start of the test, the object should be in thermal equilibrium with the air in the test chamber.

Ice Accretion Phase The climate room should simulate natural conditions for icing as closely as possible. Ice should be built up on insulators energized at normal service voltage, since the accumulation patterns on unenergized insulators or those with temporary overvoltage are not realistic. Also, under energized conditions, electrical phenomena such as corona discharges, leakage current at the surface of ice and icicles, water-drop elongation, and ionic wind also influence the general characteristics of ice [Farzaneh and Laforte, 1992]. Corona discharge and leakage current have a heating effect, whereas the elongation of water drops and ionic winds have a cooling effect [Farzaneh et al., 1990]. The interaction of these phenomena influences the characteristics of icicles considerably. Moreover, these electrical parameters vary as a function of voltage polarity and this also affects the morphology of ice and icicles [Farzaneh and Laforte, 1993].

The experimental conditions should be adjusted to form glaze ice with icicles, as this type of ice is associated with the highest probability of flashover on energized insulators. Parameters such as air temperature and wind velocity, as well as both the vertical and horizontal components of precipitation

TABLE 7-5: Summary of the Recommended Ice Deposit Parameters

Ice Deposit Parameter	Recommended Value
Type	Glaze ice with icicles
Thickness	5–30 mm on rotating cylinder
Freezing water flux	$60 \pm 20 \, \text{L/h/m}^2$
Water conductivity	100 μS/cm @ 20 °C
Air temperature	–5 °C to –15 °C
Wind speed (if used)	3–5 m/s
Precipitation direction	$45° \pm 10°$
Applied voltage	Service voltage stress

Source: Farzaneh et al. [2003].

intensities should be properly controlled and kept constant. The recommended ice deposit parameters are summarized in Table 7-5.

It may take some adjustment of parameters to grow glaze ice with icicles. The applied water droplet size and wind speed should be adjusted to give a water droplet direction of about $45° \pm 10°$ from top to bottom, compared to a vertical axis. A horizontal wind velocity of about 10 km/h (3 m/s) allows vertical icicles to form when using freezing water droplet size of less than 100 μm. If larger size water droplets are used, wind velocity up to 5 m/s may be necessary to carry the droplets to the surface of the test object.

The water should be supercooled when hitting the insulator surface. If there is no runoff water, no icicles will form. If the temperature is too cold, the water droplets impinging on the test object will be completely frozen and hard or soft rime may grow. A comprehensive review of the roles of temperature, wind speed, and other factors on artificial ice accretion is found in Farzaneh and Kiernicki [1995].

In the absence of field data, the conductivity of applied water feeding the spray system should be adjusted to $\sigma_{20} = 100$ μS/cm measured at 20 °C, in common with standards for wet tests [IEEE Standard 4, 1995]. This conductivity should be achieved by adding sodium chloride to the deionized water feeding the icing spray system. Conductivity of the applied water may be adjusted up or down on the basis of data gathered by each concerned utility in the field. Values of σ_{20} exceeding 300 μS/cm could give high intensities of leakage current and discharge activity, leading to premature insulator flashover during ice accretion before the test is completed. As well, very high freezing water conductivity can reduce the rate of ice accretion and cause an unrealistically high water drip rate from the test insulator. In these cases, the water flow can be cycled (e.g., 2 seconds on, 4 seconds off) rather than continuous. Heat tracing will be essential to keep the rain nozzles from freezing when there is no water flow.

The average thickness of ice should be measured in the applied water exposure zone on both rotating and fixed cylinders, 25–30 mm in diameter and

600 mm in length, installed near the test insulator. The longitudinal axis should be horizontal, at each end of the test specimen, to receive the same general wetting as the test insulator. The thickness of ice on fixed and rotating cylinders should be based on the requested reliability and determined by icing return periods for a given area and conditions. The ice accretion level corresponding to the onset of bridging of insulator sheds by icicles is often critical and should be measured.

A preparation period is needed between the end of the ice accretions at subzero temperature and the moment when test voltage is applied for flashover voltage evaluation. At least three ice thickness measurements and photographs should be taken on each reference cylinder.

Melting Phase   The procedure may be adjusted according to two approaches: *icing regime* or *melting regime*.

- The icing regime corresponds to flashover performance tests carried out shortly after ice accretion is completed, while a water film is still present on the ice surface. In such a case, the preparation period is short, about 2–3 minutes.
- The melting regime corresponds to flashover performance tests carried out by hardening the ice, and then warming it slowly until a water film forms on the ice surface. In such a case, the hardening period is about 15 minutes and the warming phase can take several hours.

The icing regime method is quicker but the melting regime test is more representative of natural conditions. At one laboratory, the two test methods give similar results [Farzaneh, 2000]. With additional experience it may be possible to formulate general advice.

To harden the ice prior to a melting phase, voltage is turned off, wind speed continues, and air temperature remains the same as during the icing period, to equalize insulator and ice temperatures with ambient.

During the melting sequence, immediately after the hardening sequence, while service voltage is applied, the air temperature is increased progressively from subzero to melting level at a rate of 2–3 °C/hour above −2 °C. During this sequence, the service voltage should be applied to the test insulator. A slow rate is needed to minimize premature ice shedding. Ambient and dew point temperatures, leakage current, and drip water conductivity may be monitored during the melting sequence. The most important decision here is the identification of the "critical moment." The critical moment in the melting phase is characterized by one or more of the following:

- Presence of a water film on the ice surface.
- Change in gloss (reflectivity) of the ice surface.
- Initiation of water droplet ejection from icicles.
- Increase in leakage current to values over 15 mA.

Evaluation Phase   At the end of the ice accretion phase for the icing regime test, or the critical moment of the melting regime test, the voltage should be increased or reduced at a rate of 3% of service voltage per second, until the test voltage is achieved. The test voltage is then maintained at a constant level until there is a flashover, a 15-minute withstand, or a significant shedding of the ice deposit. The time to flashover and records of leakage current activity may be helpful in ranking the relative performance of insulators.

Icing performance should be established using test series with predetermined severity to establish V_{WS}, the maximum withstand voltage, or V_{50}, the median flashover voltage. A minimum of five flashover tests at the same ice exposure are needed to establish V_{WS}, and ten are required to obtain V_{50} [IEC 60507, 1991].

7.6.6. Recommended Cold-Fog Test Method

Generally, there are three phases to a cold-fog test method: preparation, fog accretion, and repeated evaluation during a melting phase [Chisholm et al., 1996].

Preparation Phase   This phase is similar to the one described for icing tests. Insulators are always precontaminated to one of the standard ESDD levels in IEEE Standard 4 [1995] using a slurry of salt and kaolin.

Fog Accretion Phase   The cold room is chilled to –4 °C, and then cold fog is produced using water with a controlled conductivity of approximately 300 μS/cm and a median droplet diameter of 10 μm. A chilled-mirror hygrometer or other calibrated reference method is used to monitor the dew point of the cold room. The insulator is energized at service voltage stress during cold-fog accretion.

Melting and Evaluation Phase   With service voltage still applied, when the dew point is within 2 °C of ambient temperature and a dense fog is visible, cold-room temperature is increased from –4 °C to –2 °C in no less than 1 hour. Airflow should be maintained at about 10 km/h with less than 10% turbulence intensity. The temperature rise of the test chamber should be 0.6 °C/h, from –2 °C to +1 °C in 5 hours. The dew point should remain within 2 °C of ambient temperature during this sequence and test voltage should be kept constant, as in the accretion period. At regular intervals, as shown in Figure 7-20, the supply voltage is increased and held to establish 5-minute withstand levels in steps of 5% of the normal service voltage. Once a flashover is recorded, the normal supply voltage is restored. If flashover occurs between intervals, the supply voltage is reduced by a 5% step.

The declining flashover voltage with increasing temperature is extrapolated to give the intercept at 0 °C.

7.7. ICE FLASHOVER TEST RESULTS

The test results presented in this section focus mainly on the performance of standard cap-and-pin and station post insulators, but also describe early results obtained with polymer insulators. In this section, the data from outdoor icing tests are reviewed, and then laboratory results are classified into four categories—very light, light, moderate, and heavy icing.

The icing test methods used by the various researchers and engineers have been described in Section 7.6.4 and will not be repeated. With the continuous progress in developing realistic test methods, interpretation of older test results can sometimes be difficult. In particular, an appropriate melting phase with slow rate of temperature rise is needed to develop the maximum ice surface conductivity at the critical moment that typically defines the minimum flashover levels.

7.7.1. Outdoor Test Results

The performance of power system insulators at constant line voltage was described in Section 7.4.

Kawai [1970] used an outdoor test area to create ice under service voltage on strings of up to 25 disk insulators. A fog nozzle with a low precipitation-like mist was used to make "hard, dry" ice at night at a temperature of about –9 °C. Also, tests were carried out with ice mixed with misty rain at –6 °C to –1 °C. The ice was allowed to build up and reach equilibrium with melting from leakage current. This gave a heavy ice thickness of up to 25 mm on the insulators, with icicles that fully bridged many disks in the middle of the string. The flashover tests were carried out under sunrise melting conditions with the hard, dry ice, while no melting phase was needed for flashover with the mixed ice and rain (Table 7-6).

Kawai concluded that the effect of insulator shape was not important, and that the flashover voltages on the different types of insulator were mainly a function of connection length.

7.7.2. Laboratory Tests with Very Light Icing

The most comprehensive study of service-voltage flashover tests using a cold-fog test was reported by Chisholm et al. [1996]. A specialized chamber and test method, described in Section 7.6, were commissioned to ensure that temperature and dew point would be coupled together during a temperature rise period. Test results from insulators ranging from 44-kV porcelain posts to 500-kV station posts and insulator strings with 3.3-m dry arc distance were reported by Chisholm [2007].

The most significant difference in cold-fog flashover tests, compared to clean-fog tests, is the difference in duration of leakage current activity. Under cold-fog conditions, the duration of activity is extremely short—less than 10

TABLE 7-6: Results of Outdoor Flashover Tests on EHV Insulator Strings

Typical Accretion	146 mm x 254 mm	146 mm x 267 mm	172 mm x 318 mm	1195 mm x 206 m
Number of units	25, double string	19	19	3
Dry arc distance (mm)	3650	2774	3258	$1195 \times 3 = 358$
Leakage distance (mm)	$292 \times 25 = 7302$	$432 \times 19 = 8204$	$508 \times 19 = 8636$	$2098 \times 3 = 872$
Hard dry ice flashover voltage (in kV) and (Gradient in kV/m$_{\text{dry arc}}$)	288 (79)	260 (94)	288 (88)	n/a
Ice, misty rain flashover voltage (in kV) and (Gradient in kV/m$_{\text{dry arc}}$)	246 (67)	n/a	254 (78)	254 (71)

Source: Kawai [1970].

seconds—compared to the 15–30-minute duration of arcing and dry-band growth in clean-fog tests. This is illustrated with a series of tests on rigid post-type porcelain insulators with 680-mm dry arc distance and a leakage distance of 1940 mm. Four insulators of the same shape were tested in horizontal orientation at the same time. Two posts had semiconducting glaze and two had conventional glaze. Insulators were precontaminated with dilute sulfuric acid, mixed with 40 g/L kaolin, to give an ESDD of 0.35 mg/cm^2 (350 µg/cm^2) in each test.

The nature of leakage current activity in Figure 7-24 for cold-fog conditions near 0 °C is completely different from the behavior at $T_a = -5$ °C. Visual observations confirm that the arcing and dry-banding processes at −5 °C are somewhat similar to those observed in clean-fog tests. Bursts of leakage current lasting less than 10 seconds occur at 1- or 2-minute intervals, with low levels of rms and peak-to-peak levels in between. During the bursts, peak values of

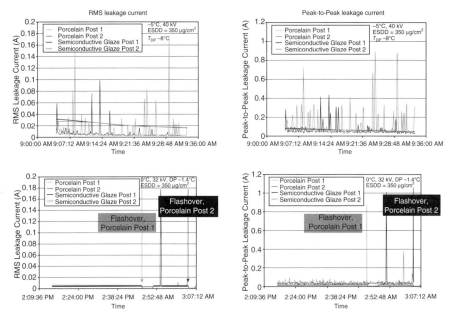

Figure 7-24: Leakage currents on precontaminated post insulators at −5 °C and 0 °C ambient in cold-fog test.

$800\,\text{mA}_{\text{p-p}}$ make large contributions to the 10-second average rms values as well.

In contrast, porcelain post insulators at 0 °C in fog are quiescent, with negligible leakage activity for periods of up to 35 and 40 minutes in Figure 7-24. When activity occurs, it either proceeds directly to flashover (marked by the vertical lines) or leads to a single burst of leakage activity, lasting between 1 and 10 seconds. Cold-fog flashovers in this case occurred on both porcelain posts at 32.0 and 32.4 kV, and these values give an average unified specific creepage distance at flashover of $(1940\,\text{mm}/32.2\,\text{kV}_{1-g}) = 60\,\text{mm/kV}$. The ESDD of $0.35\,\text{mg/cm}^2$ is the dividing line for heavy and very heavy pollution levels, and the IEC 60815 standard suggests that 44–55 mm/kV USCD would be appropriate for withstanding sodium chloride contamination. The hygroscopic and ionic behavior of the sulfuric acid precontamination at 0 °C may partially explain why insulators near acid-gas cleanout operations are at greater risk of flashover in freezing conditions.

Another aspect illustrated in Figure 7-24 is the difference in leakage current between the semiconducting glaze insulators and the porcelain posts of the same shape, size, and orientation. For the semiconducting glaze units, the peak-to-peak current was measured to be 3 ± 0.3 times the rms current whenever the high-voltage supply was on, in close agreement with the expected value of 2.83 for a pure sine wave. The complete suppression of leakage activity on the semiconducting glaze insulators means that, for practical purposes, they do not flash over in cold-fog conditions. These and other aspects of semiconducting glaze performance are discussed fully in Chapter 9.

Overall, the cold-fog performance of conventional glaze ceramic insulators is a function of leakage distance and follows the same empirical relation with ESDD as that observed for clean-fog tests. Figure 7-25 shows two relations, one fitted for deep-skirt cap-and-pin apparatus insulators with extensive sheltered creepage distance:

$$E_{50} = \begin{cases} \text{Pin} & 18.6 \cdot \left(\dfrac{ESDD}{1\,\text{mg}/\text{cm}^2} \right)^{-0.36} \\[2em] \text{Post} & 12.7 \cdot \left(\dfrac{ESDD}{1\,\text{mg}/\text{cm}^2} \right)^{-0.36} \end{cases} \tag{7-7}$$

The other expression is fitted to results for post insulators and includes the data observed for strings of 23 suspension disks or bells.

The expression for cold-fog flashover gradient E_{50} of post insulators and a 500-kV string is identical to that fitted to clean-fog test results by Baker et al. [2008]. This has led to the treatment of cold-fog flashover simply as an interesting variant of the conventional clean-fog flashover problem.

7.7.3. Insulators with Light Ice Accretion

Erven [1988] reported results of more than 200 icing tests on 500-kV station post insulators. The test method, called CIFT and described previously, was established to reproduce line-voltage flashovers on insulators with light ice accretion and moderate precontamination. The 50% withstand voltages were

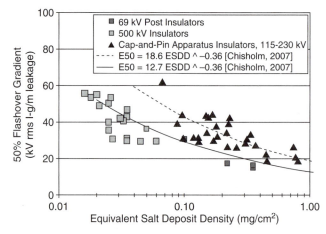

Figure 7-25: Critical flashover gradient of porcelain insulators in cold-fog test (adapted from Chisholm [2007]).

established with an up-and-down method. Figure 7-26 shows that the typical accumulation of icicles with applied water conductivity of $50\,\mu S/cm$ on a three-section station post with alternating 1-2-3 shed profiles did not bridge the space between the major sheds, and few icicles formed on the inner sheds.

Posts with uniform and alternating shed profiles were tested. Dimensions and results are shown in Table 7-7. The posts were precontaminated to $0.02\,mg/cm^2$. Critical flashover of $V_{50} = 275\,kV$ on the uniform-profile insulator gave a flashover stress of $36\,kV/m_{leakage}$ and $100\,kV/m_{dry\,arc}$. This was about 14% lower than the normal service voltage stress for these insulators.

Top section   Middle section   Bottom section

Figure 7-26: Light accretion of icicles on station post insulator.

TABLE 7-7: Comparison of Uniform and Alternating Profile Porcelain Post Performance in CIFT Conditions

Characteristic	Uniform Profile	Alternating 1-2-3 Profile
Dry arc distance (mm)	2743	2743
Leakage distance (mm)	7620	9800
Outer shed diameters top/middle/ bottom (mm)	246, 286, 321	335, 375, 400
Impulse withstand (kV)	1550	1550
Wet withstand, ANSI (kV)	620	620
Cantilever strength (kg)	2090	2090
CIFT flashover (kV) with ESDD of $0.02\,mg/cm^2$	275	335
CIFT flashover stress ($kV/m_{leakage}$)	36.1	34.3
CIFT flashover stress ($kV/m_{dry\,arc}$)	100	122

Source: Erven [1988].

Farzaneh and Melo [1994] measured ac leakge currents and flashover on short strings of 146-mm × 267-mm insulators with 356-mm leakage distance and 36-kip mechanical rating. These porcelain insulators are used in 500-kV networks in Ontario at a stress of 14 kV per disk. Rime icicles were grown in supercooled fog of 15-μm volume mean diameter and liquid water content of 0.16 g/m³ at a horizontal wind speed of 3.3 m/s. Five hours of accretion at −1.5 °C led to icicles of up to 48 mm. When liquid water content was increased to 0.27 g/m³ [Farzaneh and Melo, 1990], icicles grew to about 80 mm, which bridged about two-thirds of the shed-to-shed separation. Table 7-8 shows that freeze–thaw partitioning led to depletion of ice on the top surfaces and some enhancement of conductivity on the bottom surfaces and icicles, relative to the fixed applied water conductivity of $\sigma_{20} = 150\,\mu S/cm$.

The icicle weight after 5 hours at the high fog density was about 15 g per disk, compared with 120 g on the top surface, and 20 g on the bottom surface, for a total of 155 g per disk. The rate of change of icicle length and mass varied with the liquid water content of the fog or drizzle in the range of 0.15–0.6 g/m³ (Figure 7-27).

During melting phases after accretion of icicles in fog was completed, the rate of change of temperature had a strong effect on the conductivity of the drip water in Figure 7-28.

There is a sharp transition in the drip water conductivity–mass product in the initial phase of melting, corresponding to a low mass of highly conductive water. At this critical moment, surface conductivity is highest, the leakage current is at a maximum of 100%, and the risk of flashover for constant applied voltage is greatest. As water continues to drip away, the quantity of ions reaches the same asymptotic value for all rates of temperature rise and the leakage current level drops to about 60% of the initial value.

7.7.4. Insulators with Moderate Ice Accretion

Fujimura et al. [1979] studied the electrical performance of vertical insulators covered with ice under deenergized conditions, and then placed on

TABLE 7-8: Conductivity of Ice Accretion from Fog with σ_{20} of 150 μS/cm

Liquid Water Content (g/m³)	Icicles	Top Surface	Bottom Surface	Drip Water
	μS/cm at 20 °C			
0.16	349	76	338	290
0.22	340	94	300	277
0.27	246	70	309	327
0.40	137	46	248	325
0.53	127	50	224	290

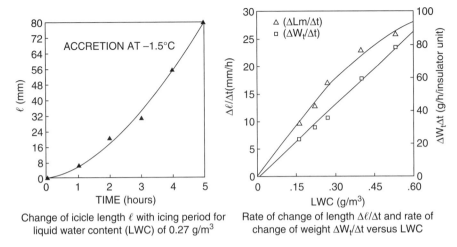

Change of icicle length ℓ with icing period for liquid water content (LWC) of 0.27 g/m^3

Rate of change of length $\Delta\ell/\Delta t$ and rate of change of weight $\Delta W_t/\Delta t$ versus LWC

Figure 7-27: Parameters of icicle growth in fog at $-1.5°$ (from Farzaneh and Melo [1990], courtesy of Elsevier).

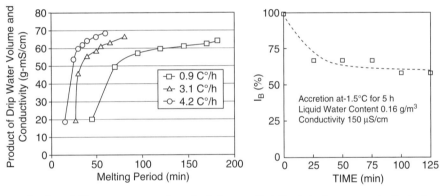

Product of drip water mass and conductivity as function of temperature rate of increase dT/dt

Variation in leakage current with duation of melting phase at dT/dt of 1.5 C°/h

Figure 7-28: Evolution of drip water ion content (mass × conductivity) and leakage current during melting phases (adapted from Farzaneh and Melo [1990]).

potential. For conditions of very light to light icing without bridging, the withstand voltage of an insulator with no or short icicles was approximately proportional to the surface leakage distance of the insulator. For moderate icing conditions, the withstand voltage of station post insulators was approximately proportional to the surface leakage distance of an icicle bridging the sheds. They defined their withstand level as maximum withstand voltage value that gave four withstands and no flashovers for the same level of ice accretion.

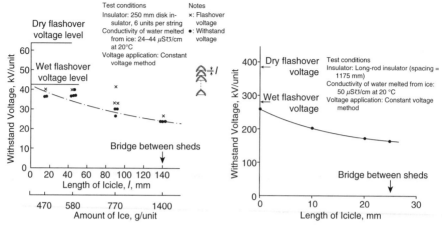

String of six 146-mm × 250-mm porcelain disks   Long-rod porcelain insulator of 1175 mm

Figure 7-29: The ac withstand voltage of insulator disks versus length of icicle (from Fujimura et al. [1979]).

Figure 7-29 shows two interesting relations. One is the amount of ice, in grams per 146-mm disk, which leads to partial or complete bridging of insulator sheds by icicles. The weight of accreted ice on the insulator string is an important variable in the calculation of the resistance of the ice layer per unit length. The second relation shows a steady decline in the flashover strength from the wet strength of 42 kV per disk to a level of about 25 kV per disk ($170 kV/m_{dry arc}$). For a long-rod insulator of 1175-mm dry arc distance, Figure 7-29 shows a withstand voltage of 180 kV, with a corresponding withstand stress of about $150 kV/m_{dry arc}$.

In both cases, the water conductivity was low, in the range of $\sigma_{20} = 24$–$50 \mu S/cm$, and the insulator surfaces were clean. The effect of a surface film of polluted water was also tested for icicle lengths ranging from 10 to 25 mm on a long-rod insulator. Figure 7-30 shows that the flashover strength can decline by a factor of 3 when the surface film conductivity exceeds 50 mS/cm, compared to the ice surface conductivity of $\sigma_{20} = 50 \mu S/cm$.

The effect of applied water conductivity was also explored and checked by measuring the conductivity of the melted ice (Figure 7-31). This also had a strong effect on the withstand voltage of station post and long-rod insulators, with flashover stress dropping below $25 kV/m_{dry arc}$ at a very high conductivity of 20 mS/m. Since it is very difficult to grow ice on energized insulators with this high level of conductivity, these test points do not represent a practical problem.

Cherney [1980] carried out icing tests using tap water accretion with $\sigma_{20} = 320 \mu S/cm$ on energized porcelain suspension strings and polymer insulators at 230 and 500 kV. The photographs of ice accretion show full bridg-

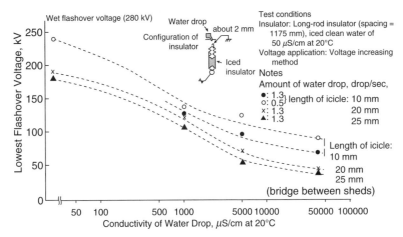

Figure 7-30: Effect of surface water conductivity on flashover of iced long-rod porcelain insulators (adapted from Fujimura et al. [1979]).

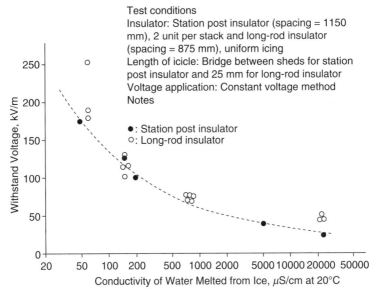

Figure 7-31: Effect of melted ice conductivity on flashover of iced station post porcelain insulators (adapted from Fujimura et al. [1979]).

ing of the polymer units but partial bridging of standard suspension disks. Accretion was measured on the four-bundle conductors and was an average of 9 mm in all tests. This is equivalent to about 5 mm on a rotating reference cylinder. All tests were carried out without a melting phase at a temperature of about –5 °C and insulators were precontaminated to an ESDD of

TABLE 7-9: Icing Flashover Test Results on 500-kV Insulators

	A	B	C	D	E
Standard Disks Type C Polymer					
			Dimensions (mm)		
Dry arc	3400	3575	3473	3500	3651
Leakage	9280	10325	8700	8800	6985
Shed spacing	55	90	55	90	146
Shed diameter	135	170/132	148	115/84	267
Number of Sheds	63	38 + 37	62	37 + 37	25
Test voltage stress ($kV/m_{dry\,arc}$) for 500 kV	93.5	89.0	91.6	90.9	87.1
Average Ice thickness on fixed cylinder	8	10	8	10	9
Number of flashovers/ total tests	3/3	1/3	2/3	1/2	1/2
Average time to flash- over (min)	101	150	92	170	116

Source: Cherney [1980, 1986].

0.4 mg/cm^2. The dimensions, ice accretion, and test performance of the 500-kV insulators are shown in Table 7-9.

Of the polymer insulators, types B and D with conical-shaped and alternating-diameter sheds had better performance than types A and C with uniform sheds, and also performed better than the string of twenty-three 146-mm × 267-mm disks. Cherney also tested 230-kV insulators. A string of fourteen standard 146-mm × 254-mm disks at a stress level of 71 kV/m$_{dry\,arc}$ flashed over in two of two tests at an ice accretion level of 8–9 mm. Generally, the flashover and withstand levels established in this test program showed that heavily polluted ice could represent an electrical danger at –5 °C.

Phan and Matsuo [1983] tested effects of several different types of ice accretion, including a milky ice of high density, called by the authors "hard rime," that is really a form of wet-grown ice. Figure 7-32 shows their relation between the thickness of accretion on a rotating monitoring cylinder and the unbridged air gap distances on short insulator strings. They found full bridging of the g_2 gap for 7–14 mm of accretion, which is roughly equivalent to 15–30 mm of ice thickness on a fixed cylinder.

Farzaneh et al. [2000] initiated studies on the model of flashover of partially bridged insulators by measuring the average air gap breakdown voltage, V_b, of air gaps in fully bridged disk insulators. Figure 7-33 shows that the break-

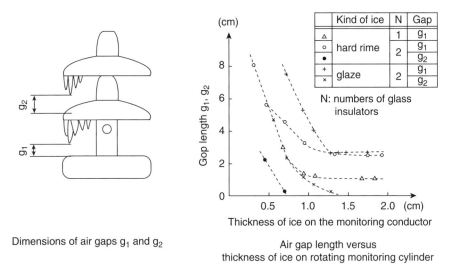

Dimensions of air gaps g_1 and g_2

Air gap length versus thickness of ice on rotating monitoring cylinder

Figure 7-32: Air gap length on standard suspension disks versus ice accretion (from Phan and Matsuo [1983]).

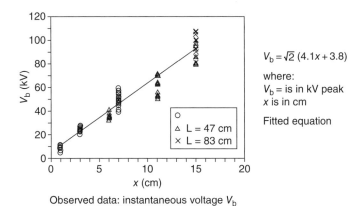

Observed data: instantaneous voltage V_b

Figure 7-33: Relation between breakdown voltage and air gap between ice surfaces (from Farzaneh et al. [2000]).

down voltage is considerable and depends only on the length of the air gap, x, and not on the length of the ice sample, L.

A number of evaluations were carried out on strings of five standard disk insulators to establish the influence of air gap length on V_{50} in the icing regime. Uniform wet-grown ice was built up off potential to a reference level of 20 mm on a rotating reference cylinder. This represents a heavy level of accretion, but the results are given here because they simulate controlled air gaps that were artificially sawn into the ice layers as shown in Table 7-10.

The initiation of a partial arc along the air gap close to the high-voltage (HV) electrode led to a redistribution of the voltage along the ice-covered insulator. The expected behavior was that additional air gaps in cases (b), (c), and (d) would automatically increase the flashover voltage, but this was not observed to be the case. In cases (b) and (c) with two air gaps, the total arc length bridged more of the ice layer than the single gap and was this closer to the critical arc length, reaching flashover at a lower voltage than case (a). In case (d), a higher voltage is needed to establish arcs across all three air gaps,

TABLE 7.10: Air Gaps in Suspension Insulator Strings and V_{50} in Icing Regime

	(a)	(b)	(c)	(d)
Icing Regime Test Up-and-Down Method [IEC 60507, 1991] $\sigma_{20} = 80\,\mu S/cm$ Accretion of 20 mm on Rotating Cylinder Five Standard 146-mm × 254-mm Porcelain Disks				
Number of air gaps (length in mm)	One (70)	Two (70, 40)	Two (70, 40)	Three (70, 40, 40)
Voltage drop across air gap(s) without partial arc (% of applied voltage)	84	75 and 6	83 and 5	79, 6, and 3
Voltage drop across air gap(s) with partial arc across air gap 1 (% of applied voltage)	15	15 and 33	15 and 31	15, 25, and 13
V_{50} (kV)	68	66	53	76

Source: Farzaneh et al. [2006b].

so flashover strength is higher. The difference between case (b) and case (c) relates to the faster rate of air gap growth due to higher arcing current in case (b), leading to higher values of V_b and V_{50}.

7.7.5. Insulators Fully Bridged with Ice

Khalifa and Morris [1967] sprayed tap or distilled water on porcelain insulators chilled to as low as $-18\,°C$ ($0\,°F$) in a freezer. The ice-coated insulators were then tested with an ac high-voltage supply having a series resistance of $100\,k\Omega$, effectively limiting short-circuit current to about $400\,mA$ at $40\,kV$. These authors noted a difference in the stability of the arc under icing conditions, compared to the dynamic behavior for contaminated surfaces. They found that currents as low as 1 mA could initiate flashover of the ice-covered insulator.

Charneski et al. [1982] tested insulator strings, porcelain posts, and polymer insulators under icing conditions including accretion of 200-$\mu S/cm$ tap water at an accretion rate between 0.5 and 2 mm per minute. Icing was continued at line voltage stress until equilibrium was reached, with few leakage current pulses above 10 mA, or until flashover. The dimensions and profiles of their insulators are given in Table 7-11.

The test results shown in Figure 7-34 were expressed as the number of flashovers as a percentage in 10 trials, against the electric stress per meter of

TABLE 7-11: Dimensions of Insulators used in Heavy Icing Tests

Parameter	A	B	C	D	E	F	G	H	I
					Dimensions (mm)				
Dry arc distance	806	1057	1057	1076	946	1187	991	1080	1143
Leakage distance	2029	2270	1718	2165	2461	2692	2359	2610	3683
Shed spacing	35	89	89	67	89	146	51	83	83
Major diameter	92	121	121	127	133	254	276	279	346

Source: Charneski et al. [1982].

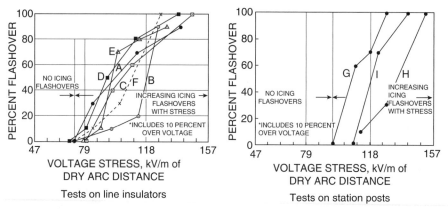

Figure 7-34: Probability of icing flashover versus stress per meter of dry arc distance (from Charneski et al. [1982]).

dry arc distance. It would seem that only insulator B offered a performance advantage over the other line suspension types. However, there was a clear improvement in performance of station post type H over both type G, with narrow shed spacing, and type I, with large diameter.

In service, Detroit Edison used the type G insulator at 345 kV with a service voltage stress of 114 kV per meter of dry arc (2.89 kV per inch) [Charneski et al. 1982]. At this stress, Figure 7-34 shows that the probability of flashover was 65%, and this matched poor in-service experience with this type of insulator. In response to the problems, a reduced service voltage stress of 104 kV per meter of dry arc was adopted by this utility. Stress levels on line insulators were lower and no flashover problems were experienced on standard disk strings of type F.

Phan and Matsuo [1983] defined a suitable leakage distance path across strings of n insulator disks as the sum of the leakage distance of a clean insulator and the $(n - 1)$ spacings between the insulators. They found a high degree of linearity between minimum flashover voltage (V_{MF}) and this dimension for short glass disk strings of up to four units, as illustrated in Figure 7-35.

Phan and Matsuo also noted that the ratio of iced to wet flashover levels decreased with increasing number of disks, from 70% for $n = 1$ to 42% for $n = 4$.

Polymer insulators were tested under icing conditions with a melting phase [Nourai and Peszlen, 1984; Nourai and Pokorny, 1986]. Heavy ice accretion was applied to an NCI with 37-mm shed-to-shed distance, 92-mm diameter, 0.7-m dry arc distance, and 1.7-m leakage distance. Leakage current was tested with a service voltage level of 80 kV. The insulator temperature lagged the chamber by a difference of 4 °C in a warming phase, controlled by the number

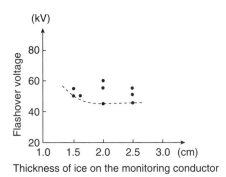

Thickness of ice on the monitoring conductor

V_{MF} of two standard glass disks as function of
ice thickness on rotating monitoring cylinder

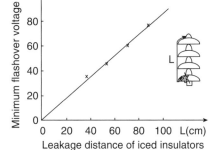

Leakage distance of iced insulators

Relation between V_{MF} and the leakage
distance as shown in figure

Figure 7-35: Minimum flashover voltage of glass disk insulators in icing conditions
(from Phan and Matsuo [1983]).

of inches that the environmental chamber door was opened. They found a
declining duration of leakage current activity with increasing insulator temperature, ranging from about 120 ac cycles (2 s) at −4 °C to less than 20 cycles
(0.33 s) near 0 °C.

As a scientific curiosity, Nourai and Pokorny [1986] also measured the rise
time of the leakage current activity on the iced surface of the polymer insulator and reported that it was faster than 500 picoseconds, the limit of their
instrumentation bandwidth.

Su and Jia [1993] addressed the problem of ice flashovers on strings of
XP-21 disks with 170-mm spacing and 280-mm diameter. Uniform and heavy
ice bridging with 70–90 µS/cm at 0 °C (σ_{20} = 120–150 µS/cm) and radial thickness of 20–30 mm was applied without service voltage. The average withstand
voltage gradient of the string dropped from 94.2 kV/m$_{dry\,arc}$ for a 9-unit (1.57-m)
string to 77.4 kV/m$_{dry\,arc}$ for 27 units (4.7 m).

Four different types of insulators were tested under heavy icing conditions by Farzaneh and Kiernicki [1995] with the results given in Table 7-12.
A thickness of 20 mm on a rotating reference cylinder, with σ_{20} = 80 µS/cm,
was used as the reference condition for each test. Flashover was
evaluated using the icing regime method, where voltage is raised immediately after ice accretion, while a film of water is still present on the ice
surface.

The service voltage stress for anti-fog, EPDM, and post type insulators corresponded to normal line-to-ground operating voltage in Québec. Ice accumulation was carried out in dry and wet regimes. Flashover levels with dry-grown
ice were much higher than service voltage stress. Vertical icicles were found
to be most severe. With a horizontal wind speed of more than 5.2 m/s, the icicle

TABLE 7-12: Parameters of Insulators and Heavy Ice Test Results

Parameter	IEEE Standard 146 mm / 254 mm	Anti-Fog 170 mm / 320 mm	EPDM	Post Type
Shed diameter (mm)	254	320	90/105	250
Shed spacing (mm)	146	170	55	51
Leakage distance (mm/unit)	305	545	2171	2525
Number of units tested	5	4	1	—
Dry arc distance of tested part (mm)	809	820	714	610
Service voltage stress (kV/m$_{dry arc}$)	62	73	71	98
Wet-grown ice E_{WS} (kV/ m$_{dry arc}$)	70	84	96	90

Source: Farzaneh and Kiernicki [1995].

angle increased to more than $20°$ off the vertical axis, and the increased air gap length resulted in a 50% increase in the flashover voltage of the string of standard disks.

Soucy [1996] carried out comparison tests of icing flashover tests on pre-contaminated EPDM, silicone, and porcelain disk insulators at SDD levels of 0.05 and 0.1 mg/cm^2. Water with $\sigma_{20} = 80\,\mu S/cm$ was used to accrete a 15-mm thickness on a rotating reference cylinder. The initial drip water at the start of the melting phase in flashover tests in Figure 7-36 was 4 times higher than the applied water when disk insulators were clean and 34 times higher when they were contaminated with an SDD of 0.05 mg/cm^2.

In spite of the remarkable change in drip water conductivity, the maximum withstand stress did not change much in the heavy icing conditions. Levels of $E_{WS} = 75.4$ and $72.2\,kV/m_{dry arc}$ were reported for the porcelain disks for clean conditions and SDD $= 0.05$ mg/cm^2, respectively. The corresponding values for both polymer insulators were $88.2\,kV/m_{dry arc}$ clean and $84.0\,kV/m_{dry arc}$ with 50 µg/cm^2. Both polymer insulators flashed over under service voltage stress of $71.4\,kV/m_{dry arc}$ during the icing period when precontaminated with SDD $= 0.1$ mg/cm^2 ($100\,\mu g/cm^2$).

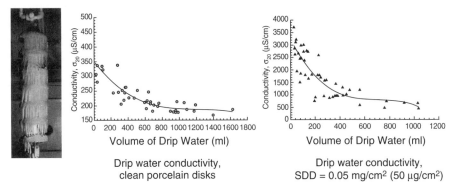

Figure 7-36: Effect of precontamination on drip water conductivity (from Soucy [1996]; courtesy of L. Soucy).

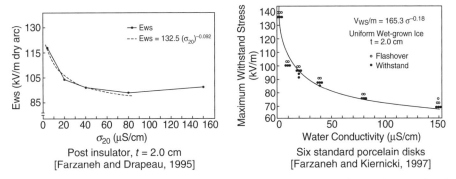

Figure 7-37: Effect of freezing water conductivity on maximum withstand stress (E_{WS}).

This work was continued and extended, using strings of six standard disks in Farzaneh and Kiernicki [1997]. The maximum withstand stress for heavy ice accretion of 20 mm decreases with increasing applied water conductivity, following empirical power-law relations that are given in Figure 7-37 for posts and standard disk insulators. In the case of the post insulator, the upturn for $\sigma_{20} = 150\,\mu S/cm$ reflects a tendency for the ice to fall off the top sheds of the post near the energized end, while ice in tests with $\sigma_{20} = 4\,\mu S/cm$ remained fully bridged.

Figure 7-38 shows that the maximum withstand stress declines with increasing thickness of wet-grown ice, measured on a rotating monitoring cylinder, up to an ice thickness of about 30 mm for post-type insulators and 25 mm for standard disks. Level-off values of 20 mm were found for glass cap-and-pin insulators [Phan and Matsuo, 1983] and 25 mm for EPDM [Farzaneh et al., 1995].

In several specialized types of high-voltage apparatus, two insulators or three bushings are located in relatively close proximity. Kuffel et al. [1999]

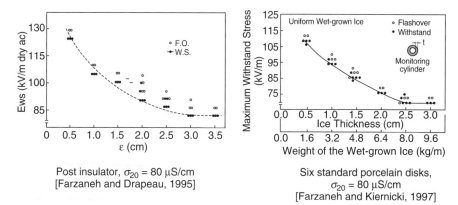

Figure 7-38: Effect of ice accretion thickness on maximum withstand stress (E_{ws}).

reported on ice tests of two parallel insulators that typically support an air-break disconnect switch, as shown in Figure 7-39. One insulator normally rotates 90° to separate a 2-m metal bus from a flange at a third insulator, not shown.

The specific insulators shown in Figure 7-39 were an experimental design, featuring a hollow fiberglass core and tapered, spiral profile of deep, widely separated silicone sheds. The polymer design had one-piece construction with a 1.84-m dry arc distance. The pair of insulators were tested by an ice layer with $\sigma_{20} = 284\,\mu S/cm$ with an average ice weight of $w = 91$ g/cm$_{dry\,arc}$, giving a severe icing stress product of $w \cdot \sigma_{20} = 26{,}000$ (g/cm) $\cdot$ ($\mu S/cm$).

With the close proximity, arcing on one insulator tended to trigger similar activity on its neighbor. In this case, the critical flashover level of $V_{50} = 135\,kV$ at +0.2 °C corresponded to a gradient of $E_{50} = 72\,kV/m_{dry\,arc}$. While they were not large, the multiple air gaps that formed under normal service voltage at the upper (energized) end of the insulators played a role in this performance.

Using an ice progressive stress (IPS) technique without a melting phase, Gutman et al. [2002] established that the flashover stress varied with the conductivity of melting water measured in their experiments. Generally, the test results with an IPS method gives flashover stresses that are considerably higher than results obtained with a melting phase because the conductivity of the ice surface never reaches its maximum value. This was the case in their study on 420-kV line insulators in Figure 7-40. The authors also reported that dripping water conductivity over a 1–3-hour period was $1.25 \cdot \sigma_{20}$, where σ_{20} was the applied water conductivity, corrected to 20 °C.

For the composite insulator, the flashover stress is fitted by $E_{50} = 560$ $(\sigma_{20})^{-0.32}$, using the observed relation that σ_{20} was 1.25 times lower than the dripping water conductivity plotted in Figure 7-40. It is not appropriate to fit the two data points for the glass insulator with a power-law curve.

Figure 7-39: Icing test on 230-kV double-column insulator supporting air-break disconnect switch.

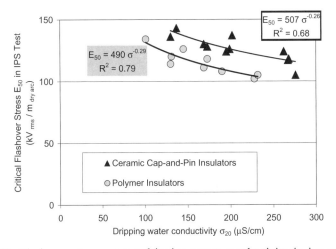

Figure 7-40: Flashover stress versus dripping water conductivity in ice progressive stress (IPS) test (data from Gutman et al. [2002]).

7.7.6. Arresters Under Heavy Icing Conditions

Kannus et al. [1998] tested surge arresters under heavy ice accretion conditions. They applied constant 68-kV line voltage to an arrester and measured internal and surface currents separately. Applied water conductivity in the range of 200–2000 µS/cm was used to create icicles. External flashover occurred when σ_{20} was greater than 500 µS/cm and a dynamic equilibrium without flashover was reached for lower values of conductivity as shown in Figure 7-41.

The leakage current inside the arrester increased to a 15-minute rms average of more than 1 mA for applied water conductivity of more than 500 µS/cm. Kannus and co-workers established a 2-mA limit for thermal instability of the metallic oxide varistor (MOV) elements in their tests at –4 °C to –9 °C.

7.7.7. Ice Flashover Under Switching and Lightning Surge

Udo [1966] noted that the switching-impulse flashover of ice-covered station post insulator stacks was as much as 50% lower than the test values for clean conditions. For ice tests on strings of standard suspension disks, the 25-unit strings had switching-surge flashover levels of +1040 and –890 kV, which were substantially the same as results in clean conditions. However, for a stack of three station posts with 205-mm diameter and 2850-mm overall length, the positive switching-impulse level fell to +750 kV from +1120 kV in clean conditions (Figure 7-42).

Tests were carried out with switching-surge waves that had a virtual front time of 120–140 µs, which differs from the standard 250-µs value used in recent test standards [IEEE Standard 4, 1995; IEC 60060, 1989].

Watanabe [1978] also carried out switching impulse tests of iced insulators, using both dry conditions and wet conditions with a water spray. A 25-unit string of standard disks was tested with a 160/1500 wave, which again differs from the standard 250/2500 wave in more recent results. Water conductivity with σ_{20} of 48 µS/cm was used. Dry flashover stresses were +360 and –380 kV/m; wet flashover stresses were +270 and –240 kV/m. These were again about one-half of the flashover stress in dry conditions.

Guerrero et al. [2005] found that, depending on the temperature, the presence of an ice layer also reduced the switching-surge flashover. At 0 °C and relative humidity of 53%, the positive (E_{50}^{+SW}) and negative (E_{50}^{-SW}) switching-surge critical flashover gradients were measured to be 562 and 711 kV/m, respectively. With a layer of ice made from rain with 80-µS/cm conductivity, these gradients both dropped to $E_{50}^{+SW} = 443\,\mathrm{kV/m}$ and $E_{50}^{-SW} = 458\,\mathrm{kV/m}$. Relative standard deviations were obtained by establishing $E_{10\%}^{+SW} = 390\,\mathrm{kV/m}$ and $E_{10\%}^{-SW} = 408\,\mathrm{kV/m}$ and then using

$$\sigma = \frac{E_{50\%} - E_{10\%}}{z}$$

Overall height A_{max} (mm)	1695	
Diameter of corona ring B (mm)	400	
Height of corona ring C (mm)	300	
Leakage distance (mm)	3739	
Dry arc distance of upper unit (mm)	800	
Dry arc distance of lower unit (mm)	500	
Major/minor shed diameters (mm)	245 / 185	
Shed spacing (mm)	60	

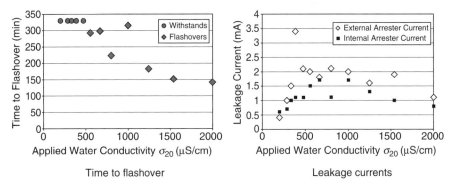

Time to flashover   Leakage currents

Figure 7-41: Dimensions of 78-kV MCOV arrester and test results with heavy icing (data from Kannus et al. [1998]).

where σ is the standard deviation and the value of z is 1.28 for a probability of 10%.

The relative standard deviations for switching-surge tests on ice-covered insulators were found to be 9.6% for positive polarity and 8.6% for negative polarity.

The relation between flashover gradient under switching impulse and conductivity of the water used to form the ice depends on polarity. The critical switching impulse stresses E_{50} found by Guerrero et al. [2005] are well fitted by the expressions

$$E_{50}^{+SW} = 497 \cdot e^{-0.0016\sigma} \quad \text{and} \quad E_{50}^{-SW} = 996 \cdot \sigma^{-0.175} \tag{7-8}$$

where E_{50} is expressed in kV/m and the conductivity σ is expressed in µS/cm. Tests were carried out at a single ice thickness, corresponding to 15 mm of

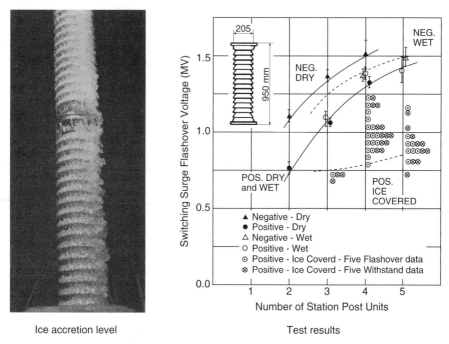

Ice accretion level   Test results

Figure 7-42: Switching-surge flashover on station post insulators (adapted from Udo [1966]).

accretion on a reference cylinder, and incorporated a melting phase. The exponent of −0.175 for the negative-polarity relation is common to the relations observed for ac flashover of iced insulators, an exponent of −0.19. However, the behavior of positive switching-surge flashover gradient does not follow this usual trend.

The relative permittivity of ice under switching impulse is much higher than that of porcelain. This results in a nonuniform distribution of the electric potential and in the increase of the electric field strength in the air gap, triggering the flashover at a relatively lower voltage value. One way to express this quantitatively is through the use of the relaxation time for the ice, which was found by Takei [2007] to change from 20 μs at 0 °C to 60 μs at −10 °C as already shown in Figure 7-5.

The values of E_{50} for lightning impulse were found to be 574 kV/m for positive polarity and 666 kV/m for negative polarity on clean post insulators. These values are consistent with normal lightning impulse flashover strength across dry arc distance of any insulator. This agreement relates to the close match of the relative permittivity of ice, with $k' = 3$ and $k'' = 0$ at the dominant lightning frequency (80–120 kHz) in Figure 7-6 to that of porcelain, leading to minimal distortion of transient electric field.

7.7.8. Effect of Diameter on ac Flashover for Heavy Icing

Watanabe [1978] tested several profiles of porcelain suspension insulators with both ac and dc, using conductivity of 20,000 Ω-cm at $0\,^{\circ}$C, equivalent to $\sigma_{20} = 83.5\,\mu$S/m. Insulators were fully bridged with ice having a weight of 2 kg per disk. Table 7-13 shows test results for ac voltage. Double strings of standard units needed 60 cm of separation to have the same minimum flashover voltage V_{MF} as single strings.

The electrical strength of the fog-type insulator string, (b), was lower than the standard string because the shed-to-shed separation of the outer edges of fog-type insulators is smaller than those of standard-type insulators. Icicles can bridge the fog-type insulators more easily as a consequence. According to the

TABLE 7-13: ac Flashover Strength of Porcelain Insulator Strings with Heavy Ice

Insulator Profile	Number of Units	String Length L (m)	Number of Strings	String Separation (cm)	V_{MF} (kV)	V_{MF}/L (kV/m)
(a)	20	2.92	1	—	430	147
	20	2.92	2	30	380	130
	20	2.92	2	40	405	139
	20	2.92	2	60	430	147
(b)	20	2.92	1	—	350	120
	20	2.92	2	40	330	113
(d)	17	2.89	1	—	405	140
	17	2.89	2	45	370	128
(e)	15	2.93	1	—	405	138
	15	2.93	2	45	415	142

Source: Watanabe [1978].

results of Watanabe [1978], increased shed-to-shed separation led to more important increases in flashover stress than reduced disk diameter.

Farzaneh et al. [2006a] also evaluated the effect of insulator diameter on minimum flashover stress. They first compared the performance of standard 146-mm × 254-mm disk insulators with a post insulator of similar diameter and 51-mm shed spacing, and an EPDM polymer insulator with alternating 90- and 105-mm sheds with 55 mm between major sheds. The values of E_{MF} were 74.2, 90.2, and 100.8 kV/m$_{dry arc}$, respectively with heavy wet-grown ice to a level of 20 mm on a rotating reference cylinder.

To decouple the influence of shed spacing and high-voltage electrode location, a series of tests were then made on the standard insulator strings. An ice layer was built up, using a mask to control the width W of the deposit as defined in Figure 7-43.

The equivalent diameter is given by $D' = 2W/\pi - 2t$, where t is the average thickness of the ice layer. This treatment is valid for $W > \pi t$. Accretion of ice gave $t = 10$ mm. Glaze ice was formed at $-12\,°C$ with a freezing water conductivity of $\sigma_{20} = 80\,\mu S/cm$. Flashover tests were carried out with evaluation in the icing regime, about 2 minutes after icing was completed, while the water film was still present on the ice surface. The tested dry arc distance was 809 mm (See Figure 7-44).

The empirical relation for the critical flashover stress $E_{50} = 141 \cdot (D')^{-0.19}$, where E_{50} is in kV/m$_{dry arc}$ and D' is in cm. This expression predicts flashover stress of $E_{50} = 76.3, 74.9,$ and 73.0 kV/m$_{dry arc}$ for the 20-mm accretion level on disks with diameters of 254, 290, and 320 mm. This 4% difference is within typical experimental errors in determining E_{50}, and also within the margin of error for the empirical fit to the observations.

Hu et al. [2007] reported test results on two different types of polymer insulators with small diameter, considered for use at 1000 kV ac. These used

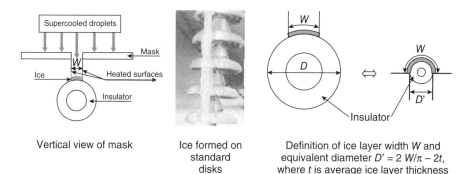

| Vertical view of mask | Ice formed on standard disks | Definition of ice layer width W and equivalent diameter $D' = 2\,W/\pi - 2t$, where t is average ice layer thickness |

Figure 7-43: Method for simulating ice accretion on different insulator diameters with same shed separation (from Farzaneh et al. [2006a]).

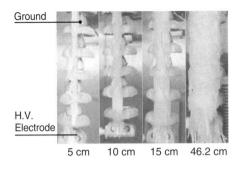

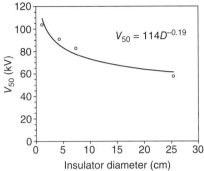

Ground

H.V.
Electrode

5 cm   10 cm   15 cm   46.2 cm

Appearance of ice layers with
W = 50, 100, 150, and 462 mm

$V_{50} = 114D^{-0.19}$

Insulator diameter (cm)

Critical flashover voltage of five standard
disks with D' of 12, 44, 75, and 254 mm

Figure 7-44: Critical flashover voltage versus ice layer equivalent diameter D' with $t = 1\,\text{cm}$.

sheds of four different diameters, with 37-mm spacing between each shed. Insulators were precontaminated to an ESDD of 0.05 mg/cm² and an NSDD of 0.30 mg/cm² prior to icing with water having $\sigma_{20} = 100\,\mu\text{S/cm}$. Glaze ice accretion was uniform because the insulators were unenergized and additionally were rotated during the accretion period at −12 °C. Table 7-14 shows the important dimensions for these insulators along with critical flashover results at an atmospheric pressure of 98.6 kPa, corresponding to an altitude of 232 m.

The critical flashover stress E_{50} was found to decline with increasing ice accretion, following a similar trend to that established in Farzaneh and Kiernicki [1997] for strings of six standard disk insulators (Figure 7-45).

The value of α, the power-law exponent relating the flashover stress to SDD, was found to be in the range of 0.186–0.211 for pressure ranging from 70 to 100 kPa. The value of n for power-law correction of flashover voltage with pressure varied from 0.40 to 0.46 for the various values of SDD.

7.7.9. dc Flashover Results for Heavy Icing

Renner et al. [1971] evaluated the possibility of ice flashover on line and station insulators using dc voltage. They concluded that the effect of polarity was not significant for heavy ice accretion. They applied tap water with a conductivity σ_{20} of about 350 μS/cm, which was much higher than their measured values of natural freezing rain and snow of $28 < \sigma_{20} < 53\,\mu\text{S/cm}$.

Their minimum flashover voltage for negative dc to the line (pin) end of the insulators was obtained when ice bridged more than 90% of the clean insulators (Figure 7-46). Positive polarity minimum flashover was only slightly higher than negative. With a 20-unit string of Type A insulators, 80% ice

TABLE 7-14: Polymer Insulator Dimensions and Icing Test Results

Typical Ice Accretion	Insulator Characteristics and Ice Severity	Profile (a)	Profile (b)
	Shed spacing (mm)	37	37
	Shed diameter (mm)	265/120/160/120/200/120/160	120/120/200/265/120/200
	Dry arc distance (mm)	3980	3980
	Leakage distance (mm)	13900	13570
	Ice Thickness on Rotating Cylinder (mm)	V_{50} in kV (standard deviation in %)	
	5	426 (4.4)	390 (5.8)
	10	344 (5.2)	317 (7.8)
	15	311 (4.6)	293 (7.2)
	20	288 (7.9)	268 (4.6)
	25	266 (4.1)	260 (5.2)
	30	267 (6.8)	260 (4.2)

Source: Hu et al. [2007].

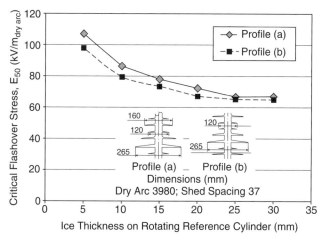

Figure 7-45: Critical flashover gradient for UHV-class polymer insulators under various icing conditions (from Hu et al., [2007]).

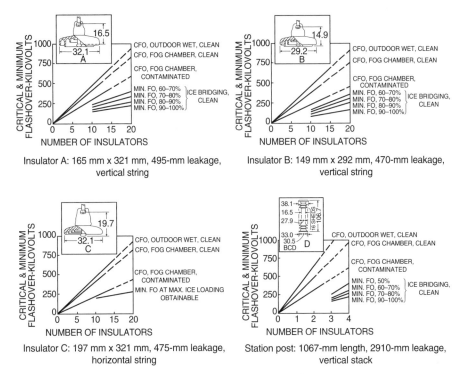

Figure 7-46: Negative dc flashover test results on insulators iced with tap water of $\sigma_{20} = 350\,\mu S/cm$ (from Renner et al. [1971]).

bridging was required to initiate a flashover at +400kV, whereas a similar specimen flashed over with 60–70% ice bridging at –400kV. The tests included accretion under normal service voltage and a "sunshine melting" phase where flashover occurred within a few minutes of dawn.

Watanabe [1978] tested the iced suspension insulators with profiles (a) and (c) in Table 7-15 with dc voltage, using applied water with resistivity of 10 or 30kΩ-cm at 0°C, corresponding to $\sigma_{20} = 167$ or 56μS/cm. His values are consolidated with those obtained by Farzaneh and Vovan [1988], Farzaneh [1991], and Farzaneh and Zhang [2000] for clean insulators, tested at $\sigma_{20} = 80–85\mu S/cm$ in icing and melting regimes as indicated in Table 7-15.

In Farzaneh and Vovan [1988] the weight of ice per 170-mm × 320-mm anti-fog disk was 1.45kg at a thickness of 20mm on a rotating reference cylinder for both polarities. The role of the melting regime was particularly strong for the negative polarity. Tests were also carried out to establish the effect of surface precontamination at two ESDD levels, 0.05 and 0.1mg/cm². (See Table 7-16).

TABLE 7-15: dc Flashover Strength of Porcelain Insulator Strings with Heavy Ice

Insulator Profile	Number of Units	Dry Arc Distance (mm)	Polarity (Regime)	σ_{20} (μS/cm)	V_{MF} (kV)	E_{MF} (kV/m)
(a) 146, 254⌀	10	1530	Positive	56	310	203
	10	1530	Negative	56	250	163
	5	809	Positive (melting)	80	72	89
	5	809	Negative (melting)	80	60	74
	14	2110	Negative	166	160	76
(c) 170, 280⌀	10	1770	Negative	56	280	158
170 mm, 320 mm	4	790	Positive (icing)	85	84	106
	4	790	Positive (melting)	82	87	110
	4	790	Negative (icing)	85	57	72
	4	790	Negative (melting)	82	75	95

Source: Adapted from Watanabe [1978], Farzaneh and Vovan [1988], Farzaneh [1991], and Farzaneh and Zhang [2000].

TABLE 7-16: V_{MF} for dc Positive and Negative Polarity for Four 170-mm × 320-mm Anti-Fog Disk Insulators Covered with 82-μS Ice

Insulator Profile	ESDD (mg/cm^2)	E_{MF} (kV/m$_{dry\,arc}$)	
		dc Positive	dc Negative
170 mm, 320 mm	0	110	95
	0.05	91	80
	0.10	110	87

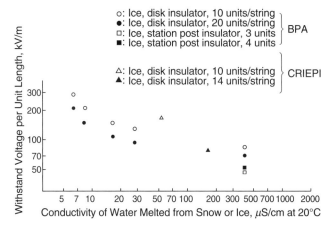

Figure 7-47: The dc negative withstand stress versus conductivity of melted ice (adapted from Fujimura et al. [1979]; data for BPA from Renner et al. [1971] and for CRIEPI from Watanabe [1978]).

The dc flashover levels had a U shape with a minimum at the ESDD level of 0.05 mg/cm^2 that was about 17% lower than the value for clean conditions. Arcing activity was much heavier for the insulators with high 0.10-mg/cm^2 (100-µg/cm^2) precontamination, leading to a type of glaze damage called "worms."

Fujimura et al. [1979] showed that the withstand voltage stress E_{WS} (kV/m$_{dry arc}$) for dc negative polarity decreased as the conductivity increased (Figure 7-47).

Farzaneh [1991] tested four anti-fog 170-mm × 320-mm disks with accretion thickness up to 25 mm on a rotating reference cylinder. Dry arc distance was 789 mm with a 680-mm connection length L. These tests included a melting regime, with a temperature rise through 0 °C after accretion of ice with $\sigma_{20} = 82\,\mu$S/cm at −12 °C. Flashover results for heavy ice accretion of 25 mm can be compared with the Watanabe values in Table 7-15.

Flashover results were also checked for light and moderate levels of accretion, giving the results for negative polarity in Figure 7-48. The percentage values of V_{MF} in Figure 7-48 are referenced to a value of 100% $E_{MF} = 72\,$kV/m$_{dry arc}$ for heavy accretion.

7.8. EMPIRICAL MODELS FOR ICING FLASHOVERS

The test results shown in the previous section cover a very wide range of ice accretion levels, with applied water conductivity also spanning a range of two orders of magnitude. It is somewhat difficult to consolidate the results without bringing them together to some common measures of stress.

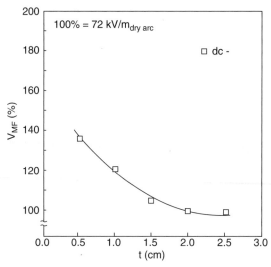

Figure 7-48: Minimum dc‾ flashover of four 170-mm × 320 mm porcelain anti-fog disks as function of ice accretion *t* on rotating reference cylinder (adapted Farzaneh [1991]).

In the case of very light icing, the insulator behavior is quite similar to that of the contaminated insulator. Flashover gradient across the leakage distance is relatively constant for insulators in the range of 1–10 m, and this gradient is a nonlinear function of ESDD. This leads to the use of a model where the surface conductivity of the leakage distance of the insulator is a function of ESDD, independent of the thickness of fog or ice accretion.

In the range of moderate to heavy icing, the insulator flashover occurs mainly across the dry arc distance. In this case, the conductivity of the applied water, augmented by some contribution of surface pollution, and the thickness of the accretion both affect the flashover stress. As a unifying concept, the product of ice conductivity and ice weight, called an *icing stress product*, gives good empirical interpolation of results and serves as a guide for the difficult task of mathematical modeling in the presence of multiple arcs.

For heavy icing, the insulator flashover across the leakage distance has a well-studied mathematical model that is almost as simple to use as an empirical model. This is described in Section 7.9.

7.8.1. The Icing Stress Product for ac Flashover Across Leakage Distance

For very light levels of ice accretion, or light levels that do not fully bridge the leakage distance, ice layer conductivity is relatively independent of ice weight on precontaminated insulators.

The influence of ice weight on layer conductivity in cold-fog conditions is negligible for a particular reason. The layer conductivity γ_s in units of μS

TABLE 7-17: Relative Contributions of Fog and Precontamination to Surface Conductivity

| Typical Fog or Cold-Fog Layer | | | Typical Pollution Layer | | |
Thickness (cm)	σ_{20} Conductivity (μS/cm)	Surface Conductivity γ_s (μS)	ESDD (mg/cm^2)	Surface Conductivity γ_s (μS)	Total Surface Conductivity γ_s (μS)
0.01	300	3	0.05	91	94
			0.2	364	367
0.05	300	15	0.05	91	106
			0.2	364	379

($M\Omega^{-1}$) is $\gamma_s = \sigma t$ and is a function only of ESDD, since every increase in t will dilute σ by the same fraction. At 20 °C, the relation $\gamma_s = 1820\,\mu$S per mg/cm^2 of ESDD [Chisholm et al., 1994] is fundamental. The majority of highly soluble ions (salt, sulfates, nitrate) are already present on the polluted insulator surface, leading to surface water with conductivity that greatly exceeds the σ_{20} of fog, which seldom exceeds 300 μS/cm as shown in Table 7-17.

For very light icing, uniform pollution level, and insulator of relatively constant diameter (such as a station post), the stretched tube that approximates the leakage distance is an appropriate approximation. Calculations of cold-fog flashover levels using this model, with a diameter of 300 mm and a quasi-water layer thickness of 1 mm, proved to be in excellent agreement with experimental results [Chisholm, discussion to Farzaneh et al., 1997]. This calculation used the approximation that $\sigma = (ESDD)/6$, where σ was in μS/cm and ESDD was in μg/cm^2. The empirical model that fits the computed flashover model on the ice cylinder closely is

$$E_{50} = 14.68 \left(\frac{ESDD}{1\,\mathrm{mg/cm}^2} \right)^{-0.358} \tag{7-9}$$

where E_{50} is the flashover stress (kV/m$_{\mathrm{leakage}}$) and $ESDD$ is in mg/cm^2.

The recommended empirical expressions [Chisholm, 2007] for critical ac flashover cap-and-pin and station post insulators based on cold-fog tests are, respectively,

$$\text{Cap and pin type:}\quad E_{50} = 18.6 \cdot (ESDD)^{-0.36} \tag{7-10}$$

$$\text{Post type:}\quad E_{50} = 12.7 \cdot (ESDD)^{-0.36} \tag{7-11}$$

where $ESDD$ is in mg/cm^2 and E_{50} is in kV$_{\mathrm{rms\ 1\text{-}g}}$/m$_{\mathrm{leakage}}$.

A direct comparison can be made between the reported flashover results for standard disk insulators [Baker et al., 2008]. The expression for clean-fog flashover of disks is

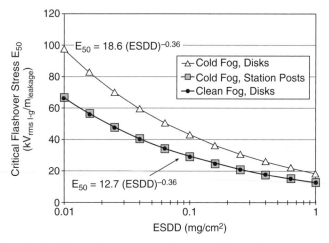

Figure 7-49: Comparison of clean-fog and cold-fog flashover stress.

$$\text{Cap and pin type:} \quad E_{50} = 12.7 \cdot (ESDD)^{-0.36} \tag{7-12}$$

Figure 7-49 shows that the clean-fog conditions are more severe than cold-fog conditions for disk insulators. However, flashover stress on station post insulators in cold-fog conditions matches the clean-fog results on ceramic disks very closely.

The close agreement again confirms that the model of icing flashover along the precontaminated leakage distance in cold-fog conditions is a manifestation of the general pollution flashover problem. The results in Figure 7-49 suggest that equivalent surface conductivity γ_e of insulators in the case of cold fog at $0\,^{\circ}\text{C}$ is about $\gamma_e = 0.4\,\gamma_s$ for the surface conductivity γ_s in the case of clean-fog wetting at $20\,^{\circ}\text{C}$. This is due mainly to the difference in equivalent conductance of sodium and chloride ions at $0\,^{\circ}\text{C}$ and $20\,^{\circ}\text{C}$, using the values in Chapter 3, Section 3.5, but is also influenced by one-sided contact of the cold-fog ice layer to the insulator surface. An "uptake factor" F is introduced later in Section 7.9.2 to describe the overall effect.

7.8.2. The Icing Stress Product for ac Flashover Across Dry Arc Distance

In contamination flashovers, the role of the resistance of the polluted layer in the Obenaus model in the propagation or extinction of surface arcing is fundamental. In the same way, the resistance of an ice, snow, or cold-fog layer is an important factor in establishing whether an arc, once initiated, will reignite and propagate in freezing conditions. If this resistance is low, the arc current will be higher. Under dc conditions, the arc will propagate further along the surface for the same electric stress. Under ac conditions, it will also take a lower voltage to reignite the arc, giving more time for extension with each half-cycle of potential.

There is a clear relation between the resistance of the ice layer and the conductivity of the water used to make up the ice. Perhaps the most significant demonstration of this effect was given by the drop in flashover voltage on line insulators found by Vucovic and Zdravkovic [1990]. They tested strings of glass disk insulators with ice formed from distilled water and from water of 8000-μS/cm conductivity. They reported a 66% reduction in flashover voltage from the former case to the latter.

In addition to the role of conductivity, the ice weight per unit length has a strong effect on the resistance of the ice or snow layer. Ice weight rather than thickness is important because each type of ice has its own density, ρ. Snow accretion, for example, has a lower density, typically $\rho = 0.3\,\text{g/cm}^3$, than glaze ice with $\rho = 0.9\,\text{g/cm}^3$ and thus needs three times the volume per unit length, v, to achieve the same mass per unit length.

For a homogeneous deposit, the weight of ice per unit length, m, can be multiplied by the conductivity of the water melted from the ice accretion, σ_{20}, to lump together all effects of ice and precontamination conductivity into a single value, called an *icing stress product* (ISP):

$$ISP = \sigma_{\text{melt water}} \cdot w = \sigma_{\text{melt water}} \cdot \rho \cdot A \tag{7-13}$$

where

- σ_{20} is the electrical conductivity (μS/cm) of the water melted from the atmospheric ice accretion,
- w is the weight of the accretion (grams per cm of dry arc distance),
- ρ is the density of the accretion, ranging from 0.3 for snow to 0.9 for ice, and A is the cross section of the accretion (cm^2), given by the volume v of the accretion (cm^3) per cm of dry arc distance.

The use of mass in the icing stress product of $\sigma_{20} \cdot w$ per unit length has proved to be a good empirical predictor of flashover performance [Farzaneh et al. 2005]. This eliminates the need to calculate the average ice cross section of the deposit. A typical sampling process for obtaining ISP from the ice accretion on a 500-kV nonceramic insulator is shown in Figure 7-50.

The first task in implementing an icing stress product model is to measure or estimate the weight per unit length of accretion on the insulator. In the test laboratory, this weight is related to the radial accretion on a reference cylinder. For 254-mm diameter ceramic disk insulator strings and 300-mm diameter station posts, the relations are expressed in Table 7-18 [Farzaneh and Kiernicki, 1997; Farzaneh et al., 2005].

w, ice weight, g per cm of dry arc distance L

times

Electrical conductivity of melted ice, σ_{20} in $\mu S/cm$

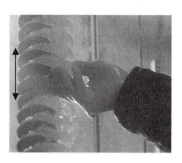

Figure 7-50: Sampling of icing stress product on nonceramic insulator.

TABLE 7-18: Relation Between Ice Accretion on Reference Cylinder and Ice Accretion on 254-mm Insulator String or 300-mm Diameter Post with Uniform Shed Profile

Ice Accretion t on Rotating Reference Cylinder (mm)	146-mm × 254-mm Standard Insulator String		300-mm Station Post Uniform 50-mm Sheds
	Ice Weight $(g/cm_{dry\,arc})$	Description	Ice Weight $(g/cm_{dry\,arc})$
5	16	Ice caps	20
10	32	Icicles start to form	40
15	48	Partial bridging	60
20	64	80% bridging	80
25	80	Maximum bridging	100

Chafiq [1995] fitted a relation of $W = 2.98\, t + 0.43$, where W was the ice weight in kg/m and t was the accumulation thickness on a rotating reference cylinder (cm) to a level of $t = 3$ cm. This is the basis of the linear relation of 3.2 g/cm ice weight per mm of accretion for standard disks in Table 7-18. Studies by Farzaneh-Dekhordi et al. [2004b] established a similar relation of 4.0 g/cm ice weight per mm of accretion of station posts fit the observations up to a level of 25 mm.

Farzaneh and Vovan [1988] established that the weight of ice per 170-mm × 320-mm anti-fog disk was 1.45 kg per disk at a thickness of 20 mm on a rotating reference cylinder for both polarities of dc. This represents a weight of $w = 85\, g/cm_{dry\,arc}$, about 33% greater than the ice accretion on standard disks under ac, where the ratio of diameter alone would give 26% more weight.

In the normal laboratory tests, the applied water conductivity σ_{20} and the conductivity of the water melted from the total ice layer are roughly equal when the insulators are clean. For precontaminated insulators, ions will diffuse from those surfaces of the insulator that receive icing into the ice layer. This will increase the conductivity of the water melted from the total ice layer, which is the value used in evaluating the icing stress product. It is relatively simple to sample an intact portion of the ice layer after a flashover test to establish the ISP value, but few researchers have made this a standard practice to date.

The following relation has been observed between flashover stress E_{50}, in kV per meter of dry arc distance, and ISP in (g/cm) · μS/cm, for transmission-class line and post insulators covered with glaze ice [Chisholm and Farzaneh, 1999; Farzaneh et al., 2005]:

$$E_{50} = 396(ISP)^{-0.19} \qquad (7\text{-}14)$$

where E_{50} is the 50% probability of flashover gradient, expressed in kV rms ac, line to ground, per meter along the ice surface (insulator dry arc distance). This equation provides a close match to experimental observations of icing flashover in Figures 7-38 and 7-51 for glaze ice accretion in the range of 5–30 mm on a rotating reference cylinder. Figure 7-51 shows that the ISP model is also suitable for moderate and heavy levels of rime ice accretion [Jiang et al., 2005].

Chen [2000] reported flashover tests on a uniformly iced, 300-mm cylinder with 17 mm of glaze thickness on an 89-mm diameter. This represents an ice deposit weight w of 51 g/cm for a glaze ice density of 0.9 g/cm^3. Chen's minimum flashover values V_{MF} in the melting regime on short ice samples were 30% higher than the values predicted with Equation 7-14.

In contrast to the results on short insulators, the results when Equation 7-14 was applied to station post insulators of 1.4–4-m length were rather better. Farzaneh-Dehkordi et al. [2004b] measured an average deposit weight of 60 g/

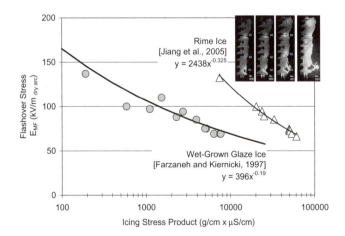

Figure 7-51: Flashover stress E_{MF} versus icing stress product for glaze ice [Farzaneh and Kiernicki, 1997; Chafiq, 1995] on 6 IEEE disk insulators and rime ice [Jiang et al, 1995] on 9 IEC disk insulators.

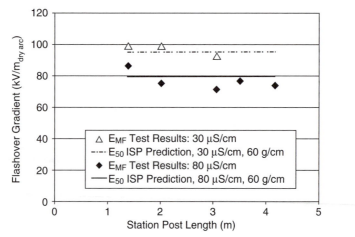

Figure 7-52: Comparison of predicted and experimental flashover gradients for station post insulators (data Farzaneh-Dehkordi et al. [2004b]).

$cm_{dry\,arc}$ on station post insulators with 50-mm shed spacing and exterior diameter that ranged from 246 to 262 mm (Figure 7-52). This deposit weight was obtained for a 15-mm accretion t on a reference cylinder, leading to a linear relation $w = 4.0\,t$ recommended in Farzaneh et al. [2005] and Table 7-18.

The predicted critical flashover stress E_{50} from Equation 7-14 for the measured ISP is constant and provides a reasonable match to E_{MF} values for the practical range of station post lengths.

The ISP can also be calculated for the case of thin ice layers, accumulated along the entire leakage distance of the insulator. Ice accumulation at typical wind speeds will be one-sided, while fog may accumulate on every surface. This situation includes the very light icing and, in particular, the cold-fog condition. In this case, the ice weight w does not play an important role in determining ISP. Every doubling of thickness t dilutes the surface contamination by a factor of 2, leading to about the same equivalent value of the surface conductivity γ associated with precontamination in Table 7-19.

Equation 7-15 [Chisholm et al., 2000; Farzaneh et al., 2005] describes the relation of flashover gradient along the entire leakage distance to the ISP for the thin ice layer:

$$E_{50} = 1196(ISP)^{-0.37} \tag{7-15}$$

where

E_{50} is the critical flashover stress in cold-fog conditions ($kV/m_{leakage}$) and,

ISP is in $(g/cm_{leakage}) \cdot (\mu S/cm)$, calculated for an ice layer thickness of $t = 1$ mm and a density of 0.9.

TABLE 7-19: Typical Pollution Severity, ISP, and Leakage Distance Flashover Stress on Line Post Insulators in Cold-Fog Conditions

Pollution Level	ESDD (mg/cm^2) $(\mu g/cm^2)$	20°C Surface Conductivity γ (μS)	Equivalent Icing Stress Product $(g/cm_{leakage} \cdot \mu S/ cm)$	Flashover Stress E_{50} Along Leakage Distance $(kV/m_{leakage})$
Very Light	<0.006(<6)	<11	<900	>96
Light	0.006–0.025 (6–25)	$11 < \gamma < 46$	900–4,000	96–55
Medium	0.025–0.1 (25–100)	$46 < \gamma < 180$	4,000–15,000	55–34
Heavy	0.1–0.4 (100–400)	$180 < \gamma < 670$	15,000–60,000	34–20
Very heavy	>0.4 (>400)	>670	>60,000	<20

The exponent in Equation 7-15 varies only slightly from the value of –0.36 in Equation 7-10 for cap-and-pin type insulators, from the correction for conductivity with salinity in thin layers of water or ice. The calculated values of flashover stress from the two expressions agree within ±3% for any ice thickness 0.005 cm < t < 1 cm covering all surfaces of strings of standard 146-mm × 254-mm disk insulators having equivalent diameter D_e = 17 cm. The equivalent diameter D_e is given by the cylinder with length equal to leakage distance whose surface area matches that of the insulators in the chain. The critical flashover stress of station posts with D_e = 47 cm is optimally described by Equations 7-11 and 7-15 for the same wide range of ice thickness.

Overall, the ISP method can be recommended as a good way to interpolate test data for specific design purposes, setting the leakage or dry arc distance, rather than as a precise method for calculating flashover strength for a particular insulator. There are better models for flashover calculations, and they appear in the next section. The specific role of the ISP approach in the insulation coordination process is in the evaluation of the joint probability of combinations of conductivity σ_{20} and accretion weight w that will cause flashover for a particular insulator dry arc distance.

7.8.3. Implementation of ISP Model for dc Flashover Under Heavy Ice Conditions

Based on Farzaneh [2000], the maximum withstand stress E_{ws} for dc positive polarity is almost the same as that for ac rms. Also, polarity had no effect on the weight of ice accretion [Farzaneh and Vovan, 1988]. These factors suggest that the ISP model for ac can also predict dc positive flashover. The withstand stress for dc negative flashover was 17% lower in the icing regime, and 30% lower in the melting regime. The ISP coefficient for this case should be reduced by the same amount:

$$E_{50dc^+} = 400(ISP)^{-0.19}$$

$$E_{50dc^-} = 340(ISP)^{-0.19}$$

(7-16)

where E_{50} is in $kV/m_{dry\,arc}$ and ISP is in $(\mu g/cm) \cdot (g/cm_{dry\,arc})$.

The dc test results from Renner et al. [1971] provide one basis for evaluating the ISP model for dc conditions under full and partial ice bridging conditions. For the most part, low values of applied water conductivity σ_{20} were sprayed on energized strings of clean 165-mm × 321-mm disks. A fully bridged weight of $w = 107\,g/cm_{dry\,arc}$ was used in the ISP model in Figure 7-53, based on results for similar disks [Farzaneh and Vovan, 1988].

For the fully bridged insulators, the values of E_{50dc^-} from Equation 7-16 fall a little below the test data but do match the slope in the range of engineering interest. Collection of the necessary data to evaluate ISP can thus be recommended as a means to interpolate results for dc tests.

7.8.4. Comparison of Ice Flashover to Wet Flashover

In Chapter 4, Figure 4-5 showed the critical wet flashover level under wet conditions to be about 450–500 kV for 2-m strings of suspension disk insulators, depending on test conditions. For tests carried out with an applied

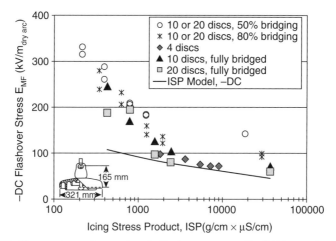

Figure 7-53: Comparison of negative dc flashover stress E_{50} ($kV/m_{dry\,arc}$) with experimental values of E_{MF} from Renner et al. [1971] and Farzaneh and Vovan [1988].

water conductivity of $100\,\mu\text{S/cm}$, Equation 7-14 can be inverted to establish the corresponding ice weight per cm of dry arc distance. For an applied water conductivity of $\sigma_{20} = 100\,\mu\text{S/cm}$, the ice weight causing critical flashover works out to only about $0.1–0.2\,\text{g/cm}$ of dry arc distance, which is an extremely thin coating, unlikely to cause full bridging. This illustrates in a crude way why icing flashovers are much more of a problem than heavy rain flashovers.

Nonuniform rain has proved to be an important flashover problem for expensive, hollow porcelain HVDC wall bushings. Depending on rain resistivity and rate, flashovers at line voltage have been reproduced in laboratory tests [Schneider and Lux, 1991]. With the accretion of ice giving a similar nonuniform distribution along the top surface of the horizontally oriented bushing, similar problems would be expected in winter conditions.

7.9. MATHEMATICAL MODELING OF FLASHOVER PROCESS ON ICE-COVERED INSULATORS

The arc discharge in series with a heavy ice layer has been studied extensively [Farzaneh et al., 1997, 2000; Farzaneh, 2000]. Generally, the electrical flashover process on heavy ice layers can be described well by adapting the Obenaus [1958] approach for dc flashover, and by taking into consideration the modifications of this model for ac arc reignition conditions [Claverie, 1971; Rizk, 1981]. This successful adaptation was based on direct laboratory measurements of arc constants, surface conductance, and reignition conditions, leading to values that differed systematically from those found for flashover on polluted surfaces. Follow-up validation of the mathematical model on full-scale insulators was also completed to verify the validity.

While the mathematical model was conceived for heavy ice layers, it became quickly apparent that it could also be used in the case of a very thin layer on the entire leakage distance of the insulator, and this was validated with full-scale test data as well. However, there are still some open questions related to the treatment of air gaps in the ice layer for light icing conditions, and there are very few flashover test data to support the validation for these cases.

This section presents a summary of the parameters for arc $V\!-\!I$ characteristics, ice layer conductivity, and reignition conditions based on a consolidation of measurements and application experience. A model for the influence of insulator surface precontamination on the ice layer conductivity is also set out. The contribution of multiple arc roots is then evaluated using a multiarc model. With this foundation, the mathematical model is then applied to each ice accretion severity, as defined by accretion t on a rotating reference cylinder in Table 7-2:

- Very light icing, $t < 1\,\text{mm}$, using the leakage distance. Ice layer conductivity in this case is dominated by the preexisting ESDD and the equivalent quasi-water layer is typically 1 mm thick. A single arc starts a rapid flash-over process. The arc path is shorter than the leakage distance.

- Light icing with $1 \leq t < 6$ mm, where the number of arcs in series may approach the number of disks in series. With 5 mm of ice, once a white arc is initiated, it proceeds to flashover in about 3 s. The arc path is shorter than the leakage distance.

- Moderate icing with $6 \leq t \leq 10$ mm, where partial ice bridging reduces the number of arcs compared to the light icing case. Generally, time to flashover with a white arc increases to the range of 10 s. The arc path follows the dry arc distance.

- Heavy icing with $t > 10$ mm, where the number of arcs reaches a minimum, one across a single air gap up to 1-m insulation and as many as three gaps for posts more than 3 m long. Time to flashover with the white arc often exceeds 30 s. The arc path follows the dry arc distance.

7.9.1. dc Flashover Modeling of Ice-Covered Insulators

The mathematical model for ice-covered insulators relies on the existence of an air gap in series with a distributed pollution layer. For constant applied dc voltage, the total supply voltage V is the sum of three components [Farzaneh and Zhang, 2000]:

$$V = V_e + A \cdot K_p x \cdot I^{-n} + IR(x) \tag{7-17}$$

where

V_e is the electrode voltage drop (V),

A and n are arc constants,

x is the position of the arc root along the leakage path (cm),

K_p is the ratio of the partial arc length in air to the arc root location x,

$R(x)$ is the resistance (Ω) along the leakage path from the location x to the ground electrode,

I is dc current (A), and

V is the dc voltage (V).

The solution of Equation 7-17 resolves into a series of U-shaped curves as x is varied from a small value (with a small arc near the high-voltage terminal) to a value of $x = L$. L is the total length of the leakage path (cm), which is approximately equal to the dry arc distance for moderate and heavy ice and is equal to the leakage distance in the case of very light icing such as cold-fog conditions. Traditionally, a uniform pollution layer $R(x) = r_p(L - x)$, with r_p in Ω/cm, was assumed as a simplification. Using this, Neumärker [1959] set the derivative of Equation 7-17 with respect to x equal to zero and obtained the critical arc length x to obtain flashover:

$$x_c = \frac{L}{n+1} \quad (7\text{-}18)$$

With the uniform pollution layer, if there is sufficient applied voltage for the arc to reach the location x_c, there is more than enough potential to complete the flashover process across the rest of the leakage distance. Typically, the value of n in the V–I characteristic of arcs is about 0.5 and this means that the value of x_c is about two-thirds of the way along the flashover path L.

By differentiating Equation 7-17 with respect to I and equating to zero, it is also possible to establish the critical current I_c and voltage V_c at flashover:

$$I_c = \left(\frac{A}{r_p}\right)^{1/(n+1)} \quad (7\text{-}19)$$

$$V_c = L \cdot A^{1/(n+1)} \cdot r_p^{n/(n+1)} \quad (7\text{-}20)$$

On a flat surface of conductive ice with $t = 5\,mm$, the arc was found to propagate either along the ice–air interface or inside the ice itself [Zhang and Farzaneh, 2000], depending somewhat on conductivity and ice temperature. The dc arc in Figure 7-54 propagates across the 280-mm distance in two main stages:

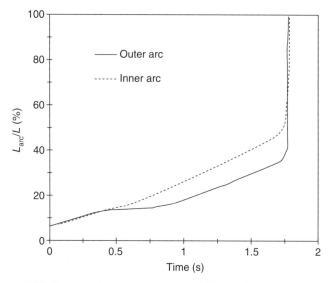

Figure 7-54: Propagation process arc at −27 kV on ice surface with σ_{20}.

- Stage 1 from 0 to 1.75 s, where the arc grows from an initial value of 0.05 L to 0.45–0.60 L, at a relatively slow speed of 0.05–0.3 m/s, for example, with 0.07 m/s in Figure 7-54.
- Stage 2, with a very rapid transition at an accelerating speed from 20 to 100 m/s across the rest of the air–ice surface once the arc reached the critical value x_c.

The electrode voltage drop is established for dc conditions by evaluating the offset in the relationship between arc voltage and arc length for a variety of current levels. The same V_e was found for static arcs and for dynamic arcs that move along the ice surface as shown in Figure 7-55.

A value of V_e = 799 V was recommended for the dc$^+$ condition in Farzaneh and Zhang [2000] along with V_e = 527 V for dc$^-$.

Unlike the value of V_e, the coefficients A and n change depending on whether the arc is static or dynamic.

After subtracting the constant value of V_e from experimental results, the arc coefficients that relate the arc gradient E_{arc} to the current are measured, plotted, and fitted as shown in Figure 7-56.

The ac flashover takes place at the peak of the ac wave. This is denoted by the use of the maximum (crest) values of voltage V_m and current I_m in later developments.

The dynamic case is of interest for flashover modeling. For the case of the dc$^+$ arc, the corresponding arc constants are A = 209 and n = 0.45, based on current (A) and arc gradient (V/cm).

The coefficient K_p in Equation 7-17 was introduced into the dc flashover model [Farzaneh and Zhang, 2000] based on the observations of arc paths on

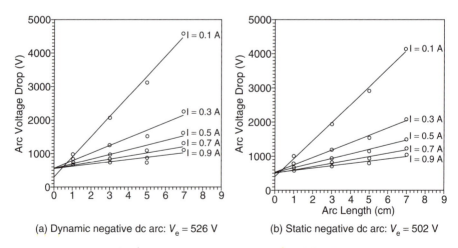

(a) Dynamic negative dc arc: V_e = 526 V   (b) Static negative dc arc: V_e = 502 V

Figure 7-55: Evaluation of arc root voltage drop V_e for static and dynamic dc$^-$ conditions (from Farzaneh et al. [1998]; courtesy of Elsevier).

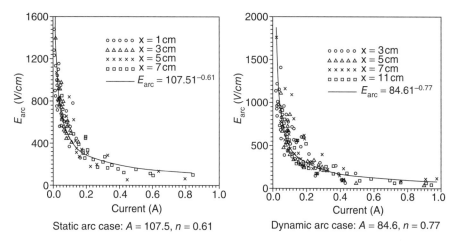

Static arc case: $A = 107.5$, $n = 0.61$ Dynamic arc case: $A = 84.6$, $n = 0.77$

Figure 7-56: Measurements for establishing arc constants A and n for dc⁻ (from Farzaneh et al. [1998]; courtesy of Elsevier).

dc flashover on ice-covered insulators. It varies between 1.2 and 1.4 for the dc case, and a median value of 1.3 was recommended for mathematical modeling of fully bridged icing. In the case of flashover under very light icing conditions, K_p is roughly equal to the ratio of dry arc to leakage distance because the partial arc develops in the air along the side of the insulator rather than following the leakage distance along the insulator surface. For this case, K_p is typically in the range of 0.3–0.5.

The final step in calculating dc flashover is the evaluation of the resistance of the ice layer in series with the arc, as already described in Section 7.3.2 of this chapter. There are two terms for this expression. The first relates to the resistance of the unbridged ice layer, described by a half-cylinder of diameter D, length L, and thickness d [Farzaneh et al., 1997; Wilkins, 1969]. The second term in Equation 7-21 adds the series resistance of N arc roots, based on the hemispherical contact of each arc root of radius r:

$$R(x) = \frac{10^6}{2\pi\gamma_e}\left[\frac{4(L-x)}{D+2t} + N\ln\left(\frac{D+2t}{4r}\right)\right] \qquad (7\text{-}21)$$

where:

$R(x)$ is the resistance (Ω) along the leakage path from the location x to the ground electrode,

γ_e is the equivalent surface conductivity (μS),

L is the length (cm) of the arc path,

x is the position (cm) of the arc root along the leakage path,

D is the insulator diameter (cm),

t is the ice thickness (cm),

N is the total number of arc roots on the ice surface, and
r is the arc root radius (cm).

The value of surface conductivity γ_e was studied for dc conditions and was found to vary with polarity and applied water conductivity σ_{20} over the range of 0–300 μS/cm as follows [Farzaneh and Zhang, 2000]:

$$\gamma_e = 0.0599\sigma_{20} + 2.59 \quad dc^- \tag{7-22}$$

$$\gamma_e = 0.082\sigma_{20} + 1.79 \quad dc^+ \tag{7-23}$$

where σ_{20} is in μS/cm and γ_e is in μS. The coefficents imply a "quasi-water" layer thickness of about 0.6–0.8 mm for highly conductive water with σ_{20} of 300 μS/cm. Even for deionized water with $\sigma_{20} < 1$ μS/cm, the chemical effects of the electrical discharge and exposure to air combine to give a nonzero level of surface conductivity. (See Figure 7-57.)

For dc arcs with $L < 1$ m, there are two arc roots, $N = 2$, representing one arc root near the line end and another near the ground end of the single ice layer [Farzaneh and Zhang, 2000].

For positive and negative arcs on the ice–air interface, Table 7-1 gave the values of arc root radius r (cm) on the ice–air surface as a function of arc current I (A):

$$r = 0.714\sqrt{I} \quad dc^- \tag{7-24}$$

$$r = 0.701\sqrt{I} \quad dc^+ \tag{7-25}$$

Since the pollution resistance along the ice surface $R(x)$ is not constant, an analytical solution such as the one proposed by Neumärker [1959] in Equation

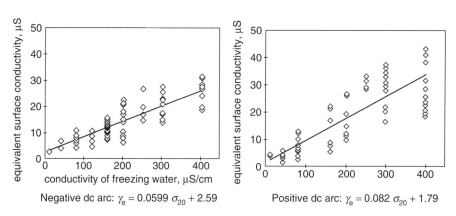

Figure 7-57: Ice surface conductivity γ_e for dc⁻ and dc⁺ conditions (from Farzaneh and Zhang [2000]; courtesy of IET).

7-20 no longer applies. A numerical approach is needed to solve Equation 7-17 using the nonuniform resistance $R(x)$ in Equation 7-21. This is done with the following process:

- The values of V_e, γ_e, A, and n for the voltage polarity are fixed.
- For each value of x along the ice surface of length L, a trial current of about 0.01 A is assumed.
- The arc root radius r corresponding to the trial current is calculated.
- The value of $R(x)$ in Equation 7-21 is calculated, using the trial radius r.
- The voltage V from Equation 7-17 is calculated for each value of x.
- The trial current I for each value of x is changed until V is minimized for this particular value of x.
- The highest of the converged values $V(x)$ is the critical flashover voltage, V_c. The corresponding arc length x_c where this maximum occurs and the current I_c that causes the flashover are also read off the spreadsheet.

For the case of five IEEE standard disk insulators with $L = 809\,\mathrm{mm}$, $D = 254\,\mathrm{mm}$, ice thickness $t = 15\,\mathrm{mm}$, and $K_p = 1.3$, the values of critical flashover stress E_{50} are calculated from this modeling process and given in Table 7-20.

The dc$^+$ flashover is predicted to be a median of 16% higher than the dc$^-$ level with the mathematical model, comparing well with the experimental value of 20%. This makes calculation of the dc$^-$ flashover voltage more important for engineering applications.

Some insight into the numerical calculation process can be gained by plotting the source voltage V as a function of x as shown in Figure 7-58.

There is a large difference between the flashover level predicted for a uniform pollution layer from Equations 7-18 to 7-20, and the dc$^+$ flashover level from the numerical model using two arc roots. While they are convenient, the analytical solutions using the derivatives in fact fail for the dc case with two arc roots in the example shown.

TABLE 7-20: Calculated Values of E_{50} for dc$^+$ and dc$^-$ Flashover of Five Standard Disk Insulators

E_{50} (kV/m$_{dry\ arc}$)		Applied Water Conductivity σ_{20} (μS/cm)				
		10	20	40	80	150
dc$^-$	Calculated	109.1	100.7	88.8	74.0	60.1
	Experimental				74	
dc$^+$	Calculated	128.7	117.1	102.3	86.2	71.9
	Experimental				89	

Source: Adapted from Farzaneh and Zhang [2000].

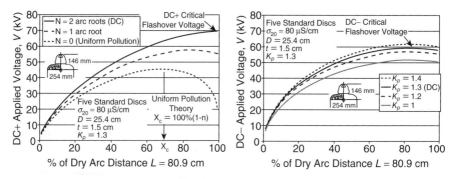

Figure 7-58: Minimum values of source voltage V to sustain dc arcs at distance x.

The influence of the arc path factor K_p in Figure 7-58 is not as strong but should still be included in the mathematical model for the heavy icing flashover calculation. These graphs also show that the mathematical model of dc⁻ flashover of five standard disk insulators is 60 kV, compared to 70 kV for the dc⁺ flashover.

7.9.2. Influence of Insulator Precontamination on dc Flashover of Ice-Covered Insulators

As mentioned in Chapter 3, the precontamination levels on HVDC insulators may be higher on the insulators with positive polarity at the line end, compared to negative polarity. The difference is at least 20% and may reach a factor of 2 or 3 in areas of low industrial contamination. This leads into the question of how to include insulator surface precontamination in the calculation of the dc⁺ and dc⁻ flashover voltage in icing conditions.

When a half-cylinder of ice forms on insulators, the deposit tends to form on the windward side and does not come into contact with a large fraction of the bottom surface of typical disks or polymer insulators. For typical shed-to-shed separation and 45° incidence angle leading to heavy incidence of freezing rain, about 20% of the ions on the top surface of insulators will migrate into the ice layer. A fraction F is defined to represent in this heavy icing case the function of partial contact of the ice to the full top surface, related to sheltering from the shed above. The fraction F for heavy icing also takes into account the effects of pollution loss through drip water.

The additional contribution of precontamination to the surface pollution can simply be modeled by evaluating the contribution of the ESDD to the applied water conductivity σ_{20}, and using this combined value in Equations 7-22 and 7-23. This involves solving for the conductivity as a function of salinity S_a using the expression in IEEE Standard 4 [1995].

The amount of salt m that migrates into and is retained in a heavy ice layer deposited on the insulator is

$$m = F \cdot ESDD \cdot A_{top} \qquad (7\text{-}26)$$

where

F is the contamination uptake factor from the insulator surface into the ice layer,

$ESDD$ is the equivalent salt deposit density (mg/cm^2),

A_{top} is the total area (cm^2) of the upward-facing insulator surfaces, and

m is the equivalent weight of salt (mg).

As an example, each standard disk insulator has 691 cm^2 of top surface area, so a string of five disk insulators would have 3455 cm^2 of surface area. An ESDD of 0.05 mg/cm^2 with an uptake factor $F = 0.2$ for heavy ice accretion would give a value of $m = 34.5$ mg.

The ice volume in contact with the insulator is given by the accretion thickness t (cm), the insulator diameter D (cm), and the dry arc or leakage distance L (cm), depending on whether or not the insulator is fully bridged. The water volume V_{water} (mL) is equal to the ice volume times the ice density of 0.9

$$V_{water} = 0.9 \cdot \frac{\pi}{4} \left((D + 2t)^2 - D^2 \right) \cdot L \qquad (7\text{-}27)$$

In the case of very light icing along the entire leakage distance, such as that produced in cold-fog conditions, it should be considered that F remains 0.4 for station posts and reduces to about $F = 0.15$ for disk insulators with less exposure. In this case, the total top and bottom surface area A_{total} (cm^2) of the insulators determines the salt mass m (mg) using $m = F \cdot ESDD \cdot A_{total}$. The water volume (cm^3) is given by $V_{water} = 0.06$ cm $\cdot A_{total}$, using a minimum value of $t = 0.07$ cm, derived from Figure 7-57, to represent the surface conductance of the quasi-water layer even though it may only be 0.01 cm thick. Also, as noted previously, the arc path factor K_p for this case is equal to the dry arc distance divided by the leakage distance.

In every case, whether for heavy or light icing, the salinity of the water $S_a = m/V_{water}$ (mg/cm^3) is then used to calculate the equivalent applied water conductivity $\sigma_{p\,20}$ (μS/cm) [IEEE Standard 4, 1995]:

$$\sigma_{p20} = 10^4 \cdot \exp\left[\frac{\ln S_a}{1.03} - \ln(5.7) \right] \qquad (7\text{-}28)$$

The total value of σ_{20} used in Equations 7-22 and 7-23 to calculate surface conductivity γ_c will be the sum of the natural precipitation conductivity $\sigma_{n\,20}$ and the additional contribution $\sigma_{p\,20}$ from the ESDD:

$$\sigma_{20} = \sigma_{n\,20} + \sigma_{p\,20} \qquad (7\text{-}29)$$

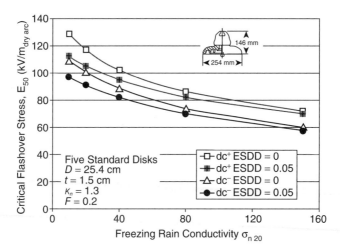

Figure 7-59: Predicted dc$^+$ and dc$^-$ flashover stress for five standard disk insulators.

For the example of five standard disks, with 1.5 cm of ice, with $F = 0.2$, an ESDD of 0.05 mg/cm^2 would raise σ_{20} from 80 to 95 μS/cm. This is predicted to have a modest effect on the critical flashover stress E_{50} of a string of five standard disk insulators under heavy icing conditions, as shown in Figure 7-59.

As might be anticipated, the effect of ESDD is greatest for the cases where the freezing rain conductivity is low, and the effect is seen to be about 10% for a value of $\sigma_{n\ 20} = 10$ μS/cm.

It is also seen that the dc$^+$ flashover stress for polluted insulators is predicted to remain well above the dc$^-$ flashover stress for clean insulators. This confirms that the evaluation of the dc$^-$ performance under icing conditions should be the first concern of experimental validation for the mathematical model.

7.9.3. ac Flashover Modeling of Ice-Covered Insulators

There are several changes needed to adapt to the arc model introduced for dc conditions in Equation 7-17 to ac conditions. The electrode voltage drop V_e term is smaller than the dc case than for ac, and its effect can be considered directly in the calculation of the other terms. The coefficient K_p is unity for ac flashover along an ice surface that bridges the dry arc distance, rather than 1.3 for the dc arc case, because the effect of "arc float" is not observed under ac in most practical cases. It is retained for use in calculation of ac flashover for very light ice conditions, where K_p becomes the ratio of dry arc to leakage distance in the mathematical model. Finally, the peak values of voltage V_m and current I_m are used in the analysis, as the flashover tends to occur at the peak of the ac wave.

The updated equation, including the reignition coefficient, becomes

$$V_m = A \cdot K_p x \cdot I_m^{-n} + I_m R(x) \qquad (7\text{-}30)$$

where

V_m is the peak of the applied ac voltage (V),

I_m is peak of the ac current (A),

A and n are arc constants,

$K_p x$ is the length of the partial arc, where x is the position (cm) of the arc root along the leakage path and K_p is the ratio of the partial arc length in air to the arc root location x, and

$R(x)$ is the resistance (Ω) along the leakage path from the location x to the ground electrode.

The values of $A = 205$ and $n = 0.56$ are recommended for ac arcs on the ice surfaces, based on the experimental results in Figure 7-60.

As for dc flashover, the ac arc may travel in two different ways: through the air over the outer surface of the ice, or inside the ice. When the arc takes the path inside the ice, the propagation speed is slower, but this somewhat random variation does not affect the flashover voltage. Arc velocity for ac flashover is lower than for dc in the first stage of propagation but the maximum velocity before flashover is higher for ac than dc in the second stage [Zhang and Farzaneh, 2000].

The expression for the equivalent surface conductivity γ_e of ice in ac conditions, using the data from Figure 7-3 [Farzaneh et al., 1997], is

$$\gamma_e = 0.0675\sigma_{20} + 2.45 \quad ac \tag{7-31}$$

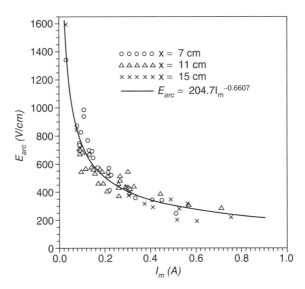

Figure 7-60: Characteristics of arc voltage gradient $E = V_m/x$ versus current on ice surfaces (from Farzaneh et al. [1997]).

where σ_{20} is the applied water conductivity (μS/cm) corrected to 20°C.

This value is used along with the number of arc roots N and the arc root radius r to calculate $R(x)$:

$$R(x) = \frac{10^6}{2\pi\gamma_e}\left[\frac{4(L-x)}{D+2t}+(N'+2N'')\ln\left(\frac{D+2t}{4r}\right)\right] \qquad (7\text{-}32)$$

where:

R(x) is the resistance (Ω) along the leakage path from the location x to the ground electrode,

γ_e is the equivalent surface conductivity (μS),

L is the length (cm) of the arc path,

x is the position (cm) of the arc root along the leakage path,

D is the insulator diameter (cm),

t is the ice thickness (cm),

N' is the total number of arcs from a metal electrode to the ice surface,

N'' is the total number of arcs across air gaps from ice surfaces to ice surfaces, and

r is the arc root radius (cm).

Equation 7-32 introduces a refinement to Equation 7-21 that is needed when describing heavy ice accumulation on long 4-m insulators, and also when evaluating performance of insulator strings that are not fully bridged by ice. The total number of arc roots N in Equation 7-21 is expressed as the sum of the number of arcs N' from metal to ice and the number of arcs N'' spanning an air gap between two ice surfaces.

The values of $N' = 2$ and $N'' = 0$ are appropriate for the calculation of dc flashover of insulators less than 1 m long.

The number of arc roots has been well studied for ac conditions on insulators up to 4 m in length [Farzaneh and Zhang, 2007]. For example, real utility insulators of 1400-mm dry arc distance tend to form two air gaps and there are three gaps on the station post with 3500-mm dry arc distance in Figure 7-61.

Only the arc roots that form on ice surfaces will cause a nonuniform current distribution along the ice surface that in turn influences the resistance of the residual ice layer. Farzaneh and Zhang [2007] evaluated the various conditions of ice gaps in series and then simplified the results into a multiarc model that accounts for the number of arcs with a single root on the ice surface (N') and the number of arcs with two roots (N''). The total number of arcs is $N = N' + N''$. The multiarc model is used in the following examples:

- In cases of partial ice bridging, where there is more than one gap to be bridged by arcing.

 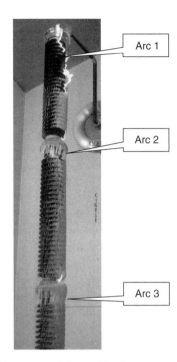

1.4-m dry arc distance: $N = 2$ 3.5-m dry arc distance: $N = 4$

Figure 7-61: Typical arcs and air gaps on station post insulators in icing conditions (from Farzaneh and Zhang [2007]).

- In cases of HV and EHV insulation, as the number of air gaps tends to increase with dry arc distance.
- To evaluate the improvement in strength when additional air gaps are deliberately introduced using insulator accessories such as booster sheds or creepage extenders.

Examples in the Farzaneh and Zhang [2007] paper with $N'' = 1$ included a 2.68-m string of standard disk insulators and a 2.70-m station post fitted with six booster sheds.

To complete the calculation of $R(x)$, the arc root radius for an outer arc path, Equation 7-32 uses the recommended value in Table 7-1 of arc root radius r (cm) in terms of peak current I_m (A) as

$$r = 0.603\sqrt{I_m} \quad ac \qquad (7\text{-}33)$$

In addition to satisfying Equation 7-30 with the appropriate constants for ac arcs, the peak voltage V must also satisfy an ac arc reignition condition that will be described later. Before showing the mathematical model, an explanation about the reignition process on the ice surface will be provided.

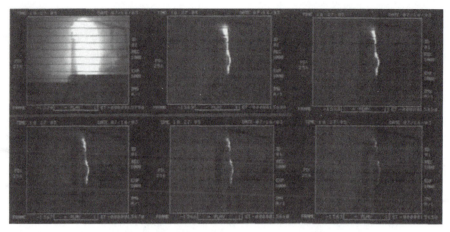

Figure 7-62: Decay of switching-surge flashover arc at 1-ms frame rate (from Guerrero [2004]; courtesy of T. Guerrero).

Under icing conditions, if the voltage is suddenly switched off, the arc does not disappear immediately. As an illustration, switching-surge impulse tests, carried out with a standard 250/2500-μs waveform, show the effect of thermal persistence after flashover. Figure 7-62 [Guerrero, 2004] shows the decay of a switching-surge discharge channel on an ice-covered insulator with 963-mm dry arc distance and 2500-mm leakage distance.

The applied impulse voltage collapses to zero in the first frame of Figure 7-62 and the generator short-circuit impulse current decays rapidly as well. The discharge channel is still visible 5 ms later. The persistence of luminance indicates that the gas is still hot, and thus susceptible to reignition and flashover at a reduced level should a second surge, or a half-cycle of ac voltage, be applied.

Under ac voltage, the current passes through zero twice in each cycle. After initiation of arcing, a local arc will extinguish and may reignite twice in each cycle. For each half-cycle of ac voltage, the arcing on thick ice surfaces extends and retreats. The equilibrium is illustrated in Figure 7-63 showing a series of photos of arcs on a 250-mm tall by 200-mm wide ice sample with an initiating gap cut into the triangle apex.

Figure 7-63 shows that the 60-Hz ac voltage reaches its peak at 4.2 ms, but also illustrates that the arc development persists after the voltage peak for another 3 ms. The presence of the arc leaves a thermal signature in the surrounding air and on the ice surface that favors arc reignition under ac voltage. This is shown, for example, by the time lag after the voltage peak and the persistence of luminosity after 9 ms.

Several researchers have studied the conditions of ac reignition for the specific case of a polluted insulator, notably Claverie [1971], Rizk [1971, 1981], and Claverie and Pocheron [1973]. The recommended adaptation of this mathematical model to the ice surface [Farzaneh et al., 1997] is that an arc reignition

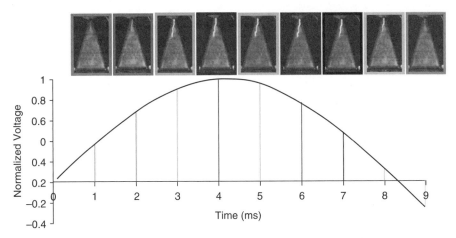

Figure 7-63: Variation of arc length on 250-mm tall by 200-mm wide ice sample under 60-Hz ac voltage: frame rate of 1 ms (adapted from Zhang and Farzaneh [2000]).

condition must be satisfied at every location of the arc root along the ice surface. This requirement is expressed by Equation 7-34:

$$V_m \geq \frac{k \cdot K_p x}{I_m^b} \tag{7-34}$$

The critical condition where the arc retains just enough thermal energy to reignite on the next half-cycle of power frequency voltage is

$$V_m = \frac{k \cdot K_p x}{I_m^b} \tag{7-35}$$

where

k and b are the arc reignition constants,

$K_p x$ is the length of the partial arc (cm) at the position x along the leakage path,

V_m is the peak value of applied voltage (V), and

I_m is the peak value of current (A).

A number of studies at the University of Québec in Chicoutimi [Farzaneh et al., 1997; Farzaneh-Dehkordi et al., 2004] have established these coefficients for ice layers at 60 Hz:

$k = 1118$ for an arc propagating upward, with natural arc buoyancy,

$k = 1300$ for an arc propagating downward, against natural arc buoyancy, and

$b = 0.5277$.

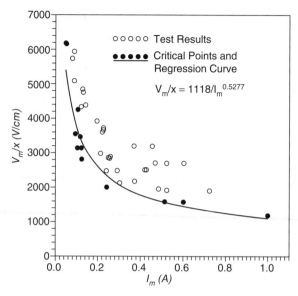

Figure 7-64: Measured values of 60-Hz ac reignition coefficient on ice surfaces (from Farzaneh et al. [1997]).

A typical set of experimental results in Figure 7-64 established k and b for an upward-propagating arc on an ice surface over the important current range from 0 to 1 A.

The thermal energy radiated by arcs tends to make them buoyant, giving, for example, the effect of a Jacob's ladder. This buoyancy has an effect on the reignition coefficients. The arc at the base of an insulator string is easier to maintain and propagate than an arc at the top of the insulator string. The physics leads to different experimental values of the arc reignition constant k that differ by about 16%.

The arc reignition process on the ice surface has coefficients that differ from those normally used for polluted surfaces, for example, $k = 800, b = 0.5$ [Claverie and Pocheron, 1973] or $k = 716, b = 0.526$ [Rizk and Rezazada, 1997]. These values for polluted surfaces at 20 °C are one-third lower than the ice surface at 0 °C. The difference is not explained simply by the effect of lower ambient temperature on the radiation of heat from the arc channel [Rizk and Rezazada, 1997], as discussed in Chapter 5. Arc propagation velocity on ice surfaces [Zhang and Farzaneh, 2000] is also somewhat slower than that observed on polluted surfaces [Ghosh and Chatterjee, 1995; Li et al., 1989] and this may be an alternate factor to temperature in explaining why the arc reignition process on iced surfaces has different coefficients from those found on polluted surfaces.

The process for calculating the ac flashover voltage follows the same general steps as that outlined for dc flashover. The peak current I_m is initialized. The corresponding arc root radius r is established using Equation 7-33. This value is used along with γ_e from Equation 7-31, modified as described for the effects of surface pollution, to compute $R(x)$ for x all along the leakage path from 0 to L. Then, Equation 7-30 is used to establish V_m.

At this stage, the calculation process for ac flashover diverges from dc. Rather than searching for the value of I_m that minimizes V_m, for most practical cases the minimum reignition condition in Equation 7-35 will establish the flashover level. A numerical search for the value of I_m that minimizes the square of the difference between Equations 7-31 and 7-35 for each value of x is efficient. With modern spreadsheets, it is possible to search for at least 100 solutions of I_m at the same time by minimizing the sum of all the squared differences.

The maximum value of V_m for all values of x establishes the peak of the ac flashover voltage. The rms line-to-ground voltage $V_m/\sqrt{2}$ is normally reported, and it is also common to calculate the critical flashover stress E_{50}:

$$E_{50} = \frac{V_m}{10 \cdot L\sqrt{2}}$$

where

E_{50} is the critical flashover stress (kV/m),

V_m is the critical flashover voltage (V), and

L is the leakage path (cm), which may be the dry arc distance across the ice surface $(L_{\text{dry arc}})$ or the leakage distance of the insulator, (L_{leakage}) depending on icing level.

The intersection between the reignition voltage curve is illustrated for two values of x along a leakage path of $L = 1\,\text{m}$ in Figure 7-65. The intersection normally takes place at a current that is higher than the current that gives a minimum in the V_m–I_m characteristic.

In calculations for extremely high conductivity or low arc length, it is possible that the inequality in Equation 7-34 may be satisfied by the minimum value of V_m, and the intersection of the reignition curve with the V_m–I_m curve would occur to the left of the minimum. These are not cases of practical interest.

7.9.4. Application Details: Flashover under Very Light Icing

Under very light icing conditions, the following parameters are recommended:

- The calculation is carried out on the basis of the leakage path L, along the leakage distance.
- The true position of the arc root x along the leakage path is corrected with a value of $K_p = 0.85$, since the arc does not follow every contour of

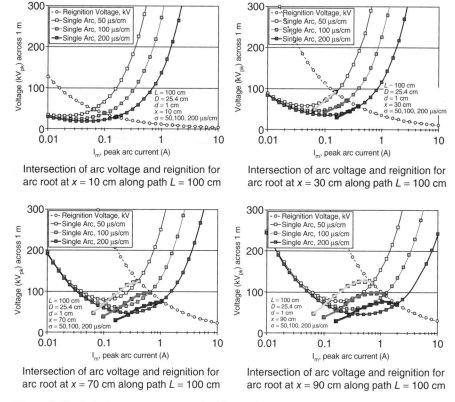

Figure 7-65: Relations among arc, reignition voltages, and arc current for arc lengths of 10–90 cm on 100-cm ice surface.

the insulator leakage distance. This factor is recommended for modeling of both strings or stacks of disk insulators and for station posts in cold-fog or light icing conditions.

- There is only one arc root on the ice surface, so $N' = 1$. As soon as this arc is initiated at any region of high electric field near a metal electrode, it extends rapidly over the insulator surface without breaking apart into multiple arcs and dry bands.

- In the calculation of equivalent surface conductivity γ_e, a quasi-water layer thickness of 0.0675 cm from Equation 7-31 is used along with the relation between conductivity and salinity in Equation 7-28. The contribution of fog accretion to the thin surface, σ_n, uses the same uptake factor F as that used for ESDD. This uptake factor is $F = 0.4$ for station posts, but falls to about $F = 0.15$ for vertical strings or stacks of disks. In both cases, upper insulator surfaces are exposed. The bottom surfaces of station posts are also nearly fully exposed, while bottom surfaces of typical disk insulators

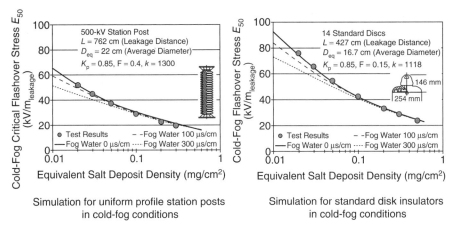

Figure 7-66: Comparison of measured cold-fog flashover stress with simulated results for very light icing.

are partially sheltered when the wind direction in nature or in the test laboratory is horizontal. As for the case with heavy icing, it is considered that half of the insulator faces the wind, and the calculation of $R(x)$ uses the same equations, with L as the leakage distance rather than the dry arc distance.

- The equivalent diameter D of the insulator in the calculation of $R(x)$ is given by the cylinder with leakage distance L that has the same surface area as the insulators. For example, with a standard disk of surface area $1599\,cm^2$ and leakage distance $30.5\,cm$, the equivalent diameter $D = (1599\,cm^2)/(\pi \cdot 30.5\,cm)$ or $16.7\,cm$.

With the recommended factors in the mathematical model, simulations of critical flashover levels for disk and post insulators in Figure 7-66 decline from about $90\,kV/m_{leakage}$ to levels of significant engineering interest, of 20–$40\,kV/m_{leakage}$ in the range of 0.05–$0.4\,mg/cm^2$ (50–$400\,\mu g/cm^2$).

The mathematical model described above provides an excellent match to the empirical fits to test results in Equations 7-10 and 7-11, especially for medium and heavy pollution levels. The effect of the conductivity of the fog itself is predicted by the model to play a role in the insulator performance for light contamination levels with ESDD below $0.04\,mg/cm^2$.

7.9.5. Application Details: Flashover under Light Icing Conditions

One difficulty in using the mathematical model for light icing conditions is that there are very few test results to validate the predictions. For example, the CIFT results in Table 7-7 compare two 500-kV post insulators with different leakage distance with the same precontamination level and applied water conductivity. Tests over a wide range of these parameters would normally be needed. This pair of test results is supplemented only by ice tests at accretion levels of $t = 0.5$ cm on a reference cylinder for clean strings of standard porcelain disks.

Flashover in light icing conditions without bridging depends mainly on the ESDD of the bottom surfaces of the insulator and the leakage distance. The ice layer forms on the top surfaces of the disks or sheds, with icicles that do not fully bridge the dry arc distance. Thus the base of the light icing calculation continues to use L as the leakage distance.

A very simple model for accretion would be to consider that any iced area reduces the leakage path. Thus the flashover stress calculated per meter of leakage distance in very light icing conditions could simply be reduced by the same fraction. However, one difficulty in this approach is evidence, such as in Figure 7-67, that the flashover path in light icing conditions often does not follow the icicles, and sometimes even finds its way between them.

If the model for very light pollution is used in the case of light icing flashovers, it seems most reasonable to change only one parameter. It is recommended that the arc path fraction K_p should be reduced from 0.85 to 0.6. This represents the influence of icicles around the skirts of the insulator, which tend to modify the arc path and bring it closer to the dry arc distance. In cases where

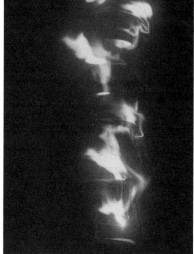

String of disk insulators   Stack of cap-and-pin apparatus insulators

Figure 7-67: Examples of flashover arcs under partial ice bridging conditions.

TABLE 7-21: Comparison of Mathematical Model and Test Results for Light Icing Conditions

Insulator Type	Applied Water Conductivity (μS/cm), ESDD (mg/cm²)	Dry Arc, Leakage Distance (mm)	Equivalent Diameter D (cm)	Calculated Flashover Stress, kV/m$_{\text{dry arc}}$ (kV/m$_{\text{leakage}}$)	Measured Flashover Stress, kV/m$_{\text{dry arc}}$ (kV/m$_{\text{leakage}}$)
Station post, alternating profile	50, 0.02	2743, 9800	26	138 (38.7)	122 (34.3)
Station post, uniform profile	50, 0.02	2743, 7620	22	115 (41.3)	100 (36.1)
Five standard Cap-and-pin disks	80, 0	809, 1525	17	106 (56.5)	113 (59.9)

icicles nearly close the gaps between sheds, the arc path fraction may drop as low as its theoretical minimum for this case, which is the dry arc distance divided by the leakage distance. (See Table 7-21.)

In each case for the application of the heavy icing model to light icing conditions, the agreement between mathematical model and experiment is imperfect. Some aspects, such as the decrease in flashover stress with increasing insulator station post diameter in the same test conditions, are modeled well. However, additional test experience with a systematic and controlled variation of ESDD and applied water conductivity would be needed to adapt the model better to this icing severity.

7.9.6. Application Details: Flashover under Moderate Icing Conditions

In the application of the mathematical model to conditions with moderate or heavy icing, it is recommended that the base of the calculation change to use the insulator dry arc distance for L. Typical polymer and post insulators with uniform sheds are fully bridged and the shed separation between cap-and-pin insulators is significantly reduced by the intrusion of icicles. The leakage path becomes the dry arc distance, and for ac conditions, $K_p = 1.0$. The effective diameter D of the ice cylinder becomes the outer diameter of the sheds, rather than an average value as used for very light or light conditions.

The number of arc roots can theoretically equal the number of air gaps in the insulator. For cap-and-pin insulators, this could equal the number of disks. Table 7-22 shows the recommendations for appropriate number of arc roots N on the ice surface, based on both insulator type and insulator system voltage.

Table 7-22 recommends the use of three arc roots for modeling the flashover performance of cap-and-pin insulators in moderate icing conditions.

TABLE 7-22: Recommended Number of Arc Roots in Multi-Arc Model

Moderate Ice Accretion Severity ($6 < t \leq 10\,\mathrm{mm}$) on Rotating Reference Cylinder	System Voltage (Phase to Phase)		
	$V \leq 110\,\mathrm{kV}$	$110\,\mathrm{kV} < V < 315\,\mathrm{kV}$	$V \geq 315\,\mathrm{kV}$
	Number of Arc Roots $N = (N' + N'')$		
Cap-and-pin insulators with 146-mm shed spacing	3	Not established	Not established
Polymer and station posts with uniform 50-mm sheds	1	2	3

Source: Adapted from Farzaneh and Zhang [2007].

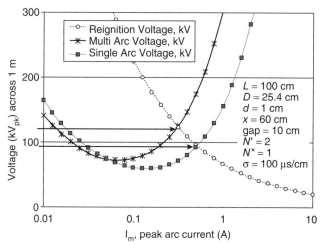

Figure 7-68: Effect of multiple arcs on balance point between arc and reignition voltage.

Under the same icing conditions, by definition, the polymer and station post sheds are fully bridged. This means that the number of partial arcs N is reduced to one for relatively short insulators. Test results [Farzaneh and Zhang, 2007] have shown that there will be up to three air gaps that form in ice layers on 4.2-m station post insulators.

The role of the multiarc model, compared to a single-arc model, is illustrated in the curve of ac reignition voltage versus V_m in Figure 7-68. For an insulator with 1-m dry arc distance, the model predicts that an ice layer with a central air gap will have a balance point of 124 kV, compared to 94 kV for the single-arc model, for the dimensions given in Figure 7-68. This difference

TABLE 7-23: Comparison of Measured and Modeled Maximum Withstand Stress for Five IEEE Standard Disk Insulators

Test Conditions	$t = 0.6$ cm	$t = 0.8$ cm	$t = 1.0$ cm
Model ($kV/m_{dry\,arc}$)	103.7	103.9	104.1
Measurement ($kV/m_{dry\,arc}$)	106	102	98

Source: Farzaneh and Kiernicki [1997].

of 30 kV is much larger than would be predicted from a fixed value of 0.95 kV per arc root on water surfaces from Wilkins and Al-Baghdadi [1971].

Validation of the numerical model for moderate icing conditions faces some of the same problems as validation for light icing. While there are more test series, as reported in Section 7.7.4 of this chapter, few of the tests included any sort of melting phase that leads to the lowest levels of flashover voltage. As an example, simulation of the results of Cherney [1980] for 320-μS/cm water and ESDD of 0.4 mg/cm^2 with a thickness $t = 0.4$ cm on a rotating reference cylinder for an insulator with $L = 340$ cm and $D = 13.5$ cm gives a result of 43 kV/m$_{dry\,arc}$, where his test results show flashover levels at $-5\,°C$ that are twice this value. A second example, simulating the tests on five clean standard disks with $\sigma_{20} = 80\,\mu$S/cm, gave the results shown in Table 7-23.

At this point, the modeling is acceptable but the measured change in flash-over level with accretion thickness is better described by an empirical icing stress product model in Equation 7-14. The refinement of the mathematical model for moderate icing conditions also needs a program of systematic testing to establish how the number of arcs increases as a function of insulator length and voltage class.

7.9.7. Application Details: Flashover Under Heavy Icing Conditions

The mathematical model for icing flashover was developed for the heavy icing condition and has proved to work well, with an acceptable level of detail for design work. The model is based on the leakage path being fully bridged, so that L becomes the dry arc distance. The equivalent surface conductivity γ_e of the halfCcylinder that covers the outer diameter D of the insulator is established only by the applied water conductivity σ_{20}, and not by the ice thickness, in the case of clean insulators. The ice thickness does affect the contribution of precontamination on γ_e.

For dc, the value of K_p is set to 1.3, and it remains $K_p = 1$ for ac because in general the ac arc does not float off the insulator surface.

In the case of heavy icing, the formation of air gaps along the ice layer depends on the voltage distribution along the insulator string during the ice accumulation process. This voltage distribution is affected by the type and length of the insulator. Generally, insulators with less than 1 m of dry arc distance (550 kV BIL) form a single air gap under line voltage during heavy icing tests. For longer insulators used in HV and EHV networks, the voltage distribution is nonuniform, particularly after icicles start to grow. Under these conditions, more than one air gap may be formed. The number, lengths, and positions of the air gaps vary for different insulator types and for different lengths of the same insulator type, as shown in the recommendations for the value of N in Table 7-24.

The performance of the multiarc model for ac flashover using the values of N from Table 7-24 is illustrated by a comparison of flashover test results with predictions for station post insulators in Figure 7-69. The predictions of a linear

TABLE 7-24: Recommended Number of Arc Roots in Multi-Arc Model

Ice Accretion Severity (t on Rotating Reference Cylinder, mm)	System Voltage (phase to phase)		
	$V \le 110\,kV$	$110\,kV < V < 315\,kV$	$V \ge 315\,kV$
	Number of Arc Roots $N = (N' + N'')$		
Heavy ($10 < t \le 15$)	2	4	6
Heavy ($t > 15$)	1	2	3

Source: Adapted from Farzaneh and Zhang [2007].

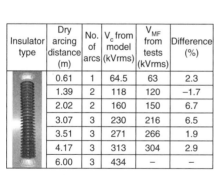

Insulator type	Dry arcing distance (m)	No. of arcs	V_c from model (kVrms)	V_{MF} from tests (kVrms)	Difference (%)
	0.61	1	64.5	63	2.3
	1.39	2	118	120	−1.7
	2.02	2	160	150	6.7
	3.07	3	230	216	6.5
	3.51	3	271	266	1.9
	4.17	3	313	304	2.9
	6.00	3	434	–	–

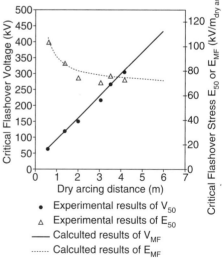

Insulator type, simulation details, and results

Comparisons: V_{50} to V_{MF}; E_{50} to E_{MF}

Figure 7-69: Comparison of experimental and calculated critical flashover level for station post insulators (adapted from Farzaneh and Zhang [2007]).

increase in V_{50} with increasing dry arc distance beyond a value of 1 m, that give a constant E_{50} stress per meter of dry arc distance, are well supported by the experimental results using 2 cm of thickness and 80-µS/cm water conductivity. Similar agreement has been found for insulators with heavy accretion of 30-µS/cm water, with and without surface precontamination [Farzaneh-Dehkordi et al., 2004b].

7.10. ENVIRONMENTAL CORRECTIONS FOR ICE SURFACES

The correction of ice flashover results for pressure is relevant to transmission lines located in high altitude. There are few substations in these regions because of the high cost of moving large equipment.

At present, the correction of icing flashover for temperature is still under development. Mathematical modeling treats the most important case, which is when the equivalent surface conductivity near the melting point is at its highest level. However, test results obtained using ice surface temperatures well below the melting point lead to significantly higher flashover levels. This is a problem involving multiple domains of electrochemistry, evaporation and sublimation rates, arc physics, and other aspects. The dissipation of power in the ice layer is one fundamental factor in this problem that has already been studied, and is reported here as an initial step toward an icing flashover model with full corrections for temperature of the ice layer.

7.10.1. Pressure Correction for Heavy Ice Tests

Tests of post and disk insulators [Farzaneh et al., 2004a] showed that the minimum ac flashover voltage V_{MF} of ice-covered insulators decreased with a decrease in atmospheric pressure. Figure 7-70 shows that the flashover strength of a heavily iced post insulator was measured to about $125\,kV/m_{dry\,arc}$ for applied water conductivity of $\sigma_{20} = 80\,µS/cm$ at standard 101.3-kPa pressure. Minimum flashover stress dropped to a level of $67\,kV/m_{dry\,arc}$ in the case with $\sigma_{20} = 250\,µS/cm$ at a pressure of 45 kPa, corresponding to high altitude of 6500 m.

Strings of disk insulators were also tested with salt and kieselguhr precontamination giving ESDD of 0.015 or 0.100 mg/cm² and NSDD of 1.0 mg/cm². The test results in Figure 7-71 show the measured minimum flashover voltage for three levels of ice accretion w, expressed in kg per m of dry arc distance.

Normally, correction for air pressure makes use of an exponent m as follows:

$$V = V_0 \left(\frac{P}{P_0} \right)^m \tag{7-36}$$

where V is the flashover voltage at ambient pressure P, P_0 is 101.3 kPa, and V_0 is the flashover voltage at pressure P_0. The exponent m depends in

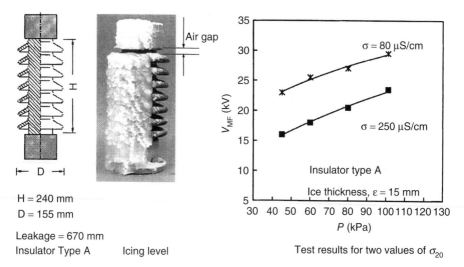

H = 240 mm
D = 155 mm

Leakage = 670 mm
Insulator Type A   Icing level   Test results for two values of σ_{20}

Figure 7-70: Effect of pressure on V_{MF} of post insulator under heavy ice conditions (from Farzaneh et al. [2004a]).

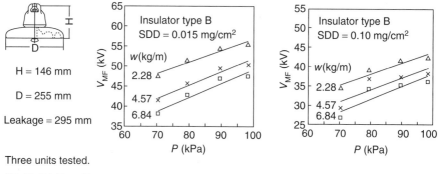

H = 146 mm

D = 255 mm

Leakage = 295 mm

Three units tested.

XP-16 disk Type B   ESDD = 0.015 mg/cm² (15 μg/cm²)   ESDD = 0.10 mg/cm² (100 μg/cm²)

Figure 7-71: Effect of pressure and ESDD on V_{MF} of disk insulators under heavy ice conditions (from Farzaneh et al. [2004a]).

general on the type of voltage stress, insulator profile, and pollution severity. Typical values for polluted insulators [Rizk and Rezazada, 1997] are in the range of $0.47 < m < 0.68$ in the range of 60–100 kPa. The value of m in icing tests also increased with increasing icing and pollution severity as shown in Table 7-25.

The variation of m for icing flashover voltage as a function of pressure is in the same range as the values obtained for pollution flashovers, with a strong tendency toward higher values of m for more severe or more conductive ice layers.

TABLE 7.25: Effect of Pollution Severity on Exponent *m* in Pressure Correction of Flashover Voltage

		Light	Heavy
Post insulator Type A 15-mm thickness	Conductivity (μS/cm) m	80 0.40	250 0.53
Three disk insulators XP-16 Type B	ESDD (mg/cm^2) m for $w = 22.8\,g/cm$ m for $w = 45.7\,g/cm$ m for $w = 68.4\,g/cm$	0.015 0.42 0.53 0.58	0.100 0.53 0.60 0.63

Source: Farzaneh et al. [2004a].

7.10.2. Arc Parameter Variation with Temperature and Pressure

Rizk and Rezazada [1997] provided the mathematical foundation to evaluate the effect of ambient temperature on the reignition condition for contaminated insulators. This effect was noted to change the reignition coefficient by about 20% in the ambient temperature range of 0–40 °C in Section 5.6.2. This does not fully explain why the reignition coefficients observed on ice surfaces ($k = 1118$ for buoyant arcs or $k = 1300$ for those starting at the top of the insulator) in Section 7.9.3 differ from the value of $k = 716$ obtained for 20 °C. It may be possible to resolve the present discrepancy through validation of the predicted arc temperature on the ice surface, using spectroscopic measurements or other experimental techniques.

7.10.3. Heat Transfer and Ice Temperature

The resistance of the ice layer is dynamic, and calculation of the resistance calls for a reasonably accurate assessment of the ice temperature.

Electrical energy from the white arc phase provides the power source. The latent heat of fusion of ice would normally give 3 mL of melt water per kJ of electrical energy. The energy furnished to four 170-mm × 320-mm porcelain anti-fog disks with 545-mm leakage distance under icing conditions was studied by Dallaire [1992] for ac and dc conditions. Energy was found to increase with ice thickness *t*, reaching a level of 600–800 kJ for arcing activity leading to withstands for $t = 25\,mm$ of accretion of applied water conductivity $\sigma_{20} = 80\,\mu S/cm$. There was a strong linear relation between the volume of melted water and the energy, with a slope of 1700 mL per 800 kJ or 2.1 mL/kJ. This implies that about 30% of the arc energy is lost through radiation, evaporation, and convection heat transfers. As an indication of power level, Tavakoli [2004] established a power dissipation during the arc propagation phase of about 300 VA/cm for cylinders and post insulators.

Chafiq [1995] also studied energy dissipation in the flashover of a string of six standard disk insulators as a function of accretion level *t*, with the results in Table 7-26.

TABLE 7-26: Energy Dissipation in Flashover of Six IEEE Standard Disk Insulators

t (mm)	Applied Water σ_{20} (μS/cm)	Drip Water σ_{20} (μS/cm)	Ice Weight w (g/cm$_{\text{dryarc}}$)	Drip Water Loss (g/cm$_{\text{dryarc}}$)	V_{MF} (kV)	E_{MF} (kV/m$_{\text{dryarc}}$)	Maximum Current (mA)	Energy (kJ)	Equivalent Arcing Period (s)
5	80	—	19	—	105	110	622	194	3
10	80	—	34	—	90	94	548	620	13
15	80	—	49	—	81	85	698	940	17
20	80	—	64	—	72	75	622	1570	35
25	80	—	79	—	66	69	668	1600	36
30	80	—	94	—	66	69	805	1800	34
20	2.9[a]	64	66	8	111[a]	137[a]	362	1000	25
20	9	86	65	11	96	100	420	1200	30
20	20	110	55	16	93	97	460	1400	33
20	40	149	57	16	84	88	510	1100	26
20	80	206	62	4	72	75	622	1600	36
20	149	229	51	13	66	69	670	1650	37

[a]Five units tested.

Source: Adapted from Chafiq [1995].

The tendency toward short arcing duration for low levels of ice accretion continues for very light accumulation levels. There are typically fewer than 10 ac cycles of leakage current activity prior to flashover in cold-fog conditions with $t < 1\,\text{mm}$ as the ice layer reaches the melting point.

7.11. FUTURE DIRECTIONS FOR ICING FLASHOVER MODELING

The mathematical model presented in Section 7.9 gives satisfactory estimates of flashover performance for very light and heavy icing severity.

Very light accumulation of ice along the entire leakage distance of an insulator with surface contamination (ESDD) leads to flashover stress E_{50} that:

- Is independent of ice thickness.
- Has a power-law exponent relation of $E_{50} \propto (ESDD)^{-0.36}$.
- Is nearly constant for leakage distance length in the range of interest for high-voltage networks.

The predictions of the mathematical model for very light icing conditions are in close agreement with test results obtained in cold-fog conditions.

Heavy accumulation of ice, bridging the entire dry arc distance of the insulator, with ice formed from applied water with conductivity σ_{20}, leads to a predicted flashover stress E_{50} that:

- Is relatively insensitive to icing thickness.
- Has a power-law exponent relation of about $E_{50} \propto (\sigma_{20})^{-0.18}$.
- Has the same power-law exponent relation with insulator diameter.
- Is also nearly constant for dry arc distance in the range of interest for high-voltage networks.

The predictions are in close agreement with test results obtained in well-controlled tests using both melting and icing regimes.

At this point, the best application advice is to use the very light icing model for light icing conditions with short icicles that do not bridge the leakage distance of the insulator, and to use the heavy icing model for moderate icing conditions on insulators that are fully bridged such as station posts and polymers with small shed-to-shed separation.

There are a number of directions that should be explored to refine the mathematical model so that it describes every icing case well. This refinement, however, should not proceed without validation of predictions that reproduce flashover problems of engineering interest, backed up by better understanding of the basic phenomena related to discharge initiation, and its propagation to a complete flashover arc.

7.11.1. Streamer Initiation and Propagation on Ice Surfaces

The process of streamer initiation and development along an ice surface has been studied with increasing sophistication, using lightning impulse voltage and ultra-high-speed optical measurements [Brettschneider et al., 2004; Farzaneh and Fofana, 2004; Ndiaye et al., 2007]. These validated a streamer onset criterion using a form factor and established that the streamer development is significantly changed by the presence of the ice layer in many ways.

The effects of several experimental parameters such as freezing water conductivity and electrode radius and form factor on the streamer inception parameters of an ice surface have been examined in all cases. Differences in the "dry" and "wet" regimes were found at temperatures well below freezing and near 0 °C, respectively.

Compared to discharge in air alone, ice–air interface discharge shows the following distinctive characteristics:

- Ice reduces the electric field needed for streamer inception.
- Ice increases the streamer propagation velocity, and the higher the surface conductivity and temperature, the greater the degree of increase.
- There is a corresponding increase in streamer currents and associated charge for each increase in streamer velocity.
- The effect of surface conductivity was in some cases stronger than the effect of the dielectric constant of the ice at lightning impulse "frequency" of about 100 kHz.

According to Ndiaye et al. [2007], the individual or combined influence of the electron emission from the ice surface by photoionization or field extraction was an important source of the excess charge that accumulated on the ice surface prior to streamer development. This deposited charge enhanced local electric fields more than permittivity of the ice and may also have led to the increased streamer propagation velocity on the ice surface.

The improved understanding and modeling of streamer initiation and propagation on the ice surface from these and additional studies will lead to better predictions of flashover on the iced surfaces of insulators.

7.11.2. Dynamics of Arc Motion on Ice Surfaces

The present mathematical model for flashover in Section 7.9 assumes that the arc is quasi-static and generally does not incorporate a condition for arc motion. The flashover calculation implies that arc propagation from a given position to the grounded end of the leakage path takes place within a small fraction of time at the instant that the applied voltage reaches its peak.

The experimental evidence has clearly shown that the arc is not fixed in one position under ac conditions, but instead grows and retreats with every half-cycle of ac. There are some retardation effects as well, that show up as a time lag between the arc position and the corresponding point of wave on applied voltage. In spite of these remarks, the static model successfully predicts flashover levels of a wide range of ice shapes and geometries, including the half-cylinder that tends to accrete on insulators in heavy icing conditions.

There has been considerable progress in studies of the dynamic behavior of the arc on the ice surface. This has been achieved with the introduction of electrical circuit elements to model capacitive loading C of the ice layer resistance and series inductance L of the arc [Tavakoli et al., 2006; Farzaneh et al., 2006a; Fofana and Farzaneh, 2007]. The RC and L/R time constants are, respectively, 13 μs and 13 ns. More important than this refinement to the circuit representation, however, is the ability of the dynamic model to predict the leakage current across the ice surface as a function of applied voltage. The arc parameters throughout the period of propagation were determined by using Mayr's equation, which is more general than the Ayrton static arc equation and has a closer link to the arc temperature and radius. The follow-up work recommended to validate this approach [Tavakoli, 2004] includes simultaneous measurements of both arc radius and arc temperature, which is presently assumed to be constant at about 4500–5000 K.

With an improved dynamic model of the arc position and current, recalculation of the instantaneous stress on the remaining portions of the insulator becomes feasible. This would lead to a better understanding of the flashover process, first on single ice layers and possibly with applications in the more difficult cases of light and moderate icing severity with several air gaps in series.

7.11.3. Dynamic Model for Ice Temperature

In Section 7.3.1, experiments showed that there was a large change in ice conductivity with ice temperature just near the freezing point. The use of a single expression relating equivalent surface conductance to applied water conductivity in Section 7.3.2 is a foundation of the success of the mathematical model in Section 7.9, but this relation was established specifically on a wet ice surface during the most important melting phase of the flashover process.

The change of surface water conductivity and water film thickness during the preflashover stage takes into account all of the heat transfer terms including conduction into the ice layer, evaporation, and radiation, as well as the heat capacity of the ice layer. This last term is very important, as it is small for thin ice layers, leading to fast times to flashover, and relatively high for thick ice layers. A dynamic model for ice surface temperature and the related change in ice layer conductivity would improve predictions of the time-to-flashover for the light and moderate icing cases, with either the present static multiarc model or the new dynamic models that are evolving now.

7.12. PRÉCIS

Freezing rain events with moderate accumulation, in the range of 6–10 mm, may not bring down phase conductors but may cause formation of a continuous ice layer along one side of many insulators. As long as temperature remains below the freezing point, this ice accumulation is usually a trouble-free insulator even though its leakage and dry arc paths are the same length. When melting eventually occurs, the ice behaves as a resistive layer with a shortened leakage path that may lead to flashover. The icing flashover process is similar in some respects to that of pollution flashover, and thus the adaptation of pollution flashover models to the heavy icing situation is particularly successful.

The flashover strength of an ice-covered insulator can be evaluated in artificial tests, using some of the same equipment and methods used in standard tests for heavy rain and contamination performance. The icing tests of main interest are those that lead to flashovers at normal service voltage. Test results show that electrical strength decreases with increasing ice weight and with increasing electrical conductivity of the melted ice. The product of these two parameters, called the icing stress product (ISP), can be used to select an appropriate dry arc distance in ways that are similar to the use of wind-on-ice maps for mechanical strength requirements.

REFERENCES

Baker, A. C., M. Farzaneh, R. S. Gorur, S. M. Gubanski, R. J. Hill, G. G. Karady, and H. M. Schneider. 2008. "Selection of Insulators for ac Overhead Lines in North America with Respect to Contamination," *IEEE Transactions on Power Delivery*, in press.

Beauséjour, Y. 1999. "Caractérization du comportement électrique d'isolateurs 735 kV en milieu naturel," in *Proceedings of the Association Francophone Pour Le Savior Colloquium on Icing*, Canada.

Bernal, J. D. and R. H. Fowler. 1933. "A Theory of Water and Ionic Solution, with Particular Reference to Hydrogen and Hydroxyl Ions," *Journal of Chemical Physics*, Vol. 1, No. 8, pp. 515–548.

Boyer, A. E. and J. R. Meale. 1988. "Insulation Flashover Under Icing Conditions on the Ontario-Hydro 500 kV Transmission Line System," in *CEA Spring Meeting*, Montreal, Canada. March.

Brettschneider, S., M. Farzaneh, and K. D. Srivastava. 2004. "Nanosecond Streak Photography of Discharge Initiation on Ice Surfaces," *IEEE Transactions on Dielectrics and Electrical Insulation*, Vol. 11, No. 3 (June), pp. 450–460.

Chafiq, M. 1995. "Comportement électriqe des isolateurs standards IEEE recouverts de glace," Mémoire de Maitrise en Sciences Appliquées, Université du Québec à Chicoutimi.

Charneski, M. D., G. L. Gaibrois, and B. F. Whitney. 1982. "Flashover Tests of Artificially Iced Insulators," *IEEE Transactions on Power Apparatus and Systems*, Vol. PAS-101, No. 8 (Aug.), pp. 2429–2433.

Chen, X. 2000. "Modeling of Electrical Arc on Polluted Ice Surfaces," PhD Thesis, Ecole Polytechnique de Montréal (Feb.).

Cherney, E. A. 1980. "Flashover Performance of Artificially Contaminated and Iced Long-Rod Transmission Line Insulators," *IEEE Transactions on Power Apparatus and Systems*, Vol. PAS-99, pp. 46–52.

Cherney, E. A. 1986. "Ice Bridging Flashover of Contaminated 500 kV Line Insulation in Freezing Rain." Ontario Hydro Report 86-80-H. April 8.

Chisholm, W. A. 1997. "North American Operating Experience: Insulator Flashovers in Cold Conditions," in *CIGRE Study Committee 33 Colloquium*, Toronto, Ontario, Canada, Paper 4.3, pp. 1–4. July.

Chisholm, W. A. 2007. "Insulator Leakage Distance Dimensioning in Areas of Winter Contamination Using Cold-Fog Test Results," *IEEE Transactions on Dielectrics and Electrical Insulation*, Vol. 14, No. 6, pp. 1455–1461.

Chisholm, W. A. and M. Farzaneh. 1999. "Une nouvelle mesure pour évaluer le comportement électrique des isolateurs recouverts de glace," Colloque *Le givrage atmosphérique et ses effets sur les équipements des réseaux électriques*, organized by CIGELE, ACFAS 67th Conference, University of Ottawa. May.

Chisholm, W. A. and Y. T. Tam. 1990. "Outdoor Insulation Studies Under CIFT Conditions, December 30–31, 1989 Storm Event." Ontario Hydro Research Division Report 90-131-H.

Chisholm, W. A., P. G. Buchan, and T. Jarv. 1994. "Accurate Measurement of Low Insulator Contamination Levels," *IEEE Transactions on Power Delivery*, Vol. 9, No. 3 (July), pp. 1552–1557.

Chisholm, W. A., K. G. Ringler, C. C. Erven, M. A. Green, O. Melo, Y. Tam, O. Nigol, J. Kuffel, A. Boyer, I. K. Pavasars, F. X. Macedo, J. K. Sabiston, and R. B. Caputo. 1996. "The Cold Fog Test," *IEEE Transactions on Power Delivery*, Vol. 11, Oct., pp. 1874–1880.

Chisholm, W. A., J. Kuffel, and M. Farzaneh. 2000. "The Icing Stress Product: A Measure for Testing and Design of Outdoor Insulators in Freezing Conditions," in *Proceedings of International Workshop on High Voltage (IWHV) 2000*, Tottori, Japan.

CIGRE Task Force 33.04.09. 1999. "Influence of Ice and Snow on the Flashover Performance o Outdoor Insulators, Part I: Effects of Ice," *Electra*, No. 187, Dec., pp. 91–111.

Claverie, P. 1971. "Predetermination of the Behaviour of Polluted Insulators," *IEEE Transactions on Power Apparatus and Systems*, Vol. PAS-90, No. 4 (July), pp. 1902–1908.

Claverie, P. and Y. Pocheron. 1973. "How to Choose Insulators for Polluted Areas," *IEEE Transactions on Power Apparatus and Systems*, Vol. PAS-92, No. 3 (May), pp. 1121–1131.

Cole, K. S. and R. H. Cole. 1941. "Dispersion and Absorption in Dielectrics I: Alternating Current Characteristics," *Journal of Chemistry and Physics*, Vol. 9, pp. 341–351.

Drapeau, J-F. 1989. "État de l'art concernant les essais et le comportement des isolateurs sous glace." Rapport scientifique, Hydro-Québec IREQ-4490. November.

Dallaire, M-A. 1992. "Contournement en courant continu et alternative des isolateures givres, precontamines ou propres," *Mémoire Maîtrise en Sciences Appliquées*, Université du Québec à Chicoutimi. January.

Erven, C. C. 1988. "500-kV Insulator Flashovers at Normal Operating Voltage," in *CEA Engineering and Operating Division Meeting*, Montreal. March.

Farzaneh, M. 1991. "Effects of the Thickness of Ice and Voltage Polarity on the Flashover Voltage of Ice Covered High-Voltage Insulators," in *Proceedings of 7th International Symposium on High Voltage Engineering*, Dresden, pp. 203–206. August 26–30.

Farzaneh, M. 2000. "Ice Accretions on High-Voltage Conductors and Insulators and Related Phenomena," *Philosophical Transactions of the Royal Society*, Vol. 358, No. 1776 (Nov.), pp. 2971–3005.

Farzaneh, M. and J-F. Drapeau. 1995. "AC Flashover Performance of Insulators Covered with Artificial Ice," *IEEE Transactions on Power Delivery*, Vol. 10, No. 2 (Apr.), pp. 1038–1051.

Farzaneh, M. and I. Fofana. 2004. "Experimental Study and Analysis of Corona Discharge Parameters on an Ice Surface," *Journal of Physics D: Applied Physics*, Vol. 37, No. 5, pp. 721–729.

Farzaneh, M. and J. Kiernicki. 1995. "Flashover Problems Caused by Ice Build-Up on Insulators," *IEEE Electrical Insulation Magazine*, Vol. 11, No. 2 (Mar./Apr.), pp. 5–17.

Farzaneh, M. and J. Kiernicki. 1997. "Flashover Performance of IEEE Standard Insulators Under Ice Conditions," *IEEE Transactions on Power Delivery*, Vol. 12, No. 4 (Oct.), pp. 1602–1613.

Farzaneh, M. and J. L. Laforte. 1992. "Effect of Voltage Polarity on Icicles Grown on Line Insulators," *International Journal of Offshore and Polar Engineering*, Vol. 2, No. 4 (Dec.), pp. 298–302.

Farzaneh, M. and J. L. Laforte. 1993. "Ice Accretion on Conductors Energized by AC or DC: A Laboratory Investigation of Ice Treeing," in *Proceedings of the Third International Offshore and Polar Engineering Conference*, Singapore, pp. 663–671. June 6–11.

Farzaneh, M. and O. T. Melo. 1990. "Properties and Effect of Freezing Rain and Winter Fog on Outline Insulators," *Cold Region Science and Technology*, Vol. 19, pp. 33–46.

Farzaneh, M. and O. Melo. 1994. "Flashover Performance of Insulators in the Presence of Short Icicles," *International Journal of Offshore and Polar Engineering*, Vol. 4, No. 2 (June), pp. 112–118.

Farzaneh, M. and M. L. Vovan. 1988. "Ice Accumulation on DC Power Transmission Line Insulators," in *Proceedings of 4th International Conference on Atmospheric Icing of Structures (IWAIS)*, Paper B5.3, pp. 301–304.

Farzaneh, M. and J. Zhang. 2000. "Modelling of DC Arc Discharge on Ice Surfaces," *IEE Proceedings on Generation, Transmission and Distribution*, Vol. 147, No. 2 (Mar.), pp. 81–86.

Farzaneh, M. and J. Zhang. 2007. "A Multi-Arc Model for Predicting AC Critical Flashover Voltage of Ice-Covered Insulators," *IEEE Transactions on Dielectrics and Electrical Insulation*, Vol. 14, No. 6 (Dec.), pp. 1401–1409.

Farzaneh, M., Y. Li, and Y. Teisseyre. 1990. "Effect of DC and AC Corona Discharges on the Accretion of Ice on H. V Conductors," in *5th International Workshop on Atmospheric Icing of Structures*, Tokyo, Paper B46, pp. 1–4. October/November.

Farzaneh, M., J. Kiernicki, and J-F. Drapeau. 1993. "AC Flashover Performance of HV Under Glaze and Rime," in *Proceedings of IEEE Conference on Electrical Insulation and Dielectric Phenomena*, Pocono Manor, pp. 499–507. October.

Farzaneh, M., J. Kiernicki, R. Chaarani, J-F. Drapeau, and R. Martin. 1995. "Influence of Wet-Grown Ice on the AC Flashover Performance of Ice-Covered Insulators," in *Proceedings of 9th International Symposium on High Voltage Engineering*, Graz, Austria, Paper No. 3176, pp. 1–4. August/September.

Farzaneh, M., J. Zhang, and X. Chen. 1997. "Modeling of the AC Arc Discharge on Ice Surfaces," *IEEE Transactions on Power Delivery*, Vol. 12, No. 1 (Jan.), pp. 325–338.

Farzaneh, M., J. Zhang, and X. Chen. 1998. "DC Characteristics of Local Arc on Ice Surface," *Atmospheric Research*, Vol. 46, pp. 49–56.

Farzaneh, M., J. Zhang, R. Chaarani, and S. M. Fikke. 2000. "Critical Conditions of AC Arc Propagation on Ice Surfaces," in *IEEE International Symposium on Electrical Insulation*, Anaheim, California, USA, pp. 211–215.

Farzaneh, M., J-F. Drapeau, C. Tavakoli, and M. Roy. 2002. "Laboratory Investigations and Methods for Evaluating the Flashover Performance of Outdoor Insulators," in *Proceedings of International Workshop on Atmospheric Icing of Structures IWAIS 2002*, Brno, Czech Republic, Session 6, pp. 1–7. June 17–20.

Farzaneh, M., T. Baker, A. Bernstorf, K. Brown, W. A. Chisholm, C. de Tourreil, J. F. Drapeau, S. Fikke, J. M. George, E. Gnandt, T. Grisham, I. Gutman, R. Hartings, R. Kremer, G. Powell, L. Rolfseng, T. Rozek, D. L. Ruff, D. Shaffner, V. Sklenicka, R. Sundararajan, and J. Yu. 2003. "Insulator Icing Test Methods and Procedures: A Position Paper Prepared by the IEEE Task Force on Insulator Icing Test Methods," *IEEE Transactions on Power Delivery*, Vol. 18, No. 3 (Oct.), pp. 1503–1515.

Farzaneh, M., Y. Li, J. Zhang, L. Shu, X. Jiang, W. Sima, and C. Sun. 2004a. "Electrical Performance of Ice-Covered Insulators at High Altitudes," *IEEE Transactions on Dielectrics and Insulation*, Vol. 11, No. 5 (Oct.), pp. 870–880.

Farzaneh-Dehkordi, J., J. Zhang, and M. Farzaneh. 2004b. "Experimental Study and Mathematical Modeling of Flashover on Extra-High Voltage Insulators Covered with Ice," *Hydrological Processes*, Vol. 18, pp. 3471–3480.

Farzaneh, M., T. Baker, A. Bernstorf, J. T. Burnham, T. Carreira, E. Cherney, W. A. Chisholm, R. Christman, R. Cole, J. Cortinas, C. de Tourreil, J. F. Drapeau, J. Farzaneh-Dehkordi, S. Fikke, R. Gorur, T. Grisham, I. Gutman, J. Kuffel, A. Phillips, G. Powell, L. Rolfseng, M. Roy, T. Rozek, D. L. Ruff, A. Schwalm, V. Sklenicka, G. Stewart, R. Sundararajan, M. Szeto, R. Tay, and J. Zhang. 2005. "Selection of Station Insulators with Respect to Ice and Snow—Part I: Technical Context and Environmental Exposure," *IEEE Transactions on Power Delivery*, Vol. 20, No. 1 (Jan.), pp. 264–270.

Farzaneh, M., J. Zhang, and C. Volat. 2006a. "Effect of Insulator Diameter on AC Flashover Voltage of an Ice-Covered Insulator String," *IEEE Transactions on Dielectrics and Electrical Insulation*, Vol. 13, No. 2 (Apr.), pp. 264–271.

Farzaneh, M., I. Fofana, I. Ndiaye, and K. D. Srivastava. 2006b. "Experimental Studies of Ice Surface Discharge Inception and Development," *International Journal of Power and Energy Systems*, Vol. 26, No. 1 (Jan.), pp. 34–41.

Farzaneh, M. (Chair), A. C. Baker, R. A. Bernstorf, J. T. Burnhan, E. A. Cherney, W. A. Chisholm, R. S. Gorur, T. Grisham, I. Gutman, L. Rolfseng, and G. A. Stewart. 2007a. "Selection of Line Insulators with Respect to Ice and Snow—Part I: Context and Stresses, A Position Paper Prepared by the IEEE Task Force on Icing Performance of Line Insulators," *IEEE Transactions on Power Delivery*, Vol. 22, No. 4 (Oct.), pp. 2289–2296.

Farzaneh, M., A. C. Baker, R. A. Bernstorf, J. T. Burnhan, E. A. Cherney, W. A. Chisholm, I. Fofana, R. S. Gorur, T. Grisham, I. Gutman, L. Rolfseng, and G. A. Stewart. 2007b. "Selection of Line Insulators With Respect to Ice and Snow—Part II: Selection Methods and Mitigation Options, A Position Paper Prepared by the IEEE Task Force on Icing Performance of Line Insulators," *IEEE Transactions on Power Delivery*, Vol. 22, No. 4 (Oct.), pp. 2297–2304.

Fikke, S. M., J. E. Hanssen, and L. Rolfseng. 1993. "Long Range Transported Pollution and Conductivity on Atmospheric Ice on Insulators," *IEEE Transactions on Power Delivery*, Vol. 8, No. 3 (July), pp. 1311–1321.

Fletcher, N. H. 1970. *The Chemical Physics of Ice*. Cambridge, UK: Cambridge University Press.

Fofana, I. and M. Farzaneh. 2007. "Application of Dynamic Model to Flashover of Ice-Covered Insulators," *IEEE Transactions on Dielectrics and Electrical Insulation*, Vol. 14, No. 6 (Dec.), pp. 1410–1417.

Forrest, J. S. 1936. "The Electrical Characteristics of 132-kV Line Insulators Under Various Weather Conditions," *IEE Proceedings*, pp. 401–423.

Forrest, J. S. 1969. "The Performance of High Voltage Insulators in Polluted Atmospheres," in *IEEE Power Engineering Society Winter Meeting*, New York. January 26–31.

Fujimura, T., K. Naito, Y. Hasegawa, and T. Kawaguchi. 1979. "Performance of Insulators Covered with Snow or Ice," *IEEE Transactions on Power Apparatus and Systems*, Vol. PAS-98, No. 10 (Oct.), pp. 1621–1631.

Ghosh, P. and N. Chatterjee. 1995. "Polluted Insulator Flashover Model for ac Voltage," *IEEE Transactions on Dielectrics and Electrical Insulation*, Vol. 2, No. 1, pp. 128–136.

Gorski, R. A. 1986. "Meteorological Summary—March 10. 1986—Multiple Outages on the Southern Ontario (Central Region) 500 kV System." Ontario Hydro Meteorology Report No. 80604-1, pp. 1–40.

Guerrero, T. 2004. "Étude Expérimentale du Contournement des Isolateurs Recouverts de Glace Sous Tensions de Foudre et Manoevre," Master of Engineering Thesis, Université du Québec à Chicoutimi. July.

Guerrero, T., M. Farzaneh, and J. Zhang. 2005. "Impulse Voltage Performance of Ice-Covered Post Insulators," in *Proceedings of 2005 Conference on Electrical Insulation and Dielectric Phenomena* (CEIDP), pp. 325–328.

Gutman, I., S. Berlijn, S. Fikke, and K. Halsan. 2002. "Development of the Ice Progressive Stress Method Applicable for the Full-Scale Testing of 420 kV Class Overhead Line Insulators," in *Proceedings of International Workshop on Atmospheric Icing on Structures*, IWAIS 2002, Brno, Czech Republic.

Hara, M. and C. L. Phan. 1979. "Leakage Current and Flashover Performance of Iced Insulators," *IEEE Transactions on Power Apparatus and Systems*, Vol. PAS-98, No. 3, pp. 849–859.

Hobbs, P. V. 1974. *Ice Physics*. Oxford, UK: Clarendon Press.

Hu, J., C. Sun, X. Jiang, Z. Zhang, and L. Shu. 2007. "Flashover Performance of Pre-contaminated and Ice-Covered Composite Insulators to Be Used in 1000 kV UHV AC Transmission Lines," *IEEE Transactions on Dielectrics and Electrical Insulation*, Vol. 14, No. 6 (Dec.), pp. 1347–1356.

IEC 60060. 1989. *High-Voltage Test Techniques. Part 1: General Definitions and Test Requirements*, 2nd ed. Lausanne, Switzerland: IEC.

IEC 62271-102. 2001. *High-Voltage Switchgear and Controlgear—Part 102: Alternating Current Disconnectors and Earthing Switches*. Lausanne, Switzerland: IEC.

IEC 60507. 1991. *Artificial Pollution Tests on High Voltage Insulators to Be Used in AC Systems*. Lausanne, Switzerland: IEC.

IEC 60815. 2008. *Selection and dimensioning of high-voltage insulators intended for use in polluted conditions*. Geneva, Switzerland: Bureau Central de la Commission Electrotechnique Internationale. October.

IEEE C37.34. 1994. *Standard Test Code for High-Voltage Air Switches*. Piscataway, NJ: IEEE Press.

IEEE Standard 4. 1995. *IEEE Standard Techniques for High Voltage Testing*. Piscataway, NJ: IEEE Press.

IEEE PAR 1783. 2008. "IEEE Guide for Test Methods and Procedures to Evaluate the Electrical Performance of Insulators in Freezing Conditions." Joint DEIS/PES Task Force on Insulator Icing, PAR 1783. January 14.

Jiang, X., S. Wang, Z. Ahang, S. Xie, and Y. Wang. 2005. "Study on AC Flashover Performance and Discharge Process of Polluted and Iced IEC Standard Suspension Insulator String," in *Proceedings of 14th International Symposium on High Voltage (ISH)*, Beijing, China. August.

Jolly, D. C. 1972. "Contamination Flashover Theory and Insulator Design," *Journal of the Franklin Institute*, Vol. 294, No. 6 (Dec.), pp. 473–500.

Kannus, K., K. Lahti, and K. Nousiainen. 1998. "AC and Switching Impulse Performance of an Ice-Covered Metal Oxide Surge Arrester," *IEEE Transactions on Power Delivery*, Vol. 13, Oct., pp. 1168–1173.

Kawai, M. 1970. "Flashover Tests at Project UHV on Ice-Covered Insulators," *IEEE Transactions on Power Apparatus and Systems*, Vol. PAS-89, No. 8 (Nov./Dec.), pp. 1800–1804.

Khalifa, M. M. and R. M. Morris. 1967. "Performance of Line Insulators Under Rime Ice," *IEEE Transactions on Power Apparatus and Systems*, Vol. PAS-86, No. 6 (June), pp. 692–697.

Kuffel, J., Z. Li, W. A. Chisholm, B. Ng, J. S. Barrett, and Y. Motlis. 1999. "Testing and Application of Composite Insulators for Lines and Substations," in *Proceedings of INMR International Conference on Insulators, Bushings and Arresters*, Barcelona, Spain.

Lee, L. Y., C. L. Nellis, and J. E. Brown. 1975. "60 Hz Tests on Ice-Covered 500 kV Insulator Strings," in *IEEE PES Summer Meeting*, San Francisco, California, Paper A75-499-4.

Li, S., R. Zhang, and K. Tan. 1989. "Measurement of the Temperature of a Local Arc Propagating," in *Proceedings of 6th International Symposium on High Voltage Engineering*, New Orleans, Louisiana, Paper No. 12.05.

Matsuda, H., H. Komuro, and K. Takasu. 1991. "Withstand Voltage Characteristics of Insulator Strings Covered with Snow or Ice," *IEEE Transactions on Power Delivery*, Vol. 6, No. 3 (July), pp. 1243–1250.

Meier, A. and W. M. Niggli. 1968. "The Influence of Snow and Ice Deposits on Supertension Transmission Line Insulator Strings with Special Reference to High Altitude Operation," in *IEE Conference Publication 44*, London, England, pp. 386–395. September.

Ndiaye, I., M. Farzaneh, and I. Fofana. 2007. "Study of the Development of Positive Streamers Along an Ice Surface," *IEEE Transactions on Dielectrics and Electrical Insulation*, Vol. 14, No. 6 (Dec.), pp. 1436–1445.

NERC (North American Electric Reliability Corporation). 1990. DAWG Database, January 1, 1989–December 31, 1989, NERC_Disturbance_Reports\dawg-89.htm.

Neumärker, G. 1959. "Verschmutzungszustand und Kriechweg," *Deutsche Akadamie*, Berlin, Vol. 1, pp. 352–359.

Nourai, A. and E. A. Peszlen. 1984. "Nonceramic Insulator Ice Flashover Tests, Part One," American Electric Power Service Corporation, Electrical Research Section Report. Dec. 28.

Nourai, A. and W. C. Pokorny. 1986. "Electrical Performance of Iced Nonceramic Insulators, Part Three—Laboratory Study," American Electric Power Electrical Research and UHV Research Section Report. January.

Obenaus, F. 1958. "Fremdschichtüberschlag und Kriechweglänge," *Deutsche Elektrotechnik*, Vol. 4, pp. 135–136.

Petrenko, P. F. and R. W. Whitworth. 1999. *Physics of Ice*. New York: Oxford University Press.

Phan, L. C. and H. Matsuo. 1983. "Minimum Flashover Voltage of Iced Insulators," *IEEE Transactions on Electrical Insulation*, Vol. EI-18, No. 6 (Dec.), pp. 605–618.

PSOD (Power System Operations Division) of Ontario Hydro. 1990. "Morning Report, January 2, 1990 covering Friday December 29, 1989 to Monday January 1, 1990."

Renner, P. E., H. L. Hill, and O. Ratz. 1971. "Effects of Icing on DC Insulation Strength," *IEEE Transactions on Power Apparatus and Systems*, Vol. PAS-90, No. 3 (May), pp. 1201–1206.

Rizk, F.A.M. 1971. "Analysis of Dielectric Recovery with Reference to Dry-Zone Arc on Polluted Insulators," *IEEE Power Engineering Society Winter Power Meeting*, Paper 71 C 134 PWR, January/February.

Rizk, F. A. M. 1981. "Mathematical Models for Pollution Flashover," *Electra*, No. 78, pp. 71–103.

Rizk, F. A. M. and A. Q. Rezazada. 1997. "Modeling of Altitude Effects on AC Flashover of High-Voltage Insulators," *IEEE Transactions on Power Delivery*, Vol. 12, No. 2 (Apr.), pp. 810–822.

Schaedlich, K. H. 1987. "Weather Conditions Associated with Insulator Flashover." Ontario Hydro Power System Operations Division (PSOD) Report.

Schneider, H. M. 1975. "Artificial Ice Tests on Transmission Line Insulators—A Progress Report," in *IEEE PES Summer Meeting*, San Francisco, Paper A75-491-1, pp. 347–353. July.

Schneider, H. M. and A. E. Lux. 1991. "Mechanism of HVDC Wall Bushing Flashover in Non-Uniform Rain," *IEEE Transactions on Power Delivery*, Vol. 6 No. 1 (Jan.), pp 448–455.

Shu, L., C. Sun, J. Zhang, and L. Gu. 1991. "AC Flashover Performance on Iced and Polluted Insulators for High Altitude Regions," in *Proceedings of 7th International Symposium on High Voltage Engineering*, Dresden, Germany, Vol. 4, Paper 43.13, pp. 303–306. August.

Soucy, L. 1996. "Effet de la fonte et de la pollution sur la tension de tenue maximale des isolateures recoverts de glace," Mémoure Maîtrise en Ingéniere, Université du Québec à Chicoutimi. October.

Su, F. and Y. Jia. 1993. "Icing on Insulator String of HV Transmission Lines and its Harmfulness," *Proceedings of Third International Offshore and Polar Engineering Conference (ISOPE)*, Singapore (June), pp. 655–658.

Takei, I. 2007. "Dielectric Relaxation of Ice Samples Grown from Vapor-Phase or Liquid-Phase Water," in *Physics and Chemistry of Ice* (W. Kuhs, ed.). Cambridge, UK: Royal Society of Chemistry, pp. 577–584.

Tavakoli, C. 2004. "Dynamic Modeling of AC Arc Development on Ice Surfaces," PhD Thesis, Université du Québec à Chicotimi. November.

Tavakoli, C., M. Farzaneh, I. Fofana, and A. Béroual. 2006. "Dynamics of AC Arc on Surface of Ice Accumulated on an Insulator String," *IEEE Transactions on Dielectrics and Electrical Insulation*, Vol. 13, No. 6 (Dec.), pp. 1278–1285.

Udo, T. 1966. "Switching Surge Sparkover Characteristics of Air Gaps and Insulator Strings Under Practical Conditions," *IEEE Transactions on Power Apparatus and Systems*, Vol. 85, No. 8 (Aug.), pp. 859–864.

Vlaar, J. 1991. "Thermal and Electrical Properties of Icicles," University of Waterloo 2B Honours Physics Report SN 88104434.

Vuckovic, Z. and Z. Zdravkovic. 1990. "Effect of Polluted Snow and Ice Accretion on High-Voltage Transmission Line Insulators," *Proceedings of 5th International Workshop on Atmospheric Icing of Structures* (IWAIS), Tokyo, Paper B4-3. October 29 to November 1.

Watanabe, Y. 1978. "Flashover Tests of Insulators Covered with Ice or Snow," *IEEE Transactions on Power Apparatus and Systems*, Vol. PAS-97, No. 5 (Sept.), pp. 1788–1794.

Wilkins, R. 1969. "Flashover Voltage of High-Voltage Insulators with Uniform Surface-Pollution Films," *Proceedings of the IEE*, Vol. 116, No. 3, pp. 457–465.

Wilkins, R. and A. A. J. Al-Baghdadi. 1971. "Arc Propagation Along an Electrolyte Surface," *Proceedings of IEE*, Vol. 118, No. 12, pp. 1886–1892.

Yoshida, S. and K. Naito. 2005. "Survey of Electrical and Mechanical Failures of Insulators Caused by Ice and/or Snow," CIGRE WG B2.03, *Electra*, No. 222.

Young, I. G. and R. E. Salomon. 1968. "Dielectric Behavior of Ice with HCl Impurity," *Journal of Chemical Physics*, Vol. 48, pp. 1635–1644.

Yu, D., M. Farzaneh, J. Zhang, L. Shu, W. Sima, and C. Sun. 2007. "Effects of Space Charge on the Discharge Process in an Icicle/Iced-Plate Electrode System Under Positive DC Voltage," *IEEE Transactions on Dielectrics and Electrical Insulation*, Vol. 14, No. 6 (Dec.), pp. 1427–1435.

Zhang, J. and M. Farzaneh. 2000. "Propagation of ac and dc Arcs on Ice Surfaces," *IEEE Transactions on Dielectrics and Electrical Insulation*, Vol. 7, No. 2 (Apr.), pp. 269–276.

CHAPTER 8

SNOW FLASHOVERS

Much of the progress in studies of snow flashover originated from research that began in Japan in the 1950s. The studies initially addressed unacceptable flashover rates on a 154-kV transmission system. Many faults were still being reported in 1981 and 1985 on the same network [Matsuda et al., 1991], and additional concerns about the vulnerability of 275- and 500-kV networks to the same problems were also expressed. More recently, for the period January 13–16, 2004, wet snow with a high concentration of sea salt accumulated on 33-kV insulators, causing widespread outages [Sugawara and Hosono, 2005]. A similar situation occurred in the 154- and 275-kV systems in the coastal Niigata Kaetsu area in the period December 22–24, 2005 [Onodera et al., 2007].

This chapter analyzes the effects of snow on the insulators, using a combination of climatology, visual observations, leakage current measurements, and results from several different physical scales that are presented in sequence.

8.1. TERMINOLOGY FOR SNOW

Dry Snow: A type of precipitation icing of 0.05–0.1-g/cm^3 density resulting in very light packs of snow accreting in various shapes and structures. Adhesion strength is weak and the accretion can easily be removed by wind or motion

Insulators for Icing and Polluted Environments. By Masoud Farzaneh and William A. Chisholm
Copyright © 2009 the Institute of Electrical and Electronics Engineers, Inc.

of the insulator. Accretion of various shapes depends on wind speed and insulator orientation and shape. When the temperature is close to zero, liquid water content of dry snow may increase, causing the accretion to fall off. If the temperature drops after accretion, adhesion strength may become very strong.

Snow: A type of precipitation (hydrometeor) in the form of hexagonal crystalline water ice, consisting of a multitude of snowflakes that fall from clouds. Since snow is composed of small ice particles, it is a granular material. It has an open and therefore soft structure, unless packed by external pressure.

Snow Cover: The accumulation of snow on the ground defined by extent, depth, duration, and water equivalent.

Snow Cover Depth: The combined total depth of both old and new snow on the ground (cm).

Snow Cover Duration: The annual number of days, or the number of days over a winter season, with snow cover.

Snow Cover Extent: The total land area covered by some amount of snow; typically reported in square kilometers (km^2).

Snow Water Equivalent: The water content obtained from melting snow. SWE is the product of depth and density: SWE (kg/m^2) = depth (m) × density (kg/m^3) or SWE (m) = depth (m) × density (kg/m^3) / density of water (kg/m^3).

Soft Rime: A type of in-cloud icing resulting in ice accretion of density 0.15–$0.3 g/cm^3$ characterized by a granular structure, feather- or cauliflower-like shape. It takes a pennant-shaped aspect against the wind on stiff objects or flexible cables. Its adherence to objects is rather weak, and it can be removed by hand.

Wet Snow: A type of precipitation icing of typically $0.3–0.8$-g/cm^3 density resulting in accretion of various shapes depending on wind speed and orientation of insulator. When the temperature is close to zero, the liquid water content of wet snow may increase, causing the accretion to fall off. If the temperature drops after accretion, adhesion strength may become very strong. Wet snow causes water drops similar to rain to form on conductors of overhead lines. If free water entirely fills the air space in the snow, it is classified as "very wet" snow.

8.2. SNOW MORPHOLOGY

Snow is solid precipitation that grows from water vapor in the atmosphere where temperature is less than $0\,°C$. Snow may fall as individual single crystals, or as an amalgamation of several crystals. Snow may melt as it descends and turn into rain, or it may be deposited on the ground as snow cover. Successive

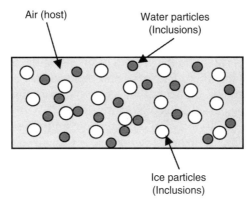

Figure 8-1: Wet snow as three-phase mixture of air, water, and ice (from Arslan et al. [2001]); courtesy of EMW Publishing).

periods of snowfall accumulation without melting periods may build up and form a snowpack.

Natural dry snow is a mixture of air and ice grains. Wet snow is distinguished by the additional presence of liquid water, as illustrated in Figure 8-1 [Arslan et al., 2001]. Wet and dry snow are fundamentally two different materials since the liquid water causes major changes in the configuration of both grains and bonds, leading to different electrical properties [Colbeck, 1997]. Wet snow is categorized by its liquid content level (high or low), while dry snow is classed by its growth rate (high or low).

Ice bonding between the grains is observed in most, but not all, types of snow. Wet snow is well bonded at low liquid water content levels from the ice-bonded clusters that form. Wet snow with high liquid water content is slushy and less cohesive because the grain boundaries are unstable.

A transitional form of snow, melt–freeze grains, can be either wet or dry. These amorphous, multicrystalline particles arise from melt–freeze cycles. They are solid within and well bonded to their neighbors.

Snow crystal growth in air is quite sensitive to small changes in various parameters such as temperature and supersaturation. The Bentley Snow Crystal Collection of the Buffalo Museum of Science is a digital library providing a high-quality collection of original and rather beautiful images of Wilson A. Bentley's original glass slide photographs of snow crystals, collected under various natural conditions.

Observations of snow crystal growth in the laboratory [Nakaya, 1954] revealed a complex dependence of crystal morphology on temperature and supersaturation. Temperature mainly determines whether snow crystals will grow into plates or columns. A higher degree of moisture supersaturation produces snow with a more complex structure. The morphology switches from columns (at $T \approx -2\,°C$) to plates (at $T \approx -15\,°C$) and back to predominantly columns (at $T \approx -30\,°C$) as temperature is decreased further. The change in crystal morphology from columns to plates at −2 °C could provide a straight-forward explanation of why the electrical conductivity of snow also peaks at this temperature.

After snow is deposited on the ground or on insulator surfaces, the particle shapes are further modified. The large and dendritic ice crystals decompose into fragments. Larger fragments then grow by absorbing small fragments. Over time, the snow particles become more rounded. With the effects of temperature gradient, gravity or wind compression, and time, large snow grains, called depth hoar, may form near the bottom of the snowpack or at the downwind interface between the snow and vertical surfaces.

8.3. SNOW ELECTRICAL CHARACTERISTICS

There are three main effects of snow on electrical insulators:

- Clean insulators may be covered with snow that contains salt or pollution.
- Contaminated insulators may be covered with clean snow.
- Clean insulators may be covered with clean snow, which then accumulates pollution.

The volume resistivity or conductivity of the snow is one of two parameters that have the strongest effect on the electrical flashover strength of an insulator that is completely covered by snow. The other parameter is the weight of snow per unit length, which is in turn a function of the snow depth, density, and liquid water content of the snow layer.

There are two main sources of electrical conductivity in snow that accumulates on insulators. First, conductivity of melted natural snow is typically found to be in the range of $\sigma_{20} = 10$–$100\,\mu$S/cm. This conductivity can be enhanced in the following ways:

- Yasui et al. [1988] show a factor-of-10 increase in median conductivity of melted snow as distance from seashore reduces from 100 to 1 km.
- Conductivity of ice formed near urban expressways, where road salting is underway, has been found to reach levels of 300–400 μS/cm, simulating marine exposure.
- Chisholm and Tam [1990] and Vuckovic and Zdravkovic [1990] have noted that surface contamination on insulators provides a dominant source of ions to thin ice layers.
- Fikke et al. [1993] show that long-range transport of man-made (anthropogenic) pollution contributes to more than 50% of the conductivity in most of their examined cases.

Figure 8-2 shows that snow can have very high conductivity, based on distance from the sea. For example, Fujimura et al. [1979] reported maximum levels of ground samples of melted snow conductivity in the range of 10,000 μS/cm at a location 30 m from the seacoast.

The median levels of snow, ice, and precipitation conductivity in Japan are considerably lower than the maximum values shown in Figure 8-2. At one typical site, measurements from 1981 to 1987 showed that there was only a 1% probability that natural precipitation would reach levels above 100 μS/cm. This suggests that most of the electrical conductivity of melted snow samples in Figure 8-2 occurred from effects of snow or insulator surface contamination. (See Table 8-1.)

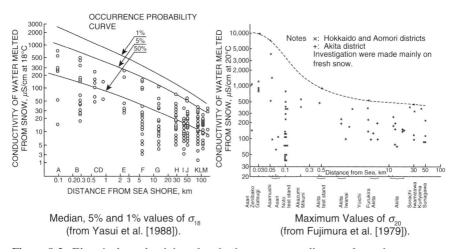

Median, 5% and 1% values of σ_{18}
(from Yasui et al. [1988]).

Maximum Values of σ_{20}
(from Fujimura et al. [1979]).

Figure 8-2: Electrical conductivity of melted snow versus distance from the sea.

TABLE 8-1: Distribution of Precipitation Conductivity and ESDD at Insulator Test Site

Probability	Conductivity of Precipitation (μS/cm) Corrected to 20°C		ESDD (mg/cm^2)	Volume Density (g/cm^3)
	Yonezawa	Ishiuchi	Yonezawa	Joetsu
Median (50%)	23	16	0.064	0.15
30%	32	21	0.072	0.18
5%	66	50	0.092	0.28
1%	102	83	0.103	0.33

Source: Takasu et al. [1988]; adapted from Yasui et al. [1988].

Matsuda et al. [1991] compared the conductivity of water melted from snow on the ground to water melted from natural ice. They found a wide range of conductivity (54–1200 μS/cm ice, 40–300 μS/cm snow corrected to 20 °C) and a factor of 2.0–2.5 that related the observations.

Figure 8-3 [Yasui et al., 1988] shows that the volume resistivity of the snow is an exponential function of the volume density in the range of natural conductivity. This is an important electrical consequence of the snow structure described in Section 8.2 of this chapter.

The relation observed between volume resistivity and density affects the use of an "icing stress product," which multiplies these terms together for a given deposit thickness to establish a single value of stress. There is no strong relation in Figure 8-3 between melted water conductivity and snow resistivity, and this suggests that further refinement of the concept may be needed.

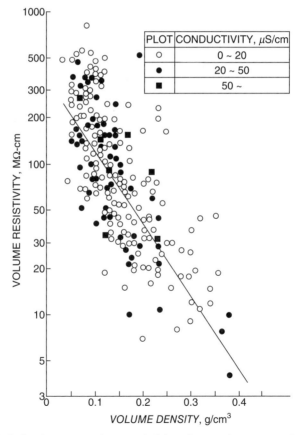

Figure 8-3: Relation between volume resistivity of natural snow samples and volume density (from Yasui et al. [1988]).

8.3.1. Electrical Conduction in Snow, dc to 100 Hz

The parameters characterizing the nature of snow are volume conductivity and density, dielectric constant, salt and water content, particle size, impurities and their nature, crystal structure, and the complex dielectric response to electric field, including effects of magnitude, polarity (for dc), and frequency (for ac).

The electric properties of snow that accumulates on an energized insulator surface also depend on geometric factors and on the nature and quality of snow contact with the electrodes.

Volume conductivity has the greatest influence in the electrical flashover characteristics of snow. In turn, this parameter is largely influenced by snow volume density, as well as by liquid water content and any electrically conductive ions in solution or inclusions. An apparatus for measuring the dc conductivity of snow, using a guard ring to mitigate surface conduction, is shown in Figure 8-4. The guard ring also ensures uniform electric field through the snow sample.

The electrical resistance of the snow deposit depends on its thickness and cross-sectional area. It is also affected by the conductivity of the deposit, σ_{snow}. The resistance of a cylindrical disk of snow with spacer thickness W and cross-sectional area A is given by $R = A/(W\sigma_{snow})$.

Many chemical impurities affect both crystal morphologies and growth rates in snow. Low concentrations of common air pollution gases and of sodium chloride from road salting can have a strong effect. Ice crystals are essentially pure water, leading to a separation of the chemical–water mixture during freezing, through purification by crystallization. This separation lowers

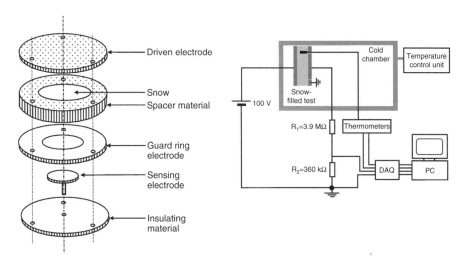

Figure 8-4: Apparatus for measuring dc conductivity of 188-cm^3 snow sample (from Farzaneh et al. [2007]).

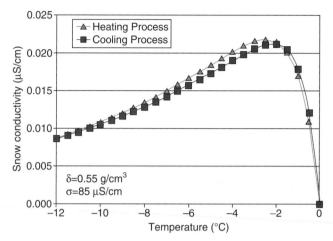

Figure 8-5: Snow conductivity variation with temperature using heating and cooling test procedures (data from Hemmatjou [2006]).

the mixture's entropy, meaning that an additional source of energy is needed to satisfy the second law of thermodynamics. This leads to a reduction in the freezing temperature of the mixture below 0 °C. The resulting effect of salt (NaCl) on the adhesion of snow to roads is well known and understood.

For low concentrations of salt in snow, leading to melted-water conductivity in the range of 40–130 μS/cm at 20 °C, the effect on freezing point depression is modest. The depression is one factor leading to a peak electrical conductivity at a snow temperature of –2 °C [Farzaneh et al., 2007; Hemmatjou et al., 2007], as shown in Figure 8-5.

The complex electrical response of the snow at 100 Hz, which is of main interest for power-frequency flashovers, was found by Takei and Maeno [2003] to be dominated by the conductivity rather than permittivity, in agreement with Farzaneh et al. [2007] and Hemmatjou et al. [2007]. Figure 8-6 shows their low-frequency test results on natural snow and artificial hoarfrost, grown in a freezer. In all cases, there is a peak in conductivity above –3 °C and below –2 °C.

Takei and Maeno [2003] also explored the activation energy, given by the slope of the conductivity curve against the inverse of absolute temperature. Their fits to the data in Figure 8-6 give activation energy of 0.7 to 1 eV, corresponding to surface ice, where a slope of 0 to 0.6 eV is normal for bulk ice. Takei and Maeno concluded that the low-frequency conductivity of the snow was caused mainly by surface conduction of ice particles.

With the implicit relation between high conductivity and low flashover voltage in the Obenaus mathematical model for polluted insulators, the results in Figures 8-5 and 8-6 suggest that the minimum electrical strength of a snow-covered insulator will also occur slightly below the freezing point.

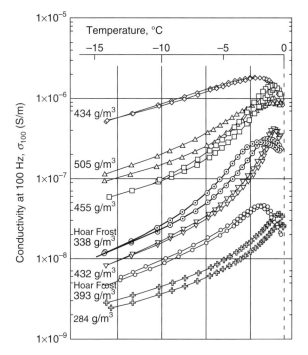

Figure 8-6: Temperature dependence of the ac conductivity (100 Hz) for snow and hoarfrost samples (from Takei and Maeno [2003]; courtesy of NRC).

8.3.2. Dielectric Behavior of Snow, 100 Hz to 5 MHz

Takei and Maeno [2003] compared the electrical conductivity of snow, artificial hoarfrost, and crushed ice over a wide frequency range, considering both the resistive and capacitive response. They reported their results as a Cole–Cole plot of the real and imaginary parts of permittivity from 50 Hz to 5 MHz in Figure 8-7.

The electrical characteristics of snow show strong temperature dependence in the range of −0.25 °C with a peak in response at 20 kHz. Also, the electrical characteristics change with annealing time at a constant temperature of −1.2 °C. Annealing changes the snow morphology over time, breaking down small crystals as the snow compacts. This process develops bonding bridges between snow particles, increasing the conductivity, and also improves connections from the measurement electrodes to the snow.

The lightning surge response of the dielectric is dominated by the behavior at 80–125 kHz, where the dielectric constant is about 1.8. The switching-surge response follows the response at 1 kHz, which is much closer to the low-frequency response and has much higher relative dielectric constant in the range of $k' = 7$–13.

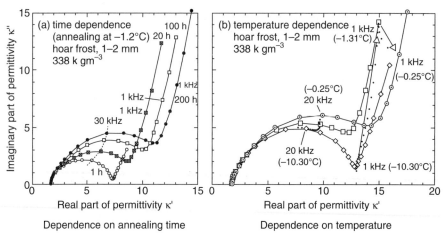

Figure 8-7: Changes in the Cole–Cole response of artificial hoarfrost snow as function of annealing time and temperature (from Takei and Maeno [2003]; courtesy of NRC).

8.3.3. Products of Electrical Discharge Activity

Partial discharge activity on snow produces heat and nitric acid. Thermal effects are most important. The heat warms the snow layer, elevating the conductivity as shown in Figures 8-5 and 8-6 until it reaches its maximum at $-2\,°C$ to $-1\,°C$. Each half-cycle of potential extends the arc to its full previous extent, and possibly a bit further. This means that regions near the arc initiation point tend to accumulate more heat than those near the tip of the arc.

Any nitric acid formed from the ozone at the arc root will raise the conductivity locally and may play a role in reducing the voltage drop at the arc root. During the flashover process, there is typically not enough time for any ions to diffuse into the liquid water layer of the bulk snow.

8.4. SNOW FLASHOVER EXPERIENCE

Meier and Niggli [1968] reported some disturbances by grounding due to snow accumulation on a 400-kV mountain line in Switzerland. In Japan, Matsuda et al. [1991] reported in January 1981 that faults on nine 154-kV transmission overhead lines occurred simultaneously due to snow covering tension insulators. (See Table 8-2.)

The role of temperature in the flashover of 275-kV line insulators was noted by Yasui et al. [1988] in a case study. There was accumulation of 1.5 m of snow on the ground, and up to 1 m of snow on insulators with a density of 0.29 g/cm^3 and a conductivity of 83 µS/cm. There were no flashovers with this snow

TABLE 8-2: Summary of Utility Experience with Flashovers in Snow Conditions

Location, Date	Ice and Contamination Type	Melting Phase	Surface Precontamination	Altitude	System Voltage
Switzerland, 1966–1967	Heavy, wet snow; 0.8-g/cm^3 density	Yes	No	High	400 kV
Norway, 1987	Contaminated snow; thin rime ice and fog	Mostly yes	Heavy	Medium	300 kV
Quebec, Canada, 1988	Contaminated snow	Yes	Low	Low	735 kV
Quebec, Canada, 1995	Snow	No, −10 °C	Low	400 m	735 kV
Okhotsk, Japan, 2004	Contaminated wet snow (sea salt) on insulator sides	No, −1 °C to −3 °C	No	Sea level	33 kV
Ontario, Canada, 2004	Contaminated snow	No, −2 °C	Heavy	Low	115 kV Bushings only
Niigata Kaetsu, Japan, 2005	Contaminated wet snow (sea salt) on insulator sides	Yes, 0 to +2 °C	No	Low	154 and 275 kV

deposit for 5 days with temperature at about −5 °C in Figure 8-8. The first flashover occurred when the temperature reached −2 °C, which is when the electrical conductivity peaks in Figure 8-5.

On April 18, 1988 at 2008, a series of six flashovers caused by heavy, wet snow accretion and fog on 735-kV station insulators in eastern Québec resulted in a major interruption of the electricity supply [Farzaneh and Kiernicki, 1997]. The interruption of flow of hydroelectric power from Churchill Falls through the Arnault substation affected export flow of 1800 MW to the United States. The weather at Sept-Isles during this period reported heavy snow showers and thunderstorms at an ambient temperature of 0 °C and a dew point of −0.2 °C. Other weather stations near the Arnaud location included Baie-Comeau (snow and fog at 1 °C ambient, 0.3 °C dew point) and Blanc Sablon (1–2 °C ambient, −2 °C to −3 °C dew point). There was a similar series of problems on the 735-kV system in 1995.

In Ontario, in January 2004, sustained winds of 50 km/h from an unfavorable direction for 24 h led to heavy accumulation of road salt at an important 115-

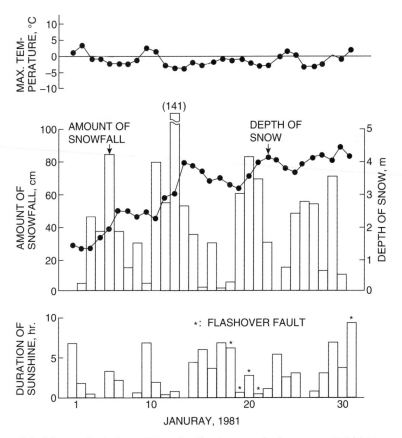

Figure 8-8: Meteorological conditions leading to snow flashovers on 275-kV transmission line (from Yasui et al. [1988]).

kV station in a major city. During a melting phase after packed snow accumulation, the only insulators that flashed over, leading to a 90-minute blackout in the downtown area, were transformer bushings.

Wieck et al. [2005, 2007] have noted that heavy accretion of snow across circuit breakers has been an operational problem on 420-kV systems. Normally, nonuniform rain, ice, or snow leads to a higher risk of flashover. Service and test photos in Figure 8-9 show the snow layers live tank circuit breakers with horizontal insulators. In this case, Figure 8-9 shows that snow is fully bridging the pole-to-pole insulation over the horizontal breaking chambers.

The most critical overvoltage situation for the breaker covered with snow will occur when connecting a line to a generator. For a short period during the synchronization, the maximum voltage over the open breaking chambers can reach 2.5 times the normal phase-to-earth voltage. The flashover problems were reproduced in outdoor and indoor testing using natural snow, as described more completely in a case study of the problem resolution in Section 8.10.

Heavy natural snow, January 2004
[Wieck et al., 2007]

Outdoor test with 20 cm of natural snow
collected from the ground
[Wieck et al., 2005]

Figure 8-9: Snow on 420-kV circuit breaker in Norway.

In 2004, the Okhotsk region of Hokkaido Island in Japan was subjected to a wet snow storm, leading to full bridging of 33-kV long-rod post insulator sheds [Sugawara and Hosono, 2005]. The accretion took place with a wind speed of about 10 m/s, and affected a line running inland about 500 m from the coast of the Okhotsk Sea. Strings of disk insulators were less affected by the wet snow exposure of 50 cm than the long-rod style with closer shed spacing.

In 2005, the Niigata Kaetsu area experienced wet snow accretion at temperatures of 0–2 °C, driven horizontally by wind speeds that peaked at 25 m/s (90 km/h) for a period of 4 hours [Onodera et al., 2007]. The snow accretion on 154- and 275-kV insulators was found to have the same ion content as seawater. While line-to-ground electrical flashovers of insulators were the main concern, there were also phase-to-phase outages related to galloping of the conductors. A total precipitation level of 33-mm water equivalent led to a buildup of 20 mm out from the outer diameter of the long-rod and cap-and-pin insulators.

8.5. SNOW FLASHOVER PROCESS AND TEST METHODS

Many parts of the world have insulators that are covered with snow for days or weeks at a time. The snow in most cases poses little additional risk of electrical flashover.

The conditions that lead to snow flashovers are relatively uncommon. They include:

- Dense deposits, corresponding to wet or compacted snow rather than dry or flaky snow.
- Cohesive and sticky deposits that adhere to insulator surfaces.
- Uniform deposits that span 60–100% of the insulator dry arc distance.

- Thick deposits that fill in the shed depths and thereby bridge the shed-to-shed dry arc distance of the insulator, or spaces between parallel insulators.
- Electrically conductive deposits that can affect the voltage distribution on the surface of the insulator and can also support reignition of ac arcing activity.
- Precontamination of insulator surfaces that diffuses into the bulk of the snow.

A suitable artificial test method will reproduce all important aspects of the conditions that lead to snow flashover.

There are two mechanisms for snow accretion. First, and more common, is the vertical accumulation on horizontal insulators. This tends to occur every winter in areas of heavy snowfall and does not represent a significant risk of flashover as long as the snow is dry and cold. The second, less common but more severe mechanism, is the accumulation of wet packed snow on one side of the insulator from the action of strong horizontal winds. This second case leads to a higher density of snow deposit that is more typical of ice accretion. In cases near the seacoast where the snow contains a high quantity of sea salt, as in Sugawara and Hosono [2005] and Onodera et al. [2007], the accumulation of wet, conductive snow can lead to widespread transmission system outages.

8.5.1. Snow Flashover Process

The snow flashover process starts when a leakage current flows on the insulator surface and through the snow layer. Initially, this current will be small, and its magnitude will be established by snow density, conductivity, and water content. Any part of the snow with high current density may start melting due to Joule heating. The current may gradually increase and become constant at levels of 50–100 mA, without the pulsations associated with leakage currents on contaminated insulators.

Luminous discharge activity signals a change in behavior. Parts of melted snow may drop away from the insulator, forming dry bands, in the same manner as contamination flashovers. This results in a nonuniform voltage distribution along the insulator. Depending on the resistance and length of the remaining snow layer, partial arcs may bridge the dry bands. Typically, the leakage current may have peak magnitudes in the range of 100–300 mA at this point and the currents are no longer continuous.

The progressive arc development and retreat on the snow and insulator surface repeats many times and may cause more snow to drop away. The arc may also separate from the exposed surface of the insulator into the air. As for contamination flashover, with sufficient stability, arcing activity in the air can progress along the snow layer until it reaches a critical length that evolves rapidly into a line-to-ground flashover arc.

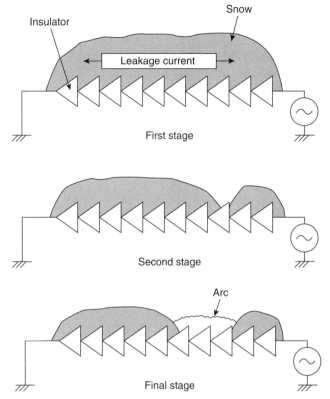

Figure 8-10: Change in state of snow on insulator string with time (from Yasui et al. [1988]).

Electrical problems under heavy snow conditions occur mainly on insulators with horizontal orientation, as shown in Figure 8-10. Risk of flashover of snow-covered horizontal insulators is increased when:

• The insulators have a large diameter, such as horizontal bushings for circuit breakers; or
• There is a possibility of retaining snow in the 0.1–1-m space between two or more insulators in parallel.

Test methods for simulating the accumulation of snow on vertically oriented insulators are still under investigation. These tests are less important in the overall risk of insulator flashovers in freezing conditions than glaze icing conditions.

8.5.2. Snow Test Methods

So far, it has proved very difficult to simulate the conditions that produce natural snow in artificial laboratories. The artificial snow produced by the ton for grooming of ski hills using snow cannons is actually a type of ice.

The development of snow flakes starts with a nucleation site. Since spontaneous freezing is difficult to achieve, injection of ice particles to serve as nucleation sites is necessary [Yasui et al., 1990]. The ice particles require a long residence time in a saturated, freezing atmosphere to develop significant accumulation with the sixfold symmetry that characterizes snowflakes. A tall vertical wind tunnel is thus needed to give sufficient residence time. Even with this approach, the volume of snow production is low and it is difficult to adapt a vertical wind tunnel to a high-voltage laboratory that is set up for testing in freezing conditions.

The second approach to simulating snow [Farzaneh and Laforte, 1998] used wet ice accretion at a temperature that gave an opaque, low-density rime ice, similar to snow in most respects. This approach was also used by Takei and Maeno [2003], who found that it was necessary to anneal the artificial accretion at −1.2 °C for many hours to obtain electrical characteristics that were similar to natural snow.

Some outdoor high-voltage facilities are located in regions where there is consistent accumulation of snow every winter. Outdoor electrical flashover tests of natural snow accretion on insulators are highly desirable, but a large number of insulators need to be exposed to obtain meaningful results. Each flashover test tends to heat and modify the snow layer, changing its electrical characteristics. Also, it is inconvenient that there are no controls over ambient temperature and wind speed, so the snow may melt or blow away before testing is completed.

The last approach of indoor tests using natural snow deposits is most suitable for test laboratories in cold regions. Natural, clean snow is gathered from the nearby environment and stored for use locally in a refrigerated chamber. The snow characteristics can be conditioned by mixing with salt using a snow blower. The volume, shape, and density of snow accretion can be controlled using forms around the insulators. The test chamber temperature controls the liquid water content of the snow and must also be carefully monitored.

8.5.3. General Arrangements for Snow Tests

In preparation for testing, these steps should be followed:

- Test insulators are selected and cleaned, and precontaminated, if required.
- Insulators are mounted to simulate the in-service electrical stress, with the proper bus or conductor size, along with any grading rings or arcing horns used in normal service.
- Appropriate electrical clearance for the test voltage, based on switching-surge criteria [IEEE Standard 4, 1995], should be provided between the insulators and any walls, floor, and ceiling or any other grounded object.
- Insulators are dried at room temperature, and cooled to the same −6 °C ± 1 °C temperature as that planned for the snow deposit process.

A high-voltage source with regulation that meets standard requirements for contamination testing [IEEE Standard 4, 1995] is needed for snow testing. In addition, depending on the dry arc distance of the test object, the source may be required to supply up to 1 A of continuous leakage current for periods of up to 60 s during the snow melting process that occurs during the final stages of the flashover test. This may require adjustment of overcurrent trip settings or other accommodations.

8.5.4. Snow Deposit Methods

The first artificial snow deposit method consists in covering the insulators with soft rime produced from very small supercooled water droplets, followed by heating the deposits by raising the air temperature, in order to increase the water content. No voltage needs to be applied during soft-rime accretion. This approach, still under investigation, is the only one recommended for simulating snow on vertical insulators.

The second method consists in covering the insulator with natural snow gathered on the ground, which can be accomplished with one of the following ways:

- The snow is loaded into a wooden form, such as the one shown in Figure 8-11, and used to cover the insulator at subzero ambient temperature; or
- Blocks of naturally accumulated snow are cut and then placed and arranged over the test insulator at subzero ambient temperature.

Snow conductivity can be increased by adding salt and mixing with a snow blower. The temperature in the test hall should typically be $-6\,°C \pm 1\,°C$ during this preparation phase. Any jig or form is maintained at the same temperature for easier removal prior to applying the test voltage.

Snow pile jig (chilled wooden form)
on EHV breaker
[Wieck et al., 2007]

Snow blocks on polymer insulator
[Hemmatjou et al., 2007]

Figure 8-11: Methods for deposit of natural snow on horizontal test objects.

TABLE 8-3: Summary of Snow Deposit Parameters Recommended by IEEE Joint Task Force [2008]

Snow Deposit Parameter	Recommended Value
Type	Natural snow
Thickness	30, 50, or 70 cm
Density	0.3 g/cm^3
Snow melted-water conductivity σ_{20}	100 μS/cm @ 20 °C by mixing snow and salt with snow blower
Air temperature	−6 °C ± 1 °C
Location of deposit	Vertically above insulator
Applied voltage	Service voltage stress

The snow layer is characterized by its height above the highest point of the test object, the total snow volume, the weight per unit dry arc distance, and electrical conductivity of water melted from the snow, corrected to 20 °C.

A third method, using a snow cannon from the ground, has also been used by Wieck et al. [2005] to build a snow-like deposit with density up to 0.3 g/cm^3 on 420-kV breakers in outdoor conditions. This was not as reliable as the use of natural snow, the method these researchers recommended in Wieck et al. [2007].

Some standard parameters have been proposed for snow tests on insulators in cases where the local snow conditions, including density, thickness, and conductivity, have not been characterized. These default conditions are listed in Table 8-3.

8.5.5. Evaluation of Flashover Voltage for Snow Tests

Withstand voltage tests for snow-covered insulators should be carried out using a constant-voltage method. A constant voltage is applied to the insulator assembly covered with snow to check the withstand voltage, until the snow falls away from the insulator assembly.

For snow tests carried out near nominal service voltage, a warming phase is needed to ensure that the conductivity of the melting snow layer reaches its maximum value at about −2 °C ± 1 °C. Once the test insulators are properly covered with snow, the ambient temperature should be progressively raised while the voltage is held constant.

After each test, the snow cover should ideally be renewed. This makes snow testing relatively expensive but ensures the best reproducibility. Some laboratories prefer to test the snow cover more than once, especially if it is in "good condition" after a withstand result. These laboratories accept the validity of the flashover level obtained at a higher test voltage on the same snow layer.

The IEC 60507 [1991] method for up-and-down testing is used to establish the test voltage, which increases or decreases by 10% depending on whether the previous test resulted in withstand or flashover, respectively. The step size may be reduced to 5% if desired for a more precise result for research purposes.

The up-and-down method with at least five "useful" tests (where the pattern of increasing voltage leads to flashover, and decreasing voltage leads to withstand) will give a test voltage that leads to three withstands and two flashovers. Snow conductivity, thickness, and density are fixed for each series.

- Maximum withstand voltage V_{WS} is the maximum level of applied voltage that gave three withstands in a series of four tests in the same conditions.
- Minimum flashover voltage V_{MF} corresponds to a voltage level that is one step (10%) higher than V_{WS} and gives two flashovers out of a maximum of three tests.
- Median flashover voltage V_{50} is established by ten "useful" tests in the up-and-down method, and the standard deviation of the ten flashover levels is also reported.

Multiple tests at the same level of snow severity are also preferred for ranking of performance. The time to flashover and records of leakage current activity should also be measured for each test, as these are important to confirm mathematical modeling.

For tests carried out with temporary overvoltage in excess of 1.5 per unit, no warming phase is necessary. It is unlikely that there would be an overvoltage of this magnitude at the same time as a melting phase. The test voltage can be applied directly. The snow will warm up under the influence of its internal leakage current. In a test process of Wieck et al. [2007], the voltage was maintained for a minimum of 1 minute and a maximum of 5 minutes. These researchers found that, provided the snow layer was completely intact and there had been no flashover on the first test, the voltage could be reapplied twice at different test levels, giving up to three test values for each snow layer.

Any artificial test method for insulators is subjected to review to see whether it meets the criteria of R^3—representative, repeatable, and reproducible—promoted by Lambeth et al. [1973]. The snow test method here is:

- Representative, as it is carried out with natural snow having a density of $0.3\,g/cm^3$ and a conductivity adjusted to $\sigma_{20} = 100\,\mu S/cm$. These are fifth-percentile values of severity in several environments. For simulating snow near the seacoast, a higher value of σ_{20} up to $4000\,\mu S/cm$ may be appropriate.
- Repeatable, with a relative standard deviation in flashover voltage of 3% when the product of snow density and melted water conductivity was

controlled in one laboratory to have a relative standard deviation of less than 10%.

· Reproducible from laboratory to laboratory, once the effects of snow deposit density, conductivity, and thickness have been normalized for each type of insulator tested. This is summarized in Section 8.7.

8.6. SNOW FLASHOVER TEST RESULTS

The results of flashover tests on snow generally show that thick layers of dense, highly conducting snow on parallel insulators may lead to flashover across the dry arc distance at normal service voltage stress.

8.6.1. Outdoor Tests Using Natural Snow Accretion

Snow accumulation on insulators with horizontal orientation has been noted to be more important than accumulation between the sheds of vertical suspension strings because the snow cover builds up more easily and persists for a much longer time without falling off.

Figure 8-12 shows typical cross sections of deposits [Fujimura et al., 1979].

The specific gravity of natural snow accumulation reached a maximum of $0.3 \, g/cm^3$, compared to levels over $0.5 \, g/cm^3$ when natural snow was deposited artificially on the insulators. The ac withstand voltage of the 6AS long-rod insulator in Figure 8-13 was considerably higher than the withstand voltage of the disk insulator strings. The withstand voltages obtained in tests with natural and artificial snow cover are in close agreement.

Tests of the flashover strength of a double tension insulator string of 12-tonne rating (26 kip) with 2.3-m string length under natural snow conditions were reported by Yasui et al. [1988]. They focused laboratory and natural-

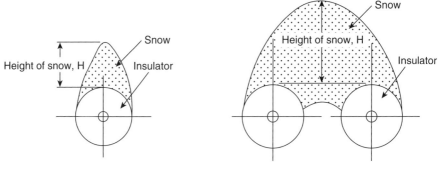

Snow deposit on single horizontal tension insulator (end view)

Snow deposit on double string of horizontal tension insulators (end view)

Figure 8-12: Conditions of snow deposit on insulators (from Fujimura et al. [1979]).

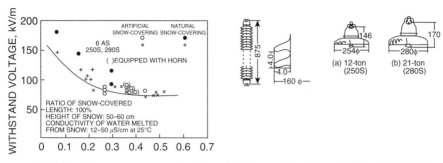

Withstand stress versus snow density   6AS long-rod insulator   250S and 280S Disks

Figure 8-13: Comparison of withstand voltage for natural and artificial deposit of natural snow (from Fujimura [1979]).

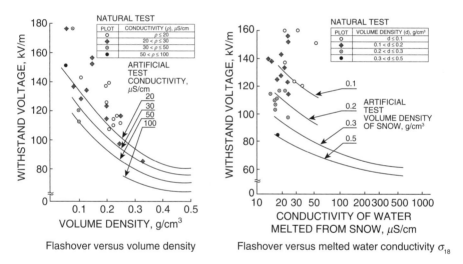

Flashover versus volume density   Flashover versus melted water conductivity σ_{18}

Figure 8-14: Flashover strength of natural deposit of natural snow accumulation on cap-and-pin insulator string (from Yasui et al. [1988]).

accretion testing on the performance of double strings of horizontal (tension) insulators, which held up to 1 m of snow at a weight of more than 20 kg per unit (1400 g per cm of dry arc distance). The authors pointed out that the temperature of maximum danger, when the snow reaches its maximum conductivity and just before snow slips from the insulator, was typically just below −1 °C.

Their results in Figure 8-14 show that the volume density and conductivity of the natural accumulation of snow both played a role in the flashover strength. Over the course of 3 weeks, snow built up on the exposed pair of horizontal insulators to a density of 0.33 g/cm³, a height of 40 cm, and a conductivity of 18 μS/cm at 18 °C, corresponding to $\sigma_{20} = 19$ μS/cm using a factor of 1.044 for the temperature coefficient of sodium chloride. With this deposit, the withstand voltage stress E_{50} was about 86 kV/m$_{dry\ arc}$.

In the long-term natural tests [Yasui et al., 1988], flashovers were observed on the insulator strings at:

- Conductivity of $\sigma_{20} = 40$ μS/cm, volume density of 0.1 g/cm³, and electric stress of 135 kV/m.
- Conductivity of 27 μS/cm, volume density of 0.16 g/cm³, and electric stress of 135 kV/m.
- Conductivity of 27 μS/cm, volume density of 0.1 g/cm³, and electric stress of 170 kV/m.

Yasui et al. [1988] concluded, in agreement with Fujimura et al. [1979], that there was a linear relation between withstand strength for natural snow accretion and for artificial deposit of natural snow, based on the summary of their data in Figure 8-15.

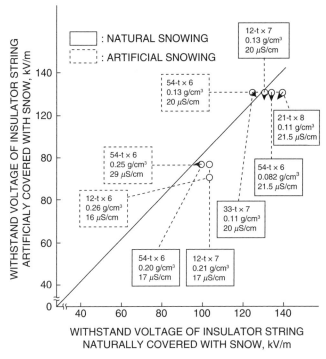

Figure 8-15: Relation between withstand voltage in natural and artificial deposit of natural snow (from Yasui et al. [1988]).

8.6.2. Outdoor Tests Using Artificial Snow Deposit

After tests with natural snow accumulation gave insufficient thickness to reproduce in-service problems, Wieck et al. [2005] adapted their test process for a 420-kV circuit breaker to make use of an artificial snow-like deposit with density up to $0.3 \, \text{g/cm}^3$. Two tests were made using a snow cannon, typically applied at ski hills, and local lake water, leaving a deposit with melted water conductivity of $\sigma_{20} = 46 \, \mu\text{S/cm}$. Three other tests used natural snow from the ground, with melted water conductivity of 3–7 $\mu\text{S/cm}$. The tests with artificial, conductive snow gave initial flashover stress in the range of 114–152 kV/$m_{\text{dry arc}}$ compared to levels of 104–122 kV/$m_{\text{dry arc}}$ for the natural snow. While outdoor air and snow temperatures were monitored, later tests by these researchers [Wieck et al., 2007] used indoor facilities with natural snow piled into jigs and better temperature control.

8.6.3. Indoor Tests Using Natural Snow Deposits

Research on snow flashovers has made effective use of natural snow, applied in a controlled volume in a test laboratory [Yasui et al., 1988]. A wooden jig, as shown in Figure 8-16, was used to enclose the snow pile. Natural snow was packed into the cooled jig to obtain the required density. The conductivity was adjusted by spraying salt solution over the snow pile. The jig was removed before electrical testing.

The more recent recommendations of the IEEE Task Force [Farzaneh et al., 2003] endorse this snow test method in most details, but the IEEE Task Force recommends the use of a snow blower to mix salt and snow to the required conductivity prior to packing into the jig. This ensures more consistent volume conductivity. In all cases, a constant-voltage method is used, with typical test duration of 20–30 minutes.

When six 250-mm diameter ceramic disk insulators were contaminated in a test laboratory and covered with natural snow, the measured surface resistance was about $20 \, \text{M}\Omega$. At a chamber temperature of +12 °C, the surface

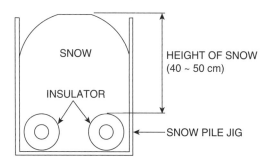

Figure 8-16: Artificial deposit of natural snow on insulators (from Yasui et al. [1988]).

resistance declined to a minimum of 2 MΩ within 20 minutes and then increased. There was a similar trend in the flashover voltage. An initial flashover level of 210 kV declined to a minimum of 110 kV within 20 minutes using a "voltage-increasing" method that tested flashover level once every 43 seconds.

A second voltage application method, where the voltage was held constant for the entire test, gave flashover levels that were about 30% lower than the voltage-increasing method. The constant-voltage method was later adopted for all testing of ice and snow flashovers in the IEEE recommendations [Farzaneh et al., 2003].

In the work of Yasui et al. [1988], the maximum applied voltage that gave withstands in four separate tests, with no flashovers, was defined as the withstand voltage. This gave a series of results shown in Figure 8-17.

The withstand voltages in Figure 8-17 can be compared directly with the results obtained using natural snow accretion in Figure 8-14. It proved to be feasible to vary the volume density and conductivity of the snow over a wide range of interest in the test lab, whereas the conductivity of melted water σ_{20} in the outdoor tests stayed in a narrow range around 20 µS/cm.

Watanabe [1978] tested double strings of 146-mm × 254-mm disk insulators with mountain snow having a conductivity of 10 µS/cm, corrected to 20 °C. For double strings of 16–25 disks, covered to a thickness of 20 cm with an unspecified density, his flashover gradient in Table 8-4 for ac and dc negative polarity was 77 kV/$m_{\text{string length}}$.

Fujimura et al. [1979] carried out their laboratory tests using natural snow that had been gathered from a mountainous area and stored indoors at –5 °C.

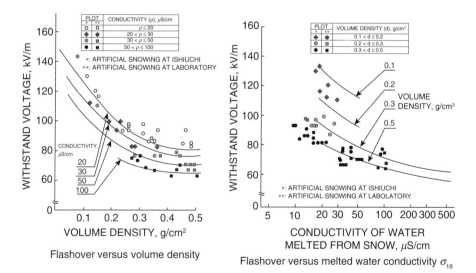

Flashover versus volume density

Flashover versus melted water conductivity σ_{18}

Figure 8-17: Flashover strength of artificial deposit of natural snow on cap-and-pin insulator string (from Yasui et al. [1988]).

TABLE 8-4: Flashover of Double Insulator Strings Covered with Mountain Snow

Insulator Type	Number of Disks	String Length (m)	Voltage Source	V_{MF} (kV)	E_{MF} (kV/m $_{length}$)
	16	2.34	ac	180	77
	25	3.65	ac	280	77
	20	2.92	dc Negative	240	82
	25	3.65	dc Negative	280	77
	20	3.30	dc Negative	270	82

Source: Watanabe [1978].

The electrical tests under ac, dc, and impulse conditions considered the height of snow, the weight of deposit per insulator spacing distance, and the electrical conductivity of the water melted from the snow. They tested long-rod insulators, station posts, and single and double strings of disk insulators.

The effect of the amount of snow, expressed in grams per unit of a 1175-mm long-rod insulator, is shown in Figure 8-18. The withstand voltage falls from 240 to 120 kV as a thick layer of clean snow is placed on the insulator.

Different insulators responded differently to snow deposits in Fujimura et al. [1979]. For a vertical station post insulator, the amount of snow between sheds was about $30 \, g/cm_{dry \, arc}$. In comparison, snow weight of about $140 \, g/cm_{dry \, arc}$ accumulated on disk insulators. The effects of the natural snow on the withstand voltage are given in Figure 8-19.

8.6.4. Snow Flashover Results for dc

In a summary of test results, CIGRE Task Force 33.04.09 [2000] noted that the dc flashover of snow-covered insulators was roughly equal to the rms value of the ac flashover strength in the same conditions. Negative polarity gave the lower withstand voltages in Figure 8-20.

8.6.5. Snow Flashover Under Switching Surge

The degree of snow cover plays a strong role in the flashover characteristics of the insulators in switching-surge conditions as well as under ac conditions. When an insulator is bridged with snow from 60% to 100% of its length,

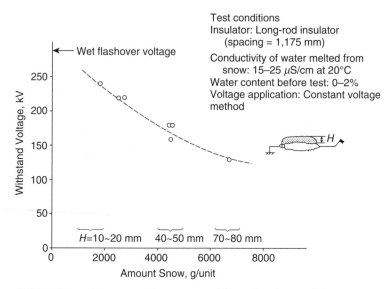

Figure 8-18: Effect of amount of snow on withstand voltage of long-rod insulator (adapted from Fujirama et al. [1979]).

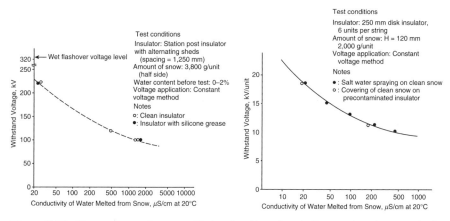

Figure 8-19: Comparison of snow withstand voltage for station post and cap-and-pin disk insulators (adapted from Fujimura et al. [1979]).

Figure 8-21 shows that the ac strength is reduced by a factor of about 2.5 from the strength without snow. Matsuda et al. [1991] showed that there is a U-shaped behavior for switching-surge flashover, with a minimum at 50–80% bridging.

Their tests also showed in Figure 8-21 that the combined effect of switching surge supcrimposed on ac stress behaved like ac stress alone, with the level being nearly constant from 60% to 100% of snow cover.

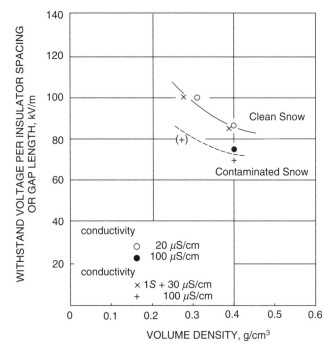

Figure 8-20: Negative polarity dc withstand voltage stress on insulator assembly as function of volume density of natural snow (from CIGRE Task Force 33.04.09 [2000]).

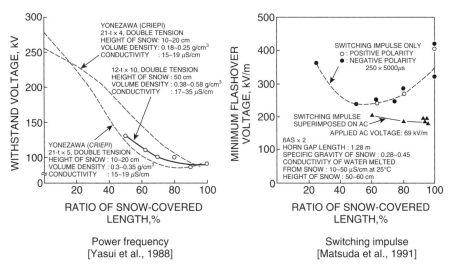

Figure 8-21: Ratio of snow-covered length to power frequency and switching impulse flashover strength.

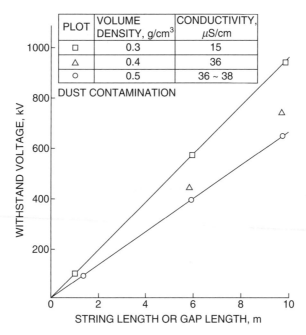

Figure 8-22: Relation of withstand voltage to insulator string length in snow flashover tests (from Yasui et al. [1988]).

8.6.6. Snow Flashover Results for Long Insulator Strings

Tests on long snow-covered insulators were carried out to test the hypothesis that flashover strength is linear with insulator dry arc distance. The constant-voltage method was used for these results. Figure 8-22 from Yasui et al. [1988] shows that a direct proportionality was found between the string length and the withstand voltage up to 10-m length, with 42 units of 54-tonne porcelain suspension disks.

The linearity of snow flashover under various types of voltage, including constant ac, temporary ac overvoltage, and switching surge, was also confirmed by the values in Table 8-5 given by Matsuda et al. [1991]. Matsuda and co-workers used snow depth of 50–100 cm with density of 0.3–0.4 g/cm^3 with a melted-snow conductivity level of 50 μS/cm at 25 °C (σ_{20} = 45 μS/cm).

8.7. EMPIRICAL MODEL FOR SNOW FLASHOVER

A unifying "icing stress product" concept was suggested in CIGRE Task Force 33.04.09 [1999] for different types of icing, at accretion levels that fully bridge the dry arc distance. This was defined as the weight of ice (grams per cm dry arc distance) times σ_{20}, the conductivity of the melted accretion (μS/cm)

TABLE 8-5: Electrical Performance of Insulator Assemblies Covered with Snow

Voltage class	154 kV	275 kV	500 kV
Insulator type	2 × 12-tonne long rod	17 × 12-tonne disk	28 × 21-tonne disk
Horn gap length (m)	1.28	2.02	4.00
ac Withstand voltage (kV)	96	152	300
Temporary ac overvoltage withstand (kV)	123	182	360
Critical switching impulse flashover (kV)	294	414	700–820
Withstand voltage (kV), SI superimposed on system ac	265	374	640–740
Critical lightning impulse flashover (kV)	445	635	1140
ac E_{WS} (kV/m_{arc})	75	75	75
ac TOV E_{WS} (kV/m_{arc})	90–95	90	90
SI E_{50} (kV/m_{arc})	230	205	175–205
SI + ac E_{50} (kV/m_{arc})	205	185	160–185
LI E_{50} (kV/m_{arc})	350	315	285

Source: Adapted from Matsuda et al. [1991].

corrected to 20 °C. This product serves as a good empirical measure for moderate to severe icing levels, leading to flashover stresses in the range of 40–150 kV per meter of dry arc distance.

8.7.1. Conversion of Test Results to Snow Stress Product

Snow density, conductivity, layer dimensions, and flashover voltage stress on pairs of horizontal insulators were reported by Yasui et al. [1988]. They tested twin strings of 25-cm diameter disk insulators that could support thick snow layers of 45 × 75-cm thickness. Their tests showed a significant loss of electrical strength, falling from 90 to 60 kV/$m_{dry\ arc}$ at high volume density.

These data were used to derive an empirical relation between flashover stress E_{WS} and "icing stress product" of snow weight (g/$cm_{dry\ arc}$) and melted snow conductivity σ_{20} (µS/cm) that was recommended for use within its range of application, from 10,000 to 100,000 (g/cm)-(µS/cm). Based on the snow cross sections, a relation of 0.1 g/cm^3 density = 338 g/$cm_{dry\ arc}$ was established.

Chisholm et al. [2000] expressed the test results for snow flashovers in Figure 8-17 with this approach, using the same icing stress product but using snow weight rather than ice weight. The stress product concept did, to a large extent, allow the definition of a single design parameter for snow flashover in the same way that the icing stress product is proving useful for ice-withstand strength analysis. The general expression that fitted these results in the range of $10,000 \leq SSP \leq 100,000$ was

$$E_{WS} = 1303(SSP)^{-0.26} \tag{8-1}$$

where

E_{WS} is the withstand flashover stress $(kV/m_{dry\ arc})$,

SSP is the snow stress product, snow weight $(g/cm_{dry\ arc})$ multiplied by σ_{20}, and

σ_{20} is the conductivity $(\mu S/cm)$ of the water melted from the snow layer, corrected to $20\,°C$.

The data from Yasui et al. [1988] are replotted in Figure 8-23 along with added values from Fujimura et al. [1979], Hemmatjou et al. [2007], and Wieck et al. [2005, 2007] for various insulator types and breaker bushings. The general

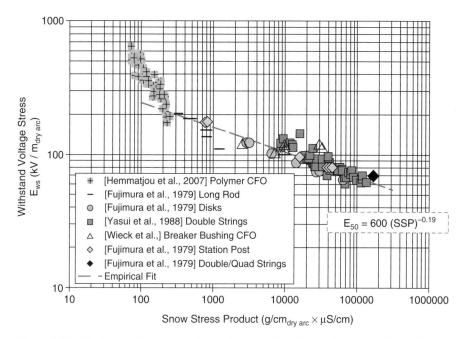

Figure 8-23: Flashover stress versus icing stress product for moderate and heavy layers of snow.

trend toward lower withstand stress E_{WS} with increasing snow stress product (SSP) is clear. Also, there is a relatively smooth transition among insulator types, ranging from very high gradients for small SSP deposits on polymer insulators and long-rod insulators, through modest levels on vertical posts and disks and into heavy snow on double horizontal strings.

Overall, the data are fitted by an updated empirical expression with the same exponent that fits the E_{WS} relation to icing stress product:

$$E_{WS} = 600(SSP)^{-0.19} \tag{8-2}$$

where

E_{WS} is the maximum flashover stress (kV per meter of dry arc distance), and

SSP is the snow stress product, given by the weight of snow ($g/cm_{dry\ arc}$) multiplied by the conductivity ($\mu S/cm$) of the melted snow corrected to $20\,^{\circ}C$.

Considering the errors in the individual data points, usually relating to the range of values for snow conductivity, the estimate of withstand stress for snow from Equation 8-2 is very simply 50% higher than the critical flashover stress if the layer was glaze ice of the same weight and conductivity.

The polymer insulator performance is relatively poorly fitted because, to some extent, the nature of the flashover changes from the interior of the snow, for high values of icing stress product, to the leakage path across the snow surface. Hemmatjou et al. [2007] reported that moderate accumulation of snow on a 15-cm diameter nonceramic insulator gave an icing stress product of about 100 (g/cm)-$(\mu S/cm)$, that actually increased the insulator strength above the typical 380-kV rms per meter of dry arc distance. For these cases, arcing at flashover followed the surface of the snow, which extended the flashover path by a factor of 2.2 compared to the dry arc distance without the snow.

The experimental values of withstand voltage versus weight of snow tend to show an asymptote at snow weight $w > 700\,g/cm$, corresponding to weight in excess of 10 kg/unit. Figure 8-24 shows that this may be a practical upper bound to the snow stress product model predictions of withstand stress using Equation 8-2.

This suggests that, in deterministic insulation coordination, the minimum withstand level of about $70\,kV/m_{dry\ arc}$ will be satisfactory for double chains of standard disk insulators for any thickness of snow at constant 0.5-g/cm^3 density and σ_{20} of $50\,\mu S/cm$ [Yasui et al., 1988].

8.7.2. Comparison of Snow Flashover to Ice and Cold Fog

Most researchers working in snow flashover have drawn conclusions about the relative severity, compared to clean-fog or salt-fog contamination flashovers.

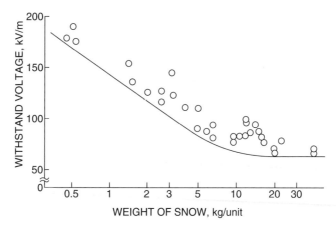

Figure 8-24: Withstand stress E_{50} (kV/m$_{dry\ arc}$) for double horizontal chains of standard disk insulators (from Yasui et al. [1988]).

For example, Watanabe [1978] compared minimum withstand voltages of 146-mm × 254-mm cap-and-pin insulator strings for salt contamination against strings with ice or snow cover for ac and dc voltages. The withstand stress of $E_{WS} = 75$ kV/m$_{dry\ arc}$ was reported for 144-µS/cm ice with a weight of 2 kg per disk or 136 g/cm. The same withstand stress was noted for 20-cm accumulation of 10-µS/cm snow on a double insulator string. This level of withstand stress was also achieved with low pollution levels of 0.02–0.03 mg/cm^2 for ac and dc, respectively. These ESDD levels are likely to be exceeded in winter conditions that have long periods without rain.

Fujimura et al. [1979] also made comparisons of the electrical performance of snow-covered insulators with the same insulators under dry, wet, or contaminated conditions (Figure 8-25). The snow by itself, with no insulator precontamination, was less dangerous than a contamination level of 0.012 mg/cm^2 (12 µg/cm^2). With precontamination, their snow-covered insulators performed better than if they were fully wetted in a fog test.

8.7.3. Comparison of Snow Flashover to Normal Service Voltage

As a brief comparison:

- Power systems with less than 250-kV system voltage are typically operated at service stress (E_{SV}) of 70–75 kV/m$_{dry\ arc}$ and would need dense, thick layers of highly conductive snow for an icing stress product of 60,000–80,000 (g/cm)-(µS/cm).
- EHV power systems are usually operated at E_{SV} of 80–90 kV/m$_{dry\ arc}$. Their double tangent insulator strings would be vulnerable to flashover with

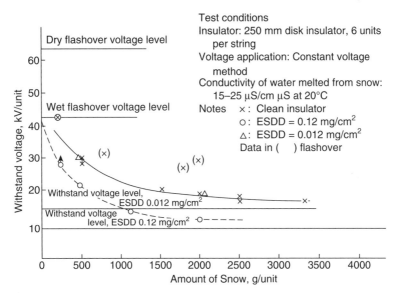

Figure 8-25: Relation of ac withstand voltage to amount of snow (from Fujimura et al. [1979]).

20,000–40,000 (g/cm)-(μS/cm), corresponding to 50-cm separation, 50-cm snow thickness, 0.3-g/cm^3 density, and 29–54-μS/cm conductivity.

- Some 500- and 735-kV systems with low levels of switching-surge over-voltages are operated at E_{SV} of 105 kV/m$_{dry arc.}$ At a snow stress product of 10,000, their insulator strings would be vulnerable to snow flashover with 50-cm thickness, 0.3-g/cm^3 density, and $\sigma_{20} = 13 \mu$S/cm.

The increase in E_{SV} with system voltage is an important factor that establishes which systems are most affected by snow accretion. Strategic EHV systems have more problems than HV systems because their levels of E_{SV} tend to be higher and they also make more use of multiple parallel insulator strings at tension (dead-end) locations. Most other power system reliability problems, such as animal and lightning outages, are far less severe on EHV lines because the basic insulation strength (dry arc distance) is larger.

8.8. MATHEMATICAL MODELING OF FLASHOVER PROCESS ON SNOW-COVERED INSULATORS

The foundation of an electrical flashover model for a snow-covered insulator rests with the successful application of the Obenaus [1958] approach to dc flashover, modified for ac reignition by Rizk [1981]. In the snow condition, as shown in Figure 8-26, the model must be adapted further. Arcing in snow as well as in air will occur. The numerical model will thus have an additional term

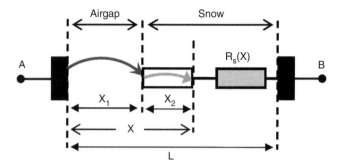

Figure 8-26: Equivalent electrical circuit of an arc propagating inside wet snow (from Hemmatjou et al [2007]).

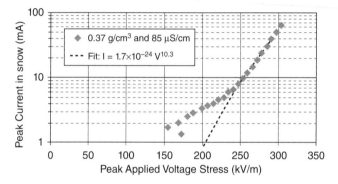

Figure 8-27: *I–V* relationship of snow layer (data from Hemmatjou [2006]).

to deal with the voltage drop of the arc in a snow layer as well as the usual voltage drop across the air gap.

where

> L is the dry arc distance (cm),
>
> x_1 is the length of the arc (cm) in air,
>
> x_2 is the length of the arc (cm) inside the snow,
>
> $x(=x_1 + x_2)$ is total arc length (cm), and
>
> $L - x$ is the length of the snow layer (cm) not bridged by the arc.

8.8.1. Voltage–Current Characteristics in Snow

The relations between arc current and voltage were established by experiments in a cylinder of snow of 300-mm length and 114-mm diameter [Hemmatjou et al., 2007]. At low current levels, the snow actually behaved in the same way as a surge arrester, as shown in Figure 8-27.

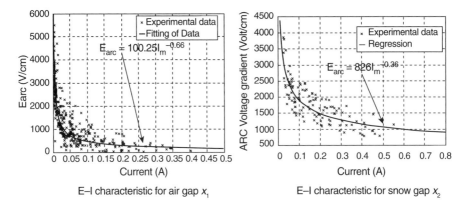

Figure 8-28: Measured voltage–current characteristic of arc in air gap x_1 and snow gap x_2 (from Hematjou et al. [2007]).

Once an arc was established, the voltage–current characteristics in regions x_1 and x_2 changed considerably. The experimental values in Figure 8-26 were established by Hemmatjou et al. [2007] in the range of peak current from 0 to 600 mA. Arc voltage gradient (V/cm) varied by a factor of 10 in air, with less dependence on current in the snow layer.

The degree of scatter in the results in Figure 8-28 is typical of the cycle-to-cycle variations normally observed during arcing. The central trend of the data for the air gap distance x_1 can be approximated by the Equation 8-3:

$$E_{arc} = A_a I_m^{-n_a} \approx 100 \cdot I_m^{-0.66} \qquad (8\text{-}3)$$

where

E_{arc} (V/cm) is the peak value of the applied voltage, divided by the air gap distance x_1

I_m (A) is the peak value of leakage current, and

A_a and n_a are constants.

The values of parameters A_a and n_a for the arc in the air between an electrode and a snow layer differ from those determined for flashover on ice surfaces, which are $A = 205$ and $n = 0.56$.

The average voltage gradient of the arc in the snow, E_{arc}(V/cm) across the distance x_2 was

$$E_{arc} = A_S I_m^{n_S} \approx 826 \cdot I_m^{-0.36} \qquad (8\text{-}4)$$

In the modified Obenaus model, the static condition of the arc in series with the resistance of the snow layer is expressed as follows:

$$V_m = V_e + A_a x_1 I_m^{n_a} + A_s x_2 I_m^{n_s} + I_m R_S(x) \tag{8-5}$$

where

V_m is the applied voltage (V),
V_e is the sum of the anode and cathode voltage drops (V),
I_m is the leakage current (A) inside wet snow,
x_1 is the length of the arc (cm) in air,
x_2 is the length of the arc (cm) inside the snow,
A_a and n_a are the arc constants in the air gap of length x_1,
A_s and n_s are the arc constants inside the snow gap x_2, and
$R_s(x)$ is the residual resistance (Ω) of the nonbridged snow layer.

The resistance of the nonbridged snow layer $R_s(x)$ is computed using

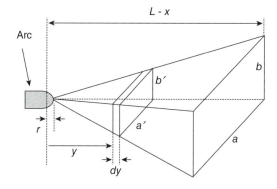

$$R_S(x) = \int_r^{L-x} \frac{1}{\sigma_e} \cdot \frac{dy}{a' \cdot b'} = \frac{(L-x)^2}{\sigma_e \cdot a \cdot b} \left(\frac{1}{r} - \frac{1}{L-x} \right) \tag{8-6}$$

The calculation of $R(x)$ relies on the equivalent conductivity of the snow, γ_e, which depends on the conductivity of the melted water σ_{20} (μS/cm at 20°C), the snow density δ (g/cm^3), and the snow temperature as shown in Figure 8-29. For the peak value of snow conductivity at -2°C, and for a limited range of application of $200 < \sigma_{20} < 800$ and $0.25 < \delta < 0.5$ from the experimental data in Figure 8-29, the equivalent volume conductivity of the snow, σ_e (μS/cm), can be estimated empirically from

$$\sigma_e = \frac{7.3\sigma_{20}\delta + 1650\delta - 1.73\sigma_{20} - 301}{1000} \tag{8-7}$$

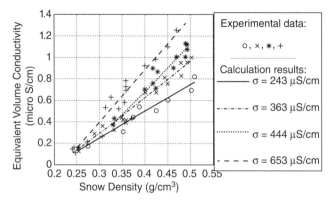

Figure 8-29: Equivalent volume conductivity of snow at $-12\,^{\circ}C$ versus density at four mean values of σ_{20} (from Hemmatjou et al. [2007]).

An empirical snow stress product model would only have the $\sigma_{20}\cdot\delta$ term in Equation 8-7, and the relative magnitudes of the other terms gives a good indication about the degree of uncertainty in the simple model.

The radius r of the arc root (cm) is an empirical function of the current I_m (A):

$$r = 0.361\sqrt{I_m} \qquad (8\text{-}8)$$

The resistance $R(x)$ in Equation 8-6 and equivalent volume conductivity σ_e are derived for a single geometry and validated for a small deposit on a short nonceramic insulator. So far, the parameters have not been able to match results for other conditions, as shown in the snow stress product graph of Figure 8-23. Additional refinement of the model can be achieved using validation with large-dimension flashover results where test parameters are well known, supplemented by laboratory experiments to establish the fundamental parameters over larger dimensions.

8.8.2. dc Flashover Voltage

Under dc conditions, the Obenaus model in Equation 8-5 includes a constant voltage term V_e for the sum of the anode and cathode voltage drops, which is typically 530 V for dc⁻ and 800 V for dc⁺ for a grounded ice layer [Farzaneh, 2000].

To use Equation 8-5, the values of V_m and x are plotted as a function of increasing I_m. The position of the arc along the snow layer will change continuously. When $x = L$, flashover occurs.

8.8.3. ac Reignition Condition and Flashover Voltage

Under ac conditions, the term V_e for anode and cathode drops in Equation 8-5 is normally set to zero and the effect is included in the arc constants. Also, there is a second necessary condition to be satisfied: the arc plasma must be sufficiently conductive and persistent to support reignition after the current crosses through zero. The arc reignition constraint was outlined by Rizk [1981] for ac flashover of polluted insulators. For an arc inside wet snow, the ac reignition constants were expressed as [Hemmatjou et al., 2007]

$$V_m \geq \frac{k_{as}x}{I_m^{b_{as}}} \tag{8-9}$$

where

k_{as} and b_{as} are the arc reignition constants, $k_{as} = 6370$ and $b_{as} = 0.486$,
I_m is the peak current (A),
V_m is the peak voltage (V), and
x is the total arc distance (cm).

Flashover is computed using the minimum value of Equation 8-9, which can be inverted to give the current needed for reignition:

$$I_m = \left(\frac{k_{as}x}{V_m}\right)^{1/b_{as}} \tag{8-10}$$

This current can be substituted into Equation 8-5 to yield

$$V_m = A_a x_1 \left(\frac{k_{as}x}{V_m}\right)^{-n_a/b_{as}} + A_s x_2 \left(\frac{k_{as}x}{V_m}\right)^{-n_s/b_{as}} + R(x)\left(\frac{k_{as}x}{V_m}\right)^{1/b_{as}} \tag{8-11}$$

where

V_m is the peak value of alternating applied voltage (V),
x is the total length of arc $(x_1 + x_2)$ (cm),
A_a and n_a are the arc constants for the air gap x_1, and
A_s and n_s are the arc constants for the arc in the snow.

When the constants A_a, n_a, A_s, n_s, k_{as}, and b_{as} and the relation $R(x)$ are known, then V_m is uniquely determined by arc lengths. The subscripts a, s, and as are used to specify the relevance of those parameters with regard to the air gap, to the arc in snow, and both of them at the same time, respectively.

8.8.4. Switching and Lightning Surge Flashover

The dominant frequency in dielectric response to switching-surge over-voltages is about 1 kHz. In tests of the high-frequency response of the real and imaginary parts of dielectric constant, or Cole–Cole plots, in Figure 8-7, this response does not deviate much from the low-frequency response.

For lightning impulse flashover, the dominant frequency is about 100 kHz. At this frequency, the snow behaves as a good dielectric with a low value of relative permittivity around 1.5–1.8, depending on morphology.

The dielectric constant of the snow at 1 kHz is relatively high compared to the insulator materials. This factor, along with its other electrical characteristics, leads in Table 8-6 to a reduction in switching-surge flashover level to 61% of the rod-to-plane value, or 57% of the value for a single insulator string with a gap factor of $k_g = 1.08$. There is approximately the same reduction in lightning impulse strength.

With the conductivity of the snow being a bit lower than ice and with similar and low values of dielectric constant at high frequency, the observed effects of snow on the lightning impulse flashover strength have not yet been explained with application of dynamic modeling.

TABLE 8-6: Comparison of Dry and Snow-Covered Impulse Flashover Levels for Insulator Strings

Insulator type	154-kV 2 × 12-tonne long rod	275-kV 17 × 12-tonne disk	500-kV 28 × 21-tonne disk
Horn gap length (m)	1.28	2.02	4.00
Dry, switching impulse E_{50} (kV/m$_{arc}$), $k_g = 1$	366	339	283
Snow, switching impulse E_{50} (kV/m$_{arc}$) (insulators)	230	205	175–205
Dry, lightning impulse E_{50} (kV/m$_{arc}$)	540	540	540
Snow, lightning impulse E_{50} (kV/m$_{arc}$)	350	315	285

Source: Data from Matsuda et al. [1991].

8.9. ENVIRONMENTAL CORRECTIONS FOR SNOW FLASHOVER

8.9.1. Pressure

At present, the same correction methods used for pressure correction of icing flashover results may be used for snow flashover. However, these corrections do not have substantial experimental support. The major concern in using the same method is that a substantial portion of arc in the snow flashover is inside the snow layer, while the arc in an icing flashover is on the air–ice surface interface.

8.9.2. Temperature

The test methods for snow flashover have a melting phase that brings the snow temperature near the melting point. The critical condition occurs at a snow temperature of −2 °C, which is the peak of electrical conductivity. For tests carried out with a melting phase, no temperature correction is necessary.

An idea about the effect of temperature on the flashover level can be inferred from the values of snow conductivity established by Hemmatjou et al. [2007]. They found that the value of γ_e at −2 °C was 1.28 times higher than the value at −12 °C. This should make less than 5% difference in the flashover voltage, which is typically within the range of experimental error in snow testing.

8.10. CASE STUDIES OF SNOW FLASHOVER

8.10.1. In-Cloud Rime Accretion: Keele Valley, Ontario

On January 17 and 18, 2004, there were three automatic circuit breaker operations on a 44-kV circuit in Ontario in a 10-hour period. The photograph in Figure 8-30 shows the trajectory of an exhaust plume from a thermal power plant, designed to use either landfill gas or natural gas. (See also Figure 8-31.)

At the times of the flashovers, the weather was a mix of snow and fog with temperature of −5 °C and dew point temperature of −6 °C to −7 °C at the nearest weather station. Wind was relatively calm from the southeast and south, shifting to the west and reaching a maximum speed of 17 km/h at the time of the third flashover. A previous flashover problem had been reported on December 23, 2003 with conditions of calm, continuous rain, fog, and drizzle and a temperature of +3 °C to +5 °C.

This case study serves to illustrate that icing can be a highly localized problem near a source of warm exhaust. The factors that could be explored to put this problem on a sound engineering basis include the following:

- The electrical conductivity of the rime ice melted from the accretion on the insulators.

Figure 8-30: Exhaust gas plume across 44-kV subtransmission line showing extent of rime accumulation on conductors. (Courtesy of Eastern Power.)

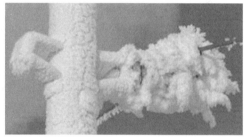

Figure 8-31: Accumulation of rime on conductors and 44-kV line post insulators from gas plume. (Courtesy of Eastern Power.)

- The density of the rime, typically about $0.3\,\text{g/cm}^3$ but possibly lower with the feathery appearance of the growth.
- The voltage stress of about 25 kV rms across 0.44 m of dry arc distance, or 58 kV/m.

With the knowledge from Takei and Maeno [2003] that hoarfrost and snow have similar electrical characteristics, it is most reasonable to treat the accretion in Figure 8-31 as a volume of snow. Inverting Equation 8-2 with service voltage stress, the SSP for withstand would be 23,000 (g/cm)-(μS/cm).

Approximating the rime in Figure 8-31 as a cube, the weight w with a density of $0.3\,\text{g/cm}^3$ would be about 600 g/cm. The conductivity σ_{20} of the rime accretion, melted and corrected to 20 °C, would need to exceed 400 μS/cm in order to explain this flashover problem. This is a conductivity that is typical of tap water in this region.

8.10.2. Temporary Overvoltage Problem: 420-kV Breaker in Norway

In the winter of 2004, Statnett in Norway found that a 90-cm snowfall was bridging the parallel insulated breaking chamber and grading capacitor insulators of their 420-kV circuit breakers with snow of a density higher than $0.3 \, g/cm^3$. The melted water conductivity from natural snow in this region was measured to be 15–20 µS/cm. The effective ac voltage across two of the circuit breaker terminals swings from ±2.5 per-unit when two parts of the power system are operating at different frequencies and are being brought into synchronization, a process that can take 60 s.

Normally, the design of hollow insulators for bushings is optimized to provide uniform radial and axial electric stress. This can be achieved with combinations of insulator geometry, capacitive (electrostatic) and resistance grading. The internal components of bushing shells have a strong effect on the flashover voltage of the outside surface. Well-designed bushings tend to have better room-temperature contamination flashovers than post insulators of the same dimensions. In service, the high operating temperature of the bushing also promotes formation of dry bands. However, in many cases, bushing designs do not consider the additional capacitance and resistance of an external ice layer.

Snow accretion on bushings will not be evenly distributed. Normally, an air gap forms at the line end, and for EHV bushings there may be two or three open areas in the pattern. The radial electric stress through the shell of the insulator is one concern from this uneven distribution. The relatively high operating temperature of the bushing may also promote melting, meaning that the highest electrical stress, and flashovers, can occur at an ambient temperature somewhat below the critical value of −2 °C that leads to maximum snow conductivity. This could be a factor in other situations where the only components that flashed over in station in icing conditions were the transformer bushings.

Tests were carried out to establish the ac flashover voltage of a standard breaker, without parallel grading capacitors, using a snow volume of $360 \, dm^3$ per horizontal bushing, corresponding to 34-cm thickness on each of a pair of 48-cm diameter bushings of 172-cm dry arc distance and 220-cm length. The snow stress product for this level of accretion is (34 cm × 48 cm × 0.3 g/cm³ × 20 µS/cm), or 9800 (g/cm)-(µS/cm). The flashover stress predicted for this snow stress product from Equation 8-2 is 105 kV per meter of dry arc distance. With two sections in series, a total flashover voltage of 360 kV can be estimated from the snow stress product approach.

The snow layers were tested three times, with changes in the appearance of the snow as shown in Figure 8-32. Flashover levels when the snow was fresh, Figure 8-32b, stayed constant with test time at 600 kV. The observed flashover level for the second voltage applications on the same snow layers was the same for short time to flashover but declined from 600 to 500 kV after 100 s of ac application. The decline with test time was more significant when the snow layer was tested a third time, falling from 550 kV at 0.1 s to 380 kV at 400 s.

(a) Snow layer before test at 360 kV

(b) Snow layer after first 1-min test

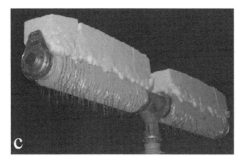

(c) Snow layer after second 1-min test

(d) Snow layer after flashover

Figure 8-32: Change in snow layer after multiple tests at 360 kV (from Wieck et al. [2007]).

Discharges near the insulator surface were measured using leakage current monitoring and ultraviolet and standard cameras. The UV equipment proved to be too sensitive once corona discharges were initiated and, instead, Figure 8-33 shows the discharges obtained using the standard video camera.

Leakage current levels across two bushings were measured to be about 500 mA at a total applied voltage of 520 kV and the current remained relatively stable over periods of 50–90 s. Leakage currents increased rapidly to more than 1500 mA within 5 seconds of flashover.

Flashover values were analyzed using the total accumulated time to flashover. The voltage values declined logarithmically with time from a value of 600 kV for 0.1-s time to flashover down to less than 400 kV for 400 s. This behavior, shown in Figure 8-34, is to be expected from the increase in snow temperature, conductivity, and melt water fraction as the electrical power of 200–300 kW is converted to heat.

The study concluded that flashover under synchronization of the two separated systems connected by this breaker was likely, and that snow should be removed from the breakers prior to connecting the generator to the system.

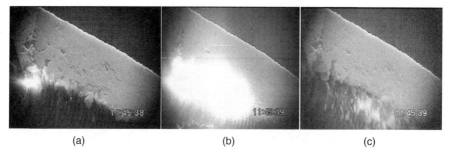

Figure 8-33: Electrical discharges in snow (a) 1 s before flashover, (b) at flashover, and (c) snow shedding after flashover (from Wieck et al. [2007]).

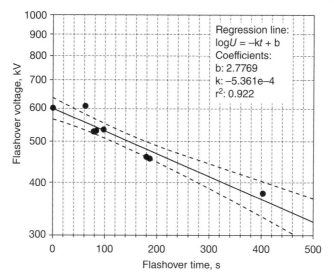

Figure 8-34: Decline in flashover voltage with accumulated test time on snow-covered circuit breaker with 2-m × 3-m insulators (adapted from Wieck et al. [2007]).

8.10.3. Snow Accretion on Surge Arresters

The effect of partial bridging by natural snow with density 0.3–0.35 g/cm³ and resistivity of 60–100 MΩ-cm in the frozen state was established by Kojima et al. [1988]. The potential distribution on a two-section 275-kV surge arrester was significantly altered with the snow covering as shown in Figure 8-35.

The change in voltage distribution along the arrester puts the full supply voltage across half the arrester, with a series resistance of the snow layer to limit power dissipation. This is an unstable situation with positive temperature coefficient of resistance for both the snow and zinc oxide materials.

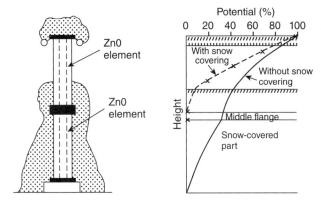

Nonuniform snow deposit   Elecric potential along vertical axis

Figure 8-35: Potential distribution along a snow-covered 275-kV zinc oxide surge arrester (from Kojima et al. [1988]).

8.11. PRÉCIS

Snow has a lower density than glaze ice, and its electrical characteristics are different as well. It takes a very thick layer of snow accumulation in the presence of pollution to lead to line voltage flashovers. Thick accumulation of snow often occurs on pairs of insulators that have horizontal orientation, so snow flashovers at service voltage have been reported mainly on dead-end structures.

So far, it has proved to be impossible to make artificial snow in a high-voltage test laboratory. Indoor test methods use jigs piled with natural snow that has been mixed with salt to simulate the local pollution environment. Arcing activity occurs inside the snow layer, adding complication to the mathematical modeling. Test results can be fitted with an empirical snow stress product model that is a good interim choice when selecting insulators.

REFERENCES

Arslan, A. N., H. Wang, J. Pulliainen, and M. Hallikainen. 2001. "Effective Permittivity of Wet Snow Using Strong Fluctuation Theory," *Progress in Electromagnetics Research*, Vol. 31, pp. 273–290.

Bentley, W. A. 1923. "The Magic Beauty of Snow and Dew," *National Geographic Magazine*, Vol. 43, pp. 103–112, with slides available at http://informatics.buffalo.edu/faculty/abbas/bms/index.htm.

Chisholm, W. A. and Y. T. Tam. 1990 "*Outdoor Insulation Studies Under CIFT Conditions, December 30–31, 1989 Storm Event*," Ontario Hydro Research Division Report, 90-131-H.

Chisholm, W. A., J. Kuffel, and M. Farzaneh. 2000. "The Icing Stress Product: A Measure for Testing and Design of Outdoor Insulators in Freezing Conditions," in *Proceedings of International Workshop on High Voltage (IWHV) 2000*, Tottori, Japan.

CIGRE Task Force 33.04.09. 1999. "Influence of Ice and Snow on the Flashover Performance of Outdoor Insulators, Part I: Effects of Ice," *Electra*, No. 187 (Dec.), pp. 91–111.

CIGRE Task Force 33.04.09. 2000. "Influence of Ice and Snow on the Flashover Performance of Outdoor Insulators, Part II: Effects of Snow," *Electra*, No. 188 (Jan.), pp. 55–69.

Colbeck, S. C. 1997. "A Review of Sintering in Seasonal Snow," US Army Cold Regions Research and Environment Laboratory (CRREL) Report 97-10, pp. 1–11.

Farzaneh, M. 2000. "Ice Accretions on High-Voltage Conductors and Insulators and Related Phenomena," *Philosophical Transactions of the Royal Society*, Vol. 358, No. 1776 (Nov.), pp. 2971–3005.

Farzaneh, M. and J. Kiernicki. 1997. "Flashover Performance of IEEE Standard Insulators Under Ice Conditions," *IEEE Transactions on Power Delivery*, Vol. 12, pp. 1602–1613.

Farzaneh, M. and J-L. Laforte. 1998. "A Laboratory Simulation of Wet Icing Buildup on H.V. Insulators," *International Journal of Offshore and Polar Engineering*, Vol. 8, No. 3 (Sept.), pp. 167–172.

Farzaneh, M., T. Baker, A. Bernstorf, K. Brown, W. A. Chisholm, C. de Tourreil, J. F. Drapeau, S. Fikke, J. M. George, E. Gnandt, T. Grisham, I. Gutman, R. Hartings, R. Kremer, G. Powell, L. Rolfseng, T. Rozek, D. L. Ruff, D. Shaffner, V. Sklenicka, R. Sundararajan, and J. Yu. 2003. "Insulator Icing Test Methods and Procedures: A Position Paper Prepared by the IEEE Task Force on Insulator Icing Test Methods," *IEEE Transactions on Power Delivery*, Vol. 18, No. 3 (Oct.), pp. 1503–1515.

Farzaneh, M., I. Fofana, and H. Hemmatjou. 2007. "Effects of Temperature and Impurities on the DC Conductivity of Snow," *IEEE Transactions on Dielectrics and Electrical Insulation*, Vol. 14, No. 1 (Feb.), pp. 185–193.

Fikke, S. M, J. E. Hanssen, and L. Rolfseng. 1993. "Long Range Transported Pollutants and Conductivity of Atmospheric Ice on Insulators," *IEEE Transactions on Power Delivery*, Vol. 8, No. 3 (July), pp. 1311–1317.

Fujimura, T., K. Naito, Y. Hasegawa, and T. Kawaguchi. 1979. "Performance of Insulators Covered with Snow or Ice," *IEEE Transactions on Power Apparatus and Systems*, Vol. PAS-98, No. 5 (Sept./Oct.), pp. 1621–1631.

Hemmatjou, H. 2006. "*Modélisation de l'Arc Électrique en Courat Alternatif à l'Intérieur de la Neige Fondante*," PhD Thesis, Université du Québec à Chicoutimi (Oct.).

Hemmatjou, H., M. Farzaneh, and I. Fofana. 2007. "Modeling of the AC Arc Discharge Inside Wet Snow," *IEEE Transactions on Dielectrics and Electrical Insulation*, Vol. 14, No. 6 (Dec.), pp 1390–1400.

IEC 60507. 1991. *Artificial Pollution Tests on High Voltage Insulators to Be Used in AC Systems*. Lausanne, Switzerland: IEC.

IEEE Standard 4. 1995. *IEEE Standard Techniques for High-Voltage Testing*. Piscataway, NJ: IEEE Press.

IEEE Joint Task Force (of DEIS and PES). 2008. "*IEEE Guide for Test Methods and Procedures to Evaluate the Electrical Performance of Insulators in Freezing Conditions*," PAR 1783. January 14.

Iwama, T., K. Kito, and K. Naito. 1982. "Ultra-High Strength Suspension Insulators and Insulator String Assemblies for UHV Transmission Line," *IEEE Transactions on Power Apparatus and Systems*, Vol. PAS-101, No. 10 (Oct.), pp. 3780–3789.

Kojima, S., M. Oyama, and M. Yamashita. 1988. "Potential Distributions of Metal Oxide Surge Arresters Under Various Environmental Conditions," *IEEE Transactions on Power Delivery*, Vol. 3, No. 3 (July), pp. 984–989.

Lambeth, P. J., J. S. T. Looms, M. Sforzini, R. Cortina, Y. Pocheron, and P. Claverie. 1973. "The Salt Fog Test and Its Use in Insulator Selection for Polluted Localities," *IEEE Transactions on Power Apparatus and Systems*, Vol. PAS-92, No. 6 (Nov.), pp. 1876–1887.

Matsuda, H., H. Komuro, and K. Takasu. 1991. "Withstand Voltage Characteristics of Insulator Strings Covered with Snow or Ice," *IEEE Transactions on Power Delivery*, Vol. 6, No. 3 (July), pp. 1243–1250.

Meier, A. and W. M. Niggli. 1968. "The Influence of Snow and Ice Deposits on Supertension Transmission Line Insulator Strings with Special Reference to High Altitude Operation," in *IEE Conference Publication No. 44*, London, England, pp. 386–395. September.

Nakaya, U. 1954. *Snow Crystals: Natural and Artificial*. Cambridge, MA: Harvard University Press.

Obenaus, F. 1958. "Fremdschichtüberschlag und Kriechweglänge," *Deutsche Elektrotechnik*, Vol. 4, pp. 135–136.

Onodera, T., H. Inukai, and T. Odashima. 2007. "Overview of Power Outage in the Niigata Kaetsu Area Caused by a Snowstorm," in *Proceedings of XII International Conference on Atmospheric Icing of Structures*, Yokohama, Japan. October 9–12.

Rizk, F. A. M. 1981. "Mathematical Models for Pollution Flashover," *Electra*, No. 78, pp. 71–103.

Sugawara, N. and K. Hosono. 2005. "Insulation Properties of Long Rod and Line Post Insulators for 33 kV Transmission Line in Wet-Snow Storm on January 2004," in *Proceedings of XI International Conference on Atmospheric Icing of Structures* (IWAIS), Montreal, Canada, pp. 197–202. June 12–16.

Takasu, K., H. Matsuda, and H. Ogawa. 1988. "Withstand Voltage Characteristics of Insulator Strings Covered with Snow," Komae Research Laboratory Report T 88026 [in Japanese].

Takei, I. and N. Maeno. 2003. "Dielectric and Mechanical Alterations of Snow Properties Near the Melting Temperature," *Canadian Journal of Physics*, Vol. 81, pp. 233–239.

Vuckovic, Z. and Z. Zdravkovic. 1990. "Effect of Polluted Snow and Ice Accretion on High-Voltage Transmission Line Insulators," in *Proceedings of 5th International Workshop on Atmospheric Icing of Structures* (IWAIS), Tokyo, Paper B4-3. October 29 to November 1.

Watanabe, Y. 1978. "Flashover Tests of Insulators Covered with Ice or Snow," *IEEE Transactions on Power Apparatus and Systems*, Vol. PAS-97, No. 5 (Sept.), pp. 1788–1794.

Wieck, H., I. Gutman, and T. Ohnstad. 2005. "Flashover Performance of a T-Shaped Circuit Breaker Under Snow Conditions," in *Proceedings of XI International Workshop on Atmospheric Icing of Structures (IWAIS)*, Montreal. June.

Wieck, H., I. Gutman, and T. Ohnstad. 2007. "Investigation of Flashover Performance of Snow-Covered Breakers," *IEEE Transactions on Dielectrics and Electrical Insulation*, Vol. 14, No. 6 (Dec.), pp. 1339–1346.

Yasui, M., K. Naito, and Y. Hasegawa. 1988. "AC Withstand Voltage Characteristics of Insulator String Covered with Snow," *IEEE Transactions on Power Delivery*, Vol. PWRD 3, No. 2 (Apr.), pp. 828–838.

Yasui, M., T. Oka, and T. Mori. 1990. "Experimental Study on Countermeasure for Snow Accretion on Power Transmission Lines," in *Proceedings of Fifth International Workshop on Atmospheric Icing of Structures* (IWAIS), Tokyo, Paper No B7-1, pp. 1–6. October 29 to November 1.

CHAPTER 9

MITIGATION OPTIONS FOR IMPROVED PERFORMANCE IN ICE AND SNOW CONDITIONS

This chapter reviews test results on a wide variety of mitigation methods, using examples drawn from both laboratory and field exposure and comparing with the empirical and theoretical models of strength.

Power system reliability under icing conditions is the combination of the reliability of many individual insulating components in parallel. Systems are designed to be tolerant of isolated faults, on a sporadic basis, but generally are more vulnerable to a series of repeated faults that may isolate several strategic lines or stations from service at the same time. We first explore the possibility of using knowledge about the climate and insulators to plan maintenance activity before, during, and after ice storms, prior to the critical period when the ice melts.

Reliability of individual insulators can also be improved in icing conditions with a number of methods. These include the use of:

- Surface coatings such as RTV silicone to decrease the surface conductivity.
- Accessories such as creepage extenders or booster sheds, added to break up icing patterns and to raise the threshold of ice accretion necessary for insulator bridging.
- More insulators, added in series mainly to increase the dry arc distance.
- Insulators with improved design, with longer leakage distance and/or greater separation between sheds, to raise the threshold of ice accretion threat level.

Insulators for Icing and Polluted Environments. By Masoud Farzaneh and William A. Chisholm
Copyright © 2009 the Institute of Electrical and Electronics Engineers, Inc.

- Semiconducting glaze, providing better performance than ceramic insulators with conventional glaze through an increased surface conductivity that mitigates any uneven electric field distribution under icing conditions.
- Polymer insulators of reduced diameter, offering benefits in areas of modest ice accretion that tend to diminish as accumulation levels increase.
- Monitoring and maintenance programs such as winter washing.

The solutions for improving contamination performance have varying degrees of efficacy in ice and snow conditions. A rough distinction is made in icing severity for electrical performance in Table 9-1 and is used to organize the mitigation options in this chapter.

Clearly, these levels of severity differ from the levels of ice accretion that would be considered severe for calculation of mechanical reliability, for example, in the ice loading criteria in Table 250-1 of IEEE/ANSI Standard C2 [2007].

Many of the methods that are effective for improving the contamination performance of insulators, such as on-demand pressure washing, increased leakage distance, RTV coatings, and semiconducting glaze insulators, are also effective for very light and light icing conditions.

The moderate icing regime is the range where insulator profile plays a dominant role in performance. Insulators with close shed-to-shed spacing are more easily bridged by icicles. Once they are fully bridged, their electrical flashover strength will be compromised.

For the heavy icing regime, countermeasures such as booster sheds and semiconducting glaze play roles that are somewhat different from the functions they perform in contaminated conditions.

Moderate amounts of snow or rime accretion have not proved to require mitigation. However, heavy accretion of snow in excess of 200 mm that fully bridges the leakage distance on standard insulator strings behaves in ways similar to the heavy icing regime if the snow is wet or is compacted by the action of high winds.

The main sections of this chapter present the mitigation options suitable for each icing regime, in an approximate order of preference. Where possible, the successes and occasional failures of the recommended mitigations are highlighted in case studies.

9.1. OPTIONS FOR MITIGATING VERY LIGHT AND LIGHT ICING

Very light icing is an electrical risk that is similar to full wetting in clean-fog conditions. The very light icing condition has a thin layer of ice on substantial portions of the insulator leakage distance, without substantial growth of ice caps or icicles on the top surfaces of insulators. The very light condition can

TABLE 9-1: Visual Classification of Ice Accretion Severity on Energized Porcelain Insulators

Very Light Ice Accretion <1 mm	Light Ice Accretion <6 mm		Moderate Ice Accretion 6–10 mm		Heavy Ice Accretion >10 mm	
Station post No icicles Thin ice layer on all surfaces	Partially bridged station post	Partially bridged 146-mm × 254-mm string	Fully bridged station post	Partially bridged 146-mm × 254-mm string	Fully bridged station post	Fully bridged 146-mm × 254-mm string
IEEE disk string No icicles Thin ice layer on all surfaces						

occur in fog with no precipitation, or with ice accumulation levels less than about 1 mm.

In approximate order of performance, the mitigation methods for very light icing are:

- Semiconducting glaze insulators.
- Increased leakage distance.
- Monitoring and washing.
- Silicone coatings on existing porcelain.
- Polymer insulators.
- Accessories.
- Increased dry arc distance with the same leakage distance.

Light icing is defined as the range where ice caps and icicles appear on the insulators, but these do not grow to an extent that fully bridges the leakage distance. In the case of small shed-to-shed distance on post insulators, this will lead to many air gaps in series. The icicles on cap-and-pin insulators may be longer and there are fewer air gaps, but each gap is much larger than those on post insulators. Generally, light icing is defined as the level that causes icicles, but does not cause bridging of station post insulators with uniform 50-mm sheds, typically an accumulation between 1 and 6 mm on a rotating reference cylinder.

In approximate order of performance, the mitigation methods for light icing are:

- Semiconducting glaze.
- Silicone coatings on existing porcelain.
- Polymer insulators.
- Increased leakage distance.
- Monitoring and washing.
- Accessories.
- Increased dry arc distance with the same leakage distance.

9.1.1. Semiconducting Glaze

The technology of ceramic insulator glazes that leads to a semiconducting surface was described as a mitigation option for pollution flashovers in Chapter 6. Commercial station posts with this option normally conduct about 1 mA at line voltage, and thus dissipate about 1 watt per kV of line-to-ground voltage. The main function of the semiconducting layer is to limit the voltage drop across dry bands, thus preventing the initial formation of arcs.

Figure 9-1 shows the difference in leakage current between the semiconducting glaze insulators and the porcelain posts of the same shape, size, and

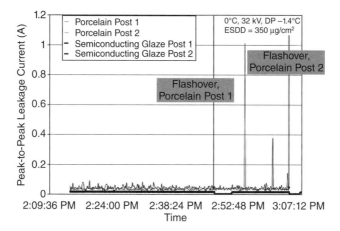

Figure 9-1: Leakage currents on standard and semiconducting glaze (RG) post insulators in cold-fog test.

orientation in cold-fog conditions near the freezing point. The posts had 680-mm dry arc and 1940-mm leakage distances and operated at 40-kV line-to-ground for a 69-kV substation. Both posts had heavy contamination levels of sulfuric acid with an ESDD of $0.35\,mg/cm^2$ ($350\,\mu g/cm^2$) to simulate effluent from a flue gas cleaning system.

For the semiconducting glaze units, the peak-to-peak current in Figure 9-1 was measured to be 3 ± 0.3 times the rms current whenever the high-voltage supply was on, in close agreement with the expected value of 2.83 for a pure sine wave. The complete suppression of leakage activity on the semiconducting glaze insulators means that, for practical purposes, they do not flash over in cold-fog conditions. If the voltage is raised to a sufficiently high level, typically well above the manufacturers' rated line voltage, thermal runaway will occur instead. As the semiconducting glaze insulators start to dissipate tens of kilowatts, thermal shock starts to break sheds off the posts.

The excellent performance of semiconducting glaze insulators in very light (cold-fog) conditions was also verified for HV and EHV applications.

For a 230-kV HV substation in a contaminated area, cold-fog tests evaluated the performance of a three-section semiconducting glaze post with 4.19-m leakage distance, 2.03-m (80 in.) overall length, and 1.65-mm dry arc distance, with a rated voltage of 135-kV line-to-ground. This post flashed over at 300 kV under a contamination level of $0.52\,mg/cm^2$ ($520\,\mu g/cm^2$) and withstood similar cold-fog tests at ESDD levels of 0.37 and $0.44\,mg/cm^2$. The maximum withstand stress E_{WS} was evaluated to be $64\,kV/m_{leakage}$ for 30 minutes at $0\,°C$.

For a group of vulnerable 500-kV EHV substations, a semiconducting glaze post with 2.87-m dry arc distance was tested in a cold-fog laboratory and the results were analyzed statistically. With 95% confidence, there was less than

2% probability of flashover for an hour in cold-fog or very light icing conditions at an ESDD of $0.15\,mg/cm^2$ $(150\,\mu g/cm^2)$. The RG porcelain post had excellent performance compared to posts with the same dry arc distance and conventional glaze.

In response to a series of 500-kV system flashovers on 1550-kV BIL post insulators [Erven, 1988], a contamination-ice-fog-temperature (CIFT) insulator test was established to simulate weather conditions in a specially designed fog chamber. This test preceded the development of standard IEEE recommendations but had many features in common with the recommendations in Farzaneh et al. [2003a], including ice accretion at normal service voltage stress with $\sigma_{20} = 50\,\mu S/cm$ at $-5\,°C$, a slow temperature rise, and melting phase as detailed in Chapter 7. Accretion of light levels of glaze ice on fixed reference cylinders was typically 10 mm, corresponding to about 5 mm on a rotating reference cylinder.

Table 9-2 shows the CIFT test results comparing a semiconducting glaze post insulator against a porcelain post with conventional glaze and alternating-shed profiles.

The test data were analyzed with the method of Barrett and Green [1994] to give the ESDD for critical flashover with an hour of exposure to melting conditions at line voltage. The semiconducting glaze post insulator gave significantly better icing test performance, while retaining the 1550-kV BIL dry

TABLE 9-2: Comparison of Contamination—Ice-Fog-Temperature (CIFT) Performance of Alternate-Shed Profile and Semiconducting Glaze (RG) Uniform-Profile 500-kV Post Insulators

Option	Rated BIL (kV)	Dry Arc Distance (m)	Shed Profile	Leakage Distance (m)	ESDD for 50% Flashover at 303 kV, 1 h (mg/cm^2)
2	1550	2.87	1-2-3 Alternating profile	9.79	0.028 ± 0.004
6	1550	2.87	Uniform profile, RG semiconducting glaze	9.4	>0.15

arc distance and dimensions. With a 95% confidence, there was less than a 10% probability of flashover of the semiconducting glaze post in light icing conditions at an ESDD of 0.15 mg/cm².

Thermal considerations, rather than cold-fog flashover performance, would be the only limit to the use of semiconducting glaze insulators in heavy pollution areas where the winter-maximum ESDD may exceed 700 μg/cm².

9.1.2. Increased Leakage Distance

With the cold-fog conditions of very light icing having much in common with contamination flashover, the well-understood solution of using lower electrical stress per meter of leakage distance is a favored alternative when insulators need to be replaced. This linearity is expressed in the use of a constant stress per meter of leakage distance, or equivalently a unified specific creepage distance (mm per kV_{l-g}) for a given site severity at all voltage levels.

For high ratios of leakage to dry arc distance, the question about whether the distance is fully effective in cold-fog conditions should be addressed. This can be analyzed using the same approach used in contaminated conditions by Baker et al. [1989] for station post insulators. The flashover stress per meter of leakage distance in a series of cold-fog tests on porcelain insulators [Chisholm, 2007] is plotted using the ratio of leakage to dry arc distance as a parameter in Figure 9-2.

Overall, in cold-fog conditions, there does not seem to be much difference in the unified specific creepage distance that causes flashover for insulator

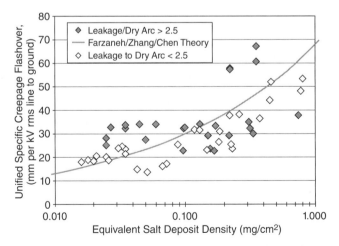

Figure 9-2: Comparison of unified specific creepage distance (USCD) flashover levels versus ESDD in cold-fog tests for high-leakage and low-leakage profiles (from Chisholm [2007]).

TABLE 9-3: Results of Cold-Fog Tests on 14-Unit String of 146-mm × 254-mm Fog-Type Insulators with 458-mm Leakage Distance per Disk

ESDD (mg/cm^2)	Critical Flashover (kV) 30 min at 0 °C	Standard Deviation of CFO (%)	E_{50} $(kV_{rms}/m_{leakage})$
0.15	219	6.0	34
0.33	213	6.7	33
0.74	170	5.4	26

Source: Chisholm [2007].

profiles that had leakage to dry arc ratios of more than 2.5, compared to those that had ratios less than this value.

For a specific comparison at high contamination levels, where the differences might be greater, cold-fog tests were carried out on 14 units of anti-fog insulators with leakage to dry arc ratio of 3.1 and with a total leakage distance of 6.41 m. The critical flashover results are shown in Table 9-3.

As a comparison, tests with ESDD in the range of 0.14–0.23 mg/cm^2 (140–230 μg/cm^2) on standard-profile 146-mm × 254-mm disk insulators with 334 mm of leakage per disk and a leakage to dry arc ratio of 2.3 gave values of E_{50} = 42–43 kV$_{rms}$/m$_{leakage}$. This suggests that the trend to reduced shed efficiency with high ratio of leakage to dry arc distance is stronger at higher pollution levels in cold-fog conditions.

At high pollution levels, the standard deviation of the CFO in the cold-fog tests on the fog-type insulators increases to levels of 5–7%, which is close to the standard deviations found in clean-fog contamination flashover tests. The cap-and-pin insulators also supported sporadic arcing activity and leakage currents at 0 °C, unlike the quiescent characteristics of porcelain post insulators in the same conditions.

The role of leakage distance in the performance of EHV station post insulators under cold-fog conditions was established by carrying out a series of laboratory tests on eight different types. Full-sized 500-kV posts were tested at ac line-to-ground voltage of up to 370 kV. Most posts had 1550-kV BIL and three sections for a total of 2.87-m dry arc distance. A four-section post with 1800-kV BIL and 3.33-m dry arc distance was also tested. The dimensions, shed profiles, and observed performance are shown in Table 9-4.

One three-section 1550-kV BIL post (option 6 in Table 9-4) had semiconducting (resistance-graded) glaze and a uniform shed profile. As discussed in the previous section, this choice exhibited superior performance in light and very light icing conditions. A second mitigation option, the use of booster sheds on a standard post with uniform profile, was tested and its results are also reported in Table 9-4 as option 7.

While several of these profiles had significantly increased spacing of major sheds, through alternating 1-2-3 or 1-2-2 profiles, none gave much advantage

TABLE 9-4: Results of Contamination-Fog-Temperature (Cold-Fog) Tests on 500-kV Station Post Insulators

Option	Rated BIL (kV)	Dry Arc Distance (m)	Shed Profile	Leakage Distance (m)	ESDD for 50% Flashover at 303 kV, 1 h (mg/cm^2)
1	1550	2.87	Uniform profile	6.56	Peak at 0.021 ± 0.004
2	1550	2.87	1-2-3 Alternating profile	9.79	0.035 ± 0.004
3	1550	2.87	1-2-2 Alternating profile	7.62	0.025 ± 0.004
4	1550	2.87	1-2-3 Alternating profile	10.22	0.035 ± 0.004
5	1550	2.87	1-2-3 Alternating profile	9.88	0.027 ± 0.003
6	1550	2.87	Uniform profile, RG semiconducting glaze	9.4	>0.15
7	1550	2.87	Option (1) with six booster sheds	8.54	0.025 ± 0.003
8	1800	3.33	1-2-3 Alternating profile	10.30	Peak at 0.045 ± 0.005

Sources: Data from Chisholm et al. [1994] and Chisholm [2007].

in flashover performance that could not be related simply to the increased leakage distance. Station posts with increased dry arc distance and 1800-kV BIL were also evaluated and the same conclusion was drawn.

The flashover and withstand results were analyzed using the statistical method described in Barrett and Green [1994] to establish the probability of flashover for 1 hour of exposure at normal 303-kV line-to-ground voltage as a function of precontamination level. A typical output from this analysis method is shown in Figure 9-3.

The test results for posts (1) and (8) in Table 9-4 showed an interesting trend, with peaks in the probability of flashover for specific values of ESDD, respectively, 0.021 and 0.045 mg/cm² (21 and 45 μg/cm²). The other 500-kV post insulators in this test series had the usual increasing probability of flashover with increasing ESDD.

The conclusion from this test program was that cold-fog performance of EHV station post insulators scales with leakage distance. This is shown using the IEC terminology of unified specific creepage distance (USCD) in Figure 9-4.

The light contamination levels below 0.06 mg/cm² in Figure 9-4 call for high USCD of at least 35 mm/kV for a single 500-kV post insulator, without neces-

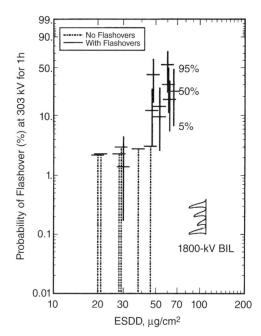

Figure 9-3: Output of statistical analysis for cold-fog (CFT) flashover test of 1800-kV BIL station post with 10.3 m of leakage distance, option (8) in Table 9-4. Vertical bars show confidence intervals (from Barrett and Green [1994]).

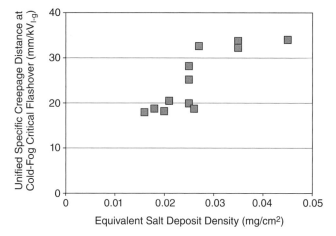

Figure 9-4: USCD for critical flashover (1 h at 303 kV) on 500-kV station post insulators with the same dry arc distance in cold-fog conditions.

sary margins for withstand versus critical flashover, many insulators in parallel, and uncertainty in ESDD. The required USCD is high because the cold-fog accretion of a typical thickness, 50 µm, tends to act as a heavy nonsoluble stabilizing deposit with NSDD of 5 mg/cm² in the flashover process [Chisholm, 2007].

9.1.3. Coating of Insulators with RTV Silicone

The use of coatings such as grease or silicone RTV can improve electrical performance in clean-fog tests. This mitigation also gives some improvement in performance under cold-fog (very light icing) and light icing conditions.

Silicone coatings increase the surface impedance of porcelain insulators in wet conditions. This increase, about six orders of magnitude above 0 °C (>100 GΩ/300 kΩ), is shown in Figure 9-5. Below freezing, the increase is not so dramatic. The silicone coating only provides about 100 MΩ of resistance instead of 100 GΩ. The resistance of the ice-coated porcelain insulator increased from 300 kΩ to over 1 MΩ. The ratio of treated to untreated surfaces was still one to two orders of magnitude.

A factor-of-10 increase in surface resistance should still be sufficient to raise the voltage needed for contamination flashover by a factor of 3. In practice, gains of about 25% in flashover strength per meter of leakage distance are gained when silicone coating is applied to porcelain insulators.

In a summary of alternatives for mitigating heavy pollution problems in winter conditions, a series of 38 cold-fog tests were carried out on 230-kV polymer post, deep-skirt apparatus insulators with and without RTV silicone

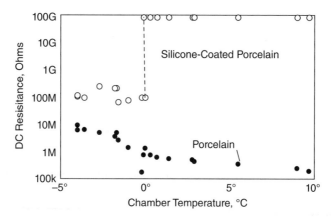

Figure 9-5: Comparison of precontaminated porcelain post insulator resistance above and below freezing with and without RTV silicone coating (from Chisholm et al. [1996]).

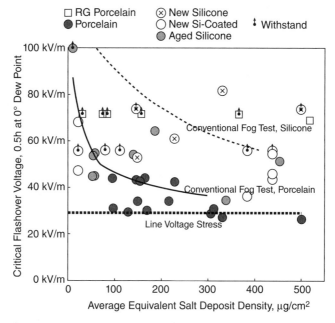

Figure 9-6: Comparison of cold-fog test E_{50} (critical flashover gradient, $kV_{rms\,1-g}/m_{leakage}$) for cap-and-pin porcelain, silicone posts, silicone-coated cap-and-pin porcelain, and RG porcelain posts [Chisholm, 1994; Chisholm, 2007].

and on resistance-graded or semiconducting glaze porcelain posts. The results are summarized in terms of E_{50} in Figure 9-6.

The tests were carried out with precontamination consisting of dry-sprayed kaolin, salt, and carbon black at a level of 20% by volume to match the local site severity. The pin-type apparatus insulators with a total of 4.8 m of leakage distance were found to have critical flashover levels of 138 kV with a standard deviation of 4.5% in cold-fog conditions, compared to the nominal insulator service voltage of 133 kV. The withstand stress for 500 insulators in parallel was 24 kV rms per meter of leakage distance on bare porcelain, which was 13% below the service voltage stress of 27.7 kV per meter of leakage distance.

The withstand stress for the large number of insulators in parallel, again using four standard deviations below the V_{50} level, improved to 5% above service voltage stress to 29 kV rms per meter of leakage distance when the same insulators were coated with RTV silicone, using the cleaning and coating processes illustrated in Chapter 6. This stress is within the experimental error of the cold-fog test method, with its standard deviation of 4% shared among all test results [Barrett and Green, 1994].

There have not been many investigations of the role of RTV silicone coatings in the flashover performance of insulators under light icing conditions. Initial concerns about the durability of the coatings at high electric field stress have been resolved with good winter service experience in a trial application in Ontario, Canada, at 500 kV. In field tests comparing a set of 3-m post insulators energized at 350-kV line to ground in a clean environment, the levels of leakage current on silicone-coated porcelain posts were consistently the lowest, even during winter flashover weather with fog and temperatures near 0 °C [Macedo, 1992].

Based on the performance of unbridged insulators with moderate ice accretion, RTV coatings for light accretion of ice of less than 6 mm on a rotating reference cylinder should perform the same as for very light conditions. This means that there will be an increase in withstand gradient from about 24 to 29 kV/m$_{leakage}$ with RTV coatings at heavy (0.1–0.4 mg/cm^2) pollution levels.

Chang and Gorur [1994] studied the rate of recovery of hydrophobicity of silicone surfaces with temperatures of –4 °C, 7 °C, 20 °C, and 28 °C. In keeping with the expected change in the viscosity of the light molecular weight fluids, the recovery rate below freezing was glacially slow, requiring weeks to achieve the same evolution of surface fluid that took hours at 28 °C.

At a 230-kV station, post insulators of uniform profile that were sprayed with an RTV silicone proved to be equally or more vulnerable to flashover from thin, highly contaminated icicles like those shown in Figure 9-7. The electrical conductivity of the water melted from accreted ice was found to be $\sigma_{20} > 400\,\mu S/cm$. The contamination source was road salt from a nearby expressway, shown in Figure 9-7.

The overall improvement in performance with silicone coatings in winter condition does not argue in favor of its selection in the design stage over other alternatives. The coatings have proved to play their most important role as a practical and long-lasting alternative when the poor winter contamination performance of an existing station must be improved. At one 230-kV station, porcelain bushings were coated with RTV silicone, and old insulators were replaced with semiconducting glaze posts as shown in Figure 9-8.

Site exposure near overhead expressways

Overall view of insulator   Detailed view

Figure 9-7: Light ice accretion leading to flashover on 230-kV post insulator in early 2003. (Courtesy of PSE&G.)

Figure 9-8: Application case study for use of RTV silicone (where necessary) and insulator replacement (where possible) for improved winter performance in highly contaminated 230-kV station (adapted from Chisholm [2005]).

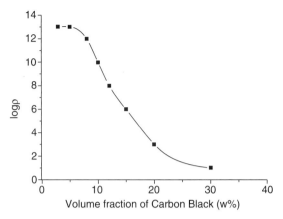

Figure 9-9: Relation of resistivity of RTV silicone (ρ, Ω-m) to volume fraction of carbon black (from Liao et al. [2007]).

The service experience on silicone-coated porcelain proved to be excellent. The carbon black that gave low flashover test results in the laboratory also melded with the silicone oil in the coating, giving surfaces that became increasingly hydrophobic over time [Chisholm, 2007] with water-drop contact angles reaching the superhydrophobic threshold of 150°. Also, Liao et al. [2007] have noted that, as the volume fraction of carbon black in the RTV increases, the resistivity ρ (Ω-m) decreases (Figure 9-9).

At a certain volume fraction, the RTV silicone may start to behave as a semiconducting layer that improves contamination performance by generating Joule heating and by improving the voltage distribution.

9.1.4. Change for Polymer

The test results for polymer insulators under cold-fog conditions gave highly mixed results. For example, tests of EPDM polymer posts with 0.93-m dry arc distance and 2.2-m leakage distance were carried out at contamination levels of 0.06–0.45 mg/cm². The lowest withstand stress of 29 kV/$m_{leakage}$ was found for an ESDD of 0.34 mg/cm². The polymer post insulators were not pursued as an option for mitigating cold-fog problems mainly because the profiles had a low ratio of leakage to dry arc distance.

However, similar cold-fog tests on 230-kV class silicone transmission line insulators with 3.8 m of leakage distance and 25 sheds gave results that improved with increasing contamination level. Insulators were tested with a dry-sprayed mix of kaolin, salt, and carbon black to simulate local pollution. Pollution levels varied in the tests from 0.15 to 1.2 mg/cm². For tests that established a critical flashover at 0.15, 0.23, and 0.33 mg/cm², the values of E_{50} actually increased from 53 kV$_{rms}$/$m_{leakage}$ to 61 and 78 kV$_{rms}$ /$m_{leakage}$, respectively [Chisholm, 1995]. At the time, the performance was attributed to cleaning of mold release, but Liao et al. [2007] offer an alternative explanation. The carbon black used as a part of the dry deposit of heavy pollution deposit in the 1995 tests may have made the silicone semiconducting as well as superhydrophobic after absorption.

Quantification of the specific effect of carbon black on the electrical properties of silicone materials in cold-fog conditions is thus recommended as an interesting area of future investigation.

9.1.5. Insulator Pollution Monitoring and Washing

Icing flashovers are more common within stations than on lines. This concentrates the risk of having a number of lines unavailable under freezing rain conditions. The weakness of station insulators is mainly a consequence of the low shed-to-shed distances, typically 50 mm for station posts compared to 120 mm from the bottom of one standard suspension disk to the top of the disk beneath it. As shown in Table 9-1, it takes a little more than 6 mm of accretion on a rotating reference cylinder to fully bridge a vertical station post insulator with uniform profile, while 10 mm or more is needed to fully bridge a string of standard ceramic transmission line disk insulators. The return period of ice-bridging problems at stations is thus considerably more frequent. This consideration focuses a large fraction of utility resources on mitigating icing performance of station insulators.

Maintenance options such as power washing are effective when buildup of insulator contamination is continuous over the course of weeks or months. In Chapter 6, this alternative for improving power system performance under conventional contamination conditions was described in considerable detail. The main point that this section will establish is that, for many areas subject to icing of any sort, the winter season is the one with the highest rate of pollution accumulation and also the one with the highest duration of exposure without natural rain washing.

The rapid rise of insulator pollution level after the start of winter road salting, and the equally steep descent when natural rain arrives in the spring, has been noted in Chapter 3. Figures 9-10 and 9-11 show a strong tendency toward a maximum level of ESDD repeated in the Toronto, Ontario, Canada area each winter.

In the ninth week in Figure 9-11, there was a remarkable drop in the pollution levels on all insulators. This corresponds to the occurrence of mixed precipitation with 15 mm of liquid rain and a lesser amount of snow. Levels built up again to a second peak on March 25 after another 2-week period without rain.

In Figure 9-11, the effect of 2 mm of winter rain was less effective at reducing the insulator ESDD, with reductions of only 20–40%. The peak ESDD levels of 30 µg/cm² for this season were one-third of the values found for the same site in the previous winter.

Insulator pollution monitoring in winter conditions should use unenergized posts or disks of the same shape, and in the same exposure, as the energized

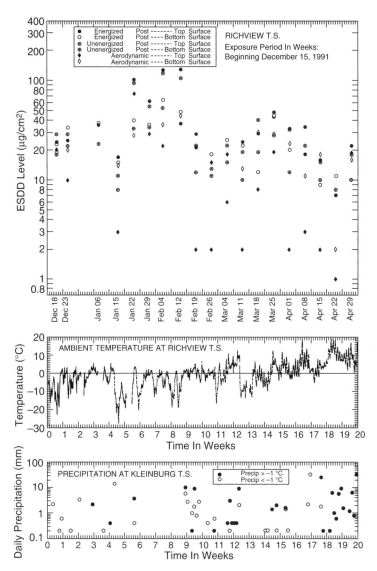

Figure 9-10: ESDD on energized and unenergized station posts and aerodynamic insulators over 1991–1992 winter season along with records of temperature and rainfall. (Courtesy of Kinectrics.)

power system insulators. Since most monitoring is carried out at transformer stations, this means that full-sized spare station posts should be used. Some considerations for monitoring in winter conditions include the following:

- Threat levels for EHV insulation may call for accurate measurement of low ESDD levels in the range of 0.015 mg/cm² (15 μg/cm²). This calls for

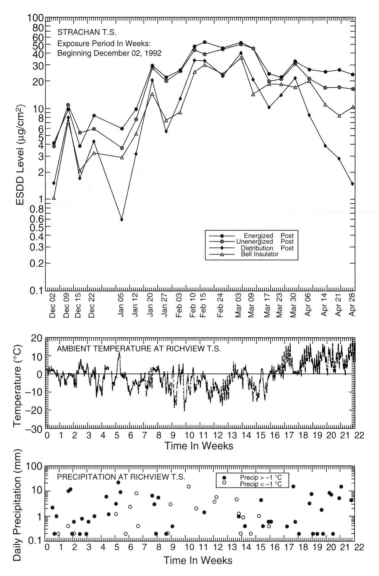

Figure 9-11: Average ESDD on energized and unenergized station posts and removable samples over 1992–1993 winter season along with records of temperature and rainfall. (Courtesy of Kinectrics.)

good training, proper procedures, and some refinements to the ESDD calculation method [Chisholm et al., 1994].

- Isopropyl alcohol can be added in any ratio up to 1:1 to keep the wash water from freezing. Tests showed that its contribution to the electrical conductivity of the wash water can be ignored. The volume of alcohol should not be included in the ESDD calculations, only the volume of water should be considered.

- Monitoring programs may use station post insulators that have been removed from service after partial shed damage. Damaged ceramic posts may have broken edges that are razor-sharp and should thus be protected with heavy reinforced duct tape that will not come loose in cold, windy conditions.

The insulator pollution levels should normally be analyzed by plotting them against the number of days without rain. The choice of what precipitation amount constitutes "rain" in winter conditions affects this analysis. In one study at five stations in Ontario, with readings similar to those shown in Figures 9-10 and 9-11, a minimum precipitation amount of 0.4 mm was found to be the best overall level, giving the linear regression results in Table 9-5.

These values were established over two winters of observations with dry periods with up to 38 days without rain. Predictions based on number of dry days form a basis of insulator maintenance in the "SMART Washing" process, but the estimates are usually supported by on-site measurements of ESDD to confirm that levels have reached critical thresholds.

Buildup of insulator surface contamination tends to be linear with duration of exposure, with some equilibrium at long exposure times. For stations where forms of pollution monitoring, such as ESDD measurements, leakage currents, or discharge activity, are deployed, it is possible to identify trigger levels for high-pressure insulator washing, prior to any icing events.

The same process described in Chapter 6, Section 6.2, should be applied for washing in freezing weather. Normally, station insulators and bushings in Canadian utilities have been washed successfully at ambient temperatures down to −10 °C. The water is usually drawn from a normal tap-water supply

TABLE 9-5: Fitted Relations Between Hours Without Rain and ESDD on Station Post Insulators for Five Locations in Ontario

Site	Equation ESDD (μg/cm^2) from Days Without Rain (D)	Regression Coefficient r^2
Strachan	ESDD = 22 μg/cm^2 + 3.1 D	0.6
Richview	ESDD = 20.5 μg/cm^2 + 0.88 D	0.5
Manby	ESDD = 9.5 μg/cm^2 + 0.80 D	0.6
Claireville	ESDD = 8.4 μg/cm^2 + 0.41 D	0.5
Milton	ESDD = 5.0 μg/cm^2 + 0.27 D	0.6
Trafalgar	ESDD = 6.6 μg/cm^2 + 0.17 D	0.2
Cherrywood	ESDD = 3.7 μg/cm^2 + 0.16 D	0.5

Source: Adapted from Chisholm et al. [1996].

through an in-line deionizing resin bed and stored in a sealed plastic tank. The wash trailer is also equipped with a high-pressure pump, pressure regulator, and hoses.

The water conductivity is maintained below $20\,\mu S/cm$ for two reasons. First, this is a level that provides high flashover strength along the water column. Second, and more important, any remaining ice on the insulator surfaces will not present an electrical danger under melting conditions. If, instead of deionized water, tap water of $300\,\mu S/cm$ is used, any icicles or ice caps that remain will function electrically as if they were made of metal. IEEE Standard 957 describes appropriate modifications to insulator washing methods for freezing conditions. Composite insulators could also benefit from winter washing but high-pressure methods that do not damage the insulators require special development and training of personnel.

9.1.6. Case Study: SMART Washing

A major utility in Ontario [Chisholm et al., 2000] reported that flashovers of station and line insulators can occur during the winter months if road salt from highway deicing activities accumulated for any significant period of time, followed up by specific winter weather conditions including fog and ice. Distance to expressways was reported to be an additional factor.

This utility implemented a "SMART Washing" program to monitor contamination trends for the purpose of initiating insulator washing at stations, starting in 1993. The SMART-Washing program allowed a gradual phase-in of improved insulation at some critical 230-kV stations and permanently postponed expensive insulator upgrades at other stations, leading to savings of about $ Cdn 17M in the 1990s. At stations where upgrades were found to be necessary, the utility did invest more than $ Cdn 10M in upgraded insulators for improved winter flashover performance at four, 115-kV stations, two 230-kV stations, and one 500-kV station.

At the same time that this mitigation program was underway, SMART Washing allowed wider flexibility in system operation under adverse freezing weather conditions for the stations that were considered less critical.

The elements of the systematic, measurable and realistically timely SMART Washing program [Chisholm et al., 2000] include:

- Monitoring of the number of days without rain.
- Estimating the ESDD, using linear regression of the historic values of ESDD against number of days without rain, for each major station.
- Verifying the predicted ESDD values with on-site measurements.
- Comparing the day-by-day ESDD levels with tolerable limits specific to each station insulator characteristics under cold-fog and icing conditions. These are, respectively, functions of the leakage and dry arc distance of the insulators.

Week-by-week measurements of contamination levels at several transformer stations showed some unexpected variations. Over several seasons, it was found that insulators remained clean until long periods of freezing weather. Good linear relations were established, station by station, between the insulator contamination levels and the number of hours without rain converting to decimal days. These relations had a fairly wide range, depending on the type of insulation and the operating voltage.

Power washing is initiated whenever:

- Freezing rain is forecast, and an ESDD level exceeds threat levels (typically $0.015–0.02 \, mg/cm^2$ or $15–20 \, \mu g/cm^2$ at $500 \, kV$), meaning the insulator cannot maintain service voltage under icing conditions.
- ESDD levels exceed somewhat higher threat levels (typically $0.025–0.03 \, mg/cm^2$ or $25–30 \, \mu g/cm^2$ ESDD at $500 \, kV$) at any time, meaning the insulator cannot maintain service voltage under cold-fog conditions.
- Once a year, only to restore good appearance as necessary, on silicone-coated, semiconducting/resistance graded, or conventional porcelain.

Some EHV stations do not reach their ESDD threat levels after 90 days without rain, while other HV stations with reduced stress per meter of leakage distance and higher ESDD threat levels may still need a wash after only 15 days.

SMART Washing is more effective for cases where small amounts of ice form naturally on the surfaces of already-contaminated insulators, with partial bridging of the dry arc distance, and not effective when there is heavy accretion of contaminated ice (from nearby road salt) deposited onto the insulators.

The equipment in the SMART Washing program consists of a 1000-psi pump and washing wands equipped with an electrical drain wire to station ground. Water tanks with 1400- or 2400-liter (300 or 500) capacity are mounted on trailers that are pulled through the station. Station post insulators as shown in Figure 9-12 are transported indoors using a fork-lift and then returned to their exposure locations once the ESDD has been established on two or three sheds.

The water used for pressure washing in winter passes through a set of four parallel deionizing columns, connected to the city water supply. Conductivity is kept below $20 \, \mu S/cm$, versus a range of $300–650 \, \mu S/cm$ for tap water in this region. The wash procedure is qualified for live-line use but, in winter, insulators are washed off-potential. Grounds are not applied.

The SMART Washing approach replaced an earlier practice of routinely washing stations every spring and autumn, and has proved to have roughly the same annual cost. Its implementation makes better use of utility resources for monitoring weather patterns and planning for contamination problems.

Station post insulator on pallet   Insulated shed for wash trailer storage
used for ESDD measurements

Wash wand for 1000-psi service

Figure 9-12: Equipment used for SMART Washing program.

A utility serving the Chicago area has a similar insulator washing program. The need for washing uses observations of bottom-surface arcing activity on 345-kV suspension insulators. Generally, arcing on one or two units is considered acceptable and insulator strings are washed if four or more units show activity. With long-term experience, routine observations are taken only at a few critical line locations near major expressways.

9.2. OPTIONS FOR MITIGATING MODERATE ICING

Moderate icing is defined as the range where the choice of insulator profile, and specifically the shed-to-shed distance, can affect the electrical performance. Different types of insulators respond differently to ice buildup. For example, moderate ice severity is defined as the level that causes bridging of station post insulators with uniform 50-mm sheds, typically about 6–7 mm on a reference cylinder, but does not cause bridging of a string of IEEE standard disks.

The weight per centimeter of dry arc distance of ice also varies, depending on whether the accumulation occurs on a post insulator or string. For standard station post insulators with uniform shed profile and 300-mm diameter, the weight is given by [Farzaneh et al., 2005]

$$Weight\left(g/cm_{dry\,arc}\right) = 4 \cdot d \qquad (9\text{-}1)$$

where d (mm) is the thickness of accretion on a rotating monitoring cylinder. The expression for a string of IEEE standard disk insulators of 254-mm diameter [Farzaneh and Kiernicki, 1997] is

$$Weight\left(g/cm_{dry\,arc}\right) = 3.2 \cdot d \qquad (9\text{-}2)$$

The measure of the "time to bridging" under constant ice accretion is also measured in the test laboratory as an indication of the response of a particular insulator profile. As soon as an insulator is fully bridged with icicles, there is a change in the leakage current, from capacitive to resistive, making this a relatively precise measure of performance.

The first factor to be considered in selecting line insulators is the local level of ice accretion. For a string of 146-mm by 254-mm ceramic disk insulators, Table 9-6 describes the evolution of ice bridging as a function of ice accretion measured on a rotating 25-mm reference cylinder.

In approximate order of effectiveness, mitigation for moderate icing should consider:

TABLE 9-6: Relation Between Ice Accretion on Reference Cylinder and Accretion on IEEE Standard Ceramic Disk Insulator String

Ice Accretion Level on Reference Cylinder (mm)	Ice Accretion Severity	Ice Accretion on 146-mm × 254-mm IEEE Standard Disk String	Ice Accretion on 300-mm Station Post Insulator
<1	Very light	Thin layer on surfaces	Thin layer on surfaces
5	Light	Ice caps on top surfaces; ~40-mm icicles span a third of the 120-mm shed spacing	Ice caps on top surfaces; ~40-mm icicles nearly span 50-mm shed spacing
10	Moderate	Ice caps on top surfaces; ~100-mm icicles span most of the shed spacing	Fully bridged with icicles
15	Heavy	Fully bridged with icicles	Spaces between icicles fill in
20	Heavy	Spaces between icicles fill in	Spaces between icicles fill in
25	Heavy	Maximum bridging	Maximum bridging

Source: Farzaneh and Kiernicki [1997].

- Increased shed spacing for the same dry arc distance.
- Increased dry arc distance.
- Semiconducting glaze.
- Booster sheds.
- Monitoring and washing.
- Polymer insulators.
- Silicone coatings.

9.2.1. Use of Profiles with Greater Shed-to-Shed Distance

The icing performance of insulator strings with alternating diameter sheds will be somewhat better because of the larger spacing and longer time needed for ice bridging. In the test laboratory, icing levels that gave full bridging of a string of standard disk insulators in Figure 9-13 left large air gaps between a large-diameter bell-shaped insulator and a standard disk beneath.

The option of alternating-diameter strings with 13 standard and 12 aerodynamic insulators was selected for 500-kV transmission lines in Ontario in the late 1980s. Figure 9-14 shows a comparison of icicle formation for insulators on the same tower, at the same height, for natural ice accretion of 10 mm.

With 550 mm of shed-to-shed spacing, it is possible to postpone ice bridging of insulator strings at a constant ice accretion rate of 5 mm/h under 400-kV service voltage for up to 7 hours [Gutman et al., 2003]. This is considerably longer than the duration of any natural ice storm and represents an upper limit of what can be achieved.

Figure 9-13: Artificial ice accretion on bell–bell–disk, bell–disk–disk, and all–disk strings of suspension insulators under similar icing conditions (from Farzaneh et al. [2007]).

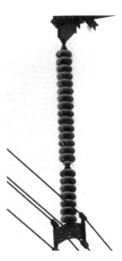

| Partial bridging alternating-diameter string of aerodynamic and standard disks [Farzaneh et al., 2007 b] | Partial bridging uniform string of standard disk insulators [Farzaneh et al., 2007 b] | 500-kV String with damaged insulator, found to be punctured on electrical inspection |

Figure 9-14: Natural ice accretion on 380-mm aerodynamic disk and 254-mm conventional suspension insulators under the same icing conditions.

Some of the same benefits of increased shed spacing can be achieved with alternating profiles on station post insulators. For example, Farzaneh and Brettschneider [2001] evaluated the ice accretion on prototype insulators with 1-2-3 alternating profile. Icing tests were carried out with glaze ice at an applied water conductivity of $\sigma_{20} = 30\,\mu$S/cm and included accumulation, hardening, and melting sequences as adopted later by the IEEE [Farzaneh et al., 2003a].

In spite of its superior ratio of leakage to dry arc distance, Alternating Profile 1 in Figure 9-15 does not perform as well as Profile 2. This is because, at the test condition of 6.5-mm accretion, all of the leakage distance of this profile was bridged by icicles.

The level of ice accretion for Alternating Profile 2 in Figure 9-15 was increased to see what level caused full bridging by icicles. This occurred at 8.5 mm, measured on a rotating reference cylinder. With this level of accretion, its flashover level fell from 312 to 237 kV. Thus the additional shed separation of 30 mm increased the "threat level" of the station by about 2 mm.

9.2.2. Increased Dry Arc Distance

Generally, under conditions of moderate icing, full bridging of icicles across the dry arc distance establishes the new leakage path length on station posts

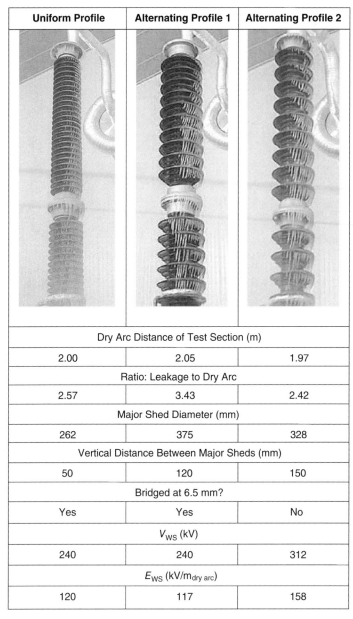

Uniform Profile	Alternating Profile 1	Alternating Profile 2
Dry Arc Distance of Test Section (m)		
2.00	2.05	1.97
Ratio: Leakage to Dry Arc		
2.57	3.43	2.42
Major Shed Diameter (mm)		
262	375	328
Vertical Distance Between Major Sheds (mm)		
50	120	150
Bridged at 6.5 mm?		
Yes	Yes	No
V_{WS} (kV)		
240	240	312
E_{WS} (kV/m$_{dry\ arc}$)		
120	117	158

Figure 9-15: Dimensions of prototype insulators tested for moderate icing conditions of 6.5-mm accretions on rotating reference cylinder.

and many polymer insulator profiles. Any increase in dry arc distance will automatically and linearly increase the flashover level for these types of insulators in moderate icing. The degree of linearity has been studied extensively and a summary is given in Section 9.3.1, dealing with heavy ice accretion.

9.2.3. Insulator Orientation

In a summary of icing test results, Kannus and Lathi [2007] tested insulator strings and noted that the electrical performance of short insulator strings with highly contaminated glaze icicles showed a significant improvement with angles of inclination $\beta > 10°$. The improvement relates to the reduction of the number of icicles that bridge the shed-to-shed distance, as shown in Figure 9-16.

A change in the insulator orientation can also be effective at preventing ice bridging under moderate icing conditions. Two sets of tests [Cherney, 1986; EPRI, 1982] ranked the electrical performance of precontaminated V-strings best among a series of competitive options. The relatively high amount of ice accretion needed to achieve full bridging for this orientation was cited as an advantage in both cases.

9.2.4. Semiconducting Glaze

A series of tests on resistance-graded posts were carried out to address a problem of flashovers at a 230-kV station with severe 0.2-mg/cm^2 (200 μg/cm^2)

Heavy accretion on string with
$\beta = 20°$

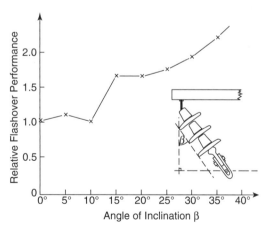

Relative electrical flashover performance

Figure 9-16: Effect of insulator string inclination on flashover performance (adapted from Kannus and Lathi [2007]).

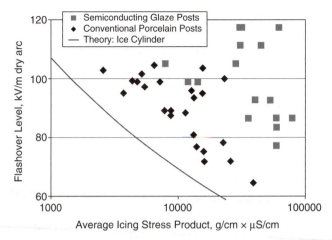

Figure 9-17: Comparison of semiconducting glaze and standard glaze porcelain posts under light icing (CIFT) conditions.

pollution and moderate icing in early 1994. A series of comparison tests again showed in Figure 9-17 that semiconducting glaze posts gave superior performance for a given dry arc distance in conditions of moderate to heavy ice accretion that did not give full icicle bridging on cap-and-pin apparatus insulators.

In a series of side-by-side tests as reported in Figure 9-17, three-section semiconducting glaze post insulators in Figure 9-18 were compared with five cap-and-pin insulators that were untreated or coated with RTV silicone [Chisholm, 1994]. Insulators were tested with precontamination levels of $0.2\,mg/cm^2$ ($200\,\mu g/cm^2$) with applied water conductivity of $\sigma_{20} = 50\,\mu S/cm$. Ice accretion to partial bridging was used, along with a melting phase including fog.

A direct comparison of conventional and semiconducting glaze insulators was also carried out on 2.0-m dry arc distance at $105\,kV/m_{dry\,arc}$ stress using an accretion of 6.5 mm on a rotating reference cylinder [Farzaneh and Brettschneider, 2001]. Their results, summarized in Figure 9-19, show a similar level of flashover stress per meter of dry arc distance on the semiconducting glaze post insulators of $135\,kV/m_{dry\,arc}$ to the values reported in Figure 9-18.

Based on the good performance at 230 kV summarized in Figure 9-18, two-piece semiconducting glaze posts with 1.84-m overall length were applied at a highly contaminated station and operated for 14 years. The performance was found to be good overall, with a significant reduction in contamination outages. There was one specific problem. On December 12, 1997, three flashovers on pairs of semiconducting glaze insulators were observed at switch positions where two insulators are located with about 0.5 m between their central axes.

A series of diagnostic tests were carried out [Chisholm, 1999] to find out why the semiconducting glaze insulators in close 508 mm (20 in.) proximity,

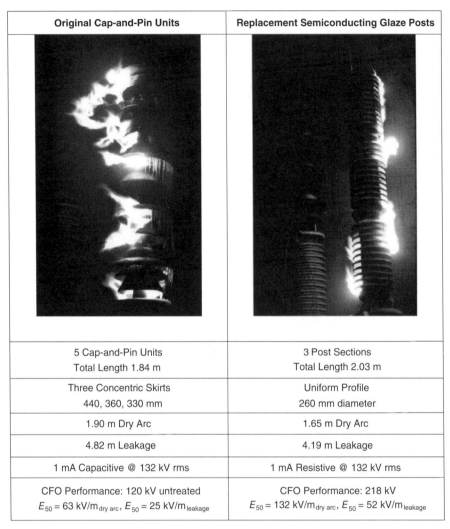

Original Cap-and-Pin Units	Replacement Semiconducting Glaze Posts
5 Cap-and-Pin Units Total Length 1.84 m	3 Post Sections Total Length 2.03 m
Three Concentric Skirts 440, 360, 330 mm	Uniform Profile 260 mm diameter
1.90 m Dry Arc	1.65 m Dry Arc
4.82 m Leakage	4.19 m Leakage
1 mA Capacitive @ 132 kV rms	1 mA Resistive @ 132 kV rms
CFO Performance: 120 kV untreated $E_{50} = 63$ kV/m$_{dry\ arc}$, $E_{50} = 25$ kV/m$_{leakage}$	CFO Performance: 218 kV $E_{50} = 132$ kV/m$_{dry\ arc}$, $E_{50} = 52$ kV/m$_{leakage}$

Figure 9-18: Comparative test results for deep-skirt apparatus and semiconducting glaze post insulators in moderate (CIFT) icing conditions (data from Chisholm [1994]).

and no others, flashed over in heavy, contaminated icing conditions. Based on insulators received from the field, the ice layer was characterized by melted water conductivity $\sigma_{20} = 403\,\mu$S/cm, with a weight of 81 g/cm dry arc distance. This is actually heavy icing, rather than moderate, and also represents an extremely high contamination level. The icing stress product of 33,000 (g/cm)-(μS/cm) is thus illustrated in Figure 9-20.

This level of contamination in the ice was 4 times higher than levels observed in field studies at other locations in the same province of Ontario, and 13 times higher than median values of $\sigma_{20} = 30\,\mu$S/cm used in Québec. Laboratory tests

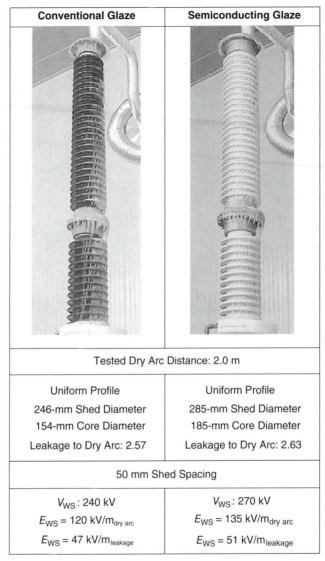

Conventional Glaze	Semiconducting Glaze

Tested Dry Arc Distance: 2.0 m	

Uniform Profile	Uniform Profile
246-mm Shed Diameter	285-mm Shed Diameter
154-mm Core Diameter	185-mm Core Diameter
Leakage to Dry Arc: 2.57	Leakage to Dry Arc: 2.63

50 mm Shed Spacing	

V_{WS}: 240 kV	V_{WS}: 270 kV
$E_{WS} = 120$ kV/m$_{dry\ arc}$	$E_{WS} = 135$ kV/m$_{dry\ arc}$
$E_{WS} = 47$ kV/m$_{leakage}$	$E_{WS} = 51$ kV/m$_{leakage}$

Figure 9-19: Comparative test results for conventional and semiconducting glaze post insulators in moderate icing conditions (from Farzaneh and Brettschneider [2001]).

reproduced the line-voltage flashover levels and also led to identification of the problem. Arcing tended to initiate between the central metal flanges of the parallel posts, as a result of unequal potential distribution when one or more of the ice layers had an air gap.

The flashover problem was mitigated by installing a cross-bond between the central flanges to equalize potentials of the two columns. This restored the performance of the double column to be identical to that of a single post.

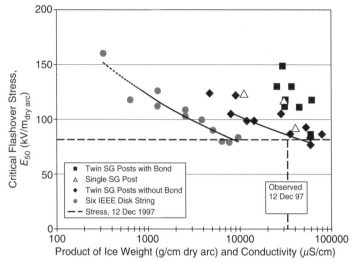

Figure 9-20: Comparison of critical flashover stress for semiconducting glaze (SG) post insulators in close proximity with heavy ice accumulation. (adapted from Chisholm [1999]).

For insulators that rotate to operate switches, a flexible lead should be fitted as shown in Figure 9-21. The bond was looped once to allow the rotating insulator to turn 90°, and was qualified for 1000 operations without fatigue of the large-diameter extra-flex copper wire.

A similar bond was adopted for parallel columns tested with heavy ice accretion at 735-kV system stress as shown in Figure 9-31.

9.2.5. Polymer Insulators

At present, polymer insulators have close shed-to-shed spacing that makes them as vulnerable to ice bridging as station post insulators. This means that their performance is not likely to be as good as cap-and-pin insulator strings in moderate icing conditions.

With a high degree of flexibility in design, and the relative success of alternating profiles on station posts, the possibility of polymer insulators with alternating-diameter sheds that increase the ice accretion level for bridging is interesting. One such profile was tested under icing accumulation conditions by Chaarani [1996] with the results in Table 9-7.

These tests were completed before the importance of the melting phase in flashover performance was defined, and represent a ranking rather than a true measure of performance. With 5 mm of accretion, neither insulator was fully bridged, and the polymer insulator performed better than the disks. With the limit of moderate accretion at 10 mm, the reduced dry arc distance per meter of connection length of the EPDM insulator puts it at a disadvantage compared to the standard disks.

Pair of insulators for air-break disconnect
switch on 508-mm (20-in.) spacing

Close-up of 4/0 extra-flex cable with single
loop to allow 90° rotation

Figure 9-21: Flexible bond between pair of stacked 230-kV semiconducting glaze (RG) station post insulators. (Courtesy of INMR.)

TABLE 9-7: Comparison of Alternating-Profile EPDM Polymer Insulator with IEEE Standard Disks in Moderate Icing Conditions

Insulator Type	Electrical Performance	Ice Accretion on Rotating Reference Cylinder	
		5 mm	10 mm
Six IEEE standard disks, 0.96-m dry arc distance, 0.88-m connection length	V_{WS} (kV)	105	90
	E_{WS} (kV/m$_{dry\ arc}$)	110	94
	E_{WS} (kV/m$_{connection}$)	120	103
EPDM, 0.71-m dry arc distance 0.945-m connection length	V_{WS} (kV)	>120	87
	E_{WS} (kV/m$_{dry\ arc}$)	>168	122
	E_{WS} (kV/m$_{connection}$)	>127	92

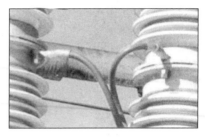

9.2.6. Corona Rings

Corona rings must be applied with care in the case of moderate to heavy icing. They can perform their function so well that ice bridging near the line end of the insulator occurs.

The selection of grading rings has a strong influence on ice accretion and ice flashover performance. A significant improvement of the electric field stress can be achieved through the use of grading rings on the line end of the insulator or on both ends in the case of EHV or polymer insulators. Figure 9-22 shows the uniform icing that was achieved at the line end of a laboratory test on a 500-kV class nonceramic insulator, resulting from the use of a corona ring. However, the grading of electric field to produce uniform icing does not necessarily lead to improved flashover performance. Tests with no corona ring on a 500-kV NCI produced such a large gap in the ice that flashover strength was increased compared to the tests with the corona ring in place.

9.2.7. Condition Monitoring Using Remote Thermal Measurements

The role of thermal monitoring in icing conditions has some specialized applications when insulators are partially bridged with ice. The applications relate more to cases where one or more sections of a multisection insulator or arrester have been fully bridged, while other sections remain partially bridged.

Under partial ice-bridging conditions, the nonuniform stress on surge arresters may lead to an internal temperature rise due to the high voltage

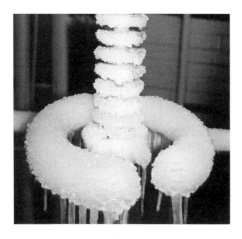

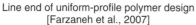

Line end of uniform-profile polymer design
[Farzaneh et al., 2007]

Line end of alternating-
profile polymer design

Figure 9-22: Uniform ice accretion on line end of 500-kV polymer insulators with corona ring and four-conductor bundle.

drops on the ice-free parts. The temperature rise of arresters inside the ice-free sections will tend to equalize along the entire length in 10–30 minutes, depending on the arrester size and construction. The heat can melt the accretion on the ice-coated parts. However, in the meantime, the ice-free parts of multisection arresters may be overstressed into thermal runaway. Infrared monitoring of arrester temperature can be used to inspect the thermal stability of the metallic oxide material, especially for 345 and 500-kV class arresters that have been prone to this failure mechanism. Similar problems were reported for snow accretion on two-section 275-kV arresters by Kojima et al. [1988].

The remedial actions when rapid temperature rise is observed on arresters are limited, as the icing events occur when flashovers and switching are likely events and voltage surges need to be controlled to prevent equipment damage. If redundant lines are available, it may be possible to deenergize arresters that are showing excessive temperature rise. These can be allowed to cool and then cleaned. It is risky to remove the ice with high-pressure washing or steam cleaning because these processes may cause thermal shock damage to the hot ceramic housings. It may also be possible to knock the ice off the arresters with live-line tools, but again the risk of damage to sheds is elevated.

Thermal monitoring is also recommended under icing conditions for stations that have been upgraded with semiconducting glaze insulators that have built up a high level of surface contamination. Proper operation of the glaze in freezing conditions may lead to temperature rise on the ice-free parts. This temperature should generally stay below +20°C. If a higher temperature is observed, there is an elevated risk of thermal runaway and damage to ceramic sheds. Aspects of this problem are similar to the multisection arrester problem, and if a high temperature rise is found, the remedial action is the same. The insulators should be deenergized, allowed to cool, and cleaned.

9.2.8. Silicone Coatings

The role of silicone coatings in moderate icing conditions depends greatly on whether the insulators are fully bridged with icicles. If they are not fully bridged, the bottom surfaces of the insulators continue to play an important role. Shed profiles with deep skirts such as cap-and-pin apparatus styles show a significant improvement of up to 30% when coated with RTV silicone.

In Figure 9-23, the icicles around the outer skirts of each insulator grew to more than 300 mm in length at a temperature of −5°C. The icicle growth on these insulators tended to follow the lines of electric field, and actually grew into the wind of 3 m/s. The water conductivity was 50 μS/cm and the insulators were precontaminated to about 0.2 mg/cm^2 (200 μg/cm^2). The flashover arc is seen to pass through or around the icicles in this case, and the arc hugs the leakage distance of the bottom sheds of each three-skirt insulator.

With tests at a median precontamination level of 186 μg/cm^2, the maximum withstand gradient across the 4.82-m leakage distance of the five insulators

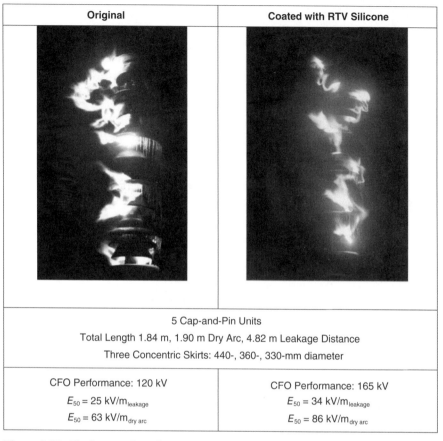

Original	Coated with RTV Silicone

5 Cap-and-Pin Units Total Length 1.84 m, 1.90 m Dry Arc, 4.82 m Leakage Distance Three Concentric Skirts: 440-, 360-, 330-mm diameter	
CFO Performance: 120 kV $E_{50} = 25$ kV/m$_{\text{leakage}}$ $E_{50} = 63$ kV/m$_{\text{dry arc}}$	CFO Performance: 165 kV $E_{50} = 34$ kV/m$_{\text{leakage}}$ $E_{50} = 86$ kV/m$_{\text{dry arc}}$

Figure 9-23: Flashover of stack of five cap-and-pin apparatus insulators, without and with RTV silicone coating, under moderate ice accretion. (data from Chisholm [1994]).

was found to be $29 \text{kV/m}_{\text{leakage}}$. This was the same value that was found in cold-fog tests at a similar pollution level, suggesting that the contamination flashover process (including negligible levels of arcing activity prior to flashover) dominate the response.

The withstand gradient across the 1.9-m dry arc distance of the five cap-and-pin insulators was established to be $74 \text{kV/m}_{\text{dry arc}}$, which is equal to the system voltage stress at a normally used 5% overvoltage. The withstand stress on identical uncoated porcelain insulators in the same icing conditions was $53 \text{kV/m}_{\text{dry arc}}$. Pollution from the upper surfaces of the untreated porcelain was more readily washed into the icicles during the ice accretion phase, making them more conductive and thus more likely to flash over. Icicles removed from samples of silicone-coated porcelain tended to have the same conductivity as the applied water, about $50 \mu\text{S/cm}$.

9.3. OPTIONS FOR MITIGATING HEAVY ICING

In approximate order of effectiveness, mitigation for moderate icing should consider:

- Increased dry arc distance.
- Semiconducting glaze.
- Accessories (booster sheds).
- Monitoring and washing.
- Polymer insulators.
- Increased leakage distance with the same dry arc distance.
- Silicone coatings.

9.3.1. Increasing the Dry Arc Distance

Increased insulation length remains the most reliable way to improve performance in heavy icing conditions.

First, considering short strings of standard 146-mm × 254-mm disk insulators, Farzaneh and Kiernicki [1997] found a linear relation between the dry arc distance L and the maximum withstand level in Figure 9-24.

Linearity of flashover performance was also studied for station post insulators by Farzaneh-Dehkordi et al. [2004] at distance scales from 1.4 to 4.2 m, leading to the results in Table 9-8 and Figure 9-25.

In both cases, the electrical strength with a value of $\sigma_{20} = 80\,\mu S/cm$ is fitted by approximately the same slope, with small offsets. This is a strong indication of linear performance with dry arc distance under heavy icing conditions, and forms the basis of performance modeling using an icing stress product (ISP) as discussed in Chapter 7.

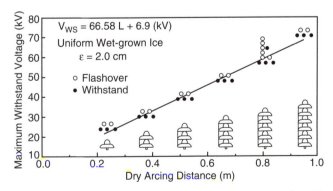

Figure 9-24: Linearity of withstand voltage under icing conditions for strings of IEEE standard disk insulators (from Farzaneh and Kiernicki [1997]).

TABLE 9-8: Flashover Performance of Uniform-Profile Station Post Insulators in Heavy Icing, 15-mm Accretion on Rotating Monitoring Cylinder with $\sigma_{20} = 80\,\mu S/cm$

Dry Arc Distance (m)	V_{MF} (kV rms)	E_{MF} (kV/m$_{dry\ arc}$)
1.39	120	86.3
2.02	150	74.3
3.07	216	70.4
3.51	266	75.8
4.17	304	72.9

Source: Farzaneh-Dehkordi et al. [2004].

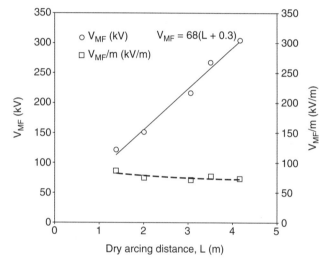

Figure 9-25: Minimum flashover voltage for 15 mm of accretion of glaze ice with $\sigma_{20} = 80\,\mu S/cm$ on station post insulators (from Farzaneh-Dehkordi et al. [2004]).

One practical limit to increases in the dry arc distance of station post insulators relates to their cantilever strength requirements. Most EHV stations have high fault currents and high reliability requirements, meaning that the post insulators must support considerable short-circuit forces. It becomes increasingly difficult to manufacture and transport large-diameter station posts. This means that a recommendation to increase dry arc distance by 25% may lead to an increase in post diameter by 33%, which in turn leads to additional ice accretion per unit length and a reduced leakage resistance. This effect of increased diameter for the same strength requirements can negate a considerable fraction of the benefit of additional dry arc distance. This can be rectified to some extent with the use of high-strength porcelain that allows a reduced post diameter for the same cantilever load rating as conventional porcelain.

9.3.2 Changing to Semiconducting Glaze

The improved performance of semiconducting glaze insulators in heavy icing conditions is mainly a result of the improved voltage grading along the insulator surface. Especially for EHV station post insulators, many sheds will be fully bridged with ice while air gaps may develop in other areas, such as the sheds near the line end as shown in Figure 9-26. The semiconducting glaze conducts enough current across these air gaps to limit voltage gradients, and thereby postpone or prevent the initiation of arcing discharges.

Conventional Glaze	Semiconducting Glaze
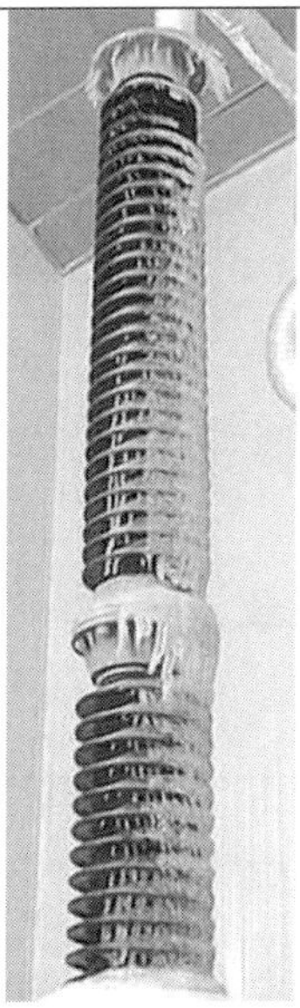	
Tested Dry Arc Distance: 2.0 m	
Uniform Profile	Uniform Profile
246-mm Shed Diameter	285-mm Shed Diameter
154-mm Core Diameter	185-mm Core Diameter
Leakage to Dry Arc: 2.57	Leakage to Dry Arc: 2.63
50-mm Shed Spacing	
V_{WS} : 195 kV	V_{WS}: 240 kV
$E_{WS} = 98$ kV/m$_{dry\ arc}$	$E_{WS} = 120$ kV/m$_{dry\ arc}$

Figure 9-26: Comparative test results for conventional and semiconducting glaze post insulators in heavy icing conditions (from Farzaneh and Brettschneider [2001]).

A direct comparison of the performance of conventional and semiconducting glaze insulators was carried out on 2.0-m dry arc distance at 105 kV/$m_{dry\ arc}$ stress using an accretion of 15 mm on a rotating reference cylinder [Farzaneh and Brettschneider, 2001]. Their results, which are part of an extensive R&D program to develop ice-resistant 735-kV insulators, are summarized in Figure 9-26. The semiconducting glaze shows a significant improvement in performance.

There is another advantage to the semiconducting glaze posts in icing conditions. In both laboratory tests and field exposure, these insulators develop larger ice-free zones than conventional glaze posts, and develop these zones earlier in the melting phase. The larger air gaps have higher flashover strength, and the quicker cascade of ice accretion reduces the time of greatest danger.

In service, the ice-free zones of 1-m extent on the 500-kV insulators in Figure 9-27 provide enough air space to withstand full ac line voltage. This large air gap can be compared with the 15-cm gap that formed at the three line-end sheds on the conventional insulator in Figure 9-26.

Semiconducting glaze disk insulators have not yet been tested under heavy icing conditions but the same benefits of arcing suppression and rapid ice loss may also lead to better performance.

As described in Section 6.7.6, the tolerance in post resistance and surface conductivity is an important specification of semiconducting glaze for three reasons: thermal runaway, matching of stacking station posts, and matching between stacking station posts. Also, monitoring of the rms current in the posts gives an excellent indication of the onset and severity of ice accretion.

Figure 9-27: Operation of 500-kV semiconducting glaze insulators under icing conditions in West Virginia. (Courtesy of Dominion Virginia Power.)

9.3.3. Adding Booster Sheds

Booster sheds have been described as an option to improve the flashover strength of contaminated insulators under heavy rain or washing conditions. These polymer slip-on accessories consist of a large-diameter C-shaped section and a binding strip that has attachment fasteners. Some designs also use a silicone rubber to seal the fastening strips.

Drapeau et al. [2002] tested 0.6-m sections of a uniform-shed porcelain station post with 0, 1 or 2 booster sheds. Using tests with an icing regime but no melting, the maximum withstand voltage stress was reported to be $102 \text{kV/m}_{dry\ arc}$ untreated, $113 \text{kV/m}_{dry\ arc}$ with one and $147 \text{kV/m}_{dry\ arc}$ with two booster sheds.

The improvement in flashover performance of single station post insulators with 655-mm diameter booster sheds was evaluated in a Cooperative Research and Development (CRD) collaboration between a utility, manufacturers, and a university [Farzaneh and Volat, 2005]. Two sections of a station post insulator were tested with a total dry arc distance of 2.7 m and a stress of 105 kV rms per meter of dry arc distance. The posts and stress levels match those used at 735-kV substations in Québec, Canada. Heavy ice tests were carried out with an applied water conductivity of 30 μS/cm and accretion of 15 or 30 mm on a rotating cylinder. The ice accretion, hardening, and melting sequences followed the IEEE Task Force recommendations defined in Farzaneh et al. [2003a]. Leakage currents during the melting phase, took a long time to evolve and never reached levels more than 20 mA prior to flashover.

The level of V_{WS} was evaluated using the process in IEC 60507 [1991] for contaminated insulators. This level increased by about 15 kV per booster shed as the number of booster sheds increased from four to six. Best results for both icing thicknesses were obtained when the booster shed positions were optimized to produce the largest air gaps. The levels of V_{WS} are compared with the value of 270 kV, normally measured for the untreated posts with heavy ice accretion of 15–30 mm (Figure 9.28).

The return period of ice accretion with thickness of more than 15 mm is relatively long in Québec; however, the tests show that even with this heavy level of accretion, the booster sheds provide an important improvement in performance. The results of about 15 kV per booster shed, consistent with those of about 20 kV per booster shed found for the 230-kV tests in Figure 9-29, correspond well with the predictions of the multiarc model for icing flashover described in Farzaneh and Zhang [2007].

In a different series of comparison tests [Chisholm, 2000], booster sheds were noted to improve the flashover performance of pairs of 230-kV posts in heavy, contaminated icing conditions. Tests were carried out at service stress levels of 76 kV per meter of dry arc distance (29.5 kV per meter of leakage distance). The booster sheds extended the dry arc distance and also introduced sheltered areas directly below, as illustrated in Figure 9-29.

Four Booster Sheds   Six Booster Sheds   Four Booster Sheds   Six Booster Sheds
15-mm ice accretion on rotating cylinder   30-mm ice accretion on rotating cylinder

V_{WS} = 330 kV V_{WS} > 350 kV V_{WS} = 285 kV V_{WS} = 315 kV

Improvement over   Improvement over   Improvement over   Improvement over
untreated: 22%   untreated: >30%   untreated: 6%   untreated: 17%

Figure 9-28: Comparative test results for conventional post insulators with 4 and 6 booster sheds in heavy icing conditions (from Farzaneh and Volat [2005]).

Insulators on 0.5-m spacing are often used for air-break disconnect switches. In these cases, it is necessary for one of the insulators to rotate freely, either 90° or for up to 30 turns. This means that it is necessary to stagger the booster sheds.

Under practically similar contaminated and heavy icing conditions, the use of two booster sheds added about 40 kV to the maximum withstand voltage of the pair of 230-kV insulators, bringing their performance from below to about 21% above service voltage.

There were a few problems noted with the performance of booster sheds in this test series. With the nonuniform accretion of ice, heavy on the top surfaces and none beneath, the electrical stress was sufficiently high that the booster sheds would either puncture, or would flash over between the neck of the booster shed and the insulator beneath, as illustrated in Figure 9-30.

Double-Column 230-kV Post **Double-Column 230-kV Post**
No Booster Sheds **With Four Booster Sheds**

ESDD: 0.025 mg/cm^2 ESDD: 0.033 mg/cm^2

σ_{20}: 500 μS/cm; w: 32 g/cm σ_{20}: 400 μS/cm; w: 33 g/cm

Icing Stress Product: Icing Stress Product:

16,000 (g/cm)(μS/cm) 13,200 (g/cm)(μS/cm)

Dry Arc Distance: 1.74 m Dry Arc Distance: 1.98 m

Service Voltage: 132 kV Service Voltage: 132 kV

V_{WS}: 120 kV V_{WS}: 160 kV

Figure 9-29: Typical test results for booster sheds on 230-kV post insulators with highly contaminated ice. (Courtesy of PSE&G.)

Typical puncture damage to booster shed Insulator glaze damage under
after heavy, conductive ice test bottom surface of failed booster shed

Figure 9-30: Flashover damage to booster sheds under heavy, contaminated icing conditions.

This type of damage occurs in the laboratory tests of insulator flashover, well above normal line voltage, and should not occur in service. However, if for any reason insulators do flash over in service, the booster sheds should be inspected for similar damage and replaced if necessary.

Double columns of station post insulators were also tested for their performance at EHV service voltage stress levels of 105 kV per meter of dry arc distance [Farzaneh et al., 2006b]. The performance of three booster sheds on each 2.7-m post in a double column of insulators, used for air-break disconnect switches, shown in Figure 9-31, was measured to be $V_{WS} = 300$ kV with 15 mm of ice accretion. This was the same withstand level found on semiconducting glaze insulators with the same profile, with a bond to equalize potential between flanges. In both cases, the value of V_{WS} was 5% above service voltage stress, and 11% above the performance of a single column of the same alternating-diameter profile and normal glaze.

Double Column of Semiconducting Glaze Station Post Insulators with Central Bond   **Double Column of Conventional Glaze Station Post Insulators with six Booster Sheds**

$V_{WS} = 300$ kV $V_{WS} = 300$ kV
Test Conditions: 15-mm accretion on rotating cylinder; $\sigma_W = 30$ μS/cm;
hardening and melting regime per IEEE recommendation [Farzaneh et al., 2003a]

Figure 9-31: Comparison of booster sheds and semiconducting glaze for double-column 2.7-m insulators used for air-break disconnect switches. (from Farzaneh et al., [2006b]).

9.3.4. Changing to Polymer Insulators

Under heavy icing conditions, Farzaneh et al. [1994] found that the flashover performance of EPDM insulators was significantly better than that of IEEE standard disk insulators. The applied water conductivity in this case was 80 μS/cm with an accretion level of 20 mm on a reference cylinder. Flashovers were tested in the icing regime similar to the method proposed in the IEEE Task Force recommendations [Farzaneh et al., 2003a]. (See Table 9-9.)

One reason why the EPDM insulators perform better relates simply to the effect of diameter, rather than any properties of the material. Figure 9-32

TABLE 9-9: Comparison of Heavy Icing Performance of EPDM Insulators and IEEE Standard Disks

Type	Mechanical Rating (kN)	Shed Diameter (Spacing) (mm)	Leakage Distance (mm)	Dry Arc Distance (mm)	V_{WS} $(kV_{rms\ l-g})$	E_{WS} $(kV/m_{dry\ arc})$
EPDM	120	90 and 105 (55)	2171	714	69	97
IEEE standard disk	120	254 (146)	305 per disk	809 (5 disks)	57	70

Source: Farzaneh et al. [1994].

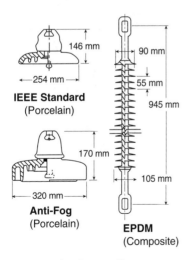

Insulator profiles

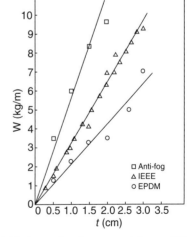

Weight of accretion (kg per m of dry arc distance) versus accretion on reference cylinder

Figure 9-32: Relation of ice accretion on insulator to accretion on reference cylinder (from Farzaneh and Kiernicki [1995]).

shows that the ice weight reached a level of 40 g/cm of dry arc distance. For the IEEE standard disk, the accretion reaches 60 g/cm. The empirical icing stress product (ISP) model from Chapter 7 suggests that the ratio between the flashover voltage gradient E_{50} will be $(40/60)^{-0.19}$ or 8% higher for the polymer insulator.

Other investigations on the effect of diameter on the electrical performance, supported by tests and modeling [Farzaneh et al., 2003b, 2006a] also support the general model that the flashover strength will increase slowly with decreasing insulator diameter. However, tests of icing flashovers during the accretion phase, such as Gutman et al. [2003], tend to favor the cap-and-pin insulator strings. The question of an ideal polymer insulator profile that has large shed spacing for high resistance to bridging, along with small diameter for good performance when fully bridged, is an important area of development.

9.3.5. Ice Monitoring Using Leakage Current

Meghnefi et al. [2007] explored the possibility of using leakage current activity as a predictor of the probability of icing flashover. They used a large station post insulator with controlled rate of icing as shown in Figure 9-33.

Generally, the leakage current during an icing episode was found to have four phases, illustrated in Figure 9-34. The initial current was capacitive in relation to the 280-kV line-to-ground voltage. After 5–10 minutes of accumulation, partial arcing was initiated. Evolutions of the magnitude of leakage current, and the third and fifth harmonic content, were used in a neural network analysis to provide on the order of 20–30 minutes of warning that flashovers would or would not be occurring during melting conditions.

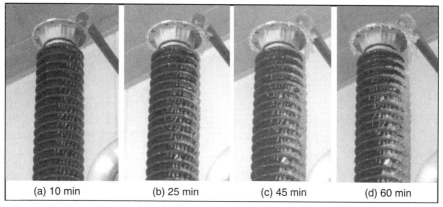

| (a) 10 min | (b) 25 min | (c) 45 min | (d) 60 min |

Figure 9-33: Ice accretion sequence on energized station post insulator (from Meghnefi et al. [2007]).

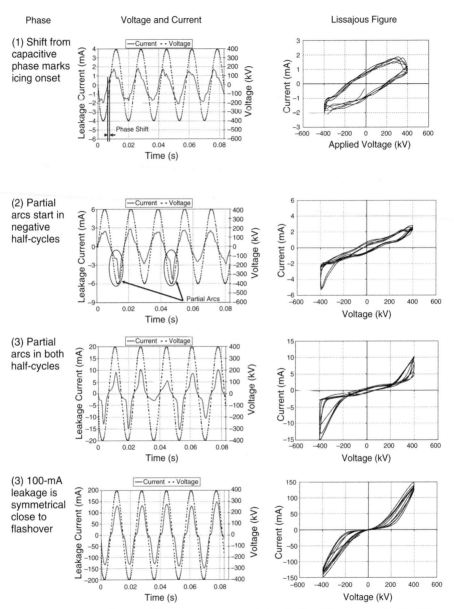

Figure 9-34: Four phases of leakage current activity during icing test with 30-µS/cm melted water conductivity and 15-mm accretion on reference conductor (data from Meghnefi et al. [2007]).

While commercial insulator monitoring systems for pollution flashover make use of leakage current magnitude, and station surge arrester monitoring systems make use of third harmonic content, these two technologies have not yet been unified into a practical system for warning of icing flashover on EHV station post insulators.

9.3.6. Ice Stripping in Freezing Weather

Two different utilities have used high-powered infrared heat lamps to melt ice on insulators and other station components. This approach has high operating costs but has proved to be economical and effective in the case of substations near waterfalls. Difficulties in this approach include the problem of replacing the lamps when they are hit by falling ice, and the need to restrict access to the area whenever deicing is in progress.

Hot water is often used to strip ice from aircraft or water systems. The method is simple but needs large quantities of water and power, and suitably clothed operators. Hot-water pressure washing was evaluated for stripping ice from insulators in laboratory conditions [Sklenicka and Vokalek, 1996]. However, this technique has not been fully developed. Factors to consider include the increased conductivity of the water with increasing temperature, the effects of wind, the high probability of flashover on a wetted ice-covered insulator, and the required minimal safety approach distances.

Glycol-based liquids are also used for deicing purposes on aircraft. However, for high ratios of glycol to water, the "slick water" becomes electrically conductive and would then present a hazard when deicing energized equipment. In common with hot-water deicing, there is an elevated risk of insulator flashover as the conductive liquid flows over the ice layer. Glycol recovery systems would also interfere with similar systems used to retain transformer oil in areas of sensitive ecology.

A more recent line of research has evaluated the use of a steam jet to strip ice from the mechanical parts of substation equipment, such as the jaws of air-break disconnect switches. Figure 9-35 shows that the method is effective, with 120 s of exposure proving to be sufficient to remove the ice from all electrically conductive fingers near the switch blade.

Figure 9-36 shows the test arrangement for stripping ice from the energized jaw of a switch, along with a practical field trial configuration.

Tests [Lanoie et al., 2005] showed that a steam jet with a mass flow rate of 135 kg/h at a pressure of 6.2 bar (90 psi) had a jet of 1–3 meters long, and a low conductivity of less than 1.5 μS/cm. The leakage current through the jet at 100 kV was less than 2.5 mA and the withstand voltage was greater than 125 kV. This method is being developed further to extend the range of application to higher voltage systems.

Standard high-pressure dry cleaning techniques such as corn blasting can be used to remove ice from insulators. This can be done safely and effectively with live-line methods by experienced operators.

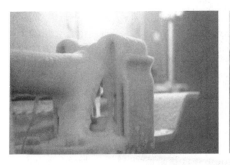 

| Jaw of air-break disconnect switch with 30 mm of ice accretion on reference cylinder | Jaw of air-break disconnect switch after 120 s of steam deicing |

Figure 9-35: Effectiveness of steam deicing for 120 s on energized switch jaw.

| Steam deicing test on energized jaw of air-break disconnect switch | Steam cleaning from live-line work platform |

Figure 9-36: Laboratory and field testing of steam deicing method on energized switches.

9.3.7. Corona Rings and Other Hardware

The corona rings were shown to play a role in the accretion of ice near the line end of polymer insulators. Normally, these will be fitted to polymer insulators to ensure long service life for a given electric stress level and contamination environment. Their role in icing performance generally needs to be established by experimental testing.

Arcing horns do not have a direct effect on the icing performance of insulators, but can serve an important auxiliary function. In cases where the insulator dry arc distance is increased to meet icing requirements, arcing horns or rod gaps can be used to restore the desired switching-surge and lightning impulse surge coordination with station equipment.

9.3.8. Increasing Shed-to-Shed Distance

An increase in shed-to-shed distance with an alternating shed profile was found to be beneficial in conditions of very light icing because it increased the overall leakage distance. The alternating profile with large shed-to-shed distance was partially effective in moderate icing because it increased the icing accumulation required to fully bridge the insulator, from 6.5 to 8.5 mm in one comparison test. For 15-mm accumulation, Drapeau et al. [2002] found uniform and alternating profiles gave identical performance per meter of post length.

Figure 9-37 shows that once the ice accretion reaches a heavy level of 15 mm on a rotating reference cylinder, there is no distinguishable difference among the performance of the various profiles. In every case, the flashover stress on the ice layer with σ_{20} of 30 µS/cm was 95–98 kV per meter of dry arc distance, which is about 7–10% below the normal 105-kV/$m_{dry\ arc}$ service voltage stress for EHV station post insulators.

9.3.9. Coating of Insulators with RTV Silicone

There is a modest improvement in performance when RTV-coated porcelain insulators are partially bridged with icicles, compared to when they are uncoated. Tests on a 0.6-m dry arc distance section of a uniform-shed post showed 11% improvement in heavy icing conditions with RTV silicone [Drapeau et al., 2002]. However, this effect is not linear, and in fact, it reversed under heavy icing conditions for tests on longer lengths of the same post insulators.

Farzaneh and Volat [2007] studied the flashover performance of a two-section uniform-shed station post insulator with 2.72-m dry arc distance, 181-mm core diameter, 275-mm shed diameter, and total leakage distance of 7 m for Hydro-Québec. Two sections of this post were tested at service voltage of 285 kV, where three sections are normally used at this stress to withstand 735-kV system voltage.

Ice bridging on silicone rubber-coated insulators occurs just as rapidly as on uncoated ones. The only difference is in the initial formation of the ice. Due to the higher hydrophobicity of the coated insulator surface, water droplets form a connected network of ice globules, whereas on uncoated insulators, water filming produces a layer of ice. Time to full bridging is nearly the same, and the ice is retained for a longer period on RTV-coated porcelain during the melting phase, essentially extending the duration of the period of greatest electrical risk in the same way that a 60-s IEC wet test gives lower flashover levels than a 10-s ANSI wet test.

In a set of 19 comparison tests, 2.7-m post insulators with and without RTV silicone coating were tested at an EHV service voltage stress of 105 kV/$m_{dry\ arc}$. The water conductivity σ_{20} was 30 µS/cm and the test followed the recommendations of the IEEE [Farzaneh et al., 2003a], including accretion of glaze ice with icicles to 15 mm on a rotating reference cylinder, hardening, and melting sequences. Figure 9-38 shows that the 7.0-m leakage distance was fully bridged by ice except near the line end.

	Standard Uniform Profile	Alternating Profile 1	Alternating Profile 2
	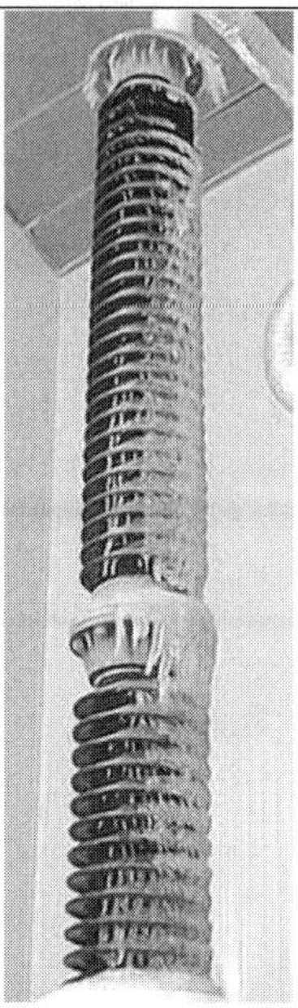	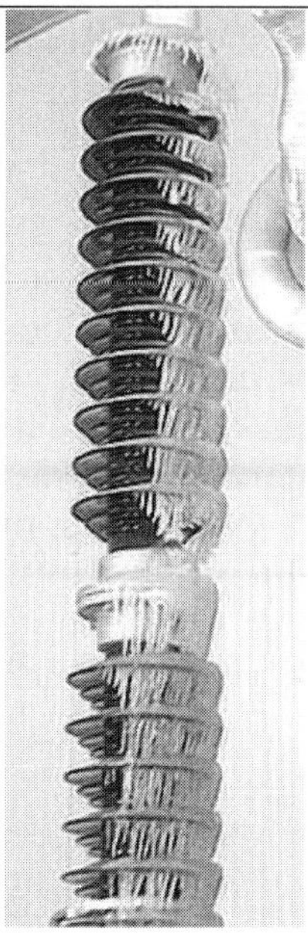	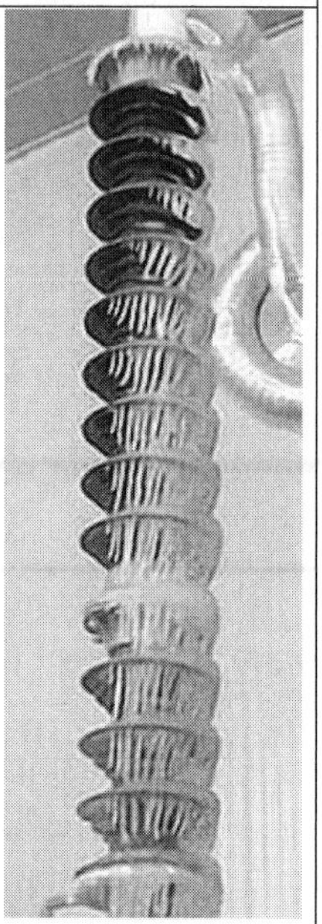
Dry Arc Distance of Test Section (m)	2.00	2.05	1.97
Ratio: Leakage to Dry Arc	2.57	3.43	2.42
Major Shed Diameter (mm)	262	375	328
Vertical Distance Between Major Sheds (mm)	50	120	150
Bridged at 15 mm?	Yes	Yes	Yes
V_{WS} (kV)	195	195	192.5
E_{WS} (kV/m$_{dry arc}$)	98	95	98

Figure 9-37: Dimensions of prototype insulators tested for heavy icing conditions of 15-mm accretion on a rotating reference cylinder (from Farzaneh and Brettschneider [2001]).

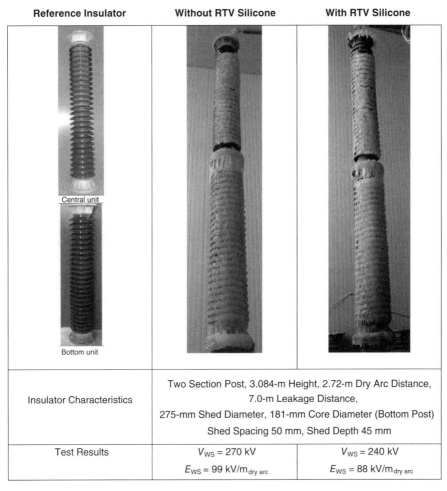

Reference Insulator	Without RTV Silicone	With RTV Silicone
Insulator Characteristics	Two Section Post, 3.084-m Height, 2.72-m Dry Arc Distance, 7.0-m Leakage Distance, 275-mm Shed Diameter, 181-mm Core Diameter (Bottom Post) Shed Spacing 50 mm, Shed Depth 45 mm	
Test Results	$V_{WS} = 270$ kV $E_{WS} = 99$ kV/m$_{dry\ arc}$	$V_{WS} = 240$ kV $E_{WS} = 88$ kV/m$_{dry\ arc}$

Figure 9-38: Comparison test of porcelain station post insulators with and without RTV silicone coating in heavy icing conditions (adapted from Farzaneh and Volat [2007]).

The withstand level V_{WS} was established using the up-and-down method prescribed in IEC 60507 [1991] and IEEE Standard 4 [1995] for evaluating polluted insulator strength in clean-fog and salt-fog tests. The level of $V_{WS} = 240$ kV for the post coated with silicone RTV was 11% lower than the level of $V_{WS} = 270$ kV for the untreated post of the same shape, in the same test conditions. This loss of electrical strength was partly attributed to the longer duration of ice retention on the rougher RTV surface.

For this reason, the use of RTV silicone coating is not recommended on post insulators with shallow shed depth, especially in areas of heavy accumulation of ice, leading to full bridging.

9.4. OPTIONS FOR MITIGATING SNOW AND RIME

There are some specific aspects to heavy snow and rime accretion that differ from ice accretion. The lower density translates generally into a reduced risk of flashover, unless the insulator arrangement allows buildup of significant thickness in the range of 0.2–1 m. This heavy accumulation is seldom found on single vertical insulator strings or posts, but often occurs when two or more insulators are placed close together in parallel, at horizontally oriented dead-end strain insulators, or on horizontal apparatus insulators and bushings of large diameter.

In rough order of preference, the mitigation options are: increased dry arc distance, profile (orientation), polymer insulators, semiconducting glaze, monitoring and washing, accessories, and silicone coatings.

9.4.1. Increased Dry Arc Distance

For reliable operation of insulators under heavy snow conditions, the most effective mitigation consists in using a sufficiently long dry arc distance, selected on the basis of withstand stress, and measured with suitable laboratory test methods, or selected on the basis of service experience under the worst snow conditions [Farzaneh et al., 2003a]. This solution has the greatest value in areas with high occurrence of snow flashovers, and where it is possible to coordinate the increased insulator BIL with other equipment.

In CIGRE Task Force 33.04.09 [2000], the recommended ac withstand voltage for snow-covered insulators scaled linearly as a function of dry arc distance. For constant ac voltage, a withstand stress of 75 kV rms per meter of dry arc distance was recommended from test results. For a 100-ms temporary overvoltage, the withstand stress of 90 kV rms per meter was reported. There were also reductions in the switching-surge and lightning-surge stresses, compared to dry, clean conditions. However, a linear increase in dry arc distance would result in a linear increase in flashover voltage under these heavy snow conditions, with conductivity of 50 µS/cm, 0.5–1-m height, and 0.3–0.4-g/cm^3 density.

9.4.2. Insulator Profile

Khalifa and Morris [1967] carried out studies of heavy rime ice accretion on transmission line insulators in research to address flashover problems that led one utility to reduce its operating voltage from 315 to 280 kV in order to keep transmission lines in service. They carried out tests at −12 °C to −7 °C (10–20 °F) on standard 146-mm × 254-mm glass or porcelain disks, anti-fog disks with 432-mm leakage, and a long-rod porcelain insulator with 2438-mm overall leakage (Figure 9.39).

The long-rod insulator had relatively poor performance, with 75-kV minimum flashover compared to 150 kV for a string of seven standard disks.

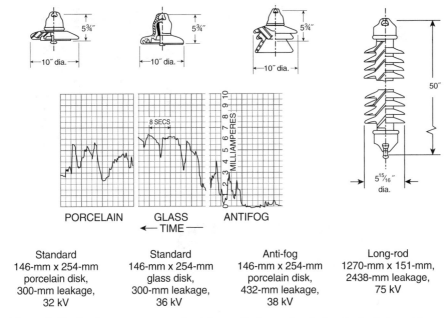

Standard	Standard	Anti-fog	Long-rod
146-mm x 254-mm porcelain disk, 300-mm leakage, 32 kV	146-mm x 254-mm glass disk, 300-mm leakage, 36 kV	146-mm x 254-mm porcelain disk, 432-mm leakage, 38 kV	1270-mm x 151-mm, 2438-mm leakage, 75 kV

Figure 9-39: Leakage currents and minimum flashover levels for single disk and long-rod insulators covered with rime ice of $35\,lb/ft^3$ at $10\text{--}20\,°F$.

This was true in spite of its greater leakage distance of 2438 mm compared to 2100 mm. The small shed-to-shed spacing of the long-rod design was easily filled with ice. The large sheltered creepage distance of their anti-fog disk gave a modest advantage over the standard disks.

9.4.3. Insulators in Parallel

Khalifa and Morris [1967] also found that rime ice could fill the space between pairs of suspension insulators, as shown in Figure 9-40. In this case, the double string flashed over at 90 kV, compared to the 150-kV level and lower leakage currents noted for the single string.

An increased separation between chains of disk insulators was recommended as a mitigation approach for heavy snow in CIGRE Task Force 33.04.09 [2000].

When the plane passing through the axes of the parallel insulator strings has vertical orientation, it is more difficult for the tension assembly to be covered with a large amount of snow. Thus in areas of heavy snow, it may be better to use vertical or braced posts at dead-end positions, with the braces oriented along the line direction, if these can meet the required mechanical duties.

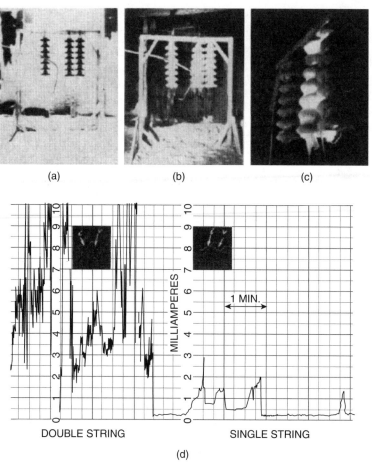

Figure 9-40: Comparison of (a) double-suspension and single-suspension strings with (b) rime ice of density $0.7\,g/cm^3$ ($45\,lb/ft^3$), (c) at $90\,kV$, with (d) leakage currents (from Khalifa and Morris [1967]).

9.4.4. Polymer Insulators

Some polymer insulators are available in mechanical ratings and dimensions that make them suitable for replacement of ceramic disk strings in horizontal configurations. The polymer insulators have smaller diameters, and each will accumulate a smaller amount of snow, giving a reduction in icing stress product and a consequent increase in flashover stress per meter of dry arc distance. In conditions of heavy snow or rime ice accretion, the increased space between the sheds of the polymer insulators will give an additional advantage when using the same yoke plates as the ceramic disks.

9.4.5. Surface Coatings

Khalifa and Morris [1967] tested an anti-sticking "Carbo" wax and grease, both used to mitigate the effects of rime ice on metal surfaces. The waxed or greased insulator strings had leakage currents of 4–5 mA compared to 2 mA on clean porcelain, in all three cases with a thick accretion of about 0.5 g/cm^3 (30 lb/ft^3) of hard rime ice. Minimum flashover actually decreased with these treatments from 38 kV for the anti-fog disk to 35 kV with the grease and 30 kV with the wax. In contrast, silicone grease gave leakage currents below 0.2 mA and an increase in minimum flashover level to 45 kV, but tended to be carried off by falling ice and degraded by the intense, long-duration white arc activity.

The general use of RTV silicone coatings to improve performance under heavy snow is not recommended without qualification testing, for the same reasons that apply to heavy icing conditions. Field observations suggest that snow may adhere better to silicone-coated porcelain than to the bare porcelain. The role of RTV coating in improving flashover strength along the leakage distance is not relevant under full snow or rime ice bridging. With few test results, additional observations of the effects of RTV coating are needed.

9.4.6. Use of Semiconducting Glaze

Practical semiconducting glaze disk insulators for transmission lines have only been commercialized recently, and their strength ratings are lower than those required for many dead-end applications. With most of the problems of heavy snow occurring on the horizontal strings of transmission lines, there has not been much motivation to test the performance of semiconducting glaze disks as a mitigation method for heavy snow.

At present, the semiconducting glaze post insulators have shown good performance in most icing conditions, but there are no test data or application experience to support their application in heavy snow conditions.

As discussed in Chapter 6, Section 6.7, semiconducting glaze insulators in close 0.5-m proximity need to be selected carefully in order to equalize the voltage between intermediate metallic flanges. Insulators in close proximity are also the configurations with the highest risk of snow accumulation, and test programs should give this aspect special attention.

9.4.7. Use of Accessories

An extension rod of about 1 m in length can be inserted between the tower arm and the tension insulator assembly. Snow tends to accumulate on the tower arms over suspension insulators. An air space reduces the chance that snow will bridge between tower arm and the insulator assembly.

Many accessories that break up the distribution of ice accumulation on vertical insulators, such as booster sheds, will also decrease the space between parallel insulators. This may leave the specific and vulnerable configurations of parallel, closely spaced insulators even more vulnerable to heavy snow accumulation.

At present, there are no accessories that are intended to break up the accumulation pattern when insulators have horizontal orientation.

9.5. ALTERNATIVES FOR MITIGATING ANY ICING

9.5.1. Doing "Nothing"

With enough advanced warning, it may be possible to configure many power systems to be more tolerant of the loss of multiple elements. These configurations may involve an increase in spinning reserve, redispatch of generation to stations that would not normally operate, or increased reliance on interconnections.

Normally, grid systems are designed with multiple lines between stations. If any one line undergoes a flashover for any reason, then the other lines can pick up their share of load and maintain service with a minimal power quality disturbance. In this way, systems are more resistant to line outages than station outages.

A large utility in Ontario operates its EHV transformer stations in a "safe posture" in winter conditions defined as "flashover weather":

- The initial air temperature is below $0\,^{\circ}$C.
- The air temperature slowly increases during the flashover weather period to at least $-1\,^{\circ}$C.
- The difference between temperature and dew point starts and remains at less than or equal to $2\,^{\circ}$C.
- Fog with visibility less than 4 km is observed at the start of the period and persists throughout.
- Fog with visibility less than 1 km occurs occasionally or continuously.
- Fog, freezing drizzle, or freezing rain causes ice accretion on exposed structures occasionally or continuously while the air temperature is below $0\,^{\circ}$C.

The "safe posture" situation was in effect for the largest loss of system minutes (4.3) from icing problems at highly contaminated 230-kV stations on December 31, 1989 after a record-breaking cold spell for the entire month.

The utility of a "safe posture" reaction depends on accurate weather forecasts, available in sufficient time to reschedule resources. While it was originally set up as a short-term solution, experience has suggested that it can be a viable option in the context of normal power system planning and operating criteria.

9.5.2. Voltage Reduction

In cases where insulators are known to be polluted and an icing event starts, or heavy icing is forecast, a second general power system operational response may be appropriate with suitable planning.

Several authors, including Hara and Phan [1979] and Cherney [1980], have suggested that a system-wide reduction in transmission voltage could be an effective icing and contamination mitigation method with limited cost. Typical staged reductions in system voltage of 5% are often practiced annually, to ensure that load flow models are up to date. These procedures can be instigated during icing conditions, either when insulators are known to have high levels of precontamination or when ice accretion reaches levels that may cause bridging.

As an example for a small station with 50 parallel insulators under full ice bridging conditions [Farzaneh et al., 2007], a reduction of the operation voltage by 5% results in a decrease of the flashover probability from $P_{n=50} = 49.5\%$ (maximum operation voltage) to $P_{n=50} = 11.7\%$. If it is possible to reduce the voltage at the affected station by 10%, the probability of any insulator flashover declines to $P_{n=50} = 1.2\%$.

9.5.3. Post-Event Inspection Using Corona Detection Equipment

Generally, the use of corona detection equipment provides sensitivity to arcing activity that is unnecessary on ice- or snow-covered insulators. Even in daylight, discharge activity is visible to the unaided eye. The role of ultraviolet equipment is more important after the ice or snow has fallen off. Inspections can identify any insulators that have been damaged by discharge activities that may have been trapped between ice layers and insulator surfaces. Inspection may also identify insulators damaged by mechanical shock due to ice shedding, sleet jump, or conductor galloping.

9.6. PRÉCIS

Mitigation options for insulator contamination problems often involve the application or use of materials that are hydrophobic. While these materials may bead water, they are not ice-phobic. There is some benefit to the use of silicone materials in freezing conditions, but it is nowhere near as effective at $0\,°C$ as it is at $20\,°C$. Since ice adheres well to polymer surfaces, the only benefit that these insulators offer when mitigating problems with heavy ice accretion is their small diameter compared to cap-and-pin insulator strings.

Options that break up the continuity of the ice layer introduce air gaps that give improved performance in moderate or heavy ice conditions. Post insulators with deep shed separation outperform uniform-profile station posts for a limited range of conditions, but can still be fully bridged once ice accretion reaches about 10 mm. Cap-and-pin insulators have a naturally larger shed-to-shed spacing that leads to fewer problems with icing that can be resolved simply by increasing the number of disks in the string. Multiple booster sheds proved to be a good solution for heavy icing performance on station post insulators. Posts with semiconducting glaze have the best all-round performance in any icing condition, along with a proven track record of field experience.

REFERENCES

Baker, A. C., L. E. Zaffanella, L. D. Anzivino, H. M. Schneider, and J. H. Moran. 1989. "A Comparison of HVAC and HVDC Contamination Performance of Station Post Insulators," *IEEE Transactions on Power Delivery*, Vol. 4, No. 2 (Apr.), pp. 1486–1491.

Barrett, J. S. and M. A. Green. 1994. "A Statistical Method for Evaluating Electrical Failures," *IEEE Transactions on Power Delivery*, Vol. 8, No. 3 (July), pp. 1524–1530.

Chaarani, R. 1996. "Détermination de la tension de tenue maximale des isolateurs composites en EPDM," Masters Thesis, University of Québec at Chicoutimi. Autumn.

Chang, J. W. and R. S. Gorur. 1994. "Surface Recovery of Silicone Rubber Used for HV Outdoor Insulation," *IEEE Transactions on Dielectrics and Electrical Insulation*, Vol. 1, No. 6 (Dec.), pp. 1039–1046.

Cherney, E. A. 1980. "Flashover Performance of Artificially Contaminated and Iced Long-Rod Transmission Line Insulators," *IEEE Transactions on Power Apparatus and Systems*, Vol. PAS-99, No. 1 (Jan.), pp. 46–52.

Cherney, E. A. 1986. "Ice Bridging Flashover of Contaminated 500 kV Line Insulation in Freezing Rain," Ontario Hydro Report 86-80-H. April 8.

Chisholm, W. A. 1994. "High-Tech Solutions to a Dirty Problem: 230-kV Flashovers at Hamilton Beach TS," Ontario Hydro Technologies Report AG94-183P. December 14.

Chisholm, W. A. 1995. "Cold-Fog Test Results on 230-kV Line Insulators for Application Near Hamilton Beach TS," Ontario Hydro Technologies Report AG-95-146-P. September 20.

Chisholm W. A. 1999. "Contaminated Ice-Cap Flashovers at Hamilton Beach TS: Diagnosis and Possible Remedial Actions," Ontario Power Technologies Report 5491-005-1995-RA-0001-R00.

Chisholm, W. A. 2000. "High Voltage Testing of Station Post Insulators Under Contaminated Ice Conditions," Kinectrics Report 8490-000-2000-RA-0001-R000 (Confidential). September 23.

Chisholm, W. A. 2005. "Ten Years of Application Experience with RTV Silicone Coatings in Canada," in *Proceedings of 2005 World Congress on Insulators Arresters and Bushings*, Hong Kong. November 27–30.

Chisholm, W. A. 2007. "Insulator Leakage Distance Dimensioning in Areas of Winter Contamination Using Cold-Fog Test Test Results," *IEEE Transactions on Dielectrics and Electrical Insulation*, Vol. 14, No. 6 (Dec.), pp. 1455–1461.

Chisholm, W. A., P. G. Buchan, and T. Jarv. 1994. "Accurate Measurement of Low Insulator Contamination Levels," *IEEE Transactions on Power Delivery*, Vol. 9, No. 3 (July), pp. 1552–1557.

Chisholm, W. A., K. G. Ringler, C. C. Erven, M. A. Green, O. Melo, Y. Tam, O. Nigol, J. Kuffel, A. Boyer, I. K. Pavasars, F. X. Macedo, J. K. Sabiston, and R. B. Caputo. 1996. "The Cold-Fog Test," *IEEE Transactions on Power Delivery*, Vol. 11, No. 4 (Oct.), pp. 1874–1880.

Chisholm, W. A., J. Kuffel, F. Kwan, and T. Kydd. 2000. "Hydro-One Smart Washing Program: The Seven-Year Itch," in *Proceedings of 2000 IEEE 9th International Conference on Transmission & Distribution Construction, Operation & Live-Line Maintenance (ESMO)*, Montreal. October 8–12.

CIGRE Task Force 33.04.09. 2000 "Influence of Ice and Snow on the Flashover Performance of Outdoor Insulators, Part II: Effects of Snow," *Electra*, No. 188 (Jan.), pp. 55–69.

Drapeau, J-F., M. Farzaneh, M. Roy, R. Chaarani, and J. Zhang. 2000. "An Experimental Study of Flashover Performance of Various Post Insulators Under Icing Conditions," in *Proceedings of IEEE Conference on Electrical Insulation and Dielectric Phenomena (CEIDP)*, Vol. 1, Victoria, British Columbia, Canada, pp. 359–364.

Drapeau, J-F., M. Farzaneh, and M. Roy. 2002. "An Exploratory Study of Various Solutions for Improving Ice Flashover Performance of Station Post Insulators," in *Proceedings of International Workshop on Atmospheric Icing of Structures (IWAIS)*, Brno, Czech Republic. June.

Electric Power Research Institute (EPRI). 1982. *Transmission Line Reference Book, 345 kV and Above*, Palo Alto, CA: EPRI.

Erven, C. C. 1988. "500 kV Insulator Flashovers at Normal Operating Voltage," in *Proceedings of Canadian Electrical Association Engineering and Operating Spring Meeting*. March.

Farzaneh, M. and S. Brettschneider. 2001. "Étude de la tension de tenue des isolateurs de postes en présence de glace atmosphérique en vue d'un choc approprié de type et configuration d'isolateurs de poste à 735 kV, Volume 1, Étude en vue du choix d'isolateurs pour le futur poste Montérégie," Report presented to Institut de Recherche d'Hydro-Québec (IREQ). September.

Farzaneh, M. and J. Kiernicki. 1995. "Flashover Problems Caused by Ice Buildup on Insulators," *IEEE Electrical Insulation Magazine*, Vol. 11, No. 2 (Mar./Apr.), pp. 5–17.

Farzaneh, M. and J. Kiernicki. 1997. "Flashover Performance of IEEE Standard Insulators Under Ice Conditions," *IEEE Transactions on Power Delivery*, Vol. 12, No. 4 (Oct.), pp. 1602–1613.

Farzaneh, M. and C. Volat. 2005. "Étude de la tension de tenue des isolateurs de postes en présence de glace atmosphérique en vue d'un choc approprié de type et configuration d'isolateurs de poste à 735 kV, Volume 3, Étude en vue de l'amélioration par l'ajout de jupes auxiliaires." Report presented to the Institut de Recherche d'Hydro-Québec (IREQ). February.

Farzaneh, M. and C. Volat. 2007. "Étude de la tension de tenue des isolateurs de postes en présence de glace atmosphérique en vue d'un choc approprié de type et configuration d'isolateurs de poste à 735 kV, Volume 8, Étude de l'influence d'un revêtement RTV appliqué à une colonne isolante standard." Report presented to Hydro-Québec TransÉnergie. August.

Farzaneh, M. and J. Zhang. 2007. "A Multi-Arc Model for Predicting AC Critical Flashover Voltage of Ice-Covered Insulators," *IEEE Transactions on Dielectrics and Electrical Insulation*, Vol. 14, No. 6 (Dec.), pp. 1401–1409.

Farzaneh, M., J. Kiernicki, and R. Martin. 1994. "A Laboratory Investigation of the Flashover Performance of Outdoor Insulators Covered with Ice," in *Proceedings of the 4th International Conference on Properties and Applications of Dielectric Materials*, Brisbane, Australia, Paper 5107, pp. 483–486. July 3–8.

Farzaneh, M., T. Baker, A. Bernstorf, K. Brown, W. A. Chisholm, C. de Tourreil, J. F. Drapeau, S. Fikke, J. M. George, E. Gnandt, T. Grisham, I. Gutman, R. Hartings, R. Kremer, G. Powell, L. Rolfseng, T. Rozek, D. L. Ruff, D. Shaffner, V. Sklenicka, R. Sundararajan, and J. Yu. 2003a. "Insulator Icing Test Methods and Procedures: A Position Paper Prepared by the IEEE Task Force on Insulator Icing Test Methods," *IEEE Transactions on Power Delivery*, Vol. 18, No. 3 (Oct.), pp. 1503–1515.

Farzaneh, M., J.-F. Drapeau, J. Zhang, M. J. Roy, and J. Farzaneh. 2003b. "Flashover Performance of Transmission Class Insulators Under Icing Conditions," in *Proceedings of Insulator News and Market Report Conference*, Marbela, Spain, pp. 315–326. August.

Farzaneh-Dehkordi, J., J. Zhang, and M. Farzaneh. 2004. "Experimental Study and Mathematical Modeling of Flashover on Extra-High Voltage Insulators Covered with Ice," *Hydrological Processes*, Vol. 18, pp. 3471–3480.

Farzaneh, M., T. Baker, A. Bernstorf, J. T. Burnham, T. Carreira, E. Cherney, W. A. Chisholm, R. Christman, R. Cole, J. Cortinas, C. de Tourreil, J. F. Drapeau, J. Farzaneh-Dehkordi, S. Fikke, R. Gorur, T. Grisham, I. Gutman, J. Kuffel, A. Phillips, G. Powell, L. Rolfseng, M. Roy, T. Rozek, D. L. Ruff, A. Schwalm, V. Sklenicka, G. Stewart, R. Sundararajan, M. Szeto, R. Tay, and J. Zhang. 2005. "Selection of Station Insulators with Respect to Ice and Snow—Part I: Technical Context and Environmental Exposure," *IEEE Transactions on Power Delivery*, Vol. 20, No. 1 (Jan.), pp. 264–270.

Farzaneh M., J. Zhang, and C. Volat. 2006a. "Effect of Insulator Diameter on AC Flashover Voltage of an Ice-Covered Insulator String," *IEEE Transactions on Dielectrics and Electrical Insulation*, Vol. 13, No. 2 (Apr.), pp. 264–271.

Farzaneh, M., C. Volat, and J. Zhang. 2006b. "Étude de la tension de tenue des isolateurs de postes en présence de glace atmosphérique en vue d'un choc approprié de type et configuration d'isolateurs de poste à 735 kV, Volume 7, Amélioriation de la tension de tenue sous glace d'une colonne double de sectionneur." Report presented to Hydro-Québec TransÉnergie. November.

Farzaneh, M. (Chair), A. C. Baker, R. A. Bernstorf, J. T. Burnhan, E. A. Cherney, W. A. Chisholm, I. Fofana, R. S. Gorur, T. Grisham, I. Gutman, L. Rolfseng, and G. A. Stewart. 2007. "Selection of Line Insulators with Respect to Ice and Snow—Part II: Selection Methods and Mitigation Options, A Position Paper Prepared by the IEEE Task Force on Icing Performance of Line Insulators," *IEEE Transactions on Power Delivery*, Vol. 22, No. 4 (Oct.), pp. 2297–2304.

Gutman, I., K. Halsan, and D. Hübinette. 2003. "Application of Ice Progressive Stress Method for Selection of Insulation Options," in *Proceedings of 2003 International*

Symposium on High Voltage (ISH-2003), Netherlands (J. Smit, ed.). Rotterdam, The Netherlands: Millpress, p. 179.

Hara, M. and C. L. Phan. 1979. "Leakage Current and Flashover Performance of Iced Insulators," *IEEE Transactions on Power Apparatus and Systems*, Vol. PAS-98, No. 3 (May), pp. 849–859.

IEC 60507. 1991. *Artificial Pollution Tests on High Voltage Insulators to Be Used in AC Systems*. Lausanne, Switzerland: IEC.

IEEE/ANSI Standard C2. 2007. *National Electrical Safety Code*. Piscataway, NJ: IEEE Press.

IEEE Standard 4. 1995. *IEEE Standard Techniques for High-Voltage Testing*. Piscataway, NJ: IEEE Press.

Kannus, K. and K. Lathi. 2007. "Laboratory Investigations of the Electrical Performance of Ice-Covered Insulators and a Metal Oxide Surge Arrester," *IEEE Transactions on Dielectrics and Electrical Insulation*, Vol. 14, No. 6 (Dec.), pp. 1357–1372.

Khalifa, M. M. and R. M. Morris. 1967. "Performance of Line Insulators Under Rime Ice," *IEEE Transactions on Power Apparatus and Systems*, Vol. PAS-86, No. 6 (June), pp. 692–697.

Kojima, S., M. Oyama, and M. Yamashita. 1988. "Potential Distributions of Metal Oxide Surge Arresters Under Various Environmental Conditions," *IEEE Transactions on Power Delivery*, Vol. 3, No. 3 (July), pp. 984–989.

Lanoie, R., D. Bouchard, M. Lessard, Y. Turcotte, and M. Roy. 2005. "Using Steam to De-ice Energized Substation Disconnect Switch," in *Proceedings of XI International Workshop on Atmospheric Icing of Structures (IWAIS)*, Montréal, Québec, pp. 353–356. June.

Liao, W., J. Zhidong, G. Zhicheng, W. Liming, J. Yang, J. Fan, Z. Su, and J. Zhou. 2007. "Reducing Ice Accumulation on Insulators by Applying Semiconducting RTV Silicone Coating," *IEEE Transactions on Dielectrics and Electrical Insulation*, Vol. 14, No. 6 (Dec.), pp. 1446–1454.

Macedo, F., Task Group Chair. 1992. "Task Group Report on 500 kV Station Insulator Retrofit Program," Ontario Hydro File 125.SIR. December 13.

Meghnefi, F., C. Volat, and M. Farzaneh. 2007. "Temporal and Frequency Analysis of the Leakage Current of a Station Post Insulator During Ice Accretion," *IEEE Transactions on Dielectrics and Electrical Insulation*, Vol. 14, No. 6 (Dec.), pp. 1381–1389.

Sklenicka, V. and J. Vokalek. 1996. "Insulators in Icing Conditions: Selection and Measures for Reliability Increasing," in *Proceedings of 7th International Workshop on Atmospheric Icing of Structures (IWAIS)*, Chicoutimi, Québec, Canada, pp. 72–76.

CHAPTER 10

INSULATION COORDINATION FOR ICING AND POLLUTED ENVIRONMENTS

Chapters 2 and 3 laid out the background data needed to estimate how rapidly pollution builds up on insulator surfaces. Chapter 4 showed how the pollution affects electrical flashover strength at room temperature in conventional clean-fog or salt-fog conditions. Chapters 7 and 8 showed that ice and snow flashovers could be modeled in similar ways to mathematical models for pollution flashover in Chapter 5. Chapter 6 explained that there are a number of different ways to improve the flashover strength in a polluted environment at room temperature. Chapter 9 showed that some of these work differently above and below the freezing point.

This final chapter closes the loops in the insulation coordination process for freezing conditions:

- Evaluating the number of flashovers related to insulator leakage distance.
- Evaluating the number of flashovers across the ice or snow layers that may bridge the insulator dry arc distance.
- Putting these risks of flashover into the overall context of substation and transmission line reliability evaluation.

The concept of insulation coordination for selecting insulator leakage distance is relatively new, compared to the processes for switching-surge and lightning performance, but can adapt many of the same standard processes.

Insulators for Icing and Polluted Environments. By Masoud Farzaneh and William A. Chisholm
Copyright © 2009 the Institute of Electrical and Electronics Engineers, Inc.

These are described first. Then, the necessary factors for quantifying the annual risk of problem incidence are developed from four climate factors in winter periods: duration of dry periods, occurrence of fog, occurrence of freezing rain and drizzle, and occurrence of heavy snow cover. The applications of the factors are illustrated using recommended methods for representing the probability distributions of precipitation conductivity and days without rain. Finally, a few utility case studies are described to show how the use of mitigation options can be of great benefit when an initial insulation coordination process has not provided the desired reliability level.

10.1. THE INSULATION COORDINATION PROCESS

Hileman [1999] very simply defined insulation coordination as the selection of the insulation strength. The IEEE Standard 1313.1 [1996] focused more specifically on the strength that would be "consistent with the expected overvoltages to obtain an acceptable risk of failure." The companion guide, IEEE Standard 1313.2 [1999] spells out how, in addition to increasing the insulator strength, overvoltage stress can be controlled by surge arresters, preinsertion resistors in circuit breakers, lightning protection (shielding and grounding), and by many other means. The IEEE 1313.2 guide also stressed the need for insulation either to be protected against temporary overvoltages or to have inherent ability to withstand typical short-duration power frequency overvoltages, when the external insulation is covered with contamination.

With many of the problems with icing flashover occurring at normal system voltage, the insulation coordination needs to be expanded somewhat to consider how the strength of air gap clearances and insulation varies with winter weather under normal service voltage stress.

The goal of the designer is to specify an optimum combination of insulation leakage and dry arc distance that achieves a desired reliability at minimum cost. Well-coordinated design practices are needed to evaluate the icing performance of both transmission lines and substations.

The composite reliability of transmission lines and substations under adverse winter conditions does not need to be perfect. An infinite mean time between failures (MTBF) from a deterministic method may be impossible or too costly to achieve and is not accepted in other evaluations of substation reliability. Risk of failure, expressed as a lightning outage rate, has also been routinely accepted in calculations of lightning performance, where the distributions of peak stroke current and footing resistance at each tower are extremely wide.

10.1.1. Classification of Overvoltage Stresses on Transmission Lines

The first and most difficult duty for insulators in icing and pollution conditions is to withstand the continuous (power frequency) voltages that are the result of normal operation of the power system.

In addition, the insulators may face a range of overvoltages, some of which may result from prior flashovers of other insulators in icing conditions. Typical problems will include:

- Temporary overvoltages, especially those caused by faults, load rejection, or line energizing after previous icing faults.
- Switching overvoltages with relatively slow-rising front time of 250 μs that are caused by line energization, fault initiation, load rejection, switching of capacitor banks, or breaking of inductive current flow, for example, by fault clearing.
- Lightning overvoltages that are caused primarily by powerful positive winter-lightning flashes that often accompany the frontal movement of freezing rain events.
- Very fast-front overvoltages that are the result of operating high-voltage disconnect switches or gas-insulated substations (GISs) to deal with icing problems.

These do not describe all the sources of overvoltage [IEEE Standard 1313.2, 1999], but instead we focus on those that are most relevant to weather-related outages.

Temporary overvoltages are described by their amplitude-duration characteristic. The per-unit (pu) system defines 1 pu as the peak of the maximum system line-to-ground voltage, which this book has defined consistently using the symbol V_m. Thus a 1.3-pu overvoltage after a fault in an effectively grounded system would ideally have a clearing time of less than 0.1 s. It may not be reasonable to design for incorrect operation of relay protection under icing conditions since this essentially describes a double contingency.

Temporary overvoltages caused by load rejection are especially important in the case of load rejection at the remote end of a long line due to the Ferranti effect. The ac voltage rise affects the apparatus at the station connected on the line side of the remote circuit breaker. Systems with long lines can have overvoltages of up to 1.5 pu at generator stations, with a duration of up to 3 s, depending on generator short-circuit power and controls. Station surge arresters generally operate to limit temporary overvoltages on ice-covered insulators above this magnitude.

Switching overvoltages tend to be higher in magnitude but shorter in duration than temporary overvoltages. The distribution of overvoltages is typically evaluated with a computer simulation and the results are then described [IEEE Standard 1313.2, 1999] using the 2% (E_2) level and a standard deviation. Line switching overvoltages are normally controlled on EHV systems with the use of preinsertion resistors on circuit breakers. This limits the E_2 values to 1.8–2.0 pu, compared to 2.8–3.0 without closing resistors. There is not much benefit to reducing the E_2 value much below 1.8, as many fault overvoltages will have similar switching-surge levels. Station surge arresters may also be selected to provide further reductions in switching-surge levels.

Some phase-to-phase insulation, such as interphase spacers and station bushings, will also be exposed to temporary phase-to-phase and switching overvoltages.

Capacitor bank energizing produces overvoltages locally, and also at line terminations, transformers, remote capacitor banks, and cables. High phase-to-phase switching overvoltages occur when ungrounded capacitor banks are energized. If at all possible, capacitor banks should not be switched under icing conditions to avoid these problems.

While it is an interesting curiosity, the lightning performance of iced insulators is not of engineering relevance as the number of hours with both icing and lightning is relatively low, on the order of less than three per year, compared to a typical exposure of 40 hours of lightning and 30 hours of icing.

Some types of icing flashover have fast current rise-times, exceeding 1-MHz system bandwidth. The ice layers may thus be susceptible to very-fast-front overvoltages from GISs or other sources. This has not proved to be an operational problem but could be a future area of investigation should the GIS technology achieve wider use.

10.1.2. High-Voltage Insulator Parameters

Following up with successful application of telegraph insulators, engineers of the first electric power systems in the 1900s and 1910s obtained reliable electrical service with the insulators of that time by limiting the voltage gradient across the dry arc distance to about 70 kV per meter of dry arc distance. With the typical 250-mm disk diameters and 150-mm cap-and-pin spacing, the insulator disks were not that much different from those used today. Each disk had approximately 10 kV (rms line to ground) per insulator disk across its terminals. With 300 mm of leakage distance, the voltage stress of 10 kV per disk corresponds to the unified specific creepage distance of 30 mm per kV rms line to ground. This stress level of 10 kV per 146-mm disk was used widely until the development of EHV transmission at voltages at or above 345 kV.

Manufacturers of ceramic disk insulators with 146-mm shed spacing supplied insulators with extended leakage distance profiles to utilities with severe contamination environments. Examples include the fog bowl, deep-rib, and multiskirt designs that were described in Chapter 6.

Within stations, the use of a few, large cap-and-pin apparatus type insulators with wider cap-to-pin spacing was common for many years. These insulators were assembled from multiple porcelain shells to give the necessary mechanical strength ratings. With the improvements in ceramic insulator technology over the years, the manufacture of post insulators of about 1-m length gave better mechanical performance and reliability with fewer joints of larger diameter. However, the post-type insulators did not provide as much sheltered leakage distance as deep-skirt apparatus types.

The choice of insulator surface material is an important aspect in the insulation coordination process for icing and polluted conditions. In Chapter 2, the use of materials such as silicone rubber as thick weather sheds over fiberglass cores was described. These insulators offer superior contamination performance in many environments when new. Thin silicone coatings were also listed in Chapter 6 as an important option to consider for improving performance of existing transmission insulators in areas of high contamination. However, Chapter 9 contained cautions that the benefits of polymer materials are not so clear for some levels of icing severity.

10.1.3. Extra-High Voltage Insulator Parameters

The voltage distribution along the insulator string becomes highly nonlinear once 20 or more disks are placed in series. This means that the ability to resist system-generated switching overvoltage no longer scales linearly with dry arc distance. When transmission voltages at or above 345 kV were engineered, it proved to be more effective to introduce other measures to calculate and control switching overvoltages than to provide longer insulator strings. The most common switching-surge control method was a preinsertion resistor fitted into circuit breakers. In the 1980s, the development of gapless metal oxide surge arresters was also effective as this type of arrester has better protective margins and higher voltage ratings than gapped silicon carbide arresters used from the 1930s.

With these switching-surge mitigation measures in place, the utilities no longer needed to respect the 10-kV per disk guideline that worked well for most high-voltage transmission lines. Typical 500-kV lines used 13–14 kV per disk (26–28 kV per meter of leakage distance) and some 735-kV lines use a stress as high as 15 kV per disk. There has been a wider range of stress on station insulators.

A typical expression for setting the clearances of electrical systems d (m) as a function of peak voltage V_{pk} (kV) is [IEEE Standard 4, 1995]

$$d \geq \frac{8 \cdot V_{pk}}{2890 \text{ kV} - V_{pk}} \tag{10-1}$$

The electrical clearance d in Equation 10-1 has a fixed margin of 15% over the critical switching-surge flashover level. This margin will vary depending on reliability requirements, but for illustration purposes, the same margin has been applied to establish the dry arc distance requirements for pollution and icing conditions in Figure 10-1.

Insulators are available with a wide ratio of specific leakage distance (the leakage distance divided by the dry arc distance), so this is used as a parameter. For standard disk insulators with a ratio of 305 mm/146 mm or 2.1, the dry arc distance (insulator length) requirements for a heavy pollution level of 0.2 mg/cm^2 completely dominate any switching-surge requirements. In this case, ceramic insulators with 3:1 or 4:1 profiles may be needed. If a system has been designed for a 1.9-pu switching-surge level, the dry arc distance requirement

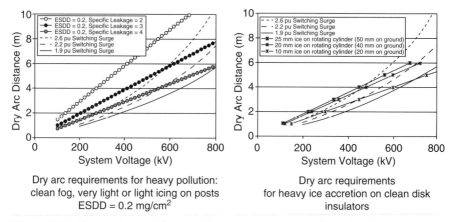

Figure 10-1: Comparison of dry arc distance requirements for switching surge, contamination, and icing conditions.

related to a 4:1 extended leakage profile is higher and thus presents the limiting design constraint.

The same comments apply to the comparison of dry arc requirements for switching-surge overvoltages and heavy icing. With the same 15% margin over critical flashover level, the dry arc distance requirement for 10 mm of radial accretion on a reference cylinder (20 mm on the ground) is more important than the switching-surge clearance for 1.9-pu overvoltage levels. For EHV systems designed for a 2.2-pu switching-surge overvoltage, 20 mm of ice represents a limiting condition.

10.1.4. Design for an Acceptable Component Failure Rate

The insulation coordination for switching-surge overvoltages benefits from having relatively low levels of standard deviation, both in the estimates of stress (E_2) and electrical strength. The strength is modeled as a Gaussian distribution that is characterized by a mean value (CFO) and a standard deviation σ_f.

In contrast to this well-defined situation, the lightning flash has a peak current that is so wide that it is necessary to take the natural logarithm of the current before fitting a median (31 kA) and evaluating a standard deviation $\sigma_{\ln(I)}$ of 0.48 [CIGRE Study Committee 33, 1991]. This wide distribution is convolved with another of the footing resistance at the base of each tower, R_f, that has an even larger value of $\sigma_{\ln(R_f)}$ that often exceeds 1.0. The case where a small lightning current strikes a tower with low R_f leads to minimal potential rise compared to the insulation level, with no probability of backflashover from the grounding system to the high-voltage conductor across the insulator dry arc distance. However, the case where a large lightning current flows through a tower with high R_f is equally probable and does often lead to a backflashover outage on the line.

TABLE 10-1: Guidelines for Accetable Lightning Performance on Transmission Lines

Typical Voltage Class	Security Requirement	Outages per 100 km per year
115 kV, 138 kV	Class C	<4
161 kV, 230 kV	Class B	<1
500 kV, 735 kV	Class A	<0.5

Lightning specialists engineer overhead lines for a desired level of reliability, expressed as a lightning outage rate per 100 km per year. Typical values of acceptable performance depend on utility, and general guidelines are shown in Table 10-1.

In contrast to overhead line design, there are no acceptable guidelines for lightning protection of cables or other types of non-self-restoring insulation. Every fault in a cable will lead to permanent damage, meaning that redundant paths are needed to preserve system reliability. The same philosophy is adopted for transmission lines that are constructed without overhead groundwire protection against lightning.

Uncertainties about lightning peak stroke current and footing resistance are not the only factors that introduce uncertainty into the calculation of lightning performance. Table 10-2 provides some indications of uncertainties in the input data that are tolerated in obtaining an engineered, nonzero but tolerable lightning performance failure rate. A different set of uncertainties exist for the contamination flashover, but the goal of a tolerable failure rate remains the same.

It is important to note that it is possible, and in fact common, to adapt some features of transmission design to the local environment. Lines may have fixed design features, such as spacing between phase conductors, but they also have variable design features that are adjusted on installation or during the long period of line operation. Some examples of these adjustments include:

- Selection in the field of tower leg lengths, body panels, or pole sizes to give the desired tower height on its foundations during line construction.
- Retensioning of transmission line conductors to restore design clearances above ground after a severe wind and ice load.
- A program of earth resistance tests, followed up by installation of additional buried electrodes at the towers with high footing resistance or high resistivity. This adaptation to the tower-to-tower soil resistivity variation is carried out either during installation or years later, during line maintenance, if the lightning performance proves unsatisfactory.
- Identification of areas with high pollution levels that lead to contamination flashover problems or premature wear-out of insulators. At the towers or locations where this occurs, it is feasible to follow up with one or more of the mitigation options described in this chapter.

TABLE 10-2: Uncertainties in the Parameters for Lightning and Contamination Performance

Parameter	Lightning	Contamination
Natural stress	Ground flash density	Type and severity of salt pollution, duration of exposure
Local stress	High soil resistivity and grounding resistance	Proximity to industrial point or line sources
Insulator parameter to be selected	Dry arc distance, clearances to towers	Leakage distance along insulator surface, shed profile
Typical electric stress	Flash current (3–200 kA) multiplied by footing impedance (3–200 Ω), across total length of insulator	Normal ac or dc service voltage, per meter of insulator leakage distance
Laboratory testing	Cannot reproduce nonlinear effects such as footing ionization, corona effects	Fully reproduces severe pollution and wetting conditions.
Typical electrical strength	540 ± 5% kV impulse voltage per meter of dry arc distance	15–45 kV ac rms line to ground per meter of leakage distance
Planning of maintenance options	Inspection and improvement of additional grounding electrodes (5–20-year cycle)	Insulator washing (1 month), cleaning (1 year), greasing (3 years), or coating (10 years)

Continuing with lightning as an example, towers in areas of high resistivity may have supplemental ground electrodes, or parts of the line may be fitted with line surge arresters. This philosophy of choosing a good design for much of a route, evaluating its performance against the desired reliability, and taking remedial actions if necessary can also be applied to insulation coordination near specific contamination sources.

10.1.5. Design for an Acceptable Network Failure Rate

Lightning is the most frequent cause of most power system faults. Only one of 10 or 20 lightning flashes has sufficient peak current magnitude to cause a backflashover. Protection systems including breakers seldom operate more than two or three times during a storm. Under icing conditions, every insulator is stressed by the same line voltage and weakened by the ice layer bridging its leakage path. Many of them may fail in a short period of time. One utility reported 57 flashovers of their critical 500-kV system in a 2-hour period, compared to a typical annual total of five lightning outages per summer. With so many breaker operations at once, common mode failures can lead to loss of protection or inability to restore power at a station after repeated faults.

The loss of any one EHV substation is a significant contingency. In cases where substations are separated by 200 km or more, utilities may be able to rely on the low probability that an icing event will affect two or more major stations at the same time. This offers some additional flexibility in the insulation coordination process.

10.2. DETERMINISTIC AND PROBABILISTIC METHODS

External insulation should be properly selected and dimensioned for an acceptable risk of flashover. It may be worthwhile to do a probabilistic, or risk-of-failure, assessment. Presently, two general approaches (deterministic and probabilistic) exist for such a purpose. The deterministic approach is used more frequently, but the more complicated probabilistic approach gives more information.

The deterministic method has been widely used for the design of many electrical and mechanical components, apparatus, and systems [CIGRE Task Force 33.04.01, 2000]. Components are then designed according to material selection and dimensioning, including both leakage distance and dry arc distance when considering design of insulators for winter and polluted environments.

The deterministic approach for selecting insulation levels is based on a worst-case analysis and includes safety factors to cover unknowns. It is assumed with this approach that there is, for example, an upper maximum of a site severity factor, such as the applied water conductivity σ_{20} that may stress the insulator when it is fully bridged with heavy accretion. This is shown as the environment "S" curve in Figure 10-2, which represents the declining cumulative probability of exceeding a particular site severity. It is also assumed in the deterministic method that the insulation strength can be described by the "W" curve that rises from zero to unity for certain withstand severity.

In this case, the "W" curve will make use of the withstand stress E_{WS}, expressed in kV/m$_{dry\ arc}$. A dry arc distance will be selected to exceed the maximum stress by a safety margin that is chosen by the designer to cover only the uncertainties in the designer's evaluation of the strength and the stress parameters.

Ice stress as a site severity in Figure 10-2 can be defined by different parameters:

- The electrical conductivity σ_{20} of precipitation alone, based on 100% bridging of the insulator sheds by icicles.
- The thickness of ice accretion alone, for a fixed level of precipitation conductivity σ_{20}.

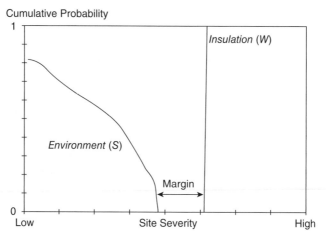

Figure 10-2: An example of deterministic coordination (adapted from CIGRE Task Force 33.04.01 [2000]).

- The icing stress product of conductivity and ice weight per meter of dry arcing distance.

The schematic in Figure 10-2 shows essentially a step change in the function W for the insulator strength as a function of site severity. This is not the case for a single insulator, but is actually rather appropriate when many insulators are considered in parallel. The flashover probability P_m of m parallel insulation points exposed to the same conditions is

$$P_m = 1 - (1 - P_1)^m \tag{10-2}$$

where P_1 is the flashover probability of a single insulation point.

Figure 10-3 from Farzaneh et al. [2005b] illustrates how the probability of flashover as a function of another index of site pollution severity, ESDD, changes when considering 10, 100, or 1000 insulators in parallel. The tendency toward a very steep change in probability of any insulator flashover validates the use of a step-function approximation of the deterministic curve W for insulation strength in Figure 10-2.

Other problems usually remain when a deterministic approach is used when selecting insulation for icing and pollution conditions.

1. The site severity S has a relatively wide distribution that may overlap the strength W with a low cumulative probability of occurrence.
2. There may be correlations among site severity values, for example, heavy accretion may tend to have lower levels of σ_{20}.
3. Different engineers may select different margins, as these can be based on engineering judgment and experience.

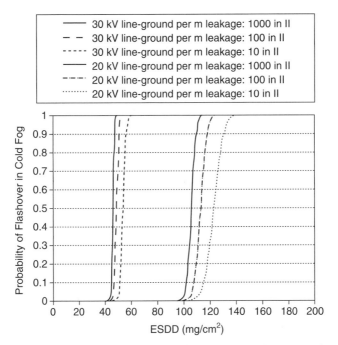

Figure 10-3: Probability of flashover versus ESDD as a function of number of insulators in parallel (∥) (from Farzaneh et al. [2005b]).

A probabilistic approach addresses the main parameters as statistical variables, defined by mean values and dispersions, as opposed to the deterministic method, where the stress and strength parameters are both assumed to be constant. The statistical dimensioning of insulators calls for the selection of the dielectric strength of an insulator, with respect to the voltage and environmental stresses, to fulfill a specific performance requirement. This is done by evaluating the risk of flashover of the candidate insulation options and selecting those with an acceptable performance.

Figure 10-4 shows the probability density function of icing stress product, along with the cumulative probability of insulator flashover, as a function of one measure of ice severity—the applied water conductivity.

The shaded area in Figure 10-4 is calculated using the following expression:

$$IFOR = \frac{N_S}{2} \int_0^{ISP_{max}} f(ISP) \left[1 - \prod_{i=1}^{m} (1 - p_i(ISP)) \right] d(ISP) \qquad (10\text{-}3)$$

where

$IFOR$ is the rate of icing flashovers per year,
N_S is the number of ice storms per year,

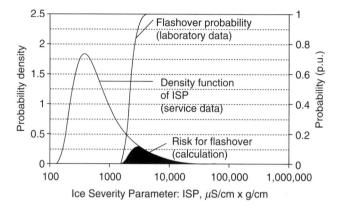

Figure 10-4: Example of probabilistic coordination (adapted from Farzaneh et al. [2005b]).

$f(ISP)$ is the probability density of icing stress product (ISP),

$p_i(ISP)$ is the probability of flashover at a particular insulator for the level of ISP,

ISP_{max} is the maximum value of precipitation water conductivity and accretion that would be expected in the environment, and

m is the number of insulators affected by the ice storm.

In the flashover calculations for icing and polluted environments, it is generally assumed that the insulators are energized at a voltage with constant amplitude, corresponding to the maximum continuous operating voltage. This differs from practice for switching-surge or lightning-surge coordination, where the voltage stress varies widely from event to event, and represents an important simplification.

The variation of the ice environmental stress $f(ISP)$ is represented by a probability density function. The ice thickness and conductivity data for this function come from local site measurements or may be estimated using the derivative of an approximation to the log-normal distribution of the ISP for a median value ISP_{median} of the form:

$$P(ISP) = \frac{1}{1 + \left(\dfrac{ISP}{ISP_{median}}\right)^{\beta}}$$

$$f(ISP) = \frac{dP(ISP)}{d(ISP)}$$

(10-4)

where

$P(ISP)$ is the probability of exceeding an icing stress product ISP,

ISP_{median} is the median value of ISP for the particular site, and

β is a fitted exponent for each site.

The distribution of ISP is in general log-normal: in other words, the logarithm of the ISP is normally distributed. This choice of grading function is common for many one-sided statistical distributions that vary over a wide range, such as lightning peak stroke current amplitude. The ISP is a product of two functions, $f(\sigma_{20})$ and $f(t)$. The function $f(\sigma_{20})$ is, for example,

$$P(\sigma_{20}) = \frac{1}{1 + \left(\dfrac{\sigma_{20}}{\sigma_{median}}\right)^{\beta}}$$

$$f(\sigma_{20}) = \frac{dP(\sigma_{20})}{d(\sigma_{20})}$$

(10-5)

where

$P(\sigma_{20})$ is the probability of exceeding a conductivity σ_{20} at a particular site,
σ_{median} is the median value of σ_{20} for the particular site, and
β is a fitted exponent for each site, with, for example, $\beta = 2.3$ for $\sigma_{median} = 25\,\mu S/cm$ in Farzaneh et al. [2007a].

The probability density function for radial accretion t of ice on conductors often uses a Weibull distribution of the form

$$f(t) = 1 - \exp\left[-\left(\frac{t}{c}\right)^{k}\right]$$

(10-6)

where c and k are fitted parameters. This distribution is complex to use as the parameters c and k are poorly defined and very sensitive. Instead, an expression of the form of Equations 10-4 and 10-5 is recommended for ice thickness:

$$P(t) = \frac{1}{1 + \left(\dfrac{t}{t_{median}}\right)^{1.3}}$$

$$f(t) = \frac{dP(t)}{dt}$$

(10-7)

where t_{median} is the median ice thickness. A mean value of $t = 3\,mm$ is recommended in CIGRE Working Group B2.16 [2006]. This choice, along with the use of an exponent $\beta = 1.3$ in Equation 10-7, corresponds to a coefficient of variation (COV) value of about 0.8 in Figure 10-5.

The variation of the COV in the Weibull distribution can be matched by selecting alternate values of median thickness t for the log-normal probability density function. Farzaneh et al [2007b] describes the calculation of ice weight w from accretion thickness t.

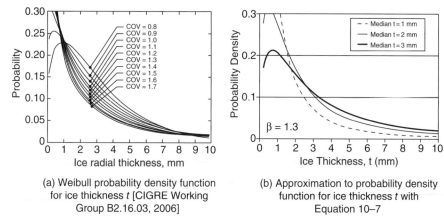

(a) Weibull probability density function
for ice thickness *t* [CIGRE Working
Group B2.16.03, 2006]

(b) Approximation to probability density
function for ice thickness *t* with
Equation 10–7

Figure 10-5: Probability density of ice radial thickness.

10.3. IEEE 1313.2 DESIGN APPROACH FOR CONTAMINATION

The IEEE guide for insulation coordination considers a wide range of stresses, including switching surges, lightning, and contamination. When contamination is present, the response of external insulation to power frequency voltages may dictate external insulation design, compared to the other risks. The specific methods used to evaluate the contamination performance of a line or station are also relevant to the same process for the winter and polluted environments.

The guide considers that insulation must withstand the maximum system voltage over the service life of the equipment. Insulation must either withstand or be protected against temporary overvoltages having a magnitude equal to the short-duration power frequency overvoltages.

Chapter 3 described in detail how precontamination on insulator surfaces can be wetted by light rain, mist, dew, or fog, at accumulation rates that do not wash away the pollution. The wet film is electrically conductive and current flows through the contamination layer as a result of normal service voltage. Dry-band arcing at areas of high current density along insulators may occur. The total insulator phase-ground voltage appears across these small dry bands, and the resulting arcs may grow outward until the insulator may eventually flash over.

In contamination conditions, insulators generally reach equilibrium between dry-band arcing and wetting. Chapter 6 described how the situation tends to change in cold-fog conditions. The thin frozen ice layer stabilizes the pollution, so that it does not wash away, leaving the insulator vulnerable to flashover just at the point that the ice layer melts.

The service voltage stress placed across the insulators is the phase-to-ground power frequency voltage, with no correction for 50 and 60 Hz.

The contamination decreases the insulators' power frequency voltage strength.

The deterministic design rule in IEEE Standard 1313.2 is to set a statistical withstand voltage ($V_{3\sigma}$) equal to the maximum phase-ground voltage (V_{TO}), which includes temporary overvoltages:

$$V_{3\sigma} = V_{50}\left[1 - 3\frac{\sigma}{V_{50}}\right] \qquad (10\text{-}8)$$

where $V_{3\sigma}$ is the withstand voltage for three standard deviations σ away from the median value V_{50}. IEEE Standard 1313.2 assumes, for contamination, that the relative standard deviation of flashover σ/V_{50} is about 10%.

Generally, industrial pollution, including road salting, is more important for winter conditions than the sea-salt exposure. The process for insulation coordination with sea-salt conditions makes use of the fact that the wetting and pollution arrive at the same time. In the winter conditions of major concern, pollution accumulates over dry periods of days or weeks, and the problems tend to occur when wetting occurs at the freezing point. However, several countries such as Japan and Norway have experienced problems with wind transport of sea-salt pollution in snow to areas far from the coast, so this strong source of ions cannot be dismissed in all cases.

Table 10-3 shows the classification of industrial pollution severity using ESDD.

The test results for contamination flashover in Chapter 4 showed that the insulation flashover and withstand stress was relatively constant for a wide range of leakage distance. This means that the use of a critical flashover stress E_{50} in kV/m$_{\text{leakage}}$ is appropriate.

For transmission lines, IEEE Standard 1313.2 [1999] recommends the insulator selection in terms of kV per meter of connection length. This unit scales with the ratio of leakage distance to dry arc distance, so that a 1.02-m connec-

TABLE 10-3: CIGRE and IEEE Classifications of Contamination Site Severity

	ESDD (mg/cm²)	
Site Severity	CIGRE	IEEE
None	0.0075–0.015	—
Very light	0.015–0.03	0–0.03
Light	0.03–0.06	0.03–0.06
Average / moderate	0.06–0.12	0.06–0.10
Heavy	0.12–0.24	>0.10
Very heavy	0.24–0.48	—
Exceptional	>0.48	—

Source: IEEE Standard 1313.2 [1999].

TABLE 10-4: Deterministic Recommendations for Transmission Line Insulator Selection for Contamination Conditions

Parameter	Pollution Level (IEEE 1313.2 Classification)			
	Very Light	Light	Moderate	Heavy
ESDD (mg/cm^2)	0.03	0.06	0.10	0.40
E_{50}, standard disks (kV/m$_{connection}$)	87 (99)	68 (82)	59 (75)	49 (66)
E_{50}, high leakage disks (kV/m$_{connection}$)	91–99	74–88	64–82	56–73

System Voltage (kV)	Number of Standard Disk I Strings (V Strings) for Site Pollution Severity			
	Very Light	Light	Moderate	Heavy
138	6 (6)	8 (7)	9 (7)	11 (8)
161	7 (7)	10 (8)	11 (9)	13 (10)
230	11 (10)	14 (12)	16 (13)	19 (15)
345	16 (15)	21 (17)	24 (19)	29 (22)
500	25 (22)	32 (27)	37 (29)	44 (33)
765	36 (32)	47 (39)	53 (42)	64 (48)

Source: IEEE Standard 1313.2 [1999].

tion length (7×146 mm) corresponds in their case to a 2.04-m leakage distance. (See Table 10-4.)

The good performance of V strings in contaminated environments has also been noted for this configuration in the heavy icing environment.

10.4. IEC 60815 DESIGN APPROACH FOR CONTAMINATION

The International Electrotechnical Commission (IEC) provides international guidance for electrical technology, including testing standards and application guidance. The IEC and IEEE standards are sometimes developed in parallel and, in the best cases, such as IEC 60060 and IEEE Standard 4, important technical details are harmonized.

The advice in IEEE Standard 1313.2 [1999] and the recently adopted IEC 60815 [2008] differ in one important way. As described in Section 10.3, the IEEE considers only the equivalent salt deposit density (ESDD) when evaluating the severity of pollution. There are two factors that change in freezing conditions:

- The ESDD has less conductivity at reduced temperature.
- The insulator wetting process, without washing or removing the contamination, is vastly better in freezing conditions, and is equivalent to high NSDD.

It is possible to adapt the IEC practice to recognize these aspects and to preserve its inherent simplicity in the deterministic output of one of five suitable design stresses for insulator leakage distance. The output is expressed as a unified specific creepage distance (USCD) in units of mm per kV of rms line-to-ground system voltage.

Normally, electric stress is expressed in terms of volts per meter, or for convenience in insulation coordination, kV rms line to ground per meter of leakage distance. Typical values range from $20 kV_{rms}/m_{leakge}$ for heavily contaminated insulators in fully wetted conditions to over $60 kV_{rms}/m_{leakge}$ for clean insulators.

The 1986 version of IEC 60815 set out deterministic recommendations for insulator leakage distance on the basis of a four-level classification of the pollution environment. These were based on the specific creepage distance (SCD), expressed as the number of millimeters needed to give satisfactory service per kilovolt of line-to-line system voltage. As an example, Kawai and Sforzini [1974] recommended SCD values ranging from 12 to 40 mm per kV of line-to-line voltage in Figure 10-6.

The relations among salt-fog, clean-fog, and layer conductivity methods for testing insulators are also shown in the scales of Figure 10-6.

With insulator test voltages normally being expressed in line-to-ground voltage, a factor of $\sqrt{3}$ is often missed in the fine print when applying SCD values. This was rectified in a revised version [IEC Standard 60815, 2008] that adopted units of unified specific creepage distance (USCD) in mm per kV of line-to-ground voltage.

The increased leakage distance requirements for kaolin-coated insulators compared to other wetting methods in Figure 10-6 has evolved over time into a better understanding of the role of nonsoluble deposits in the insulator contamination flashover performance. Figure 10-7 shows a pair of curves that are recommended for application of suspension disks and long-rod insulators to classify site pollution severity in IEC Standard 60815 [2008].

The insulation coordination process in IEC 60815 is relatively simple, once the fundamental work of establishing the expected SPS has been accomplished. For each SPS class, a basic USCD is specified. The recently adopted values are listed in Table 10-5.

The electrical utility community is celebrating the recent adoption of IEC 60815 as an important additional contribution to the selection of suitable insulators on the basis of both ESDD and NSDD.

10.5. CIGRE DESIGN APPROACH FOR CONTAMINATION

The CIGRE advice for selecting insulators was consolidated in Technical Brochure 158 [2000]. A flow chart is given in Figure 10-8 to structure the approach to selection of the appropriate leakage distance and insulator material for a given situation.

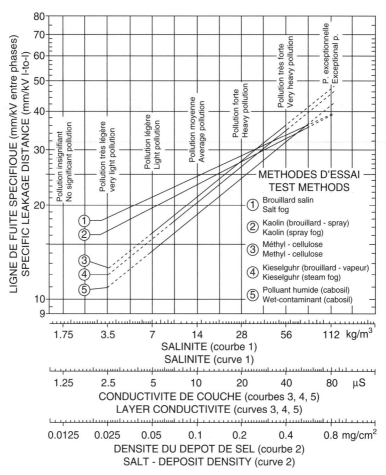

Figure 10-6: Recommended specific creepage distance for insulators in contaminated conditions (from Kawai and Sforzini [1974]; courtesy of CIGRE).

The specific leakage distance varies somewhat, depending on the test method used. Section 3.10 in Chapter 3 contained a detailed analysis of the relations among salt-fog (English) and clean-fog (IEEE) test methods. The German method uses layer conductance rather than ESDD in order to factor out the effect of wetting. (See Table 10-6.)

The provision of empirical relations between the recommended withstand levels and the local salinity, layer conductivity, or ESDD by CIGRE are essential for the basis of a probabilistic design, based on a distribution of site severity levels over a long period of exposure. The models of contamination flashover strength as a function of ESDD, taking into account insulator dimensions, configuration, and material, are derived from test results or theoretical models.

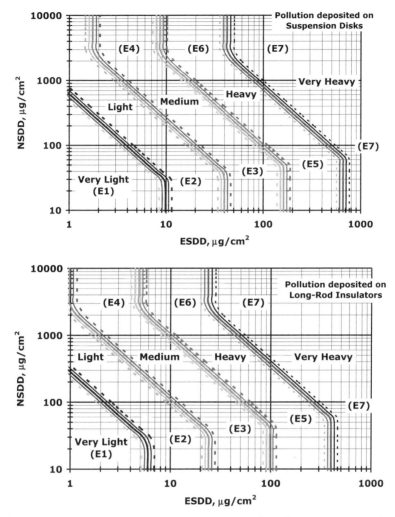

Figure 10-7: Zed-curve approximations to IEC 60815 [2008] classification of site pollution severity (SPS) for naturally exposed insulators. Different curves are under discussion for artificially contaminated insulators.

TABLE 10-5: Reference USCD as a Function of SPS Class

SPS Class	Very Light	Light	Medium	Heavy	Very Heavy
Basic USCD $(mm_{leakage}/kV)$	22	28	35	44	55
Withstand stress $(kV/m_{leakage})$	45	36	29	23	18

Source: IEC Standard 60815 [2008].

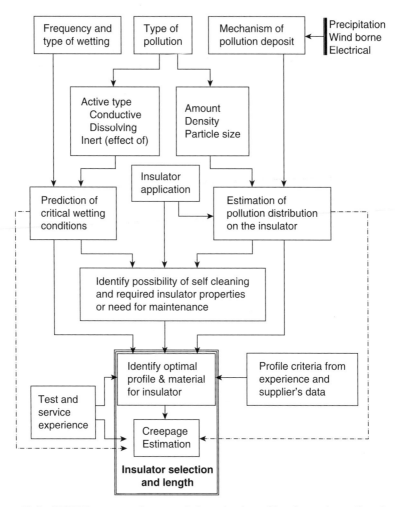

Figure 10-8: CIGRE structured approach for selection of insulators for polluted conditions (from CIGRE Task Force 33.04.01 [2000]; courtesy of CIGRE).

Similar empirical expressions (such as the icing stress product) and mathematical models for leakage and dry arc distance flashover were developed fully in Chapter 6.

The numerical models for flashover stress form the necessary bridge in the insulation coordination process between the constant line voltage and the highly variable ESDD. In winter conditions, ESDD is estimated from the number of hours or days without rain, the distance from pollution sources, the height of insulators above ground, and the wind speed and direction relative to pollution sources.

TABLE 10-6: Comparison of CIGRE and IEEE Withstand Stress (USCD) for Different Levels of Contamination

Origin	England	Germany	Japan	United States/ Canada
Method Stress	Salt fog Salinity S_a (kg/m^3)	Kieselguhr Layer conductivity γ (μS)	Fog withstand ESDD (mg/cm^2)	Clean fog ESDD (mg/cm^2)
Withstand USCD (mm/kV$_{l\text{-}g}$)	2.34 $S_a^{0.224}$	1.42 $\gamma^{0.387}$	7.14 ESDD $^{0.246}$	73 ESDD $^{0.33}$
ESDD	Unified Specific Creepage Distance, USCD (mm per kV$_{l\text{-}g}$)			
0.03	32	22	30	23
0.06	38	29	36	30
0.10	42	35	41	34
0.40	58	60	57	41

Source: IEEE Standard 1313.2 [1999].

To use the CIGRE approach for calculating the outage rates of overhead line or station insulators, a utility would need to establish the following input parameters:

- Flashover stress for the selected insulators derived from laboratory tests and interpolated with numerical models such as those in Chapter 6.
- Estimation of pollution levels on the insulators, using the guidance in the next sections.
- Number of flashover threats per year, causing the critical wetting conditions in Figure 10-8. This will be based on the occurrence of fog or icing events near 0 °C.

The experimental evidence is that insulator profile plays only a minor role in the performance of an insulator in conditions with precontamination and very light icing severity. However, profile plays a vital role in the performance in moderate icing severity, which is defined as the range where uniform sheds with 50-mm separation are fully bridged with icicles, but standard cap-and-pin suspension disks are not.

10.6. CHARACTERISTICS OF WINTER POLLUTION

Chapter 3 showed that it was possible to use existing resources for environmental monitoring to establish maps of pollution intensity. The maps of wet deposition of important pollutants such as sulfur and nitrogen provide a sound basis for calculating median precipitation conductivity σ_{20}. This value is used, along with the distribution of the intensity of icing events, to select a suitable insulator dry arc distance.

While there are fewer measurement sites, there was a good correlation between the dry and wet deposition rates of pollution. The dry deposition rate, expressed, for example, as kilograms of sulfur per hectare per year, can be converted directly into a rate of accumulation of top-surface deposit on insulator surfaces. In the absence of washing, the rate of accumulation multiplied by the exposure duration gives a general numerical basis for selecting the leakage distance of insulators.

In cold climate regions, the longest duration of exposure will likely occur in the winter. Also, the basic dry deposition rate of pollution may be supplemented with additional flux of sodium chloride depending on proximity to roads.

10.6.1. Days Without Rain in Winter

In the United States, there are extensive regions where the mean daily maximum temperature for the entire month is less than $0\,°C$. Any precipitation during these months will fall as snow or freezing rain. This means that pollution can build up on insulator surfaces without washing for extended periods. For areas in Wisconsin, northern Michigan, Minnesota, and North Dakota, indicated in Figure 10-9, there are a total of at least $(31 + 31 + 28)$ or 90 days without rain, on average, every year.

Another indicator of the duration of the winter "desert" condition is provided by the climate map in Figure 10-10 showing the number of days with snow depth greater than $25\,mm$ ($1\,in.$). The period of snow cover is also the period when insulator contamination levels may build up in the absence of rain. Snow cover also implies salt cover on roads in many urban areas, and the road salt deposits can dominate the ESDD buildup rate on exposed insulators.

Figure 10-10 shows as an example that there are about 20 states in the United States with white areas that indicate a snow duration period in excess of 84 days. These are the states where the winter periods without rain may establish the maximum contamination levels on insulators.

10.6.2. Rate of Increase of ESDD

Where it is available, the dry deposition rate of important pollution ion species, notably sulfur and nitrogen, serves as a good estimate of the background level of pollution accumulation on upward-facing insulator surfaces. As noted with the field measurements in Chapter 3, the top-surface ESDD tends to rise to values much greater than bottom-surface levels in winter conditions. Once a cleansing rain event occurs, the top-surface ESDD falls much more quickly as well.

Figure 10-9: Factor in winter pollution exposure: mean daily temperature <0 °C (data from NOAA).

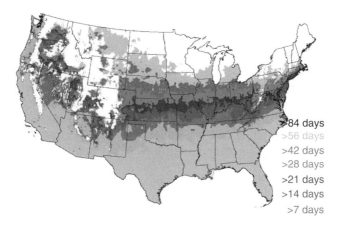

Figure 10-10: Number of days with snow depth greater than 25 mm (data from NOAA).

Spot checks, also described in Chapter 3, have also illustrated that the ion content of insulator surface contamination tends to match that found in wet deposition samples taken over the same exposure period. For the eastern coast of North America, Figure 10-11 shows the measurements of wet deposition

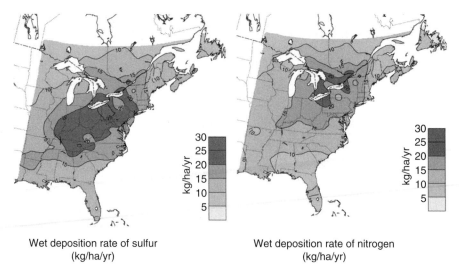

Wet deposition rate of sulfur
(kg/ha/yr)

Wet deposition rate of nitrogen
(kg/ha/yr)

Figure 10-11: Spatial distribution of wet deposition rates for sulfur and nitrogen, 1996–2001 (from Meteorological Service of Canada [2004]; courtesy of the Government of Canada).

rate, expressed in kilogram per hectare per year. For this region, Chapter 3 showed that the dry deposition could be predicted from the wet deposition rate, using a conversion factor of about 50–60%. This means that the wet deposition rate of 25 kg/ha/yr would convert to a dry deposition rate of about 13 kg/ha/yr. Using the conversion factors of 365 days per year, $10^9 \mu g/kg$, and $10^8 cm^2/ha$, the dry deposition rate converts to $0.36 \mu g/cm^2$ per day of sulfur. With a 90-day dry period, a top-surface deposit of $0.032 mg/cm^2$ ($32 \mu g/cm^2$) would be expected strictly from sulfur deposition.

A similar amount would be expected from nitrogen, with additional contributions from sodium and chloride being important near the sea and man-made salt sources. After correction for molecular weight and specific conductance of each ion, the deposition rates in Figure 10-11 can be expressed as a background level of ESDD accumulation, in the same recommended units of $\mu g/cm^2/day$.

Urban areas collect considerably more pollution than rural areas. Figures 10-12 and 10-13 compare the particulate concentrations in several sites through the United States [Meteorological Service of Canada, 2004].

The rural/urban ratio in these two figures is about 3:1. The measurements of particulate matter with 2.5-μm size ($PM_{2.5}$) particles are less reliable indicators of the rate of increase of ESDD as direct measurements of dry deposition, but there are many more sampling sites to facilitate comparisons. Thus it is reasonable to infer that the escalation rate for ESDD, based on the equivalent conductance of sulfate, nitrate, ammonia, and other ions, will also be higher in the urban regions than the background level for rural regions.

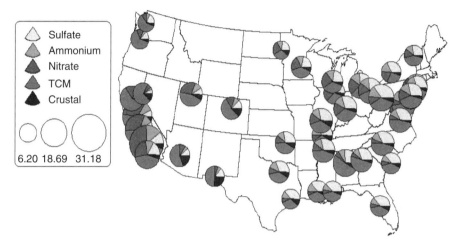

Figure 10-12: Urban ion species in PM$_{2.5}$ data for the United States. Size of pie graphs indicates average PM$_{2.5}$ concentration (from Meteorological Service of Canada [2004]; courtesy of the Government of Canada).

Figure 10-13: Rural ion species in PM$_{2.5}$ data for the United States. Size of pie graphs indicates average PM$_{2.5}$ concentration (from Meteorological Service of Canada [2004]; courtesy of the Government of Canada).

10.6.3. Effect of Road Salt

Where measurements of pollution have been carried out in urban areas in winter, the role of sodium chloride deposition as a result of nearby road salting has been strong. Typically, as shown in Chapter 3, the spray term has a distance constant of about 20m. Figure 10-14 shows that these salt contributions can

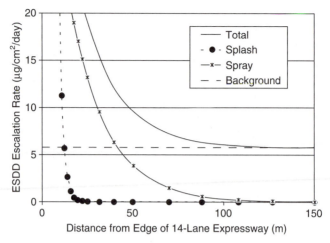

Figure 10-14: Contributions of salt spray and splash to rate of escalation of ESDD.

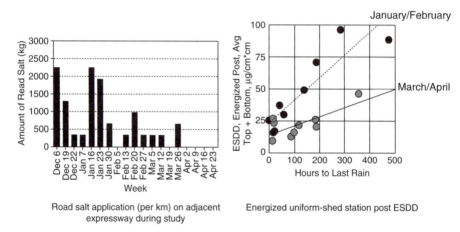

Road salt application (per km) on adjacent expressway during study

Energized uniform-shed station post ESDD

Figure 10-15: Relation of energized station post ESDD to hours without rain in winter for station near expressway (from Chisholm et al. [1993]).

double the rate of increase of ESDD for those parts of a substation at ground level that are located closer than 50 m from the edge of a 14-lane expressway.

The contribution of salt to the rate of increase of ESDD is very strong in the peak road-salting season of January and February.

It is always best to have direct measurements of the rate of increase of ESDD in winter, like the ones shown in Figure 10-15, when designing a new station or selecting remediation options for an existing one. Using a parallel example, it is always best to carry out a resistivity survey prior to specifying the ground grid design of a substation. For the case of line design, the process described in this section allows a rough estimation of performance. The parallel process in calculating lightning outage rates is to use a constant value of soil resistivity along the line route, rather than obtaining measured values of resistance or resistivity at each tower base.

10.7. WINTER FOG EVENTS

Overall, the IEEE Task Force summarized the incidence of freezing fog or fog at a temperature less than $0\,°C$ in Figure 10-16.

The "normal" fog activity in an area depends on the local climate. Near the seacoast, the freezing rain and freezing fog exposure is very much a function of the distance from shore. This is seen, for example, in the high incidence of fog around the Gulf of Mexico in Figure 10-17.

The incidence of winter fog in the Northeast and Central United States and most of Canada is of greater relevance for dimensioning insulators for winter conditions. Generally, every winter month has about 3 days of dense fog along the Atlantic coast and near the Great Lakes. The probability of fog events is fairly uniform throughout the winter in most regions, meaning that any extended cold period will experience several fog events that lead to much greater risk of insulator leakage distance flashover.

In Canada, the climatology of fog for each first-order weather station is presented as a three-dimensional plot of time of day, month of year, and prob-

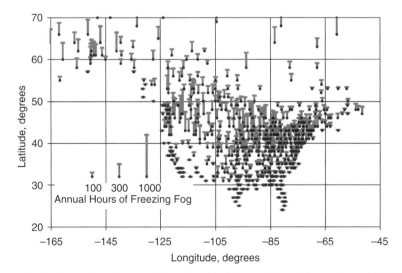

Figure 10-16: Annual incidence of freezing fog (adapted from Farzaneh et al. [2005a]).

Figure 10-17: Number of days per month with dense fog (data from NOAA).

ability of occurrence. This form of presentation in Figure 10-18 is very helpful for establishing that winter fog in the sensitive March or April periods tends to occur in the 6 hours before sunrise.

10.8. FREEZING RAIN AND FREEZING DRIZZLE EVENTS

It is pointless to evaluate the icing performance of an insulator in areas that are too warm to have significant accretion. However, a combination of light pollution and light icing in the range of 1–6-mm accretion on a reference cylinder (2–12 mm at ground level) can lead to widespread problems on EHV insulators. Any area that has a positive value in an ice-loading map for mechanical design should also consider the return period of icing and pollution events when selecting appropriate insulation for EHV systems.

10.8.1. Measurement Units

Accumulation of freezing rain is measured at ground level by meteorological services. Electrical utilities have empirical rules that relate meteorological

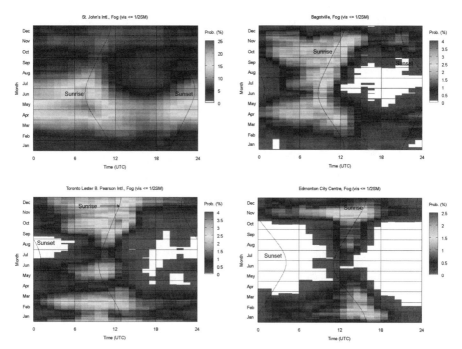

Figure 10-18: Probability of fog occurrence with visibility <800 m versus time of day and day of year for four sites in Canada (from Hansen et al. [2007]; reproduced by permission of the Government of Canada).

values to the size and weight of equivalent radial ice thickness t on conductors. A common rule-of-thumb is that the accretion on the ground will be twice as much as the radial accumulation on a conductor. There can be some deviation from this rule, as conductors will twist as ice accumulates, so accurate measurement uses the radial accretion on a 25-mm cylinder that rotates at 1 rpm. As an example, in the 1998 ice storm, the peak ice accumulation at ground level in Québec of 100 mm led to an equivalent peak radial accumulation of 75 mm on overhead groundwires and single conductors of transmission lines.

10.8.2. Frequency of Occurrence

The first indicator of occurrence of icing events consists in the climatology of the conditions that may lead to icing. In Europe, the COST 727 program has established that areas where the ambient temperature at a height of 200 m is less than 0 °C, and the cloud height is also less than 200 m, are at risk of in-

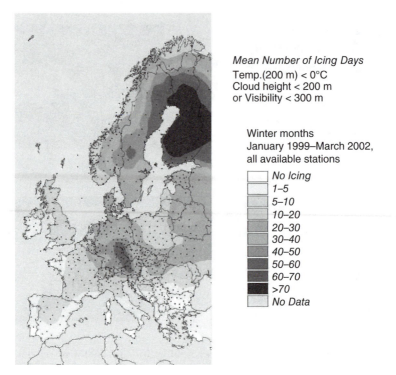

Mean Number of Icing Days
Temp.(200 m) < 0°C
Cloud height < 200 m
or Visibility < 300 m

Winter months
January 1999–March 2002,
all available stations

☐ *No Icing*
☐ *1–5*
☐ *5–10*
☐ *10–20*
☐ *20–30*
☐ *30–40*
☐ *40–50*
☐ *50–60*
☐ *60–70*
■ *>70*
☐ *No Data*

Figure 10-19: Mean number of "icing days" in Europe (from Tammelin and Fikke [2007]).

cloud icing. Figure 10-19 [Tammelin and Fikke, 2007] shows the mean annual number of icing days based on these criteria.

Cortinas et al. [2004] studied the actual occurrence of freezing precipitation in North America. They found that freezing rain, freezing drizzle, and ice pellets were associated most frequently with surface temperatures slightly less than 0°C, as shown in Figure 10-20.

Figure 10-21 shows some areas where there are a median of 40h/year of freezing rain, and also regions where there are more than 80h/year of freezing drizzle.

The design figure of greatest interest for light and moderate icing conditions is the overall number of hours of exposure to freezing rain and/or freezing drizzle. The contours of this parameter are given in Figure 10-22.

Similar climate maps of freezing precipitation for other regions and for the entire world are in preparation as technical support for the monitoring of global climate change. Like lightning at a dew point of more than +24°C, the occurrence of freezing rain in the ambient temperature range of –2°C to 0°C is a sensitive indicator as well as a risk factor.

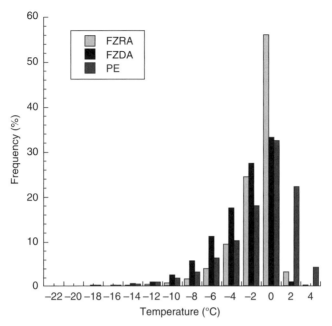

Figure 10-20: Ambient temperature at 2 m above ground level for three forms of freezing precipitation: freezing rain (FZRA), freezing drizzle (FZDZ), and ice pellets (PE) (from Cortinas et al. [2004]).

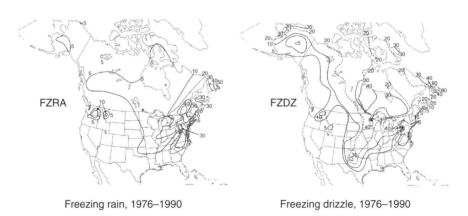

Freezing rain, 1976–1990   Freezing drizzle, 1976–1990

Figure 10-21: Median annual hours of freezing rain and freezing drizzle (from Cortinas et al. [2004]).

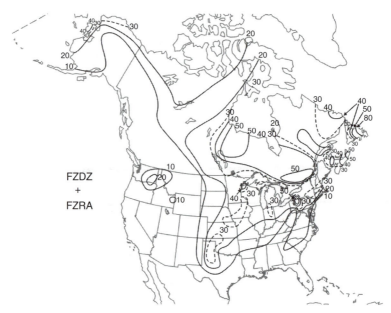

Figure 10-22: Median annual hours of freezing rain and/or freezing drizzle (from Cortinas et al. [2004]).

10.8.3. Time of Day and Time of Year of Freezing Precipitation Occurrence

Hansen et al. [2007] plotted the daily variation of freezing rain and drizzle against monthly variation and superimposed sunrise and sunset in the graphs shown in Figure 10-23. Two of the four sites show some tendency toward occurrence of freezing rain a few hours before sunrise. A similar trend was noted in the averaged climate results presented by Cortinas et al. [2004].

10.8.4. Severity of Freezing Rain Occurrence

Jones et al. [2002] evaluated the combined climatology of freezing rain events and wind speeds, to develop an icing map in Figure 10-24 that forms part of overhead power system design standards such as the National Electrical Safety Code [IEEE/ANSI Standard C2, 2007].

The study by Jones et al. [2002] is focused mainly on the accretion of ice on horizontal conductors and is thus the strongest basis for evaluating the accumulation of ice on insulators, using the same measurement method. Line design is based on the same uniform radial thickness t that will accumulate on a rotating reference cylinder, and this level will be about half of the actual accumulation of freezing rain at ground level. Thus the areas in Figure 10-24 with 6 mm ($\sim\frac{1}{4}$ in.) of accretion should use insulators that can withstand light

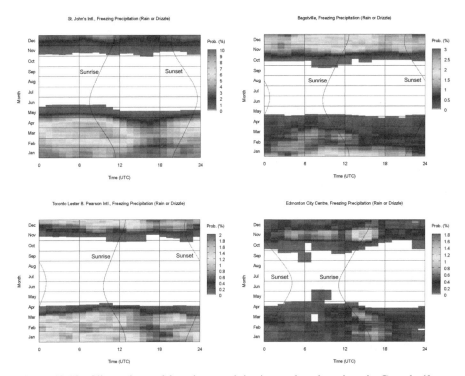

Figure 10-23: Climatology of freezing precipitation at four locations in Canada (from Hansen et al. [2007]; courtesy Government of Canada).

ice accretion, and areas noted with 13 mm ($\sim\frac{1}{2}$ in.) of accretion should use insulators that will withstand heavy icing severity.

10.8.5. Electrical Conductivity of Freezing Rainwater

Generally, the conductivity of melted samples of fresh snow and freezing rain have the same wide distribution of conductivity σ_{20} as ordinary rain. A series of 383 rain samples from measurements in Japan [Fujimura and Naito, 1982] had a median value of 15 kΩ-cm ($\sigma_{20} = 67\,\mu S/cm$) along with the distribution of values shown in Figure 10-25.

The raw distribution can be fitted with a good approximation to a cumulative log-normal distribution of σ_{20} as follows:

$$P(\sigma_{20}) = \frac{1}{1 + \left(\dfrac{\sigma_{20}}{67\,\mu S/cm}\right)^{2.3}} \tag{10-9}$$

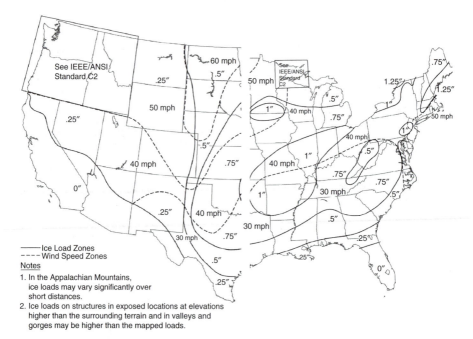

Figure 10-24: Uniform radial ice thicknesses t due to freezing rain, with concurrent 3-s gust wind speed, for a 50-year return period (adapted from Jones et al. [2002]).

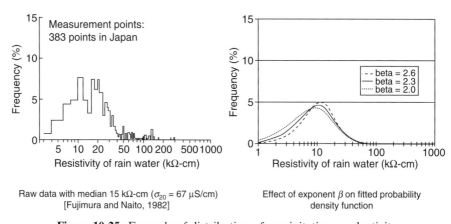

Figure 10-25: Example of distribution of precipitation conductivity.

Figure 10-25 shows that the exponent $\beta = 2.3$ in Equation 10-9 approximates the raw data best, but also illustrates that the choice of exponent is not sensitive. This is a major advantage of the form of this equation over other multiparameter grading formulas. Other areas in Japan further from the sea have median precipitation conductivity as low as $14\,\mu S/cm$, with the same value of β.

10.9. SNOW CLIMATOLOGY

With its important economic and climate roles, observations of snow have been part of standard operating procedures for meteorologists in many countries. Snow observations are important to manage the risk of air and ground transport accidents through deicing procedures. They are also used to assess the quantity of spring runoff that can be expected, both to manage the risk of floods and also to evaluate the amount of energy that is likely to become available through hydroelectric resources.

Snow is less of an electrical risk to power systems than ice, but the presence of snow also signals the presence of road salt in many urban regions. Thus aspects of snow climatology are relevant to selection of leakage distance as well as dry arc distance.

10.9.1. Standard Methods for Snow Measurements

Several elements must be considered under the heading "snow cover." These were defined in Section 8.1 as snow extent, snow depth, snow duration, and snow water equivalent.

Requirements of Global Climate Observing System (GCOS) stations call for point measurements of daily snow depth at ground station network stations. With a wide range of domestic needs for snow data, measurements are often taken hourly or more frequently with automatic snow depth sensors. Manual measurements are taken with a snow ruler or similar graduated rod, pushed carefully through all the snow and ice layers to the ground surface. Several depth measurements are averaged in open areas where the snow cover undergoes drifting and redistribution by the wind.

In remote regions, aerial snow depth markers visible from binoculars, telescopes, or aircraft may be used. These generally are vertical poles with horizontal colored crossarms mounted at fixed heights, oriented with reference to the point of observation. The depth of snow at the stake or marker may be observed from distant ground points or from aircraft by means of binoculars or telescopes. The stakes or markers are painted white to minimize the undue melting of snow immediately surrounding them.

Snow intensity is classed as *light*, *moderate*, or *heavy* depending on the accumulation rate of water equivalent i and the visibility; see Table 10-7. Depending on the air temperature, the snow may take a number of forms including snow (SN), snow pellets (SHGS), snow grains (SG), or ice crystals (IC). Solid hydrometeors that are not considered to be snow include ice pellets (PL or IP), hail (SHGR), and graupel (GS).

10.9.2 Snow Accumulation and Persistence

Snow tends to occur in the local winter when ambient temperature is below freezing at ground level. Between 30% and 40% of the Earth's land surface

TABLE 10-7: WMO Standards for Snow Observations

Classification	Precipitation Intensity, i (mm/h water equivalent)	Visibility, V (m)	Flakes
Light	$i > 1$	$V \geq 1000$	Small and spare
Medium	$1 \leq i < 5$	$400 > V \geq 1000$	Larger, more numerous
Heavy	$i \geq 5$	$V \leq 400$	Numerous of all sizes

Source: WMO [2001].

may be covered seasonally by snow. In the Northern Hemisphere, half of the land surface has snow cover in winter. Snow is highly reflective and changes the surface albedo, or response to solar input. Albedo of snow is the ratio of the reflected to the incoming solar radiation. It can be about 80% for fresh snow cover and may drop below 40% as the snow surface becomes weathered or dirty. With a large area of coverage, snow cover is an important term in calculations of the radiation balance of the Earth.

Snow also acts as a thermal and moisture sink, modifying the surface temperature and humidity. Snow cover has a number of important physical properties that exert an influence on global and regional energy, water, and carbon cycles. Snow makes up a significant fraction of the water available for agriculture, hydroelectricity, and potable water supply in many semiarid regions of the world. Changes in snow cover conditions can have serious economic and social impacts. Hence regular monitoring of snow cover, including extent, depth, and water equivalent, is a high priority activity for global climate monitoring.

Global observations of snow pack are relatively easy to make, compared to many other weather parameters. The very first image from the Television and Infrared Observation Satellite (TIROS-1) in 1960 showed snow-covered regions. With the large difference in albedo, satellite images can easily be processed to distinguish snow cover from other natural surfaces. Also, since snow persists for days, it is also feasible to average over the effects of cloud cover, which also affects the albedo. Improved vidicon and radiometer sensors have been deployed by NASA and analyzed, leading, for example, to a US National Weather Service atlas of snow cover [Matson et al., 1986].

The moderate resolution imaging spectroradiometer (MODIS) instrument is presently used to gather intensity data in 36 spectral bands. Snow cover is mapped from some of the MODIS spectral intensity data, along with a normalized difference snow index analysis and other data filters [Roy et al., 2002]. The MODIS data are available in several digital formats and resolutions [Hall et al., 2006], including those shown in Figures 10-26 to 10-28.

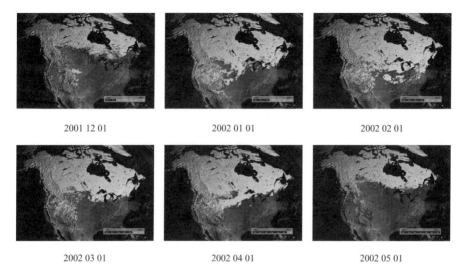

2001 12 01   2002 01 01   2002 02 01

2002 03 01   2002 04 01   2002 05 01

Figure 10-26: Typical evolution of snow cover in North America, observed from MODIS Satellite (from Hall et al. [2006]; courtesy of NASA/Goddard Space Flight Center Scientific Visualization Studio).

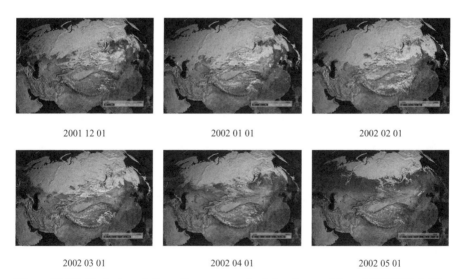

2001 12 01   2002 01 01   2002 02 01

2002 03 01   2002 04 01   2002 05 01

Figure 10-27: Typical evolution of snow cover in Asia, observed from MODIS Satellite (from Hall et al. [2006]; courtesy of NASA/Goddard Space Flight Center Scientific Visualization Studio).

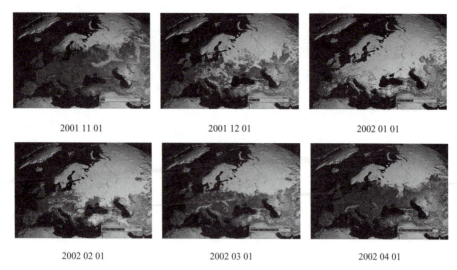

2001 11 01   2001 12 01   2002 01 01

2002 02 01   2002 03 01   2002 04 01

Figure 10-28: Typical evolution of snow cover in Europe, observed from MODIS Satellite (from Hall et al. [2006]; courtesy of NASA/Goddard Space Flight Center Scientific Visualization Studio).

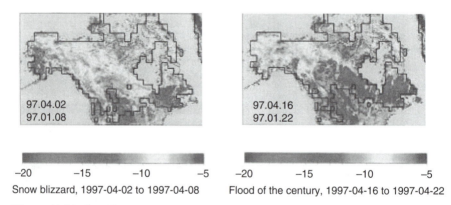

Snow blizzard, 1997-04-02 to 1997-04-08   Flood of the century, 1997-04-16 to 1997-04-22

Figure 10-29: Satellite observations of Ku-band backscatter over snow cover (from Ngheim and Tsai [2001]).

10.9.3. Snow Melting

The electrical danger of snow is greatest at ambient temperatures near the melting point, when wet snow density and electrical conductivity are elevated.

Global remote sensing methods, such as the backscatter from Ku-band radar illumination at 13.4–14.6 GHz, have been used in Figure 10-29 to establish snow depth and wetness as well as the extent [Ngheim and Tsai, 2001].

Identifying the onset of snow melting is important both for predicting floods, such as the "flood of the century" in mid-April 1997, and also for establishing when snow on electrical insulators will present the greatest risk of flashover.

10.10. DETERMINISTIC COORDINATION FOR LEAKAGE DISTANCE

The conditions of very light and light icing severity, including cold-fog events, can be treated using the same deterministic coordination methods as those described for contaminated insulators in Section 10.4

Table 10-8 shows the conversion of USCD to stress, and adds calculations of the clean-fog and cold-fog ESDD that would correspond to the 50% flashover levels for each of these values. The critical ac flashover stresses E_{50} for standard porcelain disks and posts are

$$E_{50}^{clean\ fog} = 12.8 \cdot (ESDD)^{-0.36}$$

$$E_{50}^{cold\ fog,post} = 12.7 \cdot (ESDD)^{-0.36} \tag{10-10}$$

$$E_{50}^{cold\ fog,pin\ type} = 18.6 \cdot (ESDD)^{-0.36}$$

where $ESDD$ is in units of mg/cm^2 and E_{50} is given in kV rms of line-to-ground voltage.

TABLE 10-8: Discussion Values of USCD Versus SPS in IEC 60815 for Standard Ceramic Disk Insulators Compared with Artificial Test Laboratory Results

Site Pollution Severity (SPS) Class	Very Light	Light	Medium	Heavy	Very Heavy
ESDD at NSDD of 50–60 μg/cm^2 mg/cm^2 (μg/cm^2)	<0.06 (<6)	0.02 (20)	0.08 (80)	0.2 (200)	>0.6 (>600)
USCD (mm$_{leakage}$/ kV$_{line\ to\ ground}$)	22	28	35	44	55
Line voltage operating stress (kV/m$_{leakage}$)	45.5	35.7	28.6	22.7	18.2
ESDD causing 50% flashover, ceramic insulators in clean-fog-test[a] mg/cm^2 (μg/cm^2)	0.03 (30)	0.058 (58)	0.107 (107)	0.203 (203)	0.377 (377)
ESDD causing 50% flashover, ceramic disks in cold-fog test[a] mg/cm^2 (μg/cm^2)	0.084 (84)	0.163 (163)	0.304 (304)	0.573 (573)	1.065 (1065)

[a]Test values at NSDD of 0.05–0.06 mg/cm^2 (50–60 μg/cm^2) using 40 g/L of kaolin.

For silicone-coated ceramic and silicone polymer insulators, the E_{50} stresses for cold-fog flashover increase by about 25% compared to the corresponding values for ceramic insulators. The limited evidence to date suggests that the value of E_{50} for EPDM polymer post insulators is less than or equal to the value for ceramic posts.

Tables 10-8 and 10-9 invert the E_{50} stresses for ceramic disks and station posts using suggested IEC values [IEC 60815, 2008] for the deterministic selection of USCD on disks and long-rod porcelain insulators.

For heavy pollution on ceramic disks, the IEEE recommendations [Baker et al., 2008] suggest that a USCD of 55 mm/kV will not be sufficient.

The tables do not show a consistent margin in the corresponding levels of ESDD for all pollution severity levels. The pollution level needed to cause flashover in artificial fog tests at a USCD of 22 mm/kV is much higher than suggested by the IEC classification. For high pollution levels, the cold-fog flashover stress on station posts matches well with the long-rod flashover stress in clean fog.

10.11. PROBABILISTIC COORDINATION FOR LEAKAGE DISTANCE

The process for probabilistic coordination of leakage distance has the following steps:

- Establish the desired reliability level, in outages per 100 km of line length per year for overhead lines or in mean time between failures for stations.
- Invert the probability of any of m insulators in parallel having a flashover to establish the design probability for a single insulator.

TABLE 10-9: Discussion Values of USCD Versus SPS in IEC 60815 for Ceramic Long-Rod Insulators Compared with Artificial Test Laboratory Results

Site Pollution Severity (SPS) Class	Very Light	Light	Medium	Heavy	Very Heavy
ESDD at NSDD of 50–60 µg/cm² mg/cm² (µg/cm²)	<0.003 (<3)	0.015 (15)	0.04 (40)	0.17 (170)	>0.4 (>400)
USCD (mm of leakage per kV of line-to-ground voltage)	22	28	35	44	55
Line voltage operating stress (kV/m$_{leakage}$)	45.5	35.7	28.6	22.7	18.2
ESDD causing 50% flashover, ceramic disks in cold-fog test[a] mg/cm² (µg/cm²)	0.029 (29)	0.057 (57)	0.105 (105)	0.20 (200)	0.37 (370)

[a]Laboratory test values at NSDD of 0.05–0.06 mg/cm² (50–60 µg/cm²) using 40 g/L of kaolin.

- Take an initial estimate of the required leakage distance using an empirical model for the insulators being evaluated.
- Determine the incidence of flashover weather, consisting of the number of fog, freezing rain,and/or freezing drizzle events per year that have a melting period. It is conservative to assume that every precipitation event has a melting phase, so a climate study should be carried out.
- Approximate the distribution of the number of dry days prior to flashover weather with a function $P(n_{dry})$. The data for this distribution can be read out of the same climate study for flashover weather incidence, or a suitable distribution can be constructed from observations of the duration of snow cover or days with maximum temperature below $0\,°C$.
- Calculate the ESDD corresponding to each of the dry periods. The daily rate of increase of ESDD may be obtained using site measurements taken in winter conditions, or alternately estimates can be derived from background measurements of dry or wet deposition and/or distance from expressways with road salting.
- Calculate the probability of any insulator flashover for each flashover weather event, based on the estimated ESDD, Equation 10-10 for critical flashover in cold-fog conditions, and the trial value of insulator leakage distance.
- Adjust the insulator leakage distance until the desired line or station reliability is achieved.

10.12. DETERMINISTIC COORDINATION FOR DRY ARC DISTANCE

The process for deterministic coordination of a suitable dry arc distance consists in establishing the design ice load for a 50-year return period, evaluating the quantity of ice that this represents on the insulators in grams per centimeter of dry arc distance, selecting a reference level of applied water conductivity, calculating the icing stress product, evaluating the flashover strength of the ice layer, and taking a suitable statistical margin below this strength for the maximum service voltage stress. A similar process is followed for evaluating the effects of heavy snow on horizontal insualtors, based on the average maximum annual snow cover depth.

10.12.1. Dry Arc Distance Requirements for Icing Conditions

The choice of a 50-year return period for icing events in Figure 10-24 is well established in transmission design. Longer return periods are often assumed for other substation reliability problems. For example, lightning protection of substations makes use of a MTBF of 200–500 years. A case study given later shows how one utility with a 50-year return period value of 75 mm on a fixed cylinder (37 mm on a rotating cylinder) adjusts for different return periods.

As shown in Chapter 6, the electrical effects of heavy ice accretion tend to level off for ice thickness of more than 25 mm on a rotating cylinder, corresponding to 50 mm (2 in.) on fixed cylinders. Additional ice accretion above this level does not influence selection of a suitable dry arc distance.

Selection of insulator dry arc distance without a reasonable estimate of equivalent radial ice thickness is a compromised process, similar to calculating the lightning performance of a line with a global or continental average of ground flash density rather than a local value. In the absence of precise data, a level of ice accretion $t = 15$ mm may be used for areas where previous ice-related mechanical or electrical failures have been experienced.

The quantity of ice w that accumulates on a vertical insulator is a function of its diameter and shed spacing. For deterministic models, the half-circumference of the insulator, multiplied by the ice thickness t and a glaze ice density of 0.9 g/cm^3, is a good choice for this weight w in units of $\text{g/cm}_{\text{dry arc}}$.

One default value of ice or snow conductivity is the same value that is used for rain conductivity in wet tests: $\sigma_{20} = 100 \,\mu\text{S/cm}$. This is a practical choice for insulation testing in icing conditions [Farzaneh et al., 2003] and also represents a conservative choice for natural conductivity at the 5% probability level [Farzaneh et al., 2005a, 2007a]. See Table 10-10.

Rural areas may be designed with a level of $\sigma_{20} = 30 \,\mu\text{S/cm}$ if there are no sources of pollution or road salt nearby. See Table 10-11.

If the local distribution of precipitation conductivity is known, it should form the basis of a probabilistic coordination of dry arc distance, using the process described later in Section 10.13.

10.12.2. Dry Arc Requirements for Snow Conditions

For deterministic selection of a suitable insulator dry arc distance, the quantity of snow that accumulates on horizontal insulators is established by the average

TABLE 10-10: Recommended E_{WS} for Icing Conditions with $\sigma_{20} = 100 \,\mu\text{S/cm}$

Radial Ice Accretion t, mm (inches)	6 (0.25)	13 (0.5)	19 (0.75)	25 (1.0)	32 (1.25)
w, 100-mm polymer, $\text{g/cm}_{\text{dry arc}}$	9	18	27	36	45
w, 254-mm disks/posts, $\text{g/cm}_{\text{dry arc}}$	23	46	68	91	114
E_{50}, 100-mm polymer, $\text{kV/m}_{\text{dry arc}}$	109	95	88	84	80
E_{50}, 254-mm disks/posts, $\text{kV/m}_{\text{dry arc}}$	91	80	74	70	67
E_{WS}, 100-mm polymer, $\text{kV/m}_{\text{dry arc}}$	95	83	77	73	70
E_{WS}, 254-mm disks/posts, $\text{kV/m}_{\text{dry arc}}$	79	69	64	61	58

TABLE 10-11: Recommended E_{WS} for Icing Conditions with $\sigma_{20} = 30\,\mu S/cm$

Radial Ice Accretion t, mm (inches)	6 (0.25)	13 (0.5)	19 (0.75)	≥25 (≥1.0)
w, 100-mm polymer, g/cm$_{\text{dry arc}}$	9	18	27	36
w, 254-mm disks/posts, g/cm$_{\text{dry arc}}$	23	46	68	91
E_{50}, 100-mm polymer, kV/m$_{\text{dry arc}}$	137	120	111	105
E_{50}, 254-mm disks/posts, kV/m$_{\text{dry arc}}$	115	100	93	88
E_{WS}, 100-mm polymer, kV/m$_{\text{dry arc}}$	119	104	97	91
E_{WS}, 254-mm disks/posts, kV/m$_{\text{dry arc}}$	100	87	81	77

TABLE 10-12: Recommended E_{WS} for Snow Conditions with $\sigma_{20} = 30\,\mu S/cm$

Snow Thickness, cm (inches)	10 (4)	20 (8)	30 (12)	50 (20)
w, 100-mm polymer, g/cm$_{\text{dry arc}}$	30	60	90	150
w, 254-mm disk string, g/cm$_{\text{dry arc}}$	76	152	229	381
w, 700-mm twin disk string, g/cm$_{\text{dry arc}}$	210	420	630	1050
E_{WS}, 100-mm polymer, kV/m$_{\text{dry arc}}$	165	144	134	121
E_{WS}, 254-mm disk string, kV/m$_{\text{dry arc}}$	138	121	112	102
E_{WS}, 700 mm twin disk string, kV/m$_{\text{dry arc}}$	114	100	92	84

annual maximum snow cover depth, D_{max}, multiplied by the insulator diameter. The quantity is more important for double strings of disk insulators, with separation that is less than D_{max}, as the entire space between the chains also fills with snow. The weight of the snow per cm of dry arc distance is calculated from the snow cross-sectional area (cm^2) and the density of the snow, $0.3\,g/cm^3$.

The default value of snow conductivity should represent the median value of precipitation conductivity, and a value of $\sigma_{20} = 30\,\mu S/cm$ is typical in many regions. Table 10-12 shows that the E_{WS} requirement for heavy snow accretion on clean horizontal insulators is not as onerous as the requirement for heavy ice accretion on vertical insulators in Table 10-10.

This suggests that an adequate selection of dry arc distance for ice conditions will also be suitable for the likely amounts of snow that will accumulate on horizontal insulators. In Japan, the median value of precipitation conductivity in Figure 10-25 was $\sigma_{20} = 67\,\mu S/cm$ in an area affected by problems. When this value is used in the snow stress product model, $E_{WS} = 600\,(SSP)^{-0.19}$, the recommended levels of withstand gradient fall by about 15%. See Table 10-13.

For cases where the local distribution of precipitation conductivity is known, however, it can and should form the basis of a probabilistic coordination of dry arc distance rather than a median value.

TABLE 10-13: Recommended E_{WS} for Snow Conditions with $\sigma_{20} = 67\,\mu S/cm$

Snow Thickness, cm (inches)	10 (4)	20 (8)	30 (12)	50 (20)
E_{WS}, 100-mm polymer, kV/m$_{dry\ arc}$	141	124	115	104
E_{WS}, 254-mm disk string, kV/m$_{dry\ arc}$	118	104	96	87
E_{WS}, 700 mm twin disk string, kV/m$_{dry\ arc}$	98	86	79	72

10.13. PROBABILISTIC COORDINATION FOR DRY ARC DISTANCE

The probabilistic coordination of dry arc distance follows the same general logic as for selecting the leakage distance. The process for evaluating threats from ice and snow are considered separately.

- Establish the desired reliability level, in outages per 100 km of line length per year for overhead liens or in mean time between failures for stations.
- Invert the probability of any of m insulators in parallel having a flashover to establish the design probability for a single insulator.
- Take an initial estimate of the required dry arc distance using an empirical icing stress product model for the insulators being evaluated.
- Determine the distribution of icing thickness t. It is reasonable to assume that every freezing rain storm will have a melting phase.
- Determine the distribution of precipitation conductivity σ_{20}. At this point, the distribution of σ_{20} is assumed to be independent of t.
- Approximate the distribution of the number of dry days prior to freezing rain ice storms in the fall–winter and winter–spring seasons.
- Calculate the ESDD that corresponds to each of the dry periods using a linear escalation rate.
- For those ice storms that lead to moderate or heavy ice accretion, with radial thickness $t \geq 6$ mm or approximately 12 mm on the ground, calculate the critical flashover level of an exposed insulator using the mathematical model in Chapter 6 on the basis of the ESDD and σ_{20} distributions.
- Adjust the insulator dry arc distance until the desired line or station reliability is achieved.

There is an alternative approach that may be suitable for substation designs to a specific mean time between failures.

- A design probability for failure of any of m insulators in parallel can be used to establish the acceptable probability of failure for a single insulator, P_{design}.

- The design probability for ESDD, σ_{20}, and t will each be given by the cube root of the acceptable probability of failure for the single insulator, so that $P(\text{ESDD}) \cdot P(\sigma_{20}) \cdot P(t) = P_{\text{design}}$.
- Approximations to the log-normal distributions to $P(\text{ESDD})$, $P(\sigma_{20})$, and $P(t)$ should be inverted to establish the design levels of ESDD, σ_{20}, and t.
- The level of E_{50} should be calculated using the mathematical model for the type of insulator being considered. If the level of thickness t falls below the 6-mm level for moderate icing, the leakage distance calculation should be carried out.
- An insulator that meets the insulation requirements with a statistical margin corresponding to a 5% standard deviation should be selected.

The design probability level of 0.00001 leads to values of $P(\text{ESDD})$, $P(\sigma_{20})$, and $P(t)$ of about 2.1%. With $\sigma_{\text{median}} = 25.4\,\mu\text{S/cm}$, the probability equation would be inverted to give

$$\sigma_{\text{design}} = \sigma_{\text{median}} \left(\frac{1 - P(\sigma_{20})}{P(\sigma_{20})} \right)^{1/2.3} = 25.4 \left(\frac{1 - 0.021}{0.021} \right)^{1/2.3} = 133\,\mu\text{S/cm} \qquad (10\text{-}11)$$

Engelbrecht et al. [2005] show distributions of ESDD observations at 7 locations in 3 countries. These data can all be described by Equation 10-11, using the site-specific values of $\text{ESDD}_{\text{median}}$ to predict $\text{ESDD}_{\text{design}}$.

The same principles should be used for coordination of the dry arc distance of multiple-string insulators in horizontal (dead-end) configurations. For the single disks and single polymer insulators in horizontal positions, the snow accumulation thickness will seldom reach levels of twice the insulator diameter, so this calculation may not be of engineering significance.

10.14. CASE STUDIES

10.14.1. Ontario 500 kV

A large utility built a number of 500-kV stations with 1800-kV BIL (91 kV/m dry arc stress) in the 1960s. This utility obtained reliable performance in areas where there were several moderate ice storms per year, coupled with low pollution.

With improvements in switching-surge control, this utility reduced the dry arc distance in its new stations in the 1970s, with a choice of 1550-kV BIL giving a stress of $106\,kV/m_{dry\ arc}$. With ac service voltage stress of $106\,kV/m_{dry\ arc}$, $46\,kV/m_{leakage}$, and the equivalent USCD of $22\,mm/kV_{l\text{-}g}$, they experienced a number of winter flashover problems with this dry arc distance, starting in 1982.

After study through the late 1980s and early 1990s, and considering the new requirement to have trouble-free stations in parallel while their 1550-kV insulators were flashing over, this utility adopted station post insulation levels of 1800-kV and 2050-kV BIL, giving a reduced stress level of as low as $80\,kV/m_{dry\ arc}$, in areas of moderate contamination near expressways.

With existing stations, this utility faced a more difficult decision. Its major 500-kV station was located in an area relatively distant from pollution sources, and it had simply too many critical 500-kV circuits to consider an extended outage to rebuild with larger dry arc distance. After a series of double contingency studies, it was decided that loss of any one 500-kV station could, with planning, be tolerated during flashover weather. A program of weather and pollution monitoring was put in place help this utility operate around problems.

The first operational use of weather and contamination monitoring together occurred in late January 1994. Light ice accretion, reaching a maximum of 6 mm on ground level, was measured at six weather stations around southern Ontario over an 11-hour period of freezing drizzle and fog on January 24, 1994. This deposit melted the next day. Starting at 1500 on January 27, 1994, there were nine more hours of freezing drizzle, fog, and some freezing rain with temperature rising from $-5\,°C$ to $0\,°C$, and dew point rising from $-5.7\,°C$ to $-0.9\,°C$. The probability of flashovers at only one of the 500-kV stations had been anticipated on the basis of its predicted ESDD of more than a trigger level of $0.012\,mg/cm^2$ ($12\,\mu g/cm^2$) on deenergized samples. This was selected as the "expendable" 500-kV station and the others were operated in normal fashion.

With wind speeds of 30 km/h, most of the 22 different flashovers on the 230-kV system during this event occurred on lines, with one failure of a 192-MVAR capacitor bank at a 230-kV station. The problems ended with two 500-kV system flashovers, one at the predicted station and one on a line nearby, as the temperature changed from $-0.8\,°C$ (at 2100 hours) to $+0.4\,°C$ (at 2200 hours) at lower wind speed. This success was the first operational use of ESDD as well as weather observations in managing the risk of winter flashovers on the 500-kV system.

A program of ESDD monitoring integrated with winter washing at stations with a manageable number of insulators was instituted and has served as a model for operation since then. Winter washing criteria considered both the number of days without rain and on-site ESDD measurements to confirm whether insulators have reached their individual threat levels. This utility settled on the use of units of $\mu g/cm^2$ for ESDD, rather than mg/cm^2, in its training and reporting programs. This was done partly to achieve effective and error-free communication of values using whole numbers, and partly to reinforce that a high degree of precision is needed in the ESDD measurements when evaluating some EHV flashover levels.

The ESDD and climate monitoring program was exercised fully in 2003 with a freezing rain event that led to flashovers at the cleanest of the 500-kV stations, but no other. Records show that there had been 26 days without rain, leading to estimated ESDD levels of about $0.015\,\text{mg/cm}^2$ ($15\,\mu\text{g/cm}^2$) at this station in an area of light pollution. Winter (SMART) washing had been carried out in advance at other stations, based on the number of dry days and the types of improved insulators that had been installed over time. Sporadic freezing rain started on February 3, 2003 at 1600 and ended when ambient and dew point temperatures reached $\pm0.1\,°\text{C}$ about 5 hours later. Repeated flashovers occurred near midnight and led to the complete operational loss of the 500-kV station for about 4 hours. The flashovers occurred on station post insulators with ac service voltage stress of $106\ \text{kV/m}_{\text{dry arc}}$, $46\,\text{kV/m}_{\text{leakage}}$, and the equivalent USCD of $22\ \text{mm/kV}_{\text{l-g}}$. This was a severe consequence but that had been part of the planning assumptions more than 10 years earlier.

This case study serves to illustrate that climate and insulator contamination monitoring, coupled with winter maintenance, can be integrated successfully into the operation practices of a large utility.

10.14.2. Ontario 230 V

In the period from 1989 to 1992, 16 freezing rain events in southern Ontario were studied with precipitation sampling, on-site observations, and automatic weather stations to establish the most important factors in the insulator flashover process. These studies were initiated to find the root cause of 500-kV flashovers, but they also proved useful in establishing why certain, and not other, freezing rain and winter fog events led to a total of 7.5 system-minutes of load loss (half a year's annual reliability target budget) with 230-kV flashovers on two days.

All flashover weather periods included temperatures trending upwards from below to above freezing, a dew point that was within $2\,°\text{C}$ of ambient, fog, and ice accretion at ground level of at least $5\,\text{mm}$. Transmission outages were reported during 9 of these 16 periods. Two events, including 35 of 44 outages in early 1989, were attributed to galloping rather than insulation flashover based on observations and phase-to-phase fault current records. See Table 10-14.

The field studies in southern Ontario were active during an event that led to a load loss of 4.3 system-minutes (518 MW for 120 min) mainly from 230-kV station insulator flashovers [PSOD, 1990; NERC, 1990]:

- The first element that faulted was a 192-MVAR 230-kV capacitor bank, shortly after freezing rain started (December 30, 1989 at 2238 hours).
- Some 230-kV lines started having line-to-ground faults in polluted areas near a coal-fired plant and steel mills (December 31, 1989 at 0000–0051).

TABLE 10-14: Performance of Power System in Ontario During Flashover Weather, 1989–1992

Year	1989							1990			1991				1992	
Month	Jan.	Feb.		Mar.			Dec.	Feb.	Mar.	Dec.	Jan.	Feb.	Mar.	Nov.	Feb.	
Day	25	2	21	4	5	17	31	16	8	3	9	18	3	29	14	18
Hours to last rain	234	184	180	126	18	300	790	34	336	120	246	785	276	118	544	65
115-kV/230-kV Outages	2	2	0	5	*34*	*1*	19	*>30*	0	0	0	0	0	0	0	0
500-kV Outages	0	0	0	0	0	0	3	2	0	0	0	0	0	0	7	2

Note: *Italic font* indicates mostly galloping outages.

- Other 230-kV lines and stations near urban expressways started having line-to-ground faults from 0153 to 0310 hours.
- A single 500-kV line fault occurred at 0553 hours during the melting period.
- Another 192-MVAR capacitor bank was damaged and removed from service at the end of the storm.

Field studies were also active on February 16, 1990 but not in the region of Hamilton, Ontario that was affected by 230-kV station flashovers after a long-duration low-intensity freezing drizzle exposure. Flashovers on bushings and apparatus insulators caused more than $ Cdn 1M of insulator damage and led to a system impact of 3.2 system-minutes, with direct effects on a major steel plant.

Light ice accretion, reaching a maximum of 6 mm, was measured at six weather stations around southern Ontario over an 11-hour period of freezing drizzle and fog on January 24, 1994. This deposit melted the next day. Starting at 1500 hours on January 27, 1994, there were nine more hours of freezing drizzle, fog, and some freezing rain with temperature rising from $-5\,°C$ to $0\,°C$, and dew point rising from $-5.7\,°C$ to $-0.9\,°C$. Twenty-two different flashovers occurred on 230-kV line and station insulators. Problems started at 1600 hours with galloping on 230-kV lines resulting from sustained 30-km/h wind, but there was also a 192-MVAR capacitor bank failure at a 230-kV station. The problems ended with two signature 500-kV system flashovers, one at a station and one on a line, as the temperature changed from $-0.8\,°C$ (2100 hours) to $+0.4\,°C$ (2200 hours) at lower wind speed.

After the January 27, 1994 event, weather in southern Ontario turned cold, leading to 26 days without rain. Some preventive washing of station insulators was carried out over this period. However, with 2 hours of freezing drizzle and fog on February 23, 1994 and temperature rising to $-0.8\,°C$ at a dew point of $-2.1\,°C$ the next morning, there were repeated flashover problems at 230-kV stations in an area of high industrial contamination. This was the "last straw" in a difficult winter for the same sensitive industrial customer affected in February 1990. The utility response led to a program to test and apply retrofit solutions for the 230-kV station insulator flashover problems [Chisholm, 2005].

The icing events that affected the customer had accumulations that are classed as light or very light. The site pollution level was among the highest in the province, with dustfall reaching up to $9\,g/m^2$ per month. Sources of salt pollution from an overhead expressway and carbon black were also present.

A wide range of mitigation options were considered to improve the winter contamination performance. These were evaluated with the use of a test program using two stages—one to ensure that the option would perform well in cold-fog conditions and another, more difficult test series, to reproduce cold-fog conditions including an icing phase.

The treatment of the 900 insulators at the problem station used a combination of:

- Replacement of station insulators with semiconducting glaze (RG) posts of the same dry arc distance.
- Cleaning of transformer and breaker bushings, followed immediately by application of an RTV silicone coating.
- Replacement of selected ceramic line insulators with polymer insulators of two different types.
- Doing nothing to insulators with sufficient leakage distance.
- Continuation of an ongoing winter washing program, supplemented by on-site measurements of ESDD and temperature rise on semiconducting glaze posts.
- Building an enclosure around an entire capacitor bank, which previously had simply been switched off potential every fall to avoid winter flashover problems.

The treatments have all been effective at reducing the contamination flashover rate at this station.

10.14.3. Newfoundland and Labrador Hydro

Much of the background data developed in a study of improved mechanical reliability for a 230-kV transmission system in Newfoundland [Haldar, 2006] can be applied to illustrate the process for selecting a suitable dry arc distance.

Figure 10-30 shows that the transmission line traverses the Avalon Peninsula, which actually geologically is a piece of Africa left over from the time when the continents separated. The salt pollution exposure of this line will vary widely with distance from shore. There were four major line failures in the study area, in 1970, 1984, 1988, and 1994, from heavy accretion of glaze ice to a radial thickness of up to 50 mm (2 in.).

The original 50-year design ice thickness of $t_{50\,yr} = 25.4$ mm for this region suggested an annual thickness of $t_{1\,yr} = 9$ mm when using a coefficient of variation of $C_V = 0.7$ and the expression $t_{50\,yr} = t_{1\,yr}\,(1 + 2.59\,C_V)$. This model was updated by Haldar [2006] to have a mean annual ice thickness of $t_{1\,yr} = 21.6$ mm with a $C_V = 0.96$. An expression to convert the 10-year maximum ice thickness on a fixed cylinder, $t_{10\,yr} = 50$ mm, into other return periods is:

$$t_{X\,yr} = t_{10\,yr}\left[1 - 0.78 \cdot C_{V10}\left\{2.30 + \ln\left(\ln\left(\frac{X}{X-1}\right)\right)\right\}\right] \qquad (10\text{-}12)$$

where

$t_{X\,yr}$ is the maximum ice thickness on a fixed cylinder for return period X, and

C_{V10} is the coefficient of variation for the 10-year return period.

The revised values of $t_{25\,yr} = 63$ mm and $t_{50\,yr} = 75$ mm on fixed cylinders are among the most severe in the world.

Figure 10-30: Location of study area for 230-kV transmission line evaluation (from Haldar [2006]; courtesy A. Haldar).

10.15. PRÉCIS

Four main climate factors establish the stresses that electrical insulation will face in winter conditions:

- The number of days without rain, which tends to have a maximum value in the winter season that behaves the same way as desert exposure in arid regions.
- The occurrence of winter fog at the end of an average or long period without rain.
- The occurrence of freezing rain, leading to light accumulation on contaminated surfaces or heavy accumulation on clean surfaces.
- The occurrence of heavy snow, leading to large accumulation on horizontal insulators.

The risk factor of number of days without rain is greater near sources of electrically conductive pollution from commercial operations, such as sulfates from coal-fired electric power plants. The risk factor is multiplied when lines or stations are located near sources of winter pollution, such as roads and highways that are salted for deicing and urban areas with many domestic heating furnaces.

The risk factor for the occurrence of winter fog can be elevated for insulators located downwind of cooling towers, dams, or other sources of moisture. Generally, the natural occurrence of freezing rain and snow varies widely with region, as illustrated, for example, by ice loading maps presently used only for mechanical design.

The processes for selecting appropriate insulators for the winter conditions break down into two main tasks:

- Establishing the maximum (deterministic) or design (probabilistic) level of insulator surface precontamination (ESDD), and choosing a leakage distance that gives the desired margin of strength over the service voltage stress, essentially treating the occurrence of cold fog as a severe form of nonsoluble deposit density that stabilizes the pollution and provides well-controlled wetting.

- Working with the existing design guidelines for maximum ice load on conductors, to select a suitable insulator dry arc distance. Icing maps for transmission line design are based on the equivalent uniform radial accretion, which is also the basis for insulator testing based on accumulation t on rotating reference cylinders. This ice loading is used along with a probabilistic distribution of precipitation conductivity and, in the general case, an estimate of the likely level of surface precontamination, based on the average number of days without rain prior to the freezing rain events over a long observation period. Once these parameters are established, models of the electrical strength of ice layers are used to choose a dry arc distance that gives the desired margin of strength over the service voltage stress.

REFERENCES

Baker, A. C., M. Farzaneh, R. S. Gorur, S. M. Gubanski, R. J. Hill, G. G. Karady, and H.M.Schneider.2008."Selection of Insulators for ac Overhead Lines in North America with Respect to Contamination," *IEEE Transactions on Power Delivery*, in press.

Chisholm, W. A. 2005. "Ten Years of Application Experience with RTV Silicone Coatings in Canada," in *Proceedings of 2005 World Congress on Insulators, Arresters and Bushings*, Hong Kong. November 27–30.

Chisholm, W. A., T. Jarv, and Y. Tam. 1993. "Determination of Insulator Contamination Levels from Environmental Measurements," Ontario Hydro Research Division Report 92-264-K. April 16.

CIGRE Study Committee 33, Working Group 01. 1991. *Guide to Procedures for Estimating the Lightning Performance of Transmission Lines*, Technical Brochure 63. Paris: CIGRE.

CIGRE Task Force 33.04.01. 2000. *Polluted Insulators: A Review of Current Knowledge*, Technical Brochure 158. Paris: CIGRE.

CIGRE Working Group B2.16, Task Force 03. 2006. *Guidelines for Meteorological Icing Models, Statistical Methods and Topographical Effects*, Technical Brochure 291. Paris: CIGRE.

Cortinas, J. V., B. C. Bernstein, C. C. Robbins, and J. W. Strapp. 2004. "An Analysis of Freezing Rain, Freezing Drizzle, and Ice Pellets Across the United States and Canada: 1976–90," *Weather and Forecasting*, Vol. 19, Apr., pp. 377–390.

Engelbrecht, C. S., I. Gutman, and R. Hartings. 2005. "A Practical Implementation of Statistical Principles to Select Insulators with respect to Polluted Conditions on Overhead ac Lines," *Proceedings* of *IEEE PowerTech 2005*, St. Petersburg, Russia, paper 129. June 27–30.

Farzaneh, M., T. Baker, A. Bernstorf, K. Brown, W. A. Chisholm, C. de Tourreil, J. F. Drapeau, S. Fikke, J. M. George, E. Gnandt, T. Grisham, I. Gutman, R. Hartings, R. Kremer, G. Powell, L. Rolfseng, T. Rozek, D. L. Ruff, D. Shaffner, V. Sklenicka, R. Sundararajan, and J. Yu. 2003. "Insulator Icing Test Methods and Procedures: A Position Paper Prepared by the IEEE Task Force on Insulator Icing Test Methods," *IEEE Transactions on Power Delivery*, Vol. 18, No. 3 (Oct.), pp. 1503–1515.

Farzaneh, M., T. Baker, A. Bernstorf, J. T. Burnham, T. Carreira, E. Cherney, W. A. Chisholm, R. Christman, R. Cole, J. Cortinas, C. de Tourreil, J. F. Drapeau, J. Farzaneh-Dehkordi, S. Fikke, R. Gorur, T. Grisham, I. Gutman, J. Kuffel, A. Phillips, G. Powell, L. Rolfseng, M. Roy, T. Rozek, D. L. Ruff, A. Schwalm, V. Sklenicka, G. Stewart, R. Sundararajan, M. Szeto, R. Tay, and J. Zhang, 2005a. "Selection of Station Insulators with Respect to Ice or Snow—Part I: Technical Context and Environmental Exposure, A Position Paper Prepared by the IEEE Task Force on Icing Performance of Station Insulators," *IEEE Transactions on Power Delivery*, Vol. 20, No. 1, pp. 264–270.

Farzaneh, M., T. Baker, A. Bernstorf, J. T. Burnham, T. Carreira, E. Cherney, W. A. Chisholm, R. Christman, R. Cole, J. Cortinas, C. de Tourreil, J. F. Drapeau, J. Farzaneh-Dehkordi, S. Fikke, R. Gorur, T. Grisham, I. Gutman, J. Kuffel, A. Phillips, G. Powell, L. Rolfseng, M. Roy, T. Rozek, D. L. Ruff, A. Schwalm, V. Sklenicka, G. Stewart, R. Sundararajan, M. Szeto, R. Tay, and J. Zhang. 2005b. "Selection of Station Insulators with Respect to Ice or Snow—Part II: Methods of Selection and Options for Mitigation, A Position Paper Prepared by the IEEE Task Force on Icing Performance of Station Insulators," *IEEE Transactions on Power Delivery*, Vol. 20, No. 1, pp. 271–277.

Farzaneh, M., A. C. Baker, R. A. Bernstorf, J. T. Burnham, E. A. Cherney, W. A. Chisholm, R. S. Gorur, T. Grisham, I. Gutman, L. Rolfseng, and G. A. Stewart. 2007a. "Selection of Line Insulators with Respect to Ice and Snow, Part I: Context and Stresses, A Position Paper Prepared by the IEEE Task Force on Icing Performance of Line Insulators," *IEEE Transactions on Power Delivery*, Vol. 22, No. 4, pp. 2289–2296.

Farzaneh, M., A. C. Baker, R. A. Bernstorf, J. T. Burnham, E. A. Cherney, W. A. Chisholm, I. Fofana, R. S. Gorur, T. Grisham, I. Gutman, L. Rolfseng, and G. A. Stewart. 2007b. "Selection of Line Insulators with Respect to Ice and Snow—Part II: Selection Methods and Mitigation Options, A Position Paper Prepared by the IEEE Task Force on Icing Performance of Line Insulators," *IEEE Transactions on Power Delivery*, Vol. 22, No. 4 (Oct.), pp. 2297–2304.

Fujimura, T. and K. Naito. 1982. "Electrical Phenomena and High Voltage Insulators," in *Proceedings of International Symposium on Ceramics*, Bangalore, India. November.

Haldar, A. 2006. "Upgrading of a 230kV Steel Transmission Line System Using Probabilistic Approach," in *Proceedings of 9th International Conference on Probabilistic Methods Applied to Power Systems*, KTH, Stockholm, Sweden. June 11–15, 2006.

Hall, D. K., G. A. Riggs, and V. V. Salomonson. 2006. Updated daily. MODIS/Terra snow cover 8-day L3 global 0.05 deg CMG V005, 2001-12-01 to 2002-05-01. Boulder, Colorado USA: National Snow and Ice Data Center. Digital media.

Hansen, B., I. Gultepe, P. King, G. Toth, and C. Mooney. 2007. "Visualization of Seasonal–Diurnal Climatology of Visibility in Fog and Precipitation at Canadian Airports," in *Proceedings of 16th Conference on Applied Climatology, 87th Annual Meeting of the American Meteorological Society*, San Antonio, Texas, January 14–18. Available at http://collaboration.cmc.ec.gc.ca/science/arma/climatology/.

Hileman, A. R. 1999. *Insulation Coordination for Power Systems*. Boca Raton, FL: CRC Press, Taylor & Francis Group.

IEC 60815. 2008. *Selection and Dimensioning of High-Voltage Insulators Intended for Use in Polluted Conditions*. Geneva, Switzerland: Bureau Central de la Commission Electrotechnique Internationale. October.

IEEE Standard 1313.1. 1996. IEEE Standard for Insulation Coordination—Definitions, Principles and Rules. Piscataway, NJ: IEEE Press.

IEEE Standard 1313.2. 1999. *IEEE Guide for the Application of Insulation Coordination*. Piscataway, NJ: IEEE Press.

IEEE/ANSI Standard C2. 2007. *National Electric Safety Code, 2007 Edition*. Piscataway, NJ: IEEE Press.

IEEE Standard 4. 1995. *IEEE Standard Techniques for High Voltage Testing*, Piscataway, NJ: IEEE Press.

Jones, K., R. Thorkildson, and N. Lott. 2002. "The Development of a U. S. Climatology of Extreme Ice Loads," National Climatic Data Center Technical Report 2002-01.

Kawai, M. and M. Sforzini. 1974. "Problems Related to the Performance of UHV Insulators in Contaminated Conditions," *CIGRE 1974 Session*, Paper 33–19. August 21–29.

Matson, M., C. F. Roeplewski, and M. S. Varnadore. 1986. *An Atlas of Satellite-Derived Northern Hemisphere Snow Cover Frequency*. Washington, DC: National Weather Service.

Meteorological Service of Canada. 2004. *Canada–US Transboundary Particulate Matter Science Assessment*. Atmospheric and Climate Science Directorate, Science Assessment and Integration Branch Report.

NERC (North American Electric Reliability Corporation). 1990. DAWG Database, January 1, 1989–December 31, 1989, NERC_Disturbance_Reports\dawg-89.htm.

Nghiem, S. V. and W-Y. Tsai. 2001. "Global Snow Cover Monitoring with Spaceborne Ku-Band Scatterometer," *IEEE Transactions on Geoscience and Remote Sensing*, Vol. 39, No. 10 (Oct.), pp. 2118–2134.

PSOD (Power System Operations Division) of Ontario Hydro. 1990. "*Morning Report, January 2, 1990 covering Friday December 29, 1989 to Monday January 1, 1990.*"

Roy, D. P., J. S. Borak, S. Devadiga, R. E. Wolfe, M. Zheng, and J. Descloitres. 2002. "The MODIS Land Product Quality Assessment Approach," *Remote Sensing of Environment*, Vol. 83, pp. 62–76.

Tammelin, B. and S. M. Fikke. 2007. "Measuring and Forecasting Atmospheric Icing on Structures," *Proceedings of COST 727 Meeting*, Brussels (June 6).

WMO. 2001. "Recommendations of WMO Expert Meeting on Automation of Visual and Subjective Observations (Trappes/Paris, France, 14–16 May 1997) and the Working Group on Surface Measurements (Geneva, Switzerland, 27–31 August 2001)."

APPENDIX A

MEASUREMENT OF INSULATOR CONTAMINATION LEVEL

Equivalent salt deposit density (ESDD) is the standard method for reporting the electrical conductivity of an arbitrary pollution deposit with unknown composition. It can be expressed in mg/cm^2 for areas of high pollution, while units of μg/cm^2 are more appropriate for risks to EHV transmission networks.

ESDD levels on top and bottom surfaces of insulators are usually different. The bottom-surface levels rise slowly with time and moderate rain does not clean the pollution off. The top-surface levels rise quickly with time, and small 0.4-mm amounts of winter rain can wash most of this pollution away. The electrical performance of the insulator depends more on the top-surface ESDD in icing conditions and on the overall value in cold-fog conditions at 0 °C.

ESDD is normally measured with a rag-wipe method. A clean cloth is rinsed several times in deionized water, with a minimum volume of 0.5–1 mL per cm^2 of surface area. The top surface area of a 254-mm diameter insulator disk is 300–500 cm^2 and the bottom surface, depending on profile, can be as high as 1000 cm^2. If the temperature is below 0 °C, ethanol can be added to the wash water without affecting the results, as long as the volume of the water is noted beforehand.

A specified volume of water is measured, using the level part of the surface as shown in Figure A-1. This volume is transferred to a sealable container such as a plastic or glass jar. The electrical conductivity of the wash water is measured, with the rag in place, prior to wiping the desired surface (Figure A-2).

Insulators for Icing and Polluted Environments. By Masoud Farzaneh and William A. Chisholm
Copyright © 2009 the Institute of Electrical and Electronics Engineers, Inc.

Figure A-1: Water volume of 400 mL, showing meniscus effect in graduated cylinder.

Figure A-2: Rag is inserted in wash water prior to measuring electrical conductivity.

Normally this reading should be less than 1 μS/cm, although a correction can be made if deionized water is not available.

When the surface deposit has been wiped clean, the rag is put back into the wash water. The conductivity will typically increase from the initial value by 10–100 μS/cm, and it is this change in conductivity that is most relevant, rather than the initial or final values.

It is good practice to maintain a control sample of a rag and wash water that is transported, handled, and exposed the same way as the rags used to wash insulators. The deionized water will pick up carbonate ions from the air, and this will give a false increase in conductivity. Evaporation has the same effect, although at a slower rate. The wash water samples should thus be transported in sealed containers. If a considerable quantity of deionized water is lost, the remaining water will have an elevated conductivity. These practical factors tend to make it more difficult to obtain accurate values of low ESDD. For EHV insulators, in particular, special procedures are recommended [Chisholm et al., 1994].

The change in conductivity with temperature is an important factor, especially when extrapolating results to freezing temperatures. Ramos et al. [1993] noted that this is true both for precontaminated insulators and for those exposed to salt fog. Since the wash water is a mixture of ions in unknown concentrations and proportions, it is prudent to measure the conductivity above and below 20 °C, and to use linear interpolation to read out the conductivity corrected to 20 °C. Instruments that have an automatic temperature correction factor (such as 2.2% per °C) will have an error of about 0.5–1.1% per °C if the wash water is acidic. This feature should be disabled by setting the correction factor to zero.

As an alternative to the preferred practice of interpolation between two temperatures, and if the ion content is known to be sodium chloride (e.g., from exposure to sea salt), IEC 60507 [1991] and IEEE Standard 4 [1995] offer an expression for correcting the measured conductivity σ_T for temperature T:

$$\sigma_{20} = \sigma_T [3.2 \times 10^{-8} \cdot T^4 - 1.096 \times 10^{-5} \cdot T^2 + 1.0336 \times 10^{-3} \cdot T^2$$
$$- 5.1984 \times 10^{-2} \cdot T + 1.7088] \tag{A-1}$$

where

σ_T is the conductivity of the NaCl solution at temperature T,

T is the solution temperature in °C, and

σ_{20} is the conductivity of the NaCl solution corrected to 20 °C.

As an introductory comment, the series resistance of the pollution layer plays a fundamental role in the electrical flashover of the insulator under contaminated conditions. This resistance varies inversely with conductivity. Figure A-3 shows that the electrical conductivity of NaCl solutions near the freezing point is about 60% of the level at 20 °C. This means that the series resistance of a wet pollution layer at 5 °C is 60% higher than the resistance at 20 °C.

Once a value for the wash water conductivity at 20 °C, σ_{20}, has been interpolated, the international practice for converting this value into ESDD is to calculate the salinity S_a. Unfortunately, there are many changes of units in this process. Table A-1 hopefully simplifies the process.

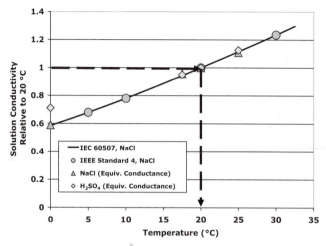

Figure A-3: Temperature correction of sodium chloride solution conductivity for ESDD measurements in IEC 60507 and IEEE Standard 4.

TABLE A-1: Conversion of Rag Wipe Results to Equivalent Salt Deposit Density Using IEC 60507 and IEEE Standard 4 Method

Insulator Surface Area, A (cm²)	Volume of Distilled Water, V (cm²)	Electrical Conductivity at 20°C	σ_{20}, Electrical Conductivity (S/m)	$S_a = (5.7 \cdot \sigma_{20})^{1.03}$ Salinity (kg/m³)	ESDD, $S_a \cdot V/A$ in mg/cm² (µg/cm²)
500	500 mL = 500 cm²	100 µS/cm	0.01	0.0523	0.0523 (52)
1000	500 cm²	200 µS/cm	0.02	0.114	0.057 (57)
500	1 L = 1000 cm²	50 µS/cm	0.005	0.256	0.512 (51)
1000	1000 mL	100 µS/cm	0.01	0.0523	0.0523 (52)
500	500 mL	1 mS/cm	0.1	0.560	0.560 (560)
1000	500 mL	2 mS/cm	0.2	1.144	0.572 (572)
500	1000 mL	0.5 mS/cm	0.05	0.274	0.549 (549)

The ESDD values are relatively insensitive to wash water volume and surface area for most common ions. Ramos et al. [1993] noted that the solubility of CaSO₄?2H₂O found in high deposit density at some sites in Mexico was far less than solubility of NaCl and that at least 2000 mL of wash water was needed to give adequate results.

For precision work on EHV insulators, it is necessary to correct for the ESDD associated with the conductivity of the control sample of wash water. This baseline value is computed, using the same algorithm, and subtracted from each of the values obtained for the insulator surfaces.

Normally, the top and bottom surfaces of individual suspension disks are swabbed separately. For long-rod or station post insulators, it can be practical to take a series of conductivity measurements as the top and bottom surfaces of each shed are wiped. These results give the distribution of pollution along the insulator string or post, which can be moderately influenced by line potential and may be heavily influenced if booster sheds or creepage extenders are used.

The ESDD level on the uppermost surface of the top insulator may be different from the levels found on other upward-facing surfaces. Rust stains from hardware, bird streamers, wind flow, exposure to rain, and many other factors are different. Normally, the ESDD on the top surface of the top insulator should be segregated.

REFERENCES

Chisholm, W. A., P. G. Buchan, and T. Jarv. 1994. "Accurate Measurement of Low Insulator Contamination Levels," *IEEE Transactions on Power Delivery*, Vol. 9, No. 3 (July), pp. 1552–1557.

IEEE Standard 4. 1995. *IEEE Standard Techniques for High Voltage Testing*. Piscataway, NJ: IEEE Press.

IEC 60507. 1991. *Artificial Pollution Tests on High-Voltage Insulators to Be Used on a.c. Systems*. Geneva, Switzerland: Bureau Central de la Commission Electrotechnique Internationale.

Ramos, G. N., M. T. Campillo, and K. Naito. 1993. "A Study on the Characteristics of Various Conductive Contaminants Accumulated on High-Voltage Insulators," *IEEE Transactions on Power Delivery*, Vol. 8, No. 4 (Oct.), 1842–1850.

APPENDIX B

STANDARD CORRECTIONS FOR HUMIDITY, TEMPERATURE, AND PRESSURE

Safety standards for electrical clearance, such as the IEEE/ANSI Standard C2, make use of a guard clearance gradient of 1.854 mm per kV of insulator basic impulse level, corresponding to a well-documented flashover gradient of 540 kV/m [Hileman, 1999]. In many areas, however, it is appropriate to apply corrections for air pressure and humidity. These corrections can change the electrical strength by 20% in some conditions.

B.1. OVERALL TREND IN UNIFORM FIELDS

Empirically, Meek and Craggs [1978] note that the Dakin curve of uniform field breakdown voltage V_{UF} (kV) versus the product of pressure p and separation d, as shown earlier in Figure 4-1, can be described by

$$V_{UF} = 2.44\delta d + 2.065\sqrt{\delta d}$$

$$\delta = \frac{p}{101.3 \text{ kPa}} \cdot \frac{293 \text{ K}}{T_{amb} + 273 \text{ K}}$$

(B-1)

where

V_{UF} is the uniform field breakdown strength (kV),
d is the electrode separation (mm),

Insulators for Icing and Polluted Environments. By Masoud Farzaneh and William A. Chisholm
Copyright © 2009 the Institute of Electrical and Electronics Engineers, Inc.

T_{amb} is the ambient temperature (°C), and

p is the ambient pressure (kPa, 1 atm = 1.013 bar = 101.3 kPa).

The expression in Equation B-1 holds for small gaps at temperatures up to 1200 °C. In these tests, a hot spot on the electrode gave the same result as heating the entire gap. A change in absolute humidity from the reference level of 11 g/m³ to dry conditions can reduce V_{UF} by 4% for small gaps, and more for large gaps.

B.2. STANDARD PROCESS FOR CORRECTING TEST LAB RESULTS

IEEE Standard 4 [1995] and IEC 60060 [1989] provide a common method to correct high-voltage test results for variations in atmospheric conditions, based on the relative air density term δ in Equation B-1.

The standards provide reference relations between the peak of the 60-Hz voltage wave that causes flashover of a rod-to-rod gap and the gap spacing d. For the *standard reference atmosphere* of 20 °C ambient, 101.3 kPa (760 mm Hg) pressure, and absolute humidity of 11 g/m³, the rod-to-rod flashover voltage in Table 11 of IEEE Standard 4 [1995] expressed as the peak of 60-Hz voltage, is

$$V_{ac\,pk} = 22 \text{ kV} + d \cdot 500 \text{ kV/m} \tag{B-2}$$

for a gap spacing d of more than 0.02 m and less than 2.2 m. The expression that gives the positive and negative polarity dc flashover is

$$V_{dc\,pk} = 2 \text{ kV} + d \cdot 534 \text{ kV/m} \tag{B-3}$$

where 0.25 m < d < 2.5 m.

The flashover values are corrected using the known values of:

T_a, the ambient temperature (°C).

T_{DP}, the dew point temperature (°C). Note that the dew point temperature depression (°C) is the difference $(T_a - T_{DP})$.

P, the ambient pressure (mbar). Note that to convert pressure in kPa to mbar, multiply by 10; for example, 101.3246 kPa = 1013.246 mbar = 760 mm Hg = 1 atmosphere.

The methods of Goff and Gratch [1946] and Goff [1957] are used to obtain absolute humidity, h/δ, expressed in g/m³. In IEEE Standard 4 [1995], these are expressed as a psychometric chart relating wet and dry bulb temperatures to absolute humidity with a limited temperature range that does not cover freezing conditions. Outdoors, normally the ambient and dew point temperatures are reported. With modern spreadsheets, including the one provided with this text, it is possible to implement the full Goff–Gratch model for absolute humidity calculations numerically without needing to refer to a graph.

E_{WS} is the US standard atmospheric pressure near sea level in mbar at an average temperature of 288 K (59 °F) in midlatitudes, 1013.246 mbar.

The *saturation vapor pressure* of water, E_W, is calculated as a function of dew point temperature T_{DP}:

$$E_W = e_{BP} \cdot 10^Z$$

$$Z = a\left(\frac{T_S}{T} - 1\right) + b\log_{10}\left(\frac{T_S}{T}\right) + c(10^{d(1-T/T_S)} - 1) + f(10^{h(T_S/T-1)} - 1) \quad \text{(B-4)}$$

$$T = T_{DP} + 273.14$$

where

E_W is the saturation vapor pressure of water (mbar),

e_{BP} = is the reference pressure (1013.246 mbar),

Z is an intermediate variable,

$a = -7.90298, b = 5.02808, c = -1.3816 \times 10^{-7}, d = 11.344, f = 8.1328 \times 10^{-3}, h = -3.49149,$

$T_S = 373.14$ K,

T is the absolute dew point temperature (K), $T_{DP} + 273.14$, with $T > 0.001$ K, and

T_{DP} is the dew point temperature (°C).

F_W is the correction factor for the departure of the mixture of air and water vapor from ideal gas laws, and it is given by

$$F_W = 1 + (5.92854 + 0.03740346 \cdot P + 1.971198 \times 10^{-4}\,T \cdot (800.0 - P) + 6.045511 \times 10^{-6}\,P \cdot T^2) \times 10^{-4} \quad \text{(B-5)}$$

where

P is the ambient pressure (mbar), and

T is the absolute dew point temperature (K), $T_{DP} + 273.14$.

If needed for calculating relative humidity, the *absolute vapor pressure* $P_{\text{abs vap}}$ is calculated from

$$P_{\text{abs vap}} = \frac{216.68 \cdot 1.01 \cdot E_W}{C_V \cdot T} \quad \text{(B-6)}$$

The appropriate value of C_V is taken from Table B-1.

TABLE B-1: C_v Values

$T_{DP} <$ $-30\,°C$	$-30\,°C <$ $T_{DP} < -20\,°C$	$-20\,°C <$ $T_{DP} < -10\,°C$	$-10\,°C <$ $T_{DP} < 0\,°C$	$0\,°C <$ $T_{DP} < 10\,°C$	$10\,°C <$ $T_{DP} < 20\,°C$	$20\,°C < T_{DP}$
$C_v = 1.0$	$C_v = 0.9998$	$C_v = 0.9997$	$C_v = 0.9995$	$C_v = 0.9992$	$C_v = 0.9988$	$C_v = 0.9982$

Continuing with the main Goff–Gratch method, a *saturation mixing ratio,* R_W (kg of water per kg of dry air), is calculated taking ambient pressure P (mbar) into account:

$$R_W = \frac{0.62197 F_W E_W}{P - F_W E_W} \tag{B-7}$$

The *saturation virtual temperature* T_V, defined from the Smithsonian [1986] Meteorological Tables is computed:

$$T_V = T_{DP} + (T_{DP} + 273.13) \cdot R_W \cdot \frac{E-1}{1 + R_W} \tag{B-8}$$

where

E is the apparent molecular weight of dry air ($M_d = 28.966$) over water ($M_w = 18.016$) giving 1.607795, and

T_{DP} is the dew point temperature (°C).

The density of air, D_{air} (g/m³) for an air pressure P (mbar) is

$$D_{air} = \frac{348.38 \cdot P}{T_V + 273.13} \tag{B-9}$$

The *absolute water vapor density,* h (g/m³), as used in IEEE Standard 4 [1995] is finally

$$h = D_{air} \cdot \frac{R_W}{1 + R_W} \tag{B-10}$$

The next step [IEEE Standard 4, 1995] in correction of ac electrical strength using absolute water vapor density h is to calculate the relative air density δ:

$$\delta = \frac{P}{101.3\,\text{mbar}} \cdot \frac{293\,\text{K}}{273\,\text{K} + T_{ambient}}$$

$$k = 1 + 0.012\left(\frac{h}{\delta} - 11\,\text{g/m}^3\right) \tag{B-11}$$

The factor of 0.012 for ac flashover tests in Equation B-11 is changed to 0.010 for impulse flashover and to 0.014 for dc test results.

The next step of the correction process has some inherent uncertainties. The atmospheric correction factors depend on the types of predischarge. The correction factor g for peak withstand test voltage V_B across the minimum discharge path length (dry arc distance) L is

$$g = \frac{1.1 \cdot V_B}{500 \text{ kV} \cdot L \cdot \delta \cdot k} \tag{B-12}$$

In cases where the critical flashover $V_{50\%}$ is established, its value is substituted in the numerator of Equation B-12.

The value of g varies with the type of test. Steep-front lightning surges, leading to short times to flashover, have little if any predischarge activity. With the typical 50% gradient of 822 kV/m for flashover of ceramic insulator strings at 2 μs, the value of g can exceed 1.6. On the other hand, contamination or icing flashovers tend to have significant predischarge activity in the tens or hundreds of ac cycles prior to flashover. Also, the critical flashover levels often decline to levels of 75–100 kV per meter of dry arc distance. This drops the value of g to less than 0.2.

Empirical exponents, m and w, are used in IEEE Standard 4 [1995] to describe the changing effect of atmospheric corrections, based on the predischarge coefficient g. The values remain under discussion by standardization groups, but the existing recommendation is reasonably approximated by the following calculations:

$$\Delta g = |g - 1.1|$$
$$w_m = \max(0, 1.4279 + \Delta g \cdot (-4.0689 + \Delta g \cdot (4.5352 - \Delta g \cdot 1.9693)))$$
$$w = \min(w_m, 1) \tag{B-13}$$
$$m = \begin{cases} w, & g < 1 \\ 1, & g \geq 1 \end{cases}$$

The overall correction factor for air density and humidity, K, is

$$K = k_1 k_2 = \delta^m k^w \tag{B-14}$$

The factor K in Equation B-14 is used to correct for the flashover strength of the air.

B.3. DC, AC, AND SWITCHING SURGE

In standard laboratory conditions (1013 mbar, 11.7 °C dew point, 20 °C ambient temperature), the absolute humidity is 11 g/m³. The correction factor of $K = 0.9998$ with the Goff–Gratch process means that there is a negligible

0.02% error, caused by rounding the reference atmosphere pressure from 1013.246 to 1013 mbar in IEEE Standard 4.

In low-pressure, dry conditions (900 mbar = 90 kPa, 0 °C dew point at 20 °C ambient temperature) found in some parts of the western United States and Canada, for a value of $g = 0.82$ the corresponding value is $K = 0.90$. This means that up to 11% more distance will be needed for safe electrical clearances outdoors, for the switching impulse flashovers normally used in safety testing for the live work method tested under standard conditions.

B.4. LIGHTNING SURGE

Climate records for the Great Lakes areas in the United States and Canada suggest that the occurrence of lightning activity climbs sharply when the dew point temperature exceeds 21 °C. In typical low-pressure conditions (980 mbar) with 30 °C ambient and 21 °C dew point, and using a value of $g = 1.6$ for short-duration lightning impulse flashover at 2 μs, the absolute humidity is more than 20.6 g/m³ and the correction factor of $K = 0.96$ represents about one standard deviation in typical lab test results.

B.5. FOG CONDITIONS

In clean-fog conditions, at normal pressure and saturated conditions (1013 mbar, 19 °C dew point, 20 °C ambient) with flashover at line voltage stress, the value of $g = 0.21$, and an absolute humidity of 17.2 g/m³ give an atmospheric correction factor of $K = 1.0010$ that could be neglected.

In this case, the very low value of g reduces the values of exponents m and w in Equation B-14 to less than 0.05. This was discussed in detail in Chapter 4, because the experimental data support a value closer to $m = 0.5$ for the pressure correction.

Experimentally, the presence of fog in the rod-to-plane gap has an effect on the ac flashover that becomes more important for gaps over 2 m long and for long-duration tests. Tests by Rizk et al. [1981] were carried out at absolute humiditiy of 3.1–5.6 g/m³, with the results shown in Figure B-1. There is no influence of fog up to 2-m gap length. Above this distance, the rod-to-plane sparkover voltage decreased. Rizk et al. [1981] ascribed this to the residual space charge produced during the negative half-cycle of the ac wave. Their negative switching-surge and dc sparkover voltages were also reduced in the dense fog with a liquid water content of 3 g/m³.

B.6. SPECIAL FACTORS BELOW 0 °C

The bulk of this book is concerned with icing and pollution flashovers near the freezing point of 0 °C, rather than at room temperature in clean-fog test conditions.

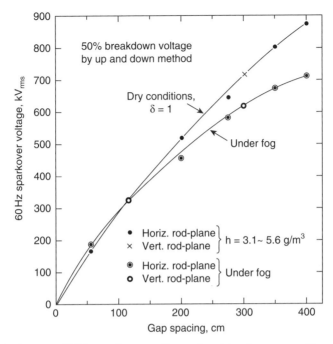

Figure B-1: Average 60-Hz sparkover voltage of rod-to-plane gap with constant 15-minute test duration using up-and-down method (from Rizk et al. [1981]).

TABLE B-2: Comparison of Electrical Flashover Processes

Normal Contamination Problems	Icing Problems
Deposit of contaminantion (few hours to a few months)	Deposit of ice or snow (about an hour)
Deposit of moisture	Formation of water film on ice surface or inside snow
Formation of dry bands on insulator surface	Formation of air gaps in ice or snow
Arc initiation over dry bands	Arc formation across air gaps
Propagation of arc and its rapid growth	Propagation of arc and its rapid growth

At present, an international standard [IEC 60815, 2008] recommends values of unified specific creepage distance (mm/kV) that are respected for all commercial outdoor insulation materials at all temperatures. These conditions are described on the left side of Table B-2.

In melting conditions during freezing rain (980 mbar, −2 °C dew point, 0 °C ambient), with a value of $g = 0.24$ for cold-fog or icing flashover at normal ac service voltage, the absolute humidity is $4.25 \, g/m^3$ and it would seem that the atmospheric correction factor $K = 0.9985$ can also be neglected.

As detailed in Chapter 9, the observed exponent m for the variation in flashover voltage with pressure follows the same trend as that found for contamination flashovers, as described for example by Rizk and Rezazada [1997]. For example, Li [2002] reports values of $0.40 < m < 0.52$ for ac voltage and $0.26 < m < 0.49$ for dc voltage of either polarity, for typical and highly contaminated ice, respectively.

REFERENCES

Goff, J. A. 1957. "Saturation Pressure of Water on the New Kelvin Temperature Scale," *Transactions of the American Society of Heating and Ventilating Engineers*, pp. 347–354.

Goff, J. A. and S. Gratch. 1946. "Low-Pressure Properties of Water from −160 to 212 °F," *Transactions of the American Society of Heating and Ventilating Engineers*, pp. 95–122.

Hileman, A. R. 1999. *Insulation Coordination for Power Systems*. Boca Raton, FL: CRC Press, Taylor & Francis Group.

IEC Standard 60060-1. 1989. *High-Voltage Test Techniques Part 1: General Definitions and Test Requirements*. Geneva, Switzerland: IEC.

IEC 60815. 2008. *Selection and Dimensioning of High-Voltage Insulators Intended for Use in Polluted Conditions*. Geneva, Switzerland: Bureau Central de la Commission Electrotechnigue Internationale. October.

IEEE/ANSI Standard C2. 2007. *National Electrical Safety Code*. Piscataway, NJ: IEEE Press.

IEEE Standard 4. 1995. *IEEE Standard Techniques for High Voltage Testing*. Piscataway, NJ: IEEE Press.

Li, Y. 2002. "Study on the Influence of Altitude on the Characteristics of the Electrical Arc on Polluted Ice Surface," PhD Thesis, Université du Québec à Chicoutimi. June.

Meek, J. M. and J. D. Craggs. 1978. *Electrical Breakdown of Gasses*. Norwich, U.K.: Wiley.

Rizk, F. A. M. and A. Q. Rezazada. 1997. "Modeling of Altitude Effects on AC Flashover of High-Voltage Insulators," *IEEE Transactions on Power Delivery*, Vol. 12, No. 2 (Apr.), pp. 810–822.

Rizk, F. A. M., Y. Beauséjour, and Hu Shi Xiong. 1981. "Sparkover Characteristics of Long Fog-Contaminated Air Gaps," *IEEE Transactions on Power Apparatus and Systems*, Vol. PAS-100, No. 11 (Nov.), pp. 4604–4611.

Smithsonian. 1986. *Smithsonian Meteorological Tables*. Washington, DC: Smithsonian Institution Press.

TERMS RELATED TO
ELECTRICAL IMPULSES

Basic Lightning Impulse Level (BIL): The crest value of a standard (1.2 × 50-μs) lightning impulse for which the insulation exhibits a 90% probability of withstand (or a 10% probability of failure) under specified conditions.

Critical Impulse Flashover Voltage (CFO): The crest value of the impulse wave that, under specified conditions, causes flashover through the surrounding medium on 50% of the applications.

Impulse: An intentionally applied transient voltage or current that usually rises rapidly to a peak value and then falls more slowly to zero.

Lightning Impulse: An impulse with front duration of up to a few tens of microseconds.

Overshoot: The value by which a lightning impulse exceeds the defined crest value.

Partial Discharge: A discharge that does not completely bridge the insulation between electrodes.

Peak Value of Impulse Voltage: The maximum value of impulses that are smooth double exponential waves without overshoot.

Response Time: A quantity that is indicative of the speed with which a system responds to changing voltage or currents.

Standard Lightning Impulse: A full lightning impulse having a front time of 1.2 μs and a virtual time to half-value of 50 μs.

Insulators for Icing and Polluted Environments. By Masoud Farzaneh and William A. Chisholm
Copyright © 2009 the Institute of Electrical and Electronics Engineers, Inc.

Standard Switching Impulse: A full switching impulse having a front time of 250 μs and a virtual time to half-value of 2500 μs.

Surge: A transient voltage or current, which usually rises rapidly to a peak value and then falls more slowly to zero, occurring in electrical equipment or networks in service.

Switching Impulse: An impulse with a front duration of some tens to thousands of microseconds.

Undershoot: The peak value of an impulse voltage or current that passes through zero in the opposite polarity of the initial peak.

Value of the Test Voltage for Lightning Impulse: The peak value when the impulse is without overshoot or oscillations.

INDEX

A

Abonneau, L., 286

Ac arc reignition
 background, 278–281
 effect of ambient temperature,
  281–282
 under icing conditions, 453–457, 458
 mathematical model, 282–283

Accuracy, defined, 157

Acid rain, 73, 79, 81

Acoustic inspection, 302–303

Acoustic noise
 in ceramic disk insulators, 251
 in dry-band arcs, 249
 in insulator partial discharge activity,
  307

Aerodynamic disks, 39, 293, 294,
 335–337

Aherne, J., 80, 81, 82

Air density
 as environmental stress on insulators,
  52
 standard correction, 224–227
 in standard process for correcting test
  lab results, 652, 654, 655

Air gap breakdown, 159–165

Air pollution, defined, 60. *See also*
 Pollution

Akbar, M., 62, 66, 69, 335, 336, 357

Akizuki, M.O., 344

Al-Baghdadi, A.A.J., 255, 463

Albedo, snow, 626

Aleksandrov, G.N., 163, 164

Algeria, insulator case study, 103

Allen, N.L., 246

Alston, L.L., 260

Alternating diameter insulators, 36, 207,
 337–338, 552–553

Amarh, F., 106, 296, 297, 298, 301

American National Standards Institute
 (ANSI), 24, 38, 43, 106, 107, 164–
 165, 211, 218, 356

Animal contacts, 53, 145, 147, 328–329,
 513

Anions, defined, 91

ANSI. *See* American National Standards
 Institute (ANSI); IEEE/ANSI
 Standard C2, 2007

Anti-fog disks, 176, 177, 195, 334–335,
 336, 340–342

Insulators for Icing and Polluted Environments. By Masoud Farzaneh and William A. Chisholm
Copyright © 2009 the Institute of Electrical and Electronics Engineers, Inc.

Applied water conductivity, defined, 364
Aquatic systems, and target loading concept, 81–82
Arai, N., 178, 193, 194
Arc discharges. *See* Violet arc; White arc
Arcing
  in artificial pollution tests, 180–181
  discharge initiation and development, 256
  dry-band, 26, 246–255
  dynamics of propagation, 261
  as factor in pollution problems, 71
  in freezing conditions, 51
  multiple arcs in series, 276–277
  V-I characteristics in free air, 256–259
  V-I characteristics on water or ice surfaces, 259–261, 262
  on wet, contaminated surfaces, 255–262
Arcing horns, 324, 331–332, 576
Arc motion, 278, 470–471
Arc reignition
  background, 278–281
  effect of ambient temperature, 281–282
  under icing conditions, 453–457, 458
  mathematical model, 282–283
Arc root radius
  in flashover modeling of ice-covered insulators, 446, 447, 452, 453, 456
  as function of leakage current, 371
  Wilkins model, 268, 269, 270, 274, 277
Argent, S.J., 6, 7
Arresters
  under heavy icing conditions, 422, 423
  snow accretion on, 524–525
Arslan, A.N., 493
Artificial pollution tests
  clean-fog tests, 180, 187–196
  cold-fog tests, 180, 199–200
  dry salt layer method, 198–199
  dust cycle method, 198
  electrical clearance in test chamber, 184–185
  heavy rain tests, 52, 180, 181, 389, 390
  natural *vs.* artificial, 180–181
  power supply characteristics, 181–184
  salt-fog tests, 180, 185–187
Artificial wetting. *See also* Precipitation, artificial

in clean-fog test procedure, 192–193
hybrid test approach for naturally polluted insulators, 197
precipitation sources for surface wetting, 139–147
ASHRAE, 1991, 130
Atmospheric icing, 2, 91, 135, 364
Ayrton, 1880s, 157

B
Baker, A.C., 49, 205, 206, 207, 208, 233, 249, 250, 283, 348, 406, 433, 535
Ball-and-socket insulators. *See* Cap-and-pin insulators
Barrett, J.S., 538, 541
Beattie, J., 69
Beauséjour, Y., 69, 106, 107, 217, 218, 219, 340, 390
Bell-shaped insulators, 338–340, 352, 552
Bentley, Wilson A., 493
Bergman, V.I., 229
Bernal, J.D., 365, 371
BIL (basic lightning impulse insulation level), 25, 50, 355, 659
Billington, R., 7, 10
Bird streamers, 53, 139, 326, 327–329, 470
Bodele Depression, Sahara Desert, 72
Booster sheds
  adding to insulators to mitigate heavy icing, 379, 453, 529, 530, 536, 537, 568–571
  background, 324–325
  for improving insulator contamination performance, 309, 324–325, 326
  vs. creep extenders, 326
Bourdages, M., 205
Boyer, A.E., 377
Brettschneider, S., 470, 553, 556, 558, 566, 567, 578
Bridging, ice
  and corona rings, 561
  and dry arc distance, 553, 555
  flashover test results, 415–421
  in moderate icing, 550, 551–556, 559, 560, 561
  and polymer insulators, 559
  role of thermal monitoring, 561–562
  and shed-to-shed spacing of insulator strings, 552–553, 554

Brighton, U.K., insulator salt-fog test observations, 175
Brunnel, D., 4
BSL (basic switching impulse insulation level), 50, 51
Burnham, J.T., 139, 314
Bursik, M., 140

C

Calcium chloride, 122, 123, 357
Canada. *See also* Hydro-Québec; Ontario Hydro
  monitoring transborder pollution, 88–89
  Ontario flashover case studies, 635–640
Canadian Electricity Association (CEA), 7, 24
Canadian Standards Association (CSA), 24, 26, 43–44
Cap-and-pin insulators, 38–39, 41, 49, 374, 485
CAPMoN program, Environment Canada, 88–89
Carberry, R.E., 324
Carbon monoxide, 60
Carreira, A.J., 55
Carrier, W.H., 130
CASTNET program, EPA, 89
Cations
  anion-cation balance, 86, 103
  background, 91–92
  defined, 91
  equivalent conductance of ions, 94
CEA (Canadian Electricity Association), 7, 24
CEIDS (Consortium for Electric Infrastructure to Support a Digial Society), 12–13
Ceramic insulators
  acoustic noise factor, 251
  classification of ice accretion levels, 374, 375
  in clean-fog testing, 190–192
  clean-fog test results, 205–210
  comparison of artificial pollution test processes, 181, 182
  defined, 31
  end fittings, 46

insulator construction, 35–44
  maintaining RTV silicone coatings, 314–315
  maintaining semiconducting glaze, 313–314
  materials for, 31, 35
  overview of types, 35–44
CFT (cold-fog-temperature) insulator tests, 394–395
Chaarani, R., 559
Chafiq, M., 436, 467, 468
Chalmers, I.D., 246
Chang, J.W., 31, 192, 541
Charneski, M.D., 382, 393, 394, 415, 416
Chatterjee, N., 260, 456
Chaurasia, D.C., 272
Chen, X., 99, 368, 437
Cheng, Y.S., 72
Cherney, E.A., 188, 193, 194, 195, 199, 229, 353, 356, 392–393, 410, 412, 463, 555
Chino, T., 332
Chisholm, W.A., 2, 11, 12, 17, 46, 74, 75, 100, 112, 113, 114, 117, 118, 122, 132, 137, 200, 232, 257, 284, 340, 370, 373, 379, 394, 402, 403, 406, 433, 437, 438, 484, 510, 535, 537, 540, 542, 543, 547, 548, 556, 557, 563, 568, 647, 1999
Chrzan, K.L., 304
Chu, S.T., 260
CIFT (contamination-ice-fog-temperature) insulator tests, 394, 395, 398, 406–407, 459
CIGRE (Conseil Internationale des Grands Resaux Electriques), 2, 607–611
CISPR (Comité International Spécial des Perturbations Radioélectriques), 251
Claverie, P., 256, 260, 262, 263, 264, 265, 266, 268, 279, 441, 454, 456
Clean-fog tests
  on alternating diameter insulator strings, 337–338
  on anti-fog disks, 334, 335
  artificial wetting processes, 192–193
  on bell-shaped insulators, 338–339
  comparison with other ceramic disk insulator test processes, 181, 182

Clean-fog tests (*cont'd*)
defined, 180
liquid pollution method, 197
precontamination process for
ceramic insulators, 188–190
precontamination process for
nonceramic insulators, 190–192
precursor, 330
rapid flashover voltage technique,
195–196
results, 205–210
validation of method, 193–195
Cleaning insulators, 69–70, 305–319,
547, 548–550, 636, 637
Clevis fittings, 38
Climate rooms
electrical clearances, 184
in icing test method, 399
Coal-burning electrical utilities
formula for time-averaged
concentration of pollution at
ground level downwind, 79–80
impact of flue gas desulfurization,
79
plants affected by EPA regulation
limits, 79
as sources of air pollution, 78–83
TVA plant design and operation
data, 80
Coating insulators, 319–324. *See also*
RTV silicone coatings
Colbeck, S.C., 493
Cold fog (freezing fog)
defined, 157
impact of RTV silicone coating on
insulators, 539–543
and increased leakage distance,
535–539
performance of semiconducting
glaze insulators, 532–535
theoretical modeling for flashover,
284–285
Cold-fog tests
background, 199–200
comparison with clean-fog tests,
403–406
comparison with other ceramic disk
insulator test processes, 181, 182
defined, 180

polymer insulator test results, 543–544
recommended method, 402
Cole, K.S., 373
Cole, R.H., 373
Cole-Cole plots, 489, 490, 519
Colomb, Michèle, 128
Concrete poles, 47
Conductance of electrolytes, 91–94
Conductivity
defined, 155–156, 364
effect of temperature on ice, 97–100
effect of temperature on liquid water,
95–97
of ice, 367–371
of ice and snow accretions, 53–54
resistance of arc root, 268–271
statistical distribution, in natural
precipitation, 110–112
temperature effects, 94–100
Conlon, W.M., 98
Conseil Iternatonale des Grands Resaux
Electriques (CIGRE), 2, 607–611
Conventional deviation of flashover
voltage
defined, 158
Cooling ponds, 61, 142–143
Cooling towers, 144–145
Corona detection equipment, for
monitoring insulator condition,
301–303
Corona rings
background, 329–331
and ice bridging, 561
impact on heavy icing performance
of polymer insulators, 576
mitigating moderate icing, 561
for polymer insulators, 329–331,
576
Correction factors, for humidity,
temperature, and pressure, 651–658
Corrosion, estimating ESDD, 106–110
Cortina, R., 176
Cortinas, J.V., 620, 621, 622
COST (Cooperation Scientifique et
Technique), 90–91, 128, 619–620
Craggs, J.D., 159, 651
Creepage distance. *See* Leakage distance
(creepage distance)
Creepage extenders, 325–327

CRIEPI (Central Research Institute of Electric Power Industry), Japan, 8, 176–178, 179
Critical flashover level, defined, 24
Critical flashover voltage, defined, 157
Critical impulse flashover voltage (CFO), defined, 659
Critical loading, 46, 81, 82
Croydon, U.K., insulator test observations, 171–174
CSA (Canadian Standards Association), 24, 26, 43–44
Current, household *vs.* transmission line, 23

D
Dakin, T.W., 160
Dallaire, M-A, 467
Darveniza, M., 252
Dawson, C.B., 141
Days without rain in winter, as pollution indicator, 612
Dead-end insulators, 34–35, 55, 313, 383
Deicing station insulators in freezing weather, 575–576
De La O, A., 191
Desert environment, typical sources of pollution deposits on power system insulators, 62, 357
De Tourreil, C., 331
Dew point temperature, 52, 53, 652, 653, 654
Directional dust deposit gauges (DDDG), 74
Discharge, defined, 158
Disruptive discharges, 24, 28, 158. *See also* Flashovers
Drapeau, J-F, 374, 383, 394, 568, 577
Drizzle, defined, 157
Drouet, M.G., 270
Dry arc distance
case studies, 635–641
defined, 25
deterministic insulation coordination, 631–634
electrical stress per meter, 51
ISP for ac flashover across, 434–439
probabilistic insulation coordination, 634–635

requirements for icing conditions, 631–632
requirements for snow conditions, 632–634
voltage stress per meter, 49
Dry-band arcing
background, 71
on contaminated surfaces, 246–255
defined, 26
electrical properties of wetted surfaces, 246–247
enlargement of dry bands, 249–250
evolution to flashover, 256
formation of dry bands, 248–249
stabilization, 256
surface impedance effects, 247
temperature effects leading to dry-band formation, 247–248
wetted layer thickness, 246–247
Dry deposition, 63, 88–89, 90, 134, 612, 614
Dry ice, 317, 403
Dry salt layer (DSL) method, 198–199
Dry snow, 481–482, 483
Duce, R.A., 72
Dust and dustfall, 72, 73, 74, 75–76, 112, 113, 119, 172, 639
Dust cycle method (DCM), 198

E
EdF (Electricité du France), 201
Eklund, A., 198
El-A Slama, S., 103, 104
Eldridge, K., 324
Electrical arcing
in artificial pollution tests, 180–181
discharge initiation and development, 256
dry-band, 26, 246–255
dynamics of propagation, 261
as factor in pollution problems, 71
in freezing conditions, 51
multiple arcs in series, 276–277
V-I characteristics in free air, 256–259
V-I characteristics on water or ice surfaces, 259–261, 262
on wet, contaminated surfaces, 255–262

Electrical conductivity. *See* Conductivity
Electrical flashovers. *See* Flashovers
Electrical puncture, 43, 44, 45
Electrical utilities
    coal-burning, 78–83
    cost of outages, 11–13
    insulator exposure test methods,
        169–178
    as sources of air pollution, 78–83
    transmission system forced outage
        rates, 7–8
Electric fields
    air gap breakdown, 159–162
    uniform *vs.* nonuniform, 159–162
Electric Power Research Institute
        (EPRI), 11, 12, 23, 164, 182, 201,
        555
Electrolytes, conductance, 91–94
Ellena, G., 277
Ely, C.H.A., 169, 175, 308, 325, 326
End fittings, 46–47
ENEL (Ente Nazionale per L'Energia
        Elettrica), Italy, 176, 201
Engelbrecht, C.S., 198, 199
Environmental factors
    in freezing conditions, 53–54
    in temperate conditions, 52–53
    typical sources of pollution deposits
        on power system insulators, 62–63
Environment Canada, CAPMoN
        program, 88, 89
EPDM (ethylene-propylene-diene
        monomer) insulators
    changing to, 572
    characteristics, 31, 44, 68, 70, 221–222
    in ice flashover tests, 417, 418, 419
    in moderate icing conditions, 559–
        560
EPRI (Electric Power Research
        Institute), 11, 12, 23, 164, 182, 201,
        555
Equivalent conductance, 93, 94
Equivalent salt deposit density (ESDD)
    case studies, 101–103
    defined, 156
    effect of road salt, 615–617
    as environmental factor, 52, 53
    estimating from environmental
        measures for corrosion, 106–110

impact of adding booster sheds to
        insulators, 325
insulator testing at Japan sites,
        177–178
measurement unit, 53
measuring insulator contamination
        level, 100, 645–649
monitoring pollution during insulator
        maintenance, 292–296
rate of increase in winter conditions,
        612–615
referring to solutions by, 100–122
traditional unit, 52, 82, 100
Error, defined, 158
Erven, C.C., 394, 406
ESDD. *See* Equivalent salt deposit
        density (ESDD)
Exposure test methods, 169–178
External insulation, defined, 158

F

Faircloth, D.C., 246
Farber, S.A., 61
Farzaneh, M., 2, 17, 53, 77, 78, 98, 99,
        115, 259, 260, 261, 262, 285, 369,
        370, 371, 377, 381, 383, 384, 385,
        386, 390, 391, 394, 397, 399, 400,
        401, 408, 409, 413, 414, 417, 418,
        419, 426, 427, 429, 430, 431, 433,
        435, 436, 437, 439, 440, 441, 442,
        443, 444, 445, 446, 447, 451, 452,
        453, 454, 455, 456, 462, 463, 464,
        465, 466, 470, 471, 487, 488, 491,
        496, 503, 504, 517, 550, 551, 552,
        553, 556, 558, 564, 566, 567, 568,
        569, 572, 573, 577, 578, 579, 580,
        600, 601, 602, 603, 2006, 2007
Farzaneh-Dehkordi, J., 437, 438, 455,
        465, 565, 2004
Ferrous end fittings, 46–47
FGH-GmbH, Germany, 201
Fiberglass poles, 47
Fifty percent disruptive discharge
        (critical flashover) voltage
    defined, 158
Fikke, S.M., 86, 383, 390, 484, 620
Filter, R., 252, 253
Flashover arcs, 326, 327, 331–332, 365,
        454, 460, 469, 494

Flashovers
  ac contamination modeling, 278–284
  ac modeling of ice-covered insulators, 450–457
  adding more insulators to improve performance, 332–333
  background, 24
  comparison of ac and dc models, 283–284
  comparison of pollution layer models, 274–276
  dc arc parameter changes, 277–278
  dc contamination modeling, 271–277
  dc modeling of ice-covered insulators, 442–450
  defined, 24, 158
  external *vs.* internal, 28–29
  field observations of insulator performance, 170–171
  future directions for modeling, 285–286
  ice *vs.* wet, 440–441ho
  icing, 363–471
  insulator form factor solution, 273–274
  multiple arcs in series, 276–277
  nonuniform pollution layer solution, 274
  pressure effects, 224–229
  snow, 481–528
  temperature effects, 229–233
  types of partial discharges, 242–246
  uniform pollution layer solution, 271–273
  winter conditions, common features, 4
Flashover weather, operating EHV transformer stations in "safe posture" during, 584
Flashunder phenomenon, 29, 46ho
Fletcher, N.H., 371
Flue gas desulfurization, 79
Fofana, I., 470, 471
Fog. *See also* Clean-fog tests; Cold fog (freezing fog); Salt-fog tests
  anti-fog disks, 195, 334–335, 340–342
  climatology, 127–128
  critical wetting conditions, 131–132
  density, 52, 124–125
  deposition on insulators, 128–129
  events in winter conditions, 617–618
  forecasting, 127
  heat balance between accretion and evaporation, 130–131
  measurement methods, 124–125
  relationship to precipitation, 127
  surface wetting by accretion, 124–132
  wetted layer conductivity, 247
Fog surging, 171, 376
Form factor
  background, 26–27, 28
  in dc pollution flashover modeling, 273–274
  defined, 26, 156
  influence in contamination performance, 219–221
Forrest, J.S., 69, 169, 171–172, 173, 174, 296, 316, 319, 340, 343, 374, 376, 390
Forward scattering spectrometer probe (FSSP), 124–125
Fowler, D.R., 81
Fowler, R.H., 365, 371
France, insulator test observations, 176
Freezing conditions
  contamination tests, 180
  environmental stresses, 53–54
  insulator washing procedures in, 316
  major electrical factors, 51
  mechanical stresses, 54–55
  purification by crystallization, 97–98
Freezing fog. *See* Cold fog (freezing fog)
Freezing rain/drizzle
  daily variations in occurrence, 622
  defined, 364
  electrical conductivity, 623–624
  frequency of occurrence, 619–622
  measuring accumulation, 618–619
  Ontario 230-kV flashover case study, 637–640
  severity of occurrence, 622–623
Friend, J.P., 59, 60
Fujimura, T., 308, 309, 392, 408, 411, 431, 485, 500, 501, 502, 504, 505, 506, 510, 512, 513, 623
Fujitaka, S., 187
Fukui, H., 329, 343

G
Gallimberti, I.I., 243, 244, 245
Garcia, R.W.S., 181
Gers (researcher), 272

Ghosh, P., 260, 456
Glass insulators, 31, 35, 36, 37, 39, 40, 41, 43. *See also* Ceramic insulators
Glaze ice (clear ice), 157, 364, 366–367
Global Climate Observing System (GCOS), 625
Global dust deposit rate, 75–76
Glycol-based liquids, as deicer, 310, 575
Gnandt, E.P., 55
Goff, J.A., 130, 652
Gonos, I.F., 213, 219, 220, 221, 273, 274
Gorski, R.A., 377
Gorur, R.S., 23, 31, 44, 192, 272, 320, 321, 324, 541
Gratch, S., 130, 652
Grayson, M., 37
Greases, use in ceramic insulator performance, 319–320, 321, 322, 583
Green, M.A., 538, 541
Grennfelt, P., 82
Guan, Z., 248, 260, 262, 267–268, 272, 276, 283
Guano guards, 327–329. *See also* Bird streamers
Gubanski, S.M., 344
Guerrero, T., 422, 423, 454
Gultepe, I., 125, 126, 127, 128
Gunther, E.W., 8
Gutman, I., 17, 18, 191, 398, 420, 421, 552, 573

H
Hadfield, P.G., 6, 7
Haldar, A., 640, 641
Hall, D.K., 626, 627, 628
Hall, J.F., 66, 67, 68
Hallett, J., 95
Hansen, B., 128, 129, 138, 619, 622, 623
Hara, H., 86, 87, 103
Hara, M., 388, 585
Harada, T., 226
Hard rime, 364, 385, 393, 400, 412–413, 583
Hayashi, M., 95, 96
Heat balance
    in clean semiconducting glaze insulators, 345–348
    in contaminated semiconducting glaze insulators, 348–349

between fog accretion and evaporation, 130–131
Heating, as factor in insulator pollution problems, 71
Heat transfer, and ice temperature, 467–469, 471
Heavy icing
    adding booster sheds to insulators to mitigate, 568–571
    changing from ceramic to polymer insulators, 572–573
    dc flashover results, 427–431
    effect of insulator diameter on ac flashover, 425–427
    flashover experience, 381–384
    flashover process, 387–388
    flashover test results, 415–421
    impact of RTV silicone coating on insulators, 577–579
    mathematical modeling of flashovers, 463–465
    options for mitigating, 564–579
    pressure correction for tests, 465–467
    removing from insulators, 575–576
    role of leakage current in monitoring, 573–575
Heavy rain tests, 52, 180, 181, 389, 390
Hemmatjou, H., 488, 497, 510, 511, 514, 515, 517, 518, 520
Hettelingh, J.-P., 82
Hewitt, G.F., 97, 246
Higashiyama, Y., 119
High voltage insulators, defined, 24
Hileman, A.R., 50, 51, 163, 592, 651
Hoarfrost, defined, 157, 364
Hobbs, P.V., 366
Hodgdon, A.D., 61
Holtzhausen, J.P., 185
Hosono, K., 481, 493, 494
Houlgate, R.G., 185
Howes, D.R., 300, 337, 338, 339
Hu, J., 426, 428
Humidity
    as environmental factor, 52
    standard correction, 224–227
Huraux, C., 255, 260
Hydro One Networks, 306
Hydrophilic insulators, 165–167

Hydrophobic materials, 31, 32, 33,
    167–168
Hydro-Québec, 134–135, 217, 383, 577
Hygroscopicity, 122

I

Ice. *See also* Icing
    accretion, classification of severity,
        374, 375, 385, 441–442, 530, 531
    adhesion strength, 47–48
    bridging, 51, 133, 363, 374, 386,
        415–421, 551, 552–555
    crystal structure, 365–366
    dry, 317, 403
    electrical characteristics, 367–373
    electrical conductivity, 367–371
    high-frequency behavior, 371–373
    measuring, 134–135
    morphology, 365–367
    polluted, 363, 367
    temperature and heat transfer,
        467–469, 471
    temperature effect on conductivity,
        97–100, 367, 368, 471
Ice bridging
    and corona rings, 561
    and dry arc distance, 553, 555
    flashover test results, 415–421
    in moderate icing, 550, 551–556, 559,
        560, 561
    and polymer insulators, 559
    role of thermal monitoring, 561–562
    and shed-to-shed spacing of insulator
        strings, 552–553, 554
Ice pellets, cleaning insulators with,
    317
Ice point temperature, defined, 53
Ice progressive stress (IPS) test method,
    398, 420–421
Ice storms, Ontario 500-kV flashover
    case study, 635–637
Ice thickness, defined, 364
Icing. *See also* Heavy icing; Light icing;
        Moderate icing; Very light icing
    classification of overvoltage stresses
        on transmission lines, 592–594
    defined, 180
    deicing station insulators in freezing
        weather, 575–576

empirical models for flashovers,
    432–441
environmental corrections, 465–469
flashover test results, 403–431
future directions for flashover
    modeling, 469–471
heavy, flashovers, 381–384
history of laboratory testing, 391–
    398
insulation coordination process,
    591–641
light, flashovers, 377–380
mathematical modeling of flashover
    process, 441–465
moderate, flashovers, 380–381
options for mitigating, 530–584
outdoor tests, 390–391, 403
recommended test method, 398–402
standardization of test method, 389
test methods, 388–402
very light, flashovers, 374–376
Icing stress product (ISP)
    and accretion density, 54
    for ac flashover across dry arc
        distance, 434–439
    for ac flashover across leakage
        distance, 432–434, 438, 439
    in calculating ice surface conductivity,
        370
    for dc flashover under heavy ice
        conditions, 439–440
    defined, 364, 432
    equation, 435
IEC (International Electrotechnical
        Commission), 24, 43, 49
IEC Standard 60060, 24, 142, 422
IEC Standard 60071, 25, 50, 159
IEC Standard 60437, 251
IEC Standard 60507, 97, 103, 105, 180,
    186, 187, 188, 189, 192, 196–197,
    224, 226, 395, 396, 397, 499, 568,
    579, 647, 648, 652
IEC Standard 60815, 49, 63, 72, 74, 77,
    82, 90, 176, 183, 217, 251, 342, 385,
    386, 606–607
IEC Standard 61109, 180, 323
IEC Standard 61245, 186
IEC Standard 62217, 180, 190, 200
IEC Standard 62271, 390

IEEE/ANSI Standard C2, 2007, 25, 54, 311, 530, 622, 624
IEEE PAR 1783, 2, 17, 389, 390
IEEE Standard 4, 1995, 24, 26, 50, 51, 96, 103, 105, 142, 159, 160, 180, 183, 184, 186, 188, 196–197, 224, 226, 246, 280, 282, 295, 400, 422, 448, 449, 496, 497, 579, 647, 648, 652
IEEE Standard 100, 10, 24
IEEE Standard 493, 12
IEEE Standard 738, 346
IEEE Standard 957, 306–307, 307, 310, 311, 312, 313, 316, 320, 354, 355
IEEE Standard 987, 2001, 353, 354
IEEE Standard 1313.1, 592
IEEE Standard 1313.2, 592, 593, 604–606
IEEE Standard 1366, 8, 9
IEEE Standard C37.34, 390
IEEE Task Force on Icing Test Methods, 391, 503, 568, 572, 617
IEEE Working Group on Insulator Contamination, 68, 131
Iliceto, F., 161
Impulse electrical stresses, 50–51
Impulses, defined, 659
Information Technology Industry Council (ITIC), 16
Infrared inspection, 303–305
Insulation
    acceptable component failure rate, 596–598
    acceptable network failure rate, 598–599
    coordination process, 16–17, 592–599
    coordination process for freezing conditions, 591–641
    deterministic coordination, 599–601
    probabilistic coordination, 601–604
    risk-of-failure assessment, 599–604
    as self-restoring, 24
Insulators
    accessories for, 324–332
    adding more to improve contamination performance, 332–333
    average diameter, 214–219
    breakdown process, 162–169
    ceramic vs. polymeric materials, 31–32
    changing from ceramic to polymer, 352–357
    changing to improved designs, 334–343
    classification, 30–35
    cleaning, 69–70, 305–319, 547, 548–550
    coating (See RTV silicone coatings)
    condition monitoring, 292–305
    construction, 35–48
    dead-end, 34–35, 55, 313, 383
    dimensions, 24–29
    electrical stresses, 48–51
    electromagnetic interference factor, 251–252
    environmental stresses, 52–54
    form factor, 26–27, 28, 219–221, 273–274
    glass, 35, 36, 37, 39, 40, 41, 43
    history of laboratory ice testing, 391–398
    ice-covered, 363–471, 530–579
    mathematical modeling of flashover process in winter conditions, 441–465, 513–519
    measuring ESDD, 100, 645–649
    measuring surface resistance, 103–105
    mechanical stresses, 54–55
    parameter effects, 211–223, 594–596
    pollution deposits, 62–94
    porcelain, 35–36, 37, 38, 39, 40–43
    semiconducting glaze, 343–352
    sheds and booster sheds, 25, 26–27, 28, 29, 324–325, 552–553, 577
    snow-covered, 481–528, 580–584
    station vs. line, 32–34
    terminology, 23–30
    test methods, 169–202, 389–390
International Electrotechnical Commission (IEC), 24, 43, 49
International Symposium on High Voltage Engineering (ISH), 2
International Workshop on Atmospheric Icing of Structures (IWAIS), 2
IPS (ice progressive stress) test method, 398, 420–421
IREQ (Research Institute of Hydro-Québec), 204–205, 217
Ishii, M., 17, 229, 230, 232, 248, 277, 278

ISO 9223 (International Organization for Standardization), 107–108, 110, 137–138
ISP. *See* Icing stress product (ISP)
Italy, ENEL (Ente Nazionale per L'Energia Elettrica), 176, 201
Italy, insulator test observations, 176
ITIC (Information Technology Industry Council), 16

J
Japan
  CRIEPI insulator test observations in marine environments, 176–178, 179
  insulator case study, 103
  Negata Kaetsu snow flashovers, 481, 491, 493
  Okhotsk snow flashovers, 491, 493
Javan-Mashmool, M., 47
Jia, Y., 398, 417
Jiang, X., 112, 223, 437
Jickells, T.D., 72, 74, 75, 76
Jing, T., 245, 246
Johnson, G., 36
Johnson, J.C., 309, 310, 312
Johnson, W., 144, 145
Johnson, W.C., 319
Jolly, D.C., 249, 250, 255, 260, 261, 270, 271, 376
Jones, K.R., 622, 624
Jun, X., 246

K
Kannus, K., 396, 422, 423, 555
Kaolin, 75, 76, 188–189, 191
Karady, G.G., 167, 214
Kawai, M., 49, 187, 206, 207, 249, 250, 299, 330–331, 333, 381, 384, 391, 403, 404
Kawamura, T., 227, 228
Khalifa, M.M., 381, 391, 415, 580, 581, 582, 583
Kiernicki, J., 2, 53, 384, 400, 417, 418, 419, 427, 435, 437, 463, 491, 551, 564, 572
Kimoto, I., 69, 86, 132
Knudsen, N., 161
Kohlrausch's axiom, 93
Kojima, S., 524, 525

Kolobova, O.I., 229
Kontargyri, V.T., 220, 340
Köppen-Geiger world Climate Classification, 2–3
Korsuncev, A.V., 268, 269, 270
Kouadri, B., 131, 248, 285
Kuffel, E., 159
Kuffel, J. ? might be E., 419

L
Laforte, J.L., 399, 496
Lakes, and target loading concept, 81–82
Lala, G.G., 128
Lambeth, P.J., 63, 70, 74, 164, 170, 175, 186, 187, 188, 193, 195, 196, 203, 205, 308, 334, 335, 349, 350, 499
Lanoie, R., 575
Lathi, K., 555
Layer conductivity, 52–53, 156
Leakage current
  case studies, 106, 107
  defined, 156
  field observations of activity, 169–170
  for monitoring insulator condition, 296–301
  pole fire factor, 252–254
  as predictor of icing flashover, 573–575
  and self wetting of insulators, 122–123
Leakage distance (creepage distance)
  background, 26–28
  and cold fog, 535–539
  defined, 26, 156
  deterministic insulation coordination, 629–630
  and electrical stresses, 49, 51
  and icing stress product, 432–434, 438, 439
  increasing in contamination tests, 211–212
  probabilistic insulation coordination, 630–631
  sheltered, 37–38
  in winter conditions, 29–30
Leakage to dry arc ratio, 211, 535–536
Lee, L.Y., 392
Lee, P.K.H., 119, 121
Leitch, J.E., 110
Le Roy, G., 165, 166, 169, 197, 203, 206, 265, 268

Li, S., 255, 456
Li, Y., 658
Liao, W., 543
Light icing
  defined, 532
  flashover experience, 377–380
  flashover process, 385–386
  flashover test results, 406–408
  impact of RTV silicone coating on
    insulators, 539–543
  mathematical modeling of flashovers,
    459–461
  options for mitigating, 530–550
Lightning
  conditions for occurrence, 656
  cost of protection, 11–12
  as major event day, 9
  and overvoltages, 593, 594
  and redundant components, 10
  as root cause of short-circuit currents
    in power systems, 53
Lightning impulse, defined, 659
Line post insulators, 34, 35, 41, 44, 45,
    439, 605–606
Liquid water content (LWC), defined,
    364. *See also* Water
Local pollution, 63, 73–74, 113, 115, 116,
    117
London, early telegraph systems, 35–36
Long-rod insulators, 25, 38, 39, 41, 44,
    45, 410–411, 500, 501, 505, 506,
    630
Looms, J.S.T., 23, 35, 37, 41, 42, 43, 63,
    69, 70, 169, 212
Los (reseaarcher), 272
Lucas insulators, 212
Lundmark, A., 113, 117, 118
Lux, A.E., 441

M
Macchiaroli, B., 209, 210
Maeno, N., 488, 489, 490, 496, 521
MAIFI (Momentary Average
    Interruption Frequency Index), 8
Major event days, 9
Manurigation, 145–147
Marine environment
  CRIEPI insulator test observations in
    Japan, 176–178, 179

typical sources of pollution deposits
  on power system insulators, 62
Martens, G., 145
Martigues, France, insulator test
  observations, 176
Matsuda, H., 382, 481, 486, 490, 506, 507,
    508, 509, 519
Matsuo, H., 393, 412, 413, 416, 417, 419
Matsuoka, R., 68, 75, 76, 188, 205, 215,
    216, 221, 222
Mauldin, F.P., 66, 67, 68
Maximum withstand voltage
  measurement method, 395–396
Mayr (researcher), 272
Mays, L., 142
Meale, J.R., 377
Meek, J.M., 159, 651
Meghnefi, F., 573, 574
Mehta, H., 8
Meier, A., 391, 490
Meier, B., 36
Meijer, E., 59
Melo, O., 53, 98, 377, 408, 409
Melted water conductivity, defined, 365
Mercure, H.P., 227, 270
Metal poles, 47
Meteorological Services of Canada, 121
Methylcellulose liquid pollution method,
    197
Metwally, I.A., 327
Mexico, insulator case study, 101–103
Meyer, L.H., 44
Meyer, M.B., 128
Mikhailov, A.A., 80, 108, 110
Milone, D.M., 187, 299
Minimum flashover voltage, defined, 158
Mintz, J.D., 252, 253
Mist, defined, 157
Mizuno, Y., 230, 231, 344, 347
Moderate icing
  defined, 550
  flashover experience, 380–381
  flashover process, 386–387
  flashover test results, 408–415
  impact of RTV silicone coating on
    insulators, 562–563
  mathematical modeling of flashovers,
    461–463
  options for mitigating, 550–563

and polymer insulators, 410–412, 559–560
role of thermal monitoring, 561–562
Moderate resolution imaging spectroradiometer (MODIS), 626, 627, 628
Momentary Average Interruption Frequency Index (MAIFI), 8
Monitoring cylinder, defined, 365
Moran, J.H., 343
Morgan, V.T., 347
Moro, F., 304
Morphology, 365–367
Morris, R.M., 381, 391, 415, 580, 581, 582, 583
Mousa, A.M., 10

N

Näcke model, 265, 266
NADP (National Atmospheric Deposition Program), 84, 85, 89
Naito, K., 2, 17, 178, 179, 186, 191, 204, 374, 623
Nakaya, U., 493
Nasser, E., 165, 272
National Electrical Safety Code. *See* IEEE/ANSI Standard C2, 2007
Natural precipitation. *See* Precipitation, natural
Ndiaye, I., 245, 246, 470
Negata Kaetsu, Japan, snow flashovers, 481, 491, 493
Nejedly, Z., 121
NEPA (National Environmental Policy Act), 78
NERC (North American Electric Reliability Corporation), 2, 13–16, 305, 373, 637
Neumärker, G., 442–443
Newfoundland, ice thickness case study, 640, 641
Nghiem, S.V., 628
Nguyen, D.H., 183
Nicholls, C.W., 212, 341
Niemeyer, L., 242, 243
Niggli, W.M., 391, 490
Nigol, O., 340, 343, 352
Nilsson, J., 82
Nitrogen dioxide, 78, 80, 81

Nitrogen oxides, 60
NOAA (National Oceanic and Atmospheric Administration), 127, 146, 613, 618
Nonceramic insulators. *See* Polymer insulators
Nondisruptive discharge (partial discharge), defined, 158
Nonsoluble deposit density (NSDD)
defined, 156
direct measurement method, 72–73
as environmental factor, 53
indirect measurement methods, 73–75
measurement case studies, 75–78
measurement unit, 53
role in insulator surface resistance, 75
Nonsustained disruptive discharge
defined, 158
Nonuniform electric fields, air breakdown in, 161–162
Norgord, D., 3
North American Electric Reliability Corporation (NERC), 2, 13–16, 305, 373, 637
Norway, 17, 18, 63, 86, 383, 491, 493, 522–524
Nottingham (researcher), 257, 258, 272
Nourai, A., 394, 416, 417
Novak, J.P., 277

O

Obenaus, F., 211, 256, 259, 260, 273, 281, 284, 441, 488, 513
Oil-filled insulators, 319, 321
Okhotsk, Japan snow flashovers, 491, 493
Olofsson, B., 113, 117, 118
Onodera, T., 481, 494
Ontario, Canada, flashover case studies, 635–640
Ontario Hydro, 377, 380
Outdoor exposure test methods, 169–178
Overshoot, defined, 659
Overvoltages
classifying icing-related stresses, 592–594
and design of insulation systems, 16–17
impulse electrical stresses, 50
temporary *vs.* switching, 593–594

Ozone, ground-level, 60
Ozone production, 254

P

Pargamin, L., 209, 210
Partial discharges. *See also* Corona
    detection equipment, for monitoring
    insulator condition
  acoustic noise factor, 251
  defined, 659
  electromagnetic interference factor,
    251–252
  near dielectric surfaces, 243–246
  nuisance factors, wetted pollution
    layers, 250–254
  ozone production factor, 254
  pole fire factor, 252–254
  types and characteristics, 242–246
Particulate matter, 78, 83–91
Passive ice monitoring (PIM), 134–135
Peak value of alternating voltage,
    defined, 158
Peak value of impulse voltage, defined,
    659
Peelo, D.F., 256, 257, 258, 259, 260
Perin, D., 312
Persian Gulf coast, Iran
  map of pollution regions, 91
  pollution deposit densities, 77
  Shariati studies, 89–90
Peszlen, E.A., 394, 416
Petrenko, P.F., 366
Peyregne, G., 261, 262
Phan, C.L., 388, 585
Phan, L.C., 393, 412, 413, 416, 417, 419
Philips, S.E., 319
Phillips, A.J., 168, 330
Pigini, A., 162, 163
Plambeck, J.A., 94
Pocheron, Y., 260, 263, 264, 279, 454,
    456
Pokorny, W.C., 394, 416, 417
Poles, electrical
  materials for, 47–48
  wooden, electrical resistance in winter
    conditions, 252–254
Pollution. *See also* Artificial pollution
    tests
  assessing severity, 63–65

background, 59–61
condition monitoring, 292–303
conductance of electrolytes, 91–94
defined, 59
deposits on insulators, 62–71
dry deposition, 63, 65, 88–89, 90, 134,
    612, 614
electrical utility sources, 78–83
formula for time-averaged
    concentration at ground level
    downwind from coal-fired power
    plant, 79–80
insoluble, electrically inert, 72–78
insulator breakdown process, 165–
    169
insulator cleaning processes and rates,
    69–70
local, 63, 73–74, 113, 115, 116, 117
long-term changes in level, 70–71
measuring short-term accumulation,
    65–68
mitigation options for improved
    insulator performance, 291–357
site monitoring methods, 63–65
soluble, electrically conductive, 78–94
standards for tests, 180
summaries of test methods, 201–202,
    389–390
wet deposition, 65, 83–91, 113, 118,
    611–614
Pollution layer
  residual resistance, 262–271
  series resistance, 262–268
Pollution monitoring
  insulator contamination measurement,
    292–296
  using corona detection equipment,
    301–303
  using leakage current, 296–301
  using remote thermal monitoring,
    303–305
Polymer insulators. *See also* EPDM
    (ethylene-propylene-diene
    monomer) insulators
  aging tests, 180
  artificial pollution tests, 181
  background, 352–353
  changing to under contaminated
    conditions, 352–357

changing to under heavy icing conditions, 572–573
in clean-fog testing, 192–193
corona rings for, 329–331, 576
in desert environment, 357
effect of ultraviolet radiation, 53
end fittings, 46–47
flashover test results under cold-fog conditions, 543–544
flashover test results under heavy icing conditions, 416–417
flashover test results under moderate icing conditions, 410–412
interchangeability with ceramic insulators, 355–357
long-term performance in contaminated conditions, 354–355
maintaining, 314–315
materials for, 31–32, 33, 44–46, 200–201
mitigating moderate icing, 559–560
mitigating snow and rime, 582
short-term experience in contaminated conditions, 353–354
summary of pollution test methods, 201, 202
Poole, C.D., 249, 250, 261, 270, 271
Porcelain insulators, 35–36, 37, 38, 39, 40–43, 425, 430. *See also* Ceramic insulators
Posch, M., 82
Post insulators, 35. *See also* Line post insulators; Station post insulators
Potvin Air Management Consulting, 74
Powell, D.G., 343
Power supplies, characteristics for artificial pollution tests, 181, 183–184
P-percent disruptive discharge voltage (flashover voltage), defined, 158
Precipitation, artificial
bird streamers, 53, 139, 326, 327–329, 470
cooling pond overspray, 142–143
cooling tower drift effluent, 144–145
cooling tower overspray, 145
dam spray, 139–141
irrigation with recycled water, 141–142
manurigation, 145–147

surface wetting, 137–147
tower paint, 139
Precipitation, natural. *See also* Rain; Snow
frozen or solid, 133
liquid, 133
measurement methods, 133–136
statistical distribution of conductivity, 110–112
supercooled, 133
surface wetting by, 132–138
types, 132, 133
Precision, defined, 158
Pressure washing. *See* Washing insulators
Protected leakage (creepage) distance, defined, 156
PSOD (Power Systems Operations Division, Ontario Hydro), 380, 637
Puncture strength, 45

Q

Québec, Canada. *See* Hydro-Québec

R

Radial electrical stress, 49–50
Radioactive contamination, 60–61
Rag-wipe method of measuring ESDD, 100, 645–649
Rahal, A.M., 255, 260
Rain. *See also* Heavy rain tests
climatology, 137–138
defined, 157
effects of washing on surface conductivity, 137
effects of washing on surface pollution level, 69–70
electrical conductivity of rainwater, 52
pollution concentration in droplets, 136
Rain gauges, 134
Ramos, G.N., 101–102, 229, 230, 648
Random error, defined, 158
Raoult's law, 122
Raraty, L.E., 47, 48
Reignition. *See* Arc reignition
Reliability, electrical system
achieving with redundant components, 10

Reliability, electrical system (*con'd*)
background, 6
cost of outages, 11–13
measures, 6–9
regulation, 16
responsibility, 13–14
role of maintenance, 10–11
role of National Electric Reliability
Council, 13–16
stakeholders, 13–14
Remote thermal measurements
role in monitoring icing conditions,
561–562
role in monitoring insulator condition,
303–305
Renner, P.E., 391, 392, 427, 429, 440
Renowden, J.D., 299
Research Institute of Hydro-Québec
(IREQ), 204–205, 217
Residual resistance of pollution layer,
262–271
Response time, defined, 659
Return period (of icing events), defined,
365
Rezazada, A.Q., 228, 280, 281, 282, 283,
456, 466, 467
Richards, C.S., 123, 296, 299
Rime ice, 54, 364, 367, 368, 437, 482,
580–583
Riquel, G., 176
Rizk, F.A., 62, 183, 185, 205, 211, 228,
256, 265, 266, 269, 271, 278, 280,
281, 284, 357, 441, 454, 456, 466,
467, 513, 518, 656, 657
Road salt
effect on ESDD, 615–617
vehicles as source, 112–122
Robinson, R.A., 96
Roy, D.P., 626
RTV silicone coatings
background, 321–324
for improved insulator performance in
heavy icing conditions, 577–579
for improved insulator performance in
heavy snow and rime ice conditions,
583
for improved insulator performance
in moderate icing conditions, 562–
563

for improved insulator performance in
very light and light icing conditions,
539–543
and insulator washing, 314–315, 323,
324
and leakage current mitigation, 300
Rudakova, V.M., 229
Rumeli, A., 261, 262
Russia, transmission line NSDD case
study, 77–78

S
Sahin, S., 123
SAIDI (System Average Interruption
Duration Index), 8, 9
SAIFI (System Average Interruption
Frequency Index), 8
Salama, M.A., 129
Salley, D.T., 144, 145
Salomon, R.E., 367
Salt-fog tests
Brighton, England observations, 175
comparison with other ceramic disk
insulator test processes, 181, 182
defined, 180
description of method, 52, 185
quick flashover voltage technique,
187
results, 203–205
Sangkasaad, S., 343
Scale factor, defined, 159
Schaedlich, K.H., 376, 377
Schmuck F., 331
Schneider, H.M., 186, 204, 212, 292, 300,
324, 337, 338, 339, 341, 441
Scrubbers. *See* Flue gas desulfurization
Self-restoring insulation, 24, 159
Self-wetting of contaminated surfaces,
122–123
Semiconducting glaze insulators
application experience, 352
clean, heat balance, 345–348
close proximity considerations,
351–352
contaminated, heat balance, 348–349
in heavy icing conditions, 566–567
mitigating moderate ice, 555–559
need for heavy snow test data, 583
on-line monitoring with, 349–350

performance in cold-fog conditions, 532–535

role of power dissipation in fog and cold-fog accretion, 350–351

technology, 343–345

Series resistance

    mathematical functions, 265–268

    pollution layer observations, 262–264

    role in electrical flashover, 647

Sforzini, M., 206, 207

Shaft diameter, defined, 25

Shapiro, C.A., 145, 147

Shariati, M.R., 77, 90

Sharp, R., 314, 315

Sheds, insulator, 25, 26–27, 28, 29, 552–553, 577. *See also* Booster sheds

Shinoda, A., 303, 352

Shu, L., 382

Silicone coatings. *See* RTV silicone coatings

Sklenicka, V., 575

Slick water, 310, 575

SMART washing program, for insulator maintenance, 547, 548–550, 637

Smog, 60, 157, 316

Snow

    accumulation and persistence, 625–628

    background, 482–484

    case studies of flashover, 520–525

    climatology, 625–629

    conditions leading to flashover, 493–494

    defined, 482

    dielectric behavior, 489–490

    dry, 481–482, 483

    effect on insulators, 481–528

    electrical characteristics, 484–490

    electrical conduction, 487–489

    empirical model for flashover, 508–513

    environmental corrections for flashover, 520

    flashover experience, 490–493

    flashover process, 494–495

    flashover test methods, 495–500

    flashover test results, 500–508

    flashover under switching surge, 505–508

    impact of RTV silicone coating on insulators, 583

indoor tests using natural deposits, 503–505

    mathematical modeling of flashover process on insulators, 513–519

    melting, 628–629

    morphology, 482–484

    options for mitigating, 580–584

    outdoor tests, 500–502, 503

    partial discharge activity, 490

    sources of conductivity, 484–486

    standard measurement methods, 625

    wet *vs.* dry, 483

Snow contamination test, defined, 180

Snow cover

    defined, 482, 625

    MODIS observations, 626–628

Snow cover depth, 482, 625

Snow water equivalent, defined, 482

Sodium chloride, 91, 122, 123. *See also* Road salt

Sodium hydroxide (NaOH), 92, 122, 123

Soft rime, 385, 393, 400, 482, 497

Soil contamination, 60, 81, 82

Soucy, L., 418

Southern California Edison, insulator case study, 309–310, 311

Spangenberg, E., 176

Srivastava, K.D., 10

Standard lightning impulse, defined, 659

Standards, voluntary, 1, 5, 7, 13, 16, 24

Standard switching impulse, defined, 660

Station post insulators, 26, 29, 32–34, 35, 40, 45, 208, 292–293, 294, 295, 342–343, 351, 385, 544, 544–548, 551, 575–576

STE (Society of Telegraph Engineers), 35

Steinmetz, C., 257

Stokes, R.H., 96

Stokes-Einstein equation, 94

Stonkus, D.J., 353

Streamers. *See* Bird streamers

Stresses, on insulators

    electrical, 48–51

    environmental, 52–54

    mechanical, 54–55

Su, F., 398, 417

Sugawara, N., 481, 493, 494

Suginuma, Y., 254

Sulfur dioxide, 60, 78, 80–81
Sumnu, S.G., 123
Sundararajan, R., 139, 272, 328
Supercooled, defined, 365
Surface coatings. *See* RTV silicone coatings
Surface conductivity, defined, 156
Surge arresters
  under heavy icing conditions, 422, 423
  snow accretion on, 524–525
Surges, defined, 660
Su Zhiyi, S., 68
Swift (researcher), 272
Switchgear, 30, 32, 34, 390, 392
Switching impulse, defined, 660
Switching surges
  and gap factor, 51
  ice flashovers under, 422–424, 454
  and insulation coordination, 51
  mitigation measures, 595–596
  snow flashover under, 505–508
Systematic error, defined, 159
System Average Interruption Duration Index (SAIDI), 8, 9
System Average Interruption Frequency Index (SAIFI), 8

T
Tabor, D., 47, 48
Takasu, K., 103, 105, 485
Takei, I., 372, 373, 424, 488, 496, 521
Tam, Y.T., 112, 379, 484
Tammelin, B., 620
Taniguchi, Y., 62, 65, 66
Target loading hypothesis, 70, 81–82
Tavakoli, C., 266, 471
Telegraph systems, 35–36
Television and Infrared Observation Satellite (TIROS-1), 626
Temperature
  dew point *vs.* ambient, 52
  dew point *vs.* ice point, 53
  effect on ice conductivity, 97–100, 367, 368, 471
  effect on liquid water conductivity, 95–97
  effects on pollution flashover, 229–233

ice point, 53
overview of impacts on electrical conductivity, 94–100
Tennessee Valley Authority, 79, 80
Tests. *See* Artificial pollution tests
Textier, C., 131, 248, 285
Thermal measurements
  role in monitoring icing conditions, 561–562
  role in monitoring insulator condition, 303–305
Thermal pollution, 61
Thomas, F.W., 79, 80
Tikohdeev, N.N., 229
Todd, D.K., 142
Tonoko, 75, 188–189, 191, 193, 201
Topalis, F.V., 213, 219, 220, 221, 273, 274
Train, D., 183
Transfer function *[H(f)]*, defined, 159
Transmission lines
  classification of overvoltage stresses, 592–594
  nature of mechanical load, 34–35
  Russia, NSDD case study, 77–78
  traditional cap-and-pin type insulators, 38–39
Transmission towers, 47, 139
Tsai, W-Y., 628
Turner, F.J., 209, 210
2.5-Beta method, 9

U
Udo, T., 422, 424
U.K., insulator test observations, 171–175
U.K. Electricity Supply Regulations of 1937, reliability objectives, 6–7
Ulrich, H., 344
Ultrasonic microphones, 302
Ultraviolet radiation, 53, 302–303
Uncertainty, defined, 159
Undershoot, defined, 660
Unified specific creepage distance (USCD), 43, 156–157, 183, 538–539, 629, 630
Uniform electric fields, 159–161, 487
U.S. Energy Policy Act of 2005, 16
Utilities. *See* Electrical utilities

V

Value of test voltage for alternating voltage, defined, 159
Value of test voltage for lightning impulse, defined, 660
Vandalism, 53
Vapor pressure, 122, 130, 652, 653–654
Vehicle traffic
 impact of dustfall, 112
 impact of road salt, 112–122
Verhaart, H., 245
Very light icing
 defined, 530
 flashover experience, 374–376
 flashover process, 385–386
 flashover test results, 403–406
 impact of RTV silicone coating on insulators, 539–543
 mathematical modeling of flashovers, 457–459
 options for mitigating, 530–550
Violet arc, 365, 388
Viscosity, 31, 54, 82, 94, 95–97, 191
Visual pollution, 61
Vlaar, J., 99, 368
Vohl, P.E., 183
Vokalek, J., 575
Volat, C., 568, 569, 577, 579
Voltage, reducing as icing and contamination mitigation method, 585
Voltage ratio of a voltage divider, defined, 159
Voltage stress, 49, 592–594
Vovan, M.L., 429, 436, 440
Vuckovic, Z., 382, 435, 484

W

Wall thickness, insulator, 27, 28–29, 50
Washing insulators
 industry standard practices, 310–313
 methods and conditions for washing, 307–309
 natural processes, 69–70
 pressure vs. hand or abrasive methods, 316
 and RTV coatings, 314–315, 323, 324
 selecting intervals for washing, 306–307

SMART program, 547, 548–550, 637
Southern California Edison case study, 309–310, 311
 in winter, 636–637
Watanabe, Y., 382, 422, 425, 426, 429, 431, 504, 505, 512
Water. See also Icing; Rain; Washing insulators
 effect of temperature on electrical conductivity, 95–97
 freeze-thaw purification, 97–100
 pollutants that absorb, 122
 temperature dependence of viscosity, 95–97
Water pollution, defined, 60
West, H.J., 139
Wet deposition
 electrically conductive ions, 83–91
 precipitation samples, 83–91
 vs. dry deposition, 65, 88–89, 113, 118
 and winter conditions, 611–614
Wetness sensors, 134
Wet snow, 482, 628, 629
Wetted insulator surfaces
 dry-band arcing and enlargement, 249–250
 dry-band formation, 248–249
 nuisance factors from discharges, 250–254
Wetting
 artificial processes, 139–147, 192–193, 197
 as factor in pollution problems, 71
Wetting rate, 52, 255
White arc, 365, 388, 442, 467, 583
Whitworth, R.W., 366
Wieck, H., 492, 493, 497, 498, 499, 503, 523, 524
Wilkins, R., 248, 255, 265, 268, 269, 368, 370, 371, 445, 463
Wilkins model, 266, 270, 276, 283, 285
Winter conditions. See also Ice; Icing; Snow
 ESDD rate of increase, 612–615
 features of flashover events, 4
 fog events, 617–618
 insulator washing, 636–637
 Ontario flashover case studies, 635–640

Winter conditions (*cont'd*)
  pollution characteristics, 611–617
  terminology, 29–30
  typical sources of pollution deposits
    on power system insulators, 62
Withstand probability, defined, 159
Withstand voltage, defined, 159
WMO (World Meteorological
  Organization), 133, 626
Wolf, M., 122
Wood crossarms, 47, 254
Wood poles, 252–254
World Meteorological Organization
  (WMO), 133, 626
Wu, D., 68

Y
Yasuda, M., 308, 309
Yasui, M., 251, 484, 485, 486, 490, 492,
  495, 496, 500, 501, 502, 503, 504,
  507, 508, 509, 510, 511, 512

Ye, H., 198
Yeh, H.C., 72
Yoshida, S., 2, 17, 374
Young, I.G., 367
Yu, D., 387

Z
Zdravkovic, Z., 382, 435, 484
Zedan, F., 62, 66, 69, 335, 336, 357
Zhang, J., 248, 255, 370, 429, 430,
  442, 443, 444, 446, 447, 451,
  452, 453, 455, 456, 462, 464,
  568
Zhang, R., 260, 262, 267–268, 272,
  276, 283
Zimmerman, M., 319
Zoledziowski, S., 260